D1084989

The Dow Jones-Irwin Handbook of Telecommunications Management

The Dow Jones-Irwin Handbook of Telecommunications Management

James Harry Green

Dow Jones-Irwin
Homewood, Illinois 60430

TK
5102.5
.G73
1989

Fairleigh Dickinson
University Library

MADISON, NEW JERSEY

© James Harry Green, 1989

Dow Jones-Irwin is a trademark of
Dow Jones & Company, Inc.

All rights reserved. No part of this publication may be reproduced, stored in a retrieval system, or transmitted, in any form or by any means, electronic, mechanical, photocopying, recording, or otherwise, without the prior written permission of the copyright holder.

This publication is designed to provide accurate and authoritative information in regard to the subject matter covered. It is sold with the understanding that neither the author nor the publisher is engaged in rendering legal, accounting, or other professional service. If legal advice or other expert assistance is required, the services of a competent professional person should be sought.

From a Declaration of Principles jointly adopted by a Committee of the American Bar Association and a Committee of Publishers.

Project editor: Merrily Mazza
Production manager: Ann Cassady
Compositor: Carlisle Communications, Ltd.
Typeface: 11/13 Times Roman
Printer: Arcata Graphics/Kingsport

Library of Congress Cataloging-in-Publication Data

Green, James H. (James Harry)
The Dow Jones-Irwin handbook of telecommunications management / James Harry Green.
p. cm.
Bibliography: p.
Includes index.
ISBN 1–556–23030–3
1.Telecommunication systems—Management—Handbooks, manuals, etc.
2.Business—Communication systems—Management—Handbooks, manuals, etc. I. Dow Jones-Irwin. II. Title.
TK5102.5.G73 1989
651.7—dc19

88–20949
CIP

Printed in the United States of America

1 2 3 4 5 6 7 8 9 0 K 5 4 3 2 1 0 9 8

This book is dedicated to my mother, Vadura, and the memory of my father, Jim, who taught me the value of an education. Without their support this book would not have been possible.

PREFACE

Since the breakup of the Bell System in 1984, the function of telecommunications management has become increasingly important in most large organizations. No longer is it feasible to delegate the company's telecommunications requirements to a single outside source. In today's environment, with multiple products and numerous vendors, someone inside every organization must take charge of the telecommunications system. In smaller companies this person inevitably has other responsibilities because most companies can't afford full-time telecommunications management until the budget approaches about $500,000 per year. Even though an organization may be unable to justify full-time telecommunications management, professional management of the system is mandatory in virtually every company where telecommunications is more than an incidental resource.

A telecommunications system can't be effectively managed like an automobile—that is, purchased, installed, and given only occasional routine attention. The objectives of telecommunications management are to maintain a reasonable balance between cost and service in an environment that is continually changing and, each time a change occurs, to review whether the change has disrupted the cost/service balance. The manager, therefore, must frequently measure loads and grade of service and make adjustments whenever the indicators suggest it is time. Cost/service balancing requires a knowledge of common carrier tariffs and traffic engineering theory, a method of consistent data collection, and a process for making adjustments.

Telecommunications management involves far more than adjusting quantities of trunks to keep cost and service in balance. Someone must classify individuals or positions in the organization according to types of terminals or instruments, packages of features, and service restrictions.

Not only must policies be established, but records must be kept of all the variables associated with an individual's service. These include records of cable location, pair assignments, and cross connections from the point of interconnection with telephone company facilities to the individual station.

Quality control plays an important part in the telecommunications manager's job. It is possible to administer a system under the assumption that the absence of complaints indicates satisfaction, but effective management monitors for indications of service degradations exclusive of the frequency of trouble reports and before reports occur. Both switching quality, measured in terms of percentages of blocked calls, and transmission quality, measured in terms of loss, noise, and echo grade of service, must be tracked. Moreover, the telecommunications manager should know enough about transmission design to foresee and prevent circuit configurations that are likely to generate dissatisfaction. When problems do occur, receiving, recording, correcting, classifying, and analyzing trouble reports are an important part of the job.

The effective telecommunications manager is an effective planner. Planning involves forecasting usage and matching usage to capacities to foresee when shortages will occur so preventive action can be taken in time. Effective planning requires staying current with advances in technology so that, as new services and equipment enter the market, the company is positioned to take advantage of their cost-reducing potential. Effective planning also requires the manager to understand the financial structure of the company and to be familiar with economic evaluation techniques so different alternatives can be compared on the basis of their financial efficiency. In addition telecommunications managers must be capable of evaluating lease versus purchase options and deciding whether it is more economical to contract for services or to perform them with internal staff. All of these functions presume some knowledge of financial analysis techniques.

The telecommunications manager must be an effective manager of people. Most companies have at least a full-time attendant/receptionist; in larger organizations, additional staff is needed to install and maintain telecommunications services.

The telecommunications manager must be acquainted with data processing, computers, and computer communications techniques. In many companies the data processing and telecommunications jobs are assigned to the same individual, or telecommunications management may be a sideline job for the data processing manager. As the lines between computers and communications become less distinct and as distributed data processing becomes more prevalent, the effective telecommunications manager must understand data processing.

Given the variety of skills needed to manage a telecommunications system effectively, is it reasonable to expect that one person will possess the necessary knowledge? The answer is an emphatic yes. Telecommunications technology and management techniques are no more complicated than many other career fields—data processing, for example. The problem has been that a small number of companies have had a monopoly, not only on the provision of telecommunications service, but on the body of knowledge required to manage it. Only in recent years have colleges and universities begun to teach telecommunications, and much of the specialized knowledge required has not been documented as a unified collection of techniques. The need for a one-volume collection of telecommunications management techniques is the driving spirit behind this book. Of course, a book alone isn't enough to develop a skilled telecommunications manager; part of the job is intuitive and requires the sixth sense that comes with years of experience. While this book doesn't contain all the information a telecommunications manager needs, it is an effective starting place.

Effective telecommunications management requires a certain amount of quantitative analysis. I have tried to minimize the number of formulas and calculations in this book, but a few mathematical manipulations are inevitable. Fortunately, some excellent PC software is available that takes away most of the drudgery. Throughout this book I have demonstrated the use of spreadsheets to simplify the calculations. To elaborate on the method of constructing the spreadsheet is beyond the scope of this book, but enough information has been included that experienced analysts should be able to develop models to fit their applications.

James Harry Green

CONTENTS

CHAPTER 1

INTRODUCTION TO TELECOMMUNICATIONS MANAGEMENT

Like any other organizational resource, a telecommunications system must be managed. Whether your company has a large PBX or Centrex system, a small key telephone system, or a worldwide data communications network, the objectives of managing it are much the same, although the techniques will differ considerably.

A telecommunications system can be viewed as a cost center, like a large fleet of vehicles. From this point of view the objective is to contain costs. You can also look at telecommunications as a resource that enhances the productivity of people in much the same way that office automation and mechanized support systems such as cost accounting reduce manual effort. A third way of viewing a telecommunications system is as a strategic weapon that can be used to improve your company's market position. Harnessed to the company's information resource, the telecommunications system becomes a way of moving information into the hands of customers, who, it is to be hoped, will perceive greater value in dealing with your company because of the accessibility of its information. Regardless of which way your company perceives its telecommunications system, this book offers a practical way of managing it.

Before the divestiture of AT&T, telecommunications management was usually delegated to the telephone company. When there was little competition among equipment and service suppliers, a telecommunications manager's job was greatly simplified, but not necessarily enhanced. Decision making was simple; it was a matter of deciding which model of Western Electric PBX to lease, and which of a restricted range of features to equip it with. The corporate telecommunications manager, if such a position existed, was regarded as a person who ran a technical service department, usually out of the mainstream of the business.

In settling the Carterphone case of 1968, the Federal Communications Commission determined that telephone company tariffs prohibiting the interconnection of customer-owned equipment to the network were contrary to public policy. For the next 15 years, AT&T fought a rear guard action to stave off encroaching competition, but the Carterphone chink in AT&T's armor eventually ended in a collapse of the monopoly. Divestiture day, January 1, 1984, not only complicated life for a million AT&T employees, it also thrust the corporate telecommunications manager into a position of new importance. No longer were companies limited to shopping from one catalog. The problem became one of a surplus of alternatives, and new factors that had not been considered important before. Which companies will be in business to service the equipment in five years? Who offers the best long distance rates? After we put the system together, who will guarantee that it works?

If the Department of Justice's case against AT&T was the lawyer's full employment act of 1974, the spinoff of the Bell Operating Companies was the consultant's full employment act of 1984. Suddenly, a single-source commodity offered by a staid and complacent industry became a veritable minefield of alternatives. Telecommunications managers, who in the past had been contented to know what to do, were suddenly faced with the need to know how to do it. Telephone companies reviewed services that once were "free" (because they were buried in the rates) and either offered them at cost or not at all.

If the technology isn't enough to contend with, there is also a barrage of acronyms that would intimidate a New Dealer. The telecommunications manager has to fly through the flak-ridden skies of ISDN, LADT, SDN, SNA, OSI, ONA, and a host of other acronyms that once could be left to specialists. But no more. The corporate telecommunications manager is now in the same position as the transportation manager when the company obtains its first corporate jet, and finds that he or she must learn details that were once handled by travel agents or the airlines.

But where is the school that teaches the corporate telecommunications manager? Many a manager can point to painful lumps and correlate a lesson with each one. Concepts such as queueing theory and Erlangs, that may have been buried in an operations research course, suddenly spring from the irrelevant to the crucial. A few colleges and universities are beginning to teach telecommunications management, but the manager who winds up in the middle of a large corporate network may find that what is really needed is a quick graduate course.

Managing a vast telecommunications network is no easy task. The telephone companies handled it for years by breaking the jobs down into minute chores that could be taught to people with little formal education. Max Weber, the father of bureaucracy, would have been delighted to see his theories so admirably confirmed. AT&T showed that one of the most complex systems known to mankind could be managed by breaking it into bite-sized jobs, each of which was governed by a complex set of rules for designing, engineering, operating, and maintaining an unbelievably complex network.

State-of-the-art bureaucracy may have been the best way to manage the nation's telecommunications systems. History will determine whether it was when we find out how well the alternative works. Even so, consider the plight of the telecommunications manager who suddenly acquires a telecommunications system to manage along with word processing, a mini-computer, and a host of personal computers that have suddenly materialized wearing different brand names and bearing incompatible software.

What are the choices? There are three options: hire an outside management firm, turn the job over to the equipment vendor, or learn how to do it internally, preferably without suffering too many painful lumps.

Unquestionably, the first two alternatives are best for many organizations, but the most effective way to contract for any service is to have a good idea what result you expect. If an outside vendor has control of the process and also evaluates the results, you may find that an important control has been lost. This is particularly true in telecommunications, where there isn't a great deal of competition in administering systems, and those who do it may not have proven track records. Moreover, parts of the job cannot be effectively contracted. No contractor knows enough about the inner workings of a business to control use of long distance services or to allocate costs among profit centers.

This book is designed to fill the gap in the fund of information about telecommunications. It assumes that you have a reasonably good working knowledge of the technology. If you do not, the author's book, *The Dow Jones-Irwin Handbook of Telecommunications,* is recommended as a source for learning how the elements of telecommunications systems function.

This book is intended for the telecommunications manager who needs to know how to administer telecommunications systems. All three of the dimensions discussed at the opening of the chapter are covered: cost containment, productivity support, and enhancement of the organization's competitive position by managing telecommunications as part of an information resource. Each chapter describes a separate facet of telecommunications management. Not only are you shown what must be done, but you are also shown how to do it. The "how to" portion of each chapter emphasizes, where feasible, the use of personal computers and packaged software to simplify the management task. By using popular software packages such as a spreadsheet and a data base management system, you can establish a complete telecommunications management system, and design, evaluate, and control the costs of your system.

THE JOB OF TELECOMMUNICATIONS MANAGEMENT

Managing a telecommunications system, whether for a common carrier, a large private user, or even a small business or governmental organization, requires a variety of skills. The job is similar to a data processing manager's job; it requires a blend of business and technical skills. The most effective managers are equally adept at business administration techniques such as finance, accounting, economics, and statistics, the practice of which are unrelated to the technology. In this book we will discuss these concepts in enough depth to enable you to apply them to communications planning, design, and product selection.

Some managers believe that technical knowledge is of secondary importance compared to managerial skills. This school of thought suggests that good managers can manage any technology if they understand management principles. While there undoubtedly are such managers, technical knowledge is important for most managers. A telecommunications manager must deal with vendors of equipment ranging from key telephone systems to complex data communications systems, and services ranging from ordinary telephone services to packet switched data and software defined networks. Managers who venture into these fields without knowledge of the technology need either a highly competent and reliable staff, a large budget, or a lot of luck. In this book we will weigh most heavily on the side of the technology, not because it is more important than managerial ability, but because plenty of other information is available on general management techniques.

Although technical knowledge is required, a good knowledge of the technology is not enough to function effectively. In most companies the telecommunications manager's status can be enhanced only if he or she has a business orientation that far outweighs a technical orientation. A key function of the telecommunications manager is to translate technology into terms that top management can understand. To do this the manager must speak the language of both business and technology.

Planning

In telecommunications it is better to plan twice and implement once instead of the other way round. Few organizations remain static; they are expanding, contracting, moving, or simply rearranging, and the telecommunications manager's job is to plan so changes do not disrupt communications, which is the nerve center of many organizations. Telecommunications is something most users don't want to think about, and the telecommunications manager's job is to ensure that they don't have to. Hassle free communication is enhanced by an effective job of planning.

About half of this book is devoted to planning. Although only two chapters carry planning titles, all chapters support the planning process. We can identify three distinct types of telecommunications planning:

- Strategic and long range planning
- Current or short range planning
- Operational planning

The dividing line between these types of plans is not distinct. Strategic planning, as discussed in Chapter 2, is the process of supporting the company's competitive goals by opening its telecommunications and information systems to employees, customers, and the general public. As a consequence of developing strategic plans, long range plans, which are three to five years into the future, will emerge. Strategic and long range plans are not product and service specific. They represent the general direction the company intends to take to open new markets and enhance existing ones, or to improve productivity.

Current, or short range planning, is product and service specific. Chapter 6 discusses the techniques of developing requirements and converting these into specifications that are issued to vendors. Chapters 7 to 17 discuss techniques for selecting and applying equipment. This process converts long range plans into reality, and, therefore, must actively involve users' representatives.

Operational planning results in day-to-day improvements the telecommunications manager must make. It involves monitoring cost and service, and making adjustments to keep them in balance. It involves planning how to test the network, accommodating short-term service requests, and making plans for expansion of existing telecommunications equipment. It involves preparing forecasts of immediate service requirements, and when it is clear that insufficient capacity is available, feeding the forecasts into the long and short range planning process.

Operational planning is something all telecommunications managers do, if not deliberately, at least by default. Current planning is done well in some companies, and scarcely touched in others, which may lend the appearance of crisis management. Very few companies do strategic telecommunications planning, even though it may yield the greatest payoff.

Monitoring Regulatory Changes

Telecommunications costs are affected by activities in the regulatory arena, and keeping track of what is going on in the courts and commissions can be a difficult and confusing job. No future regulatory change is likely to have the impact of the decree that broke up the Bell System, but the pace of change did not stop with the agreement between AT&T and the Department of Justice. The seven operating regions that emerged from the breakup, shown in Figure 1.1, are lobbying the courts and regulatory commissions for further deregulation of their business, and for permission to offer interexchange long distance service. So far, they have been largely unsuccessful, but the regulatory process is dynamic, subject to political pressures, and may bend as additional pressure is applied. AT&T is seeking deregulation of its long distance business, and the regulators are debating about the degree to which other service providers should be regulated.

Both federal and state regulators have decreed changes in the ownership of inside wire, network channel terminating equipment, and station equipment. Long distance rates have dropped dramatically over the past few years, primarily because of changes decreed by the FCC. New tariffs are regularly being announced by local and long distance

FIGURE 1.1
The Seven Bell Operating Regions

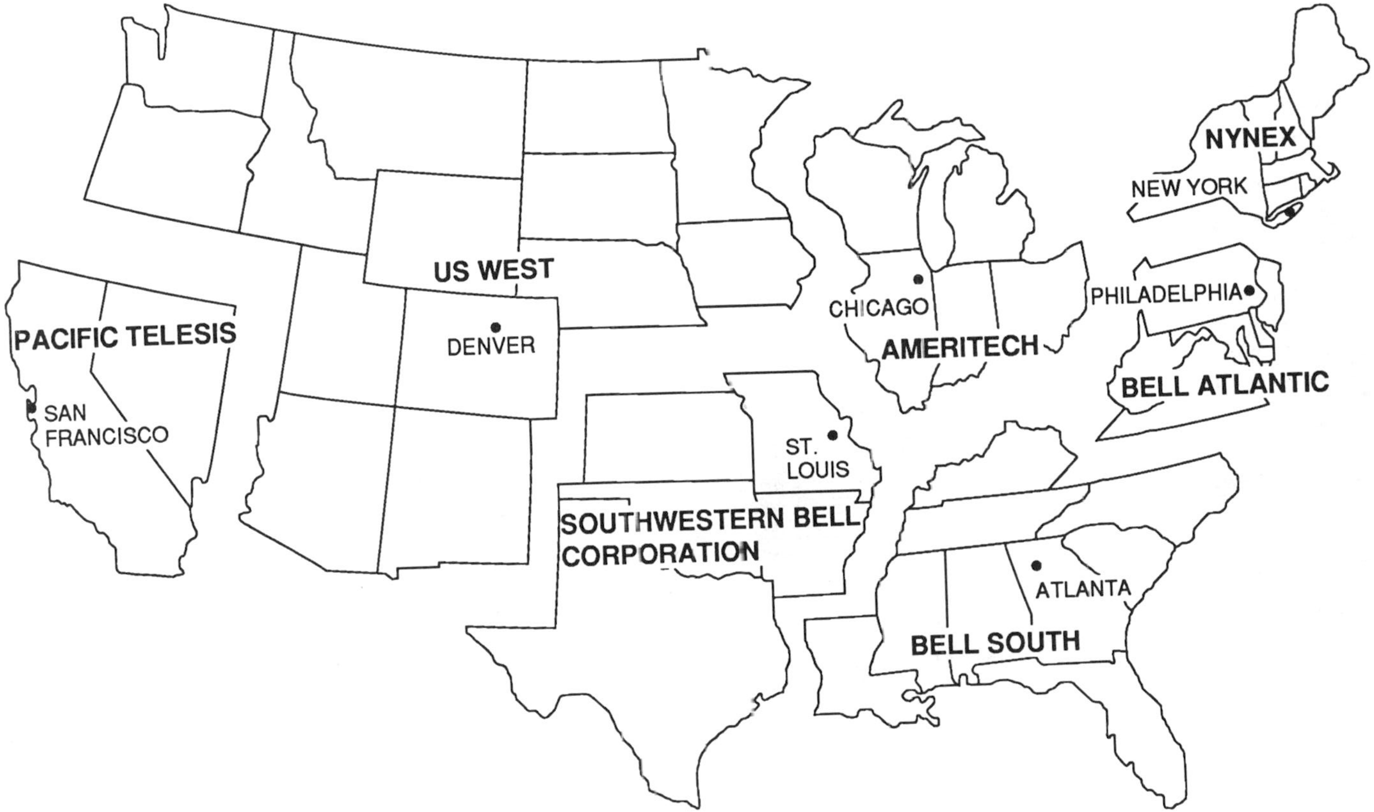

service providers. Telecommunications managers must keep informed about these changes. Decisions must continually be reexamined to see if they are still valid. Some form of information service, publications, or membership in trade associations is necessary to keep pace with regulatory change. However information is obtained, it is vital that the telecommunications manager be informed and build a plan flexible enough that it can be revised as the regulatory environment changes.

ORGANIZATIONAL ISSUES

Information takes many forms in modern organizations: voice, data, text, and video. Telecommunications is defined as the movement of information by electronic means, a definition that raises the question in many organizations of which unit takes charge. In the typical organization, voice communications came first, but only the largest companies actively managed it. Until the last few years in most organizations, voice communications was turned over to the telephone company.

Data communications started with the data processing manager, who often learned the technology as a matter of necessity. Until the last few years there has been little incentive to combine voice and data in the organizational structure, so they remained separate.

Over the last decade, the reductions in processing costs and the introduction of personal computers has given rise to a third discipline--administrative management. Because early devices were so expensive, word processing frequently came under centralized management that may also have assumed jurisdiction over telecommunications.

In some organizations the problem of who controls telecommunications has been resolved, but in others, the issue is still open. In companies where the issue is settled, it may have been resolved by centralizing all telecommunications under one head, or segmenting it among several work groups. There is no "best way" to deal with organizational issues because of differences in the structure and size of organizations. Neither centralization nor decentralization is best, but the issue must be addressed in every company.

Support Systems

Organizational issues go beyond who manages the system. A related question is whether to contract for services or to perform them with internal staff. Again, the answer is unique to the organization, but the decision should be based on criteria that are thoroughly examined. Chapter 12 deals specifically with contracting issues. Most of the remaining chapters contain enough information about product and service selection techniques and the elements of system administration that you can decide whether you want to develop internal staff or pay a somewhat higher fee for outside assistance.

In some cases, the need for outside assistance will be clear. For example, a company without internal legal staff will contract for it when it is needed in telecommunications work. Only the largest companies will decide to maintain their systems themselves, although most companies can profitably perform limited maintenance and save money compared to a maintenance contract.

Another related issue is the need for mechanized systems to monitor network performance, maintain cable and equipment inventories, support work order and assignment records, and aid in diagnosing troubles. These functions, which once were largely delegated to the telephone company, must now be performed or controlled by the using organization in today's environment of acquiring services from multiple vendors. Mechanization is also being extended to the process of testing and managing the performance of a telecommunications system, a topic that is discussed in Chapters 21 to 24.

Managing and Monitoring Service

A large percentage of the day-to-day activity of a telecommunications manager is concerned with keeping service and cost in balance. Cost in a telecommunications network is a composite of the cost of service itself and the cost of the people it is designed to support. Figure 1.2 shows conceptually how poor service can result in increasing personnel costs, but it also shows that excellent service may cost more than it is worth in terms of improved personnel performance. The telecommunications manager's job is to find how good service must be to support the legitimate demands of the users.

To illustrate this point, consider the example of local trunks from a PBX to the telephone company. If an outside trunk is always available for all users when they want one, no one will ever experience a delay in calling out. A fully connected network of this type is not economical, however, because trunks will remain idle a great deal of the time. We cannot afford to pay for such a high grade of service. We can afford to

FIGURE 1.2
Total Cost of Telecommunications Service

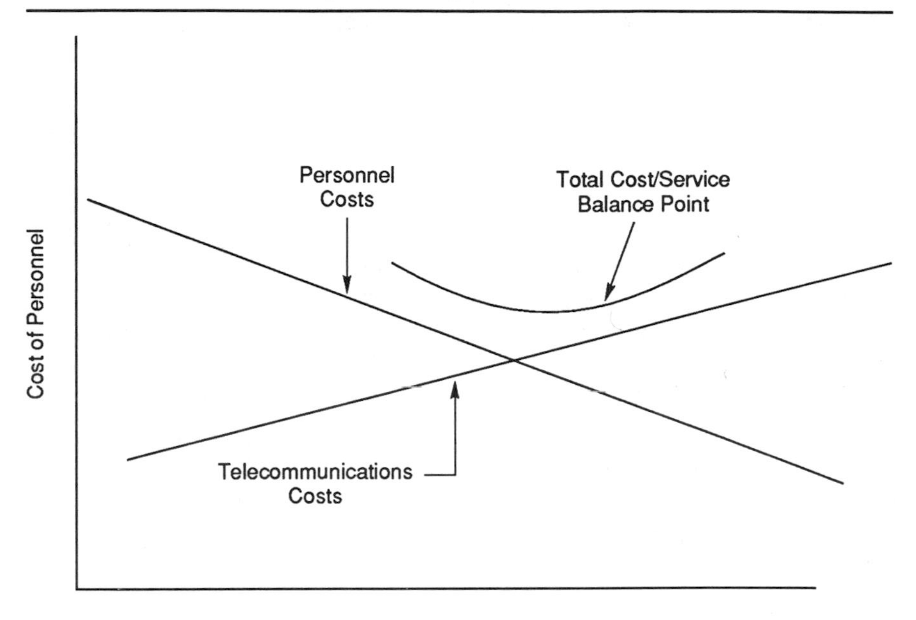

have employees and customers wait occasionally when all lines are busy, but if this happens too frequently, the cost of lost business and wasted time will offset the savings in telecommunications costs.

Formulas and tables are available and are printed in the back of this book to aid in achieving a cost/service balance, but the process isn't as simple as merely looking up numbers in a table. There are many other elements of service measurement that must be considered. Traffic tables help us determine how many circuits are required, but service objectives are required to apply them. Chapter 16 explains this process for voice networks, and Chapter 17 does the same for data networks. These chapters give you the tools you need to select the proper number of circuits, computer ports, terminals, and similar apparatus.

Simply ensuring that circuits are available when needed is not enough; users expect good transmission performance, a topic that is covered in Chapter 18 along with other service monitoring issues that are discussed in Chapter 21. They also expect the convenience of the productivity-enhancing features of a PBX. Features such as least cost routing, variable call forwarding, and message-handling facilities are discussed in Chapter 9.

Cost Control

Cost is the other side of the service coin, and there are many things telecommunications managers can do to control costs without adversely affecting service. Chapter 20 discusses the main issues. A call accounting system or station message detail register identifies the calls placed by each user. It helps control costs by allocating them to organizational units where they can be budgeted, and by enabling managers to control unauthorized use of long distance services. Other PBX features such as least cost routing, toll restriction, and queuing for low cost services, can have a dramatic affect on long distance costs. These features are discussed in Chapter 9.

When new equipment and services are selected, the use of a systematic process can improve your chances of buying the right thing the first time, and help you get the benefits of competitive procurement to avoid paying more than necessary. Chapter 7 and 8 cover procurement techniques. When major transitions take place, the opportunity for service disruption and excessive cost occurs. Chapter 14 discusses cutover management, and Chapter 15 explains how to manage telecommunications projects to avoid unnecessary costs.

The telecommunications manager is responsible for controlling other factors that will have a cost or service impact if they aren't managed. Security is an important factor. You must take measures to avoid theft of service and to prevent unauthorized access to trade secrets, private data bases, and sensitive equipment and facilities. Chapter 22 covers security on voice and data networks.

Chapter 21 discusses another aspect of cost control, telecommunications records. These are essential for clearing trouble and administering additions and rearrangements of facilities.

The Profession of Telecommunications Management

In the strictest sense of the word, telecommunications management is not a profession. It lacks a common body of knowledge, recognized standards and techniques, a recognized academic curriculum, and a professional licensing body or association, all of which tend to characterize true professions. Nevertheless, a discipline is emerging, in many cases as a matter of self-preservation on the part of the people who practice telecommunications management. Associations are being formed, and colleges and universities are beginning to offer telecommunications management curricula, but we have a long way to go. Most management techniques are learned by trial and error, and mechanized system support is far from maturity.

A major objective of this book is to offer a single collection of techniques for managing telecommunications systems in organizations of all types and sizes. Many of these techniques have been developed and used by telephone companies for managing their own networks and measuring service to their users. Other techniques such as forecasting and project management are essentially independent of the discipline they are used on, but nevertheless have specific applications to telecommunications management. Still other techniques have little or no application outside the realm of the telecommunications manager.

In most organizations, the job of telecommunications management lacks status and visibility. While the press touts the strategic value of telecommunications, top management is still in a cost-containment mode in most companies. One key to gaining the attention of top management is for the manager to possess the ability to prepare and present a credible plan with emphasis on the business as well as the technical side. The corollary of planning is execution, and, paradoxically, the telecommunications management job is often the most effective when it is the least visible.

CHAPTER 2

STRATEGIC TELECOMMUNICATIONS PLANNING

To speak of telecommunications strategic planning is to be somewhat inaccurate because a telecommunications strategic plan is actually a business plan in which telecommunications resources play a key role. As discussed in Chapter 1, telecommunications can be viewed within the organization in one of three ways:

- As a cost to be minimized
- As a tool to improve the productivity of the work force
- As a strategic weapon which, in conjunction with the company's information systems, can be used to enhance its competitive position

The planning approach depends on the nature of the enterprise. Non-competitive organizations, such as a government agency or a franchised garbage company, have little interest in strategic telecommunications planning. Their interests are more often in minimizing cost, controlling usage, and improving productivity. Other organizations have effectively examined their information systems, asked how they could be used to enhance their competitive position, and deployed a system that has repaid its cost many times over.

STRATEGIC TELECOMMUNICATIONS PLANNING

Strategic planning is a topic very much in vogue among telecommunications managers, but often not recognized or accepted by top management. To be accurate, we should refer to telecommunications planning in support of the company's strategic plan because, except for those companies whose principal business is telecommunications, the role of the telecommunications planner is to support the strategic plan of the entire business. As a supporting function, telecommunications has a vital role to play, and valuable suggestions to offer, but its role is subsidiary to the overall company mission. Therefore, we see the telecommunications and management information systems (MIS) strategic planners as a resource for the company's planning staff, with a mission of recommending ways in which the company's information base can be positioned to increase the company's effectiveness.

For purpose of this discussion, the terms "strategic" and "long range" planning can be considered synonymous. Strategic plans should be distinguished from tactical or operational plans. Operational plans are created and administered by line managers who are responsible for the day-to-day profitability and competitiveness of the enterprise. Operational plans are current, and deal with the immediate future, usually no more than two or three years out. Operational plans are concerned with details, and give managers specific tasks to perform.

By contrast, strategic plans deal with the general direction the company plans to take. The plans lack specific detail, but deal more in terms of product mix, market share, opportunities and threats, and competitive position. The strategic planner analyzes the longer range view, normally from three to seven years in the future. Some companies attempt to look even further into the future by postulating alternative scenarios, but the longer the planning range, the more likely something will render it invalid. For example, few companies predicted the breakup of the Bell System, an event that had more impact on telecommunications plans than any event since the invention of the telephone.

Assuming that the pundits of the information age, John Nasbitt (*Megatrends,* 1982) and Alvin Tofler (*The Third Wave,* 1980) are correct, we are entering an era where information is becoming a prized commodity. Since telecommunications is the electronic movement of information, the telecommunications manager has the responsibility of foreseeing the quantities of information likely to flow in the future, its destinations, and the most likely transport mechanisms. To do an effective job of supporting the enterprise's strategic plans, the telecommunications manager must keep a finger on the pulse of future trends that are likely to generate opportunities and create problems. Among these trends are:

- Marketing strategies of the major common carriers
- Sources within the company of information with a future market value
- Technological advances that are likely to affect the economic feasibility of various alternatives
- Regulatory trends that create opportunities or impose unexpected costs or restrictions
- Activities and plans of major competitors in expanding their telecommunications networks
- Financial plans and constraints within the organization

Elements of the Telecommunications Strategic Plan

This section lists and describes the contents of a telecommunications strategic plan. This outline is not rigidly formatted, but is a flexible compilation of factors that should be considered in most organizations.

The Environment

The company exists in an environment consisting of numerous external factors that affect its strategy: the financial community, customers, government, competitors, owners, and the public at large. Within the company, the telecommunications department exists in an organizational environment that affects how it acts as a support resource to improve productivity and enhance effectiveness. These forces are illustrated in Figure 2.1. The competitive environment should be discussed as an introduction to the strategic plan.

The Competitive Environment. For profit-making organizations, the competitive environment is probably the most important factor to consider in the strategic plan. The aim of the telecommunications system is to support competitiveness by making the company's workers, products, and possibly its informational data bases, accessible to its customers. Opening your company's information offers a way to distinguish its products and services from those of its competitors.

Telecommunications personnel must understand who the major competitors are, what their strengths and weaknesses are with respect to their telecommunications systems, and how your company's position can be improved *vis a vis* competitors by use of information resources.

FIGURE 2.1
The Environment of the Telecommunications Department

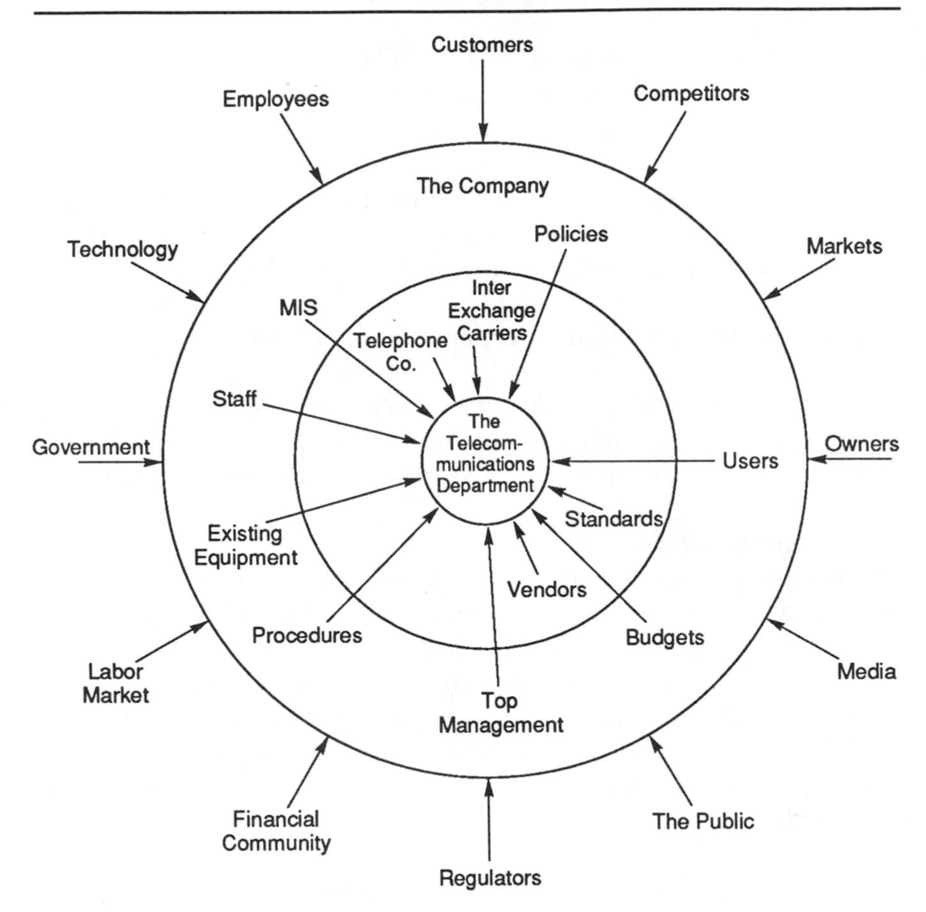

The Regulatory Environment. Without a reasonably good understanding of the regulatory process, and an interpretation of the meaning of trends emanating from the Federal Communication Commission and the state utilities commissions, the telecommunications department is in a weak position to predict costs and to be ready to take advantage of new opportunities. The strategic plan should discuss future actions you expect regulators will take, and what the impact on the organization is likely to be. For example, if local service is deregulated, as several Bell Operating Companies are pressing for, do you expect costs to increase, and if so, what action can the company take to bypass or improve utilization of local facilities? The plan should also discuss associations, and how membership and support of these can aid in supporting or opposing regulatory changes.

The Technological Environment. The pace of technological change has been staggering, generally outpacing the ability of most organizations to absorb the technology, and it shows no signs of abating. For example, 1987 was the year in which high- temperature superconductors began to demonstrate their promise. How will these affect the future of your business and when? Should your company be a leader in implementing new technology or wait until it is proven? What technologies do you expect will emerge in the future, and what will be their effect on your company? Do these technologies represent an opportunity to improve your competitive position, reduce costs, introduce new products and services, or meet customer demand more effectively?

The Vendor Environment. The last few years have brought a marked increase in the number of vendors offering telecommunications products and services. Possibly, the traditional vendor of choice no longer offers the same product and price advantage it once did. In this section of the plan you should review existing services, comment on their effectiveness in the future environment, and develop plans for strengthening or weakening ties to existing vendors. The questions of vulnerability from staying with traditional sources of supply should be examined, along with the benefits of contracting with alternate sources.

Compatibility issues should be discussed. If changes in architecture are suggested, the strategic plan should guide the organization on the timing for making changes, and the impact on existing systems that would result from shifting to a new architecture or absorbing architectural enhancements that have been announced by the vendors. For example, in this section you might discuss adopting a software defined network or changing from a traditional hierarchical data communications network to one that offers peer-to-peer communications. The suitability of your existing architecture for enabling clients to access information systems directly should be examined if this is part of the long-term strategy.

The Organizational Environment. Many companies have information and telecommunications organizations that have evolved from the early beginnings of data processing and the fully regulated telecommunications environment. The strategic plan should examine and comment on whether the current organization is appropriate for managing in the future. For example, should voice and data communications be integrated under the data processing department, a separate telecommunications department, or under an administrative services manager? Does the nature of the organization enable the manager to get information relating to future plans in sufficient time and in enough detail to enable him or her to make accurate forecasts and implementation plans? Does the organization structure give the telecommunications manager enough authority to control expenses and obtain real price competition for system enhancements? Are the MIS and telecommunications departments organizationally positioned to obtain information about future plans and trends so that they will be able to foresee requirements rather than merely to react to demand?

The Embedded Equipment Environment. All existing organizations have an embedded base of telecommunications and information processing equipment, and have a desire to retain it as long as it is economical to do so. Changing equipment often requires discarding a substantial investment, some of which may not be amortized or recoverable. To implement a strategic plan may mean making interim investments in equipment that will be retired before the end of its normal service life.

Significant changes may involve disrupting established vendor relationships, spending money for retraining and reprogramming, and rewriting policies and procedures. The adequacy of the existing base of equipment should be thoroughly discussed in the strategic plan, and recommendations for change should be supported by the economics and functionality of the overall plan. The degree to which the company is vulnerable to loss of business or productivity because of unprotected outages should also be examined, and where redundancy, uninterruptible power supplies, or other means of backup are required, these should be outlined in the plan.

The Work Force Environment. The strategic plan should examine the company's work force and comment on any changes that will be required as the strategic plan is implemented. For example, the productivity of the work force should be assessed, and where deficiencies are apparent, methods should be presented on how the telecommunications system can improve productivity. The impact of new systems should be discussed, together with plans for retraining or hiring to enable the skills of the work force to match the demands of the system. If labor relations problems or functional obsolescence problems exist, the plan should document the methods of coping with them.

The Financial Environment. Strategic plans cannot be executed without adequate financial support. Generally, the strategic plan does not deal in detailed cost analysis, but instead, describes the overall financial impact of recommended changes. The plan should explore the timing of major investments in equipment or reprogramming, and should document the expected benefits in terms of increased revenue or reduced expenses. Most plans will show the financial results expected with and without the enhancements recommended by the strategic plan. Detailed cost study techniques and plans for execution of the strategic plan are covered in Chapters 3 and 4.

The Market Environment. The strategic plan should comment on the organization's traditional markets, and describe any ways the recommended plan will affect markets. For example, will an enhanced telecommunications system open non-traditional markets? Will it significantly change the delivery system? What investments or procedural changes will be required of customers to operate under revised procedures?

The Operational Environment. A new strategic plan may be driven by geographical changes. For example, the company may be moving its headquarters, expanding its base of operations, or acquiring another company that must be integrated into the company's operations. The strategic plan should comment on how the telecommunications system will be used to integrate previously diverse operations without excessive delay or travel. This section of the plan will probably require commenting on all phases of operations with a discussion of how they will be affected by the proposed changes. If multiple networks are being integrated, the plan should comment on how they will be connected, maintained, and administered.

The Standards Environment. Telecommunications systems exist in an environment of internal and external standards. External standards include those set by major vendors and standards organizations. The standards to which the system will adhere should be explained. The plan should also discuss the internal performance standards expected from the telecommunications system including response time, blockage, service quality, and vendor support. The plan should also discuss service protection, security, and disaster recovery.

The Customer Environment. If the strategic plan involves changes in service delivery to customers, the plan should discuss how the services will be marketed to customers. If customers are being given direct access to the company's information base, the plan should discuss publicity, user training, assistance methods such as a technical assistance center, and other ways of assuring that business is not lost because users are either unaware of the availability of information or are unable to assimilate it into their operations.

Business Requirements

The telecommunications plan, as shown in Figure 2.2, is heavily influenced and, in many companies, is driven by business strategy. It must be stressed that in most businesses the telecommunications planner is not in the driver's seat, and may not be part of the overall business planning process. Knowledge of what the business plans are is of critical importance to both the strategic and short-range planning processes.

FIGURE 2.2
A View of the Telecommunications Strategic Planning Process

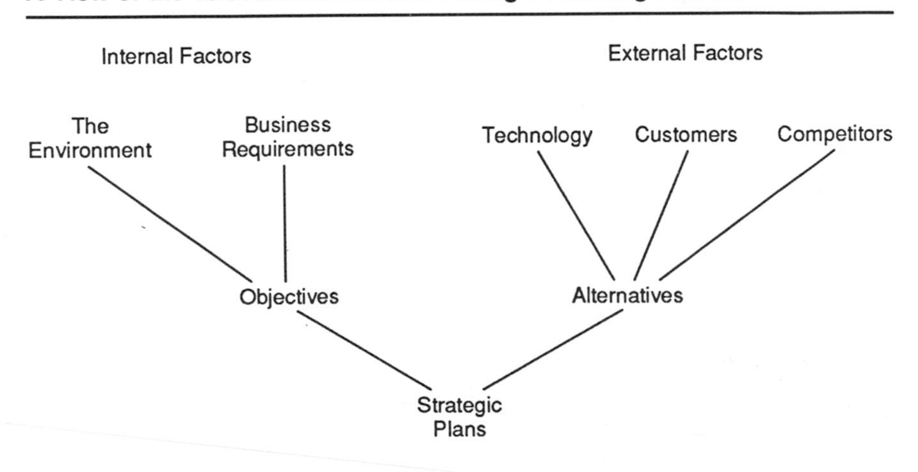

Management Information Systems Planning. The product to be delivered to the customer or employees is information; telecommunications is the means by which it is delivered. In the strategic planning process there is a close relationship between telecommunications and MIS planning. As shown in Figure 2.3, both are affected by business plans and technology, and both should be in lock-step when the final plan is prepared.

Expansion Plans. Plans for growth (or contraction) of the business have a significant effect on current and long range telecommunications plans. In strategic planning, the planner must be aware of top management's intentions to change the size of the business, to seek new markets, or to rejuvenate existing ones. Where new markets are sought, telecommunications managers should be in a position to propose ways networks can be constructed or expanded to take advantage of new opportunities.

FIGURE 2.3
Interrelatedness of Telecommunications and MIS Plans

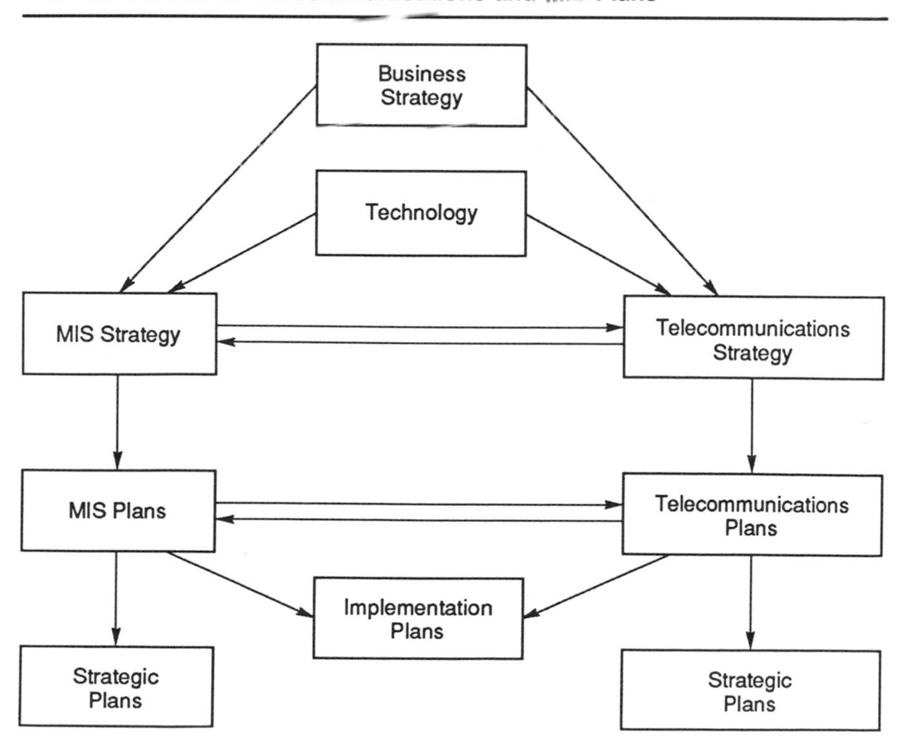

Financial Objectives. Closely related to expansion plans are plans for changes in the financial performance of the enterprise. For example, is capital expected to be in short supply, or does management plan aggressive investments to enhance financial performance?

Operating Principles. To prepare a strategic plan, you should understand the fundamental operating principles of the company. Is it risk-accepting or risk-averse? Is it liberal in extending telecommunications privileges to employees or does it exert tight controls? Is it a forward-reaching company that aggressively pursues the use of new technology, or does it have conservative management that accepts new products only after they are thoroughly tested by others? Every company has its culture, and the strategic plan that fails to support the culture will fail to win management acceptance.

Technology

Of the external factors that influence telecommunications plans, technology is one of the most important. In current planning, only presently available products are considered, but in long range planning you must consider the future thrust of technology because it is the force that makes possible plans that a few years earlier would have been impractical dreams. Several technological trends are evident today that will affect plans into the foreseeable future:

- A shift from analog to digital switching and transmission facilities
- The movement of computing resources closer to the end user
- A continually diminishing cost of bulk long-haul transmission facilities
- The compression of information into ever-smaller bandwidths
- A continuing reduction in the cost of data processing hardware
- A provision of intelligence in networks to enable them to adapt to changes in demand
- Plans of the telephone companies to deploy the Integrated Services Digital Network (ISDN)

These and other trends are reshaping information-based technologies. Strategic plans must recognize these trends, and determine what they portend for the future of the business. Other technologies such as the development of super conductors will have even more far-reaching effects, but until these developments are translated into actual products, it is impractical to consider them in strategic planning.

Customers

Any business that fails to focus on its customers is in for difficult times over the span of the strategic plan. A business must ask itself who its customers are, and how their requirements are likely to change during the time-span of the plan. Elements such as the following should be explored in the telecommunications portion of a strategic plan.

Changes in the Delivery System. Over the past few years we have seen a definite shift in the public's buying habits. More and more merchandise is being ordered over the telephone, using credit cards and package delivery services. The strategic plan must consider whether these trends will affect your company because they have such a marked impact on requirements for telecommunications services.

Changes in Customer Requirements. If the product and service needs of a major customer change, your product mix and delivery systems must change accordingly. The strategic plan should anticipate these changes even though they have not been expressed by customers. The plan should assess the impact of such changes on the need for telecommunications services.

Integration of Customers' Information Services. One of the most effective ways in which telecommunications and MIS can enhance a company's strategic position is by creating services that lock the customer in to further transactions with the company. For example, a program to allow customers access to your inventory of goods enables them to order directly from your computer, creating a tie that customers will find difficult to sever. The key is to integrate your operations with the customer's operations. For example, one company maintains inventory records for its customers and allows customers to order by using their own stock numbers.

Competitors

Strategic plans are not complete without an analysis of what the competition is likely to do in the future in response to the same influences that drive your strategic plans. Not only is it important to analyze current competition, it is also important to look at how technology or changes in the strategy of a competitor might create a threat where none exists presently. A classic example of failure to perceive a threat from a competitive product occurred in 1876–77 when Western Union Telegraph Company turned down a chance to buy the early Bell patents for $100,000. Questions such as these should be examined:

- Who are our competitors?
- What are their strengths and weaknesses?
- What are their telecommunications and MIS resources?
- Are they likely to use these resources in a way that damages our competitive position?

PLANNING FOR THE INTEGRATED SERVICES DIGITAL NETWORK (ISDN)

Depending on where you stand in the world of telecommunications, as a professional, an insider, or a user, the initials ISDN evoke a varied set of reactions. Telephone companies and many equipment vendors tout ISDN as a wave of the future. Mention ISDN to the average user, and you're apt to be met with a blank stare. Within the industry, ISDN inspires doubts and jokes ("innovations subscribers don't need", or "I sure don't know"). In this section we will discuss ISDN to give you an overview of its services and an understanding of how ISDN might fit into your company's strategic plans.

Right now there is little demand for ISDN, but this doesn't mean its detractors are correct. When desktop computers first came on the scene, there was little demand for them except for hobbyists, experimenters, and a few prescient people who could see the effect personal computers would have in the future. The same thing applies to ISDN. We don't see a great demand now because its facilities aren't available, and it may be the turn of the century, when the services are widely enough deployed for us to taste their benefits, before the impact of ISDN is apparent. The timing of ISDN is uncertain, but there is little question that it will be the way telecommunications services are delivered in the future.

ISDN Concepts

Today's telecommunications services all use separate local loops, and physical and electrical interfaces that are designed for a specific service, but that are incompatible with other types of services. Moreover, they use a variety of transmission media to connect to the telephone companies' central office. Figure 2.4 illustrates the status of today's services and interfaces.

ISDN offers a standard universal interface for switched and dedicated voice, data, and video services. The present local loop is replaced by digital channels that operate over a single pair of wires. As shown in Figure 2.5, two 64 kb/s clear channels, known as bearer or "B" channels, plus one 16 kb/s digital or "D" channel are provided. The ISDN shorthand term is "2B + D," giving each user a full 144 kb/s of bandwidth. All three

FIGURE 2.4
Existing Network Interfaces

In today's networks, each service has a separate interface to the network and a local loop dedicated to the service.

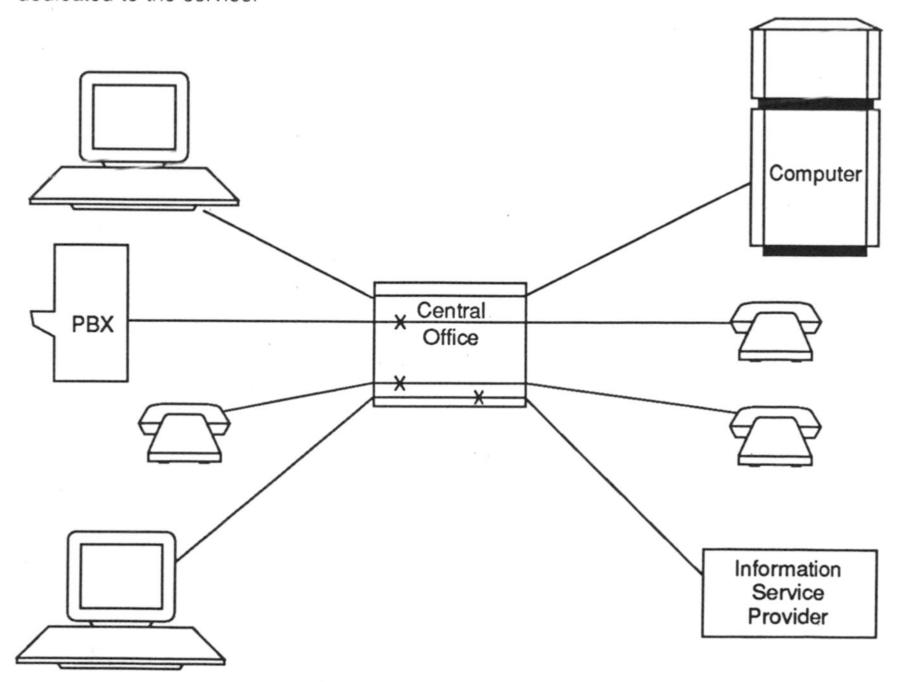

channels operate independently, which means that a user can use voice and data simultaneously over the B channels and still exchange signals with another party over the D channel.

The basic ISDN service just described is suitable for residences and small businesses, but large businesses require more bandwidth. To meet the need for more capacity, ISDN will offer a primary channel service which consists of 23 B channels and one 64 kb/s D channel. The D channel can support multiple primary channels, which means that several primary services can share the signaling capacity of one D channel.

The ISDN switch accepts information in any form from any source and directs it to the appropriate destination. As Figure 2.5 shows, this can be a voice or data circuit, a value added network, or an information provider.

ISDN not only provides access to other services, it also opens the way to a host of network services that are not practical today. For example, the following kinds of services are feasible with ISDN:

- A call waiting announcement could show the identity of the calling party, and the called party could decide to answer the call or not without interrupting the call in progress

FIGURE 2.5
ISDN Interfaces

In the ISDN, services will share common local loops and will have a uniform interface to the network.

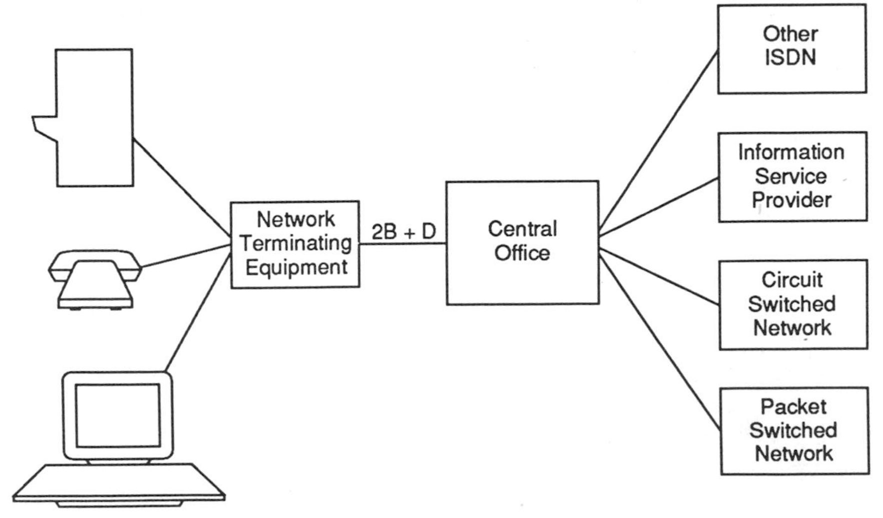

- Calls could follow traveling users to any telephone after they identified themselves to the network
- Users could establish personal feature lists that would be provided at any telephone they use

The above features are just three examples of ISDN capabilities for enabling the network to respond in a personalized rather than standardized manner to its users. The inventiveness of service designers will expand the list significantly as the capabilities of the network become clearer.

Network Signaling

In addition to the difference in transmission paths, ISDN uses a different signaling system compared to existing networks. In the present local telephone network, signaling takes place over the circuit itself. This means that the connection must be established end-to-end before the called telephone is rung. A high percentage of calls are not completed because the called telephone is busy or does not answer. Extra circuits must be provided to carry out this signaling function even though the primary purpose of the telephone system, which is carrying voice and data connections between two points, is not fulfilled.

Under ISDN, signaling will take place over a separate data network. The interoffice signaling protocols, Signaling System Number Seven (SS7), are being standardized by Consultative Committee on International Telephone and Telegraph (CCITT). By using these separate out-of-band signals, users will be able to communicate over the 16 kb/s D channel while the B channels are in use. Signals will also be exchanged between local central offices and interexchange carriers using SS7 in the same way that many interexchange carriers use out-of-band signaling between their tandem switching offices today.

Implications of ISDN for Telecommunications Managers

Despite the hoopla that has accompanied its field trials, ISDN service is not available outside trial applications and is not likely to be so within the immediate future. The first call between ISDN exchanges was placed as recently as November, 1986, which shows that the service is still very much in its infancy. The standards that exist are provisional, and because of the large number of companies, countries, and other interests that are participating in the standardization process, the overall shape of the service is likely to remain tentative until sometime in the 1990s. Even after the service is clearly defined, the transition to ISDN will be progressive and will stretch over many years. Almost certainly, existing non-ISDN telephone services will remain well into the twenty-first century. Telephone companies have enormous investments in conventional switching equipment; the bulk of the subscriber lines in the United States today are served by analog switching equipment. Even though these switches are electronically controlled, they are not ISDN compatible, nor can they easily be made so.

The first conversion of electromechanical to electronic switching equipment was made in 1965. At the time of this writing, more than twenty-three years later, a significant number of electro-mechanical switching machines still exist despite the fact that telephone companies can achieve a substantial improvement in customer service and reduction of costs by using electronic switching. By contrast, the improvements offered by ISDN service are less significant except for those few users who can make use of ISDN's bandwidth and features.

The conclusions we can draw from this discussion are these:

- The transition to ISDN will be gradual, probably spreading across at least three decades
- When ISDN central offices are placed, they will be installed where demand for the features is greatest, in areas of heavy business concentration
- Services and features will develop to meet the capabilities of ISDN, and will probably differ considerably from what is envisioned today

The key question for a telecommunications manager is what, if any, steps should be taken now to prepare for ISDN. Many products claim to be "ISDN compatible," but how can they be compatible with services and equipment that don't yet exist? The answer, of course, is that they can't, but they do support ISDN- like interfaces and services.

Network Trends

The question managers should ask is whether ISDN portends any future arrangement of communications services that would lead them to configure their networks differently than they otherwise would. To answer this question, let us analyze present trends in telecommunications services, what is driving them, and whether ISDN suggests a change in direction.

A Shift To Digital Circuits

Analog circuits are gradually being replaced by digital circuits for several reasons. First, analog switching is being replaced by digital switching because digital switches are less expensive to manufacture and give better transmission performance. Second, the nation is on the verge of an over-supply of digital circuit capacity. As numerous interexchange carriers are rushing to install fiber optic cables, the supply will inevitably drive prices down, and eventually analog circuits will be almost entirely supplanted by digital. The third force is the inherent design of digital, as opposed to analog services, which makes it more economical to purchase digital services in bulk. Digital circuits can be deployed over the local loop using existing twisted pair cable, and can interface directly with digital switches without using multiplexing equipment. The fourth reason is the continued growth of data transmission. Not only are more data communications applications emerging, but existing applications are being changed to distribute processing power closer to the end user.

These trends increase the demand for digital services, which, in most cases, are far more effective than analog services for the transmission of both data and voice. A T-1 service can be extended to a user's premises over two twisted pair wires, and can supply 24 voice channels. To distribute the same service on an analog basis would require 24 separate circuits, a coaxial cable, or the use of at least three multi-channel analog subscriber carriers. Analog services cannot compete economically, circuits are more expensive to design and maintain, and their transmission quality is inferior to digital circuits.

For these reasons, the crossover point between the cost of providing discrete analog circuits and bulk digital circuits is shifting to the point that displacement of fewer and fewer analog circuits is required to cost-justify a bulk digital facility. The individual twisted pair copper wire is still the most economical way of distributing circuits one-at-a-time to individual households, providing they are within three or four miles of a central office, but for businesses with multiple services, the economics are shifting in favor of T-1.

The trend toward digital circuits is unmistakable and inevitable. It also happens to require the same action that a manager would take in preparing for ISDN. To prepare for ISDN, managers should understand the shift toward digital services, continually search for economical applications, and design applications so voice and data can share the same transmission media. Exactly these same steps should be taken by managers who are attempting to minimize cost and maximize the effectiveness of the telecommunications services anyway, so in this case intelligently managing for today's requirements will result in future compatibility with ISDN.

Digital Switching

For the same reasons discussed under digital circuits, the tendency in PBXs and tandem switches is toward digital switching. Very few currently manufactured systems are truly analog machines, but the label "digital" does not mean compatibility with ISDN or even the digital transmission facilities discussed above. Switching systems labeled as "digital" may have one of several non-PCM switching networks; delta modulation and pulse amplitude modulation are the most common. These systems are not compatible with ISDN, and possibly more important, they are not compatible with T-1 carrier. Most digital PBXs are compatible with T-1, however, and can at least theoretically operate in an ISDN environment.

The ISDN compatibility of PBXs is probably a moot point. Most PBXs have a service life of less than ten years, and are probably fully supported by the manufacturer for no more than that. While it is important to consider the type of switching network in any PBX you purchase, ISDN compatibility is probably one of the last features to evaluate unless you are in a location where the telephone company has published plans to deploy ISDN. Several PBX manufacturers have announced plans to support ISDN interfaces. Obtaining assurance of future ISDN compatibility can do no harm, and may possibly be of future benefit, but should have little impact on most PBX selection decisions.

Network Intelligence

An important trend in networking is the deployment of intelligence in the network to re-route circuits in response to commands, changes in load, or by time of day. Network intelligence allows personalization of service such as recognizing account codes and allowing features and restrictions to follow a traveling user.

Network intelligence is employed in both private and public networks. It is not only necessary for ISDN, it is the very essence of its services.

Small private networks, for the most part, have no requirement to deploy network intelligence. As circuit prices drop, however, it becomes more important that companies develop ways to use their networks to support the company's strategic plans. The greater the degree to which a company controls its network with intelligent processors, the greater the degree to which it will be able to use ISDN services when they are available. As with other factors we have discussed in this chapter, ISDN is not the driving force, but it may eventually become the means by which your company's plans are executed.

Although ISDN is a service that managers should be aware of, its impact on near-term plans is likely to be minimal in all but a handful of companies. The evolution to ISDN will be slow; many existing systems and managers will be retired before ISDN affects them directly. Because ISDN is largely a concept at the present time, it is a factor to be understood rather than one to be managed and applied. Its primary importance to most managers lies in understanding the future directions telecommunications networks will take so you can be assured that your present plans are heading in the right direction.

OPEN NETWORK ARCHITECTURE

At the request of the FCC, the Bell Operating Companies have filed plans for Open Network Architecture (ONA). ONA should not be confused with the CCITT's Open Systems Interconnect model (OSI), which is a computer network architecture designed to permit devices from diverse manufacturers to communicate. ONA is a regulatory concept conceived by the FCC in 1986 as a way of achieving competition among regulated carriers and nonregulated providers of information and related services. Through ONA the FCC offered the telephone companies an opportunity to enter information and enhanced services markets in exchange for opening their networks to competition. When ONA becomes a reality, the FCC expects that new network features will be introduced, creating demand among users for specialized products capable of interfacing with new network offerings, thereby creating new applications.

Under ONA the capabilities of central office and local exchange services and facilities are to be unbundled. ONA is not intended to incorporate ISDN standards. The two technologies are expected to develop in parallel. Information service providers will be given the ability to provide innovative information transport, transaction, processing, retrieval, monitoring, and message delivery services to end users. ONA is expected to open the doorway to applications such as voice and electronic message, mail, and answering bureau services. Service providers will be able to use the local telephone network to provide services such as packet switching and protocol conversion, cable TV program selection, security, monitoring, and commercial alarm services and transaction-oriented services such as point-of-sale applications.

The question facing telecommunications managers is what affect ONA is likely to have on their long range planning. The answer in the near term is "very little." Compared to other countries, innovative network services tend to emerge slowly in the United States, due in large part to the pace of regulatory action. At the time of this writing, the courts have denied most requests of the Bell Operating Companies to enter information services markets, which in turn has led some observers to speculate that the BOCs will retard the pace of ONA development.

SUMMARY: INTEGRATING THE STRATEGIC PLAN

As shown in Figure 2.3, the strategic plan is an aggregation of business, MIS, and telecommunications plans and resources. Many companies that have been very successful in using their telecommunications and information resources as a competitive edge have not followed a deliberate process such as this one. They have simple perceived a need and launched a service. The successful ones such as American Airlines with its SABRE reservation system, American Hospital Supply Corporation with its ASAP system, and CitiCorp with its Global Transaction Network, are widely publicized in the media as companies that have done a strategic job of telecommunications planning. The ones that fail are seldom heard from again.

Process is no guarantee of success, and lack of process doesn't mean failure. In this chapter we make no claims that strategic telecommunications planning is a certain road to success, but with luck and a lot of good planning, it could be.

CHAPTER 3

CURRENT TELECOMMUNICATIONS PLANNING

Virtually every plan has its roots in a problem of some kind. A plan and its objectives can be understood only after you understand the nature of the problem, which is to say that the most effective planning process starts with an analysis of the underlying problems and their causes. Most of us have an unfortunate tendency to leap immediately from problem to solution, hastened along by an abundance of alternatives that promise immediate relief. The real problem is how to find the best solution in a sea of confusion, jargon, and hype.

The answer is to adopt a process that systematically carries you to the solution that most effectively addresses your specific problems, and to prepare a plan that documents the most effective solution. A telecommunications plan is normally the first step toward introducing a material change in the company's telecommunications or information management system.

In this chapter we will first discuss a problem-solving process that describes how to review the current situation in an organization's telecommunications systems and to develop alternatives for solving the problems. Next we will discuss techniques for preparing long range or fundamental plans, and carry these forward into current or short range plans. Finally, we will discuss a method of documenting and presenting the plan for higher management approval.

OVERVIEW OF THE PROBLEM SOLVING PROCESS

Whatever the profession, its specialists tend to use a consistent process for diagnosing problems and finding solutions. This section describes one approach to developing telecommunications plans, starting with a diagnosis of the most significant problems in the present system.

An excellent book on problem solving is *The Rational Manager,* by Charles H. Kepner and Benjamin B. Tregoe. Although written in 1965, the book presents a method that is still current. It is also available as a Personal Computer-based program called *Trouble Shooter,* which leads you step by step through the process and leaves you with a fully documented plan. The Kepner-Tregoe process can be simplified to the diagram shown in Figure 3.1. According to Kepner and Tregoe, a manager probes the current situation to find deviations from the level of desired performance. When managers are dissatisfied with current performance, they have, by definition, a problem. Current problems are the factors that drive the need for current plans.

Amateur and professional problem solvers alike follow the first part of the process of finding deviations and labeling them as problems, but from this point on, their techniques diverge. The amateur begins at once to develop solutions; the professional suspends judgment while refining the problem. More and more information is demanded until the nature of the problem and its causes are clearly understood. In the process, the professional begins to develop objectives.

Objectives, or goals as they are often called, are descriptions of how the situation will be when the problem has been solved. For example, an amateur might define a typical telecommunications problem as "terminal response time is too long." The objective might be "improve response time." A professional would stick with the problem until it could be defined more precisely; for example, "on 45 percent of our data transactions, terminal response time exceeds 6 seconds." With that problem stated, an objective can be developed; for example "at least 90 percent of data transactions will have a terminal response time of 3 seconds or less."

Objectives are divided into two categories, mandatory and desirable. Mandatory objectives are the conditions that an alternative must satisfy in order to be acceptable. Desirable objectives are the conditions we'd like the alternative to satisfy, but their lack won't disqualify it. For example, the objective above could be amplified to include a desirable objective: "at

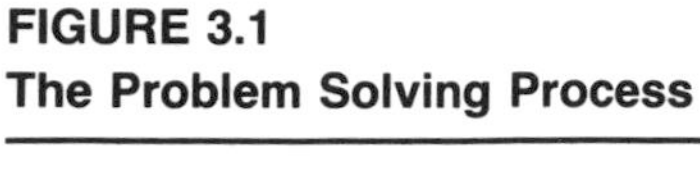

FIGURE 3.1
The Problem Solving Process

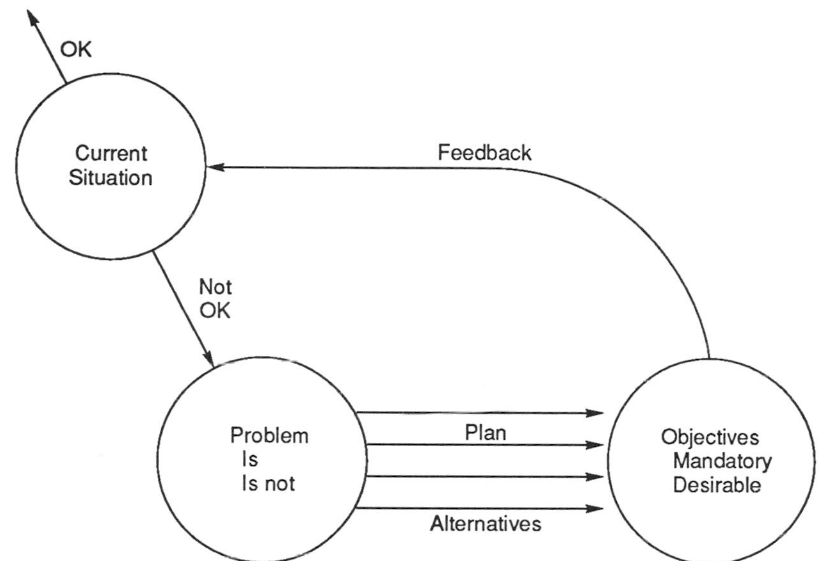

least half of the transactions should have a response time of 2 seconds or less." Alternatives that meet the requirement but fail to meet the desirable objectives can still be considered. By weighting the relative importance of desirables, different alternatives can be evaluated on an equal footing. When all the people with a stake in the outcome have agreed on the objectives, it is much easier to identify solutions.

A constructive way to view problems and objectives is to consider that they directly correspond to one another. A problem describes the situation as it exists now; an objective describes the identical situation as it will exist when the problem has been solved. Only when problems and objectives have been identified and documented in sufficient detail is it productive to begin developing alternatives. Unfortunately, this is seldom the way the planning process works in the real world. More often we choose an alternative and seek ways to justify it.

Once the objectives are clearly understood, alternatives can be developed. Later in this chapter we will discuss several methods of developing alternatives, always with the warning that easy solutions should be rejected until the problems and objectives are thoroughly understood. The alternatives are tested against the mandatory objectives, and those that fail to meet them are discarded. Of course, if objectives turn out to be unrealistic, they must be revised.

The alternatives that survive the initial screening are rated by how well they meet the weighted desirable features. The one that best complies becomes the chosen solution, and a plan is developed around that alternative. This process is described in Chapter 4, which discusses feasibility analysis.

All that remains is to implement the plan, and to review the results to ensure that expectations are being realized. This evaluation process is the feedback loop shown at the top of Figure 3.1.

THE TELECOMMUNICATIONS SYSTEM DESIGN PROCESS

Even though people may not be conscious of following the steps just outlined, the process is followed by most professionals, from doctors to architects, in diagnosing and solving a client's or patient's problem. Looking ahead to the time when the plan will be implemented, it is worthwhile at the outset to form a committee of people who can speak for their respective parts of the organization. After cutover, such a committee can become a users group that assists in verifying that the system is fulfilling its promise.

Initial Scan

A plan usually starts with a description of the present situation, determined by interviews with key people and an initial scan of documents that point to deficiencies. The sources of information that should be evaluated include the following:

- Telephone company and common carrier bills
- Review of existing equipment and network architecture
- Trouble history records
- Data transaction records
- Strategic plans
- Higher management and end user objectives

Not all of these documents will exist in all organizations, which may, of itself, be an indicator of problems that should be corrected. The objective of this phase of the analysis is to determine the source of any dissatisfaction with existing services.

Forecasting

The starting point of any telecommunications design is a forecast of requirements for lines, terminals, and trunks at cutover and through the life of the system. To paraphrase an old Mark Twain saying, forecasting is very difficult, especially with respect to the future, and it is often tempting to plan without one. Without a forecast, however, the telecommunications system may be over- or under-built, cost far more than necessary, and suffer a foreshortened life. Most systems can be expanded to a capacity limit, and when the capacity is reached, further growth requires an expensive replacement. Often, this expense can be avoided by a valid forecast.

The forecast serves as a coarse screen for filtering out alternatives that will not support requirements for the expected life of the system. For example, an organization with immediate requirements for a 75-station telephone system would consider a hybrid key telephone system, but if the forecast indicated that requirements would triple within the next five years, it would be advisable to consider only a PBX.

A telecommunications planner typically must develop forecasts for the following system elements:

- PBX stations, trunks, and ports
- Voice and data private line circuit requirements
- Front end processor and multiplexer ports
- Terminals
- Budgets
- Traffic usage
- Floor space

Ideally, a forecast should start with an analysis of historical trends. The past isn't always a good predictor of the future, but it is a good place to start. Bear in mind, however, that assuming that past trends will continue into the future has caused no end of trouble for more than one forecaster. Of course, many organizations have not kept records, and have no historical information on which to base a forecast. Chapter 5 discusses forecasting methods and techniques in more detail.

Identifying Problems and Developing Objectives

As we stated earlier, every telecommunications plan has its roots in a problem of some kind. You may be running out of capacity, moving to a new location, be dissatisfied with costs, be unhappy with the features of the present system, or be implementing a strategic plan. It is possible that no one has even verbalized the problems, or if they have, it's unlikely that all the problems have been clearly identified, and even less likely that the causes are known. Kepner and Tregoe contend that it is just as important to know what a problem is not, as to know what it is. For example, in the response time problem we discussed in the last section, it is important for a telecommunications designer to be able to state with confidence that the problem is not excessive central processing unit processing time. If this seems obvious, you might reflect on how many response time problems have been attacked by increasing modem speed, only to find that little, if any, improvement resulted.

For example, consider the multi-drop data communications circuit shown in Figure 3.2. When terminals are added to a circuit like this, the

FIGURE 3.2
Multidrop Data Circuit

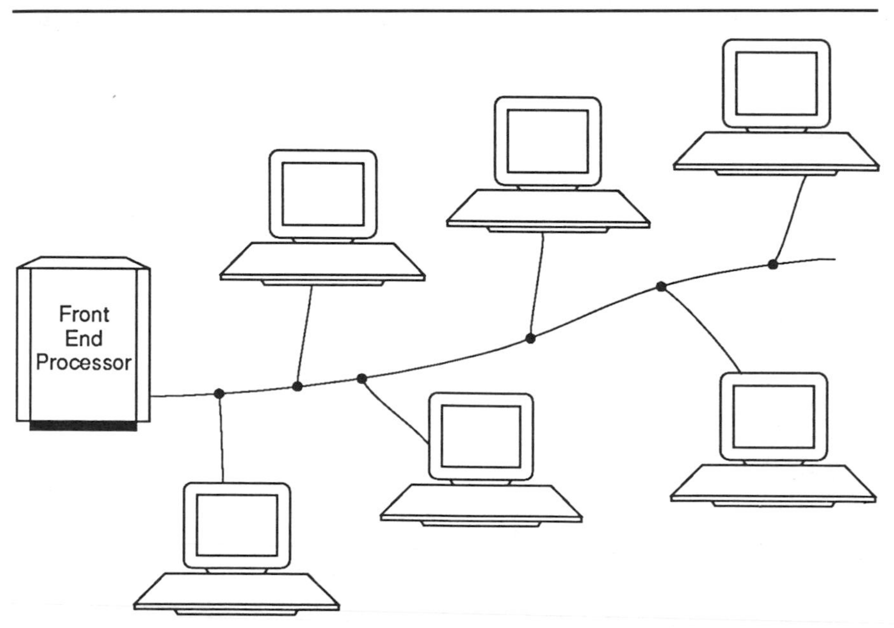

manager will eventually begin to hear complaints about terminal response time. The tempting approach is to increase modem speed. Conventional wisdom has it that doubling the modem speed cuts the response time in half, but as we will see in Chapter 17, it usually does not. As any data communications professional can tell you, the solution to response time problems lies in determining what is causing the problem. If you want to avoid wasting money on new modems or protocol changes, you must consider all the factors that could possibly cause excessive response time and find out whether they are the culprits. For example, the problem could be any of the following:

- Too many terminals on the circuit
- High error rate
- Improper block length
- Wrong polling method
- CPU delay
- Terminal operator errors
- Unrealistic expectations of the users

The list can be expanded, but the point is that attempting to solve the problem before you know what is causing it will lead you to explore blind alleys unless you're lucky enough to hit the right solution the first time. By arranging the list into two categories, what the problem is and what it is not, you can avoid a lot of wasted effort and needless expense. For example, if the problem is excessive error rate, operator error, or central processing unit blockage, raising the modem speed probably won't help much, and certainly will cost a lot of money.

It is important that the problem be stated in numerical terms if possible. For example, if the problem is that 10 percent of the transactions have a response time of 10 seconds, but 90 percent are five seconds or less, the problem may be unrealistic expectations on the part of the users. Management may have elected to pay the penalty of longer response time on 10 percent of the transactions. It is possible to design a service that never has more than three seconds response time delay, but this may drive costs too high. Objectives should be expressed in terms that management and users are willing to accept.

Objectives should be separated into mandatory and desirable categories. Mandatories are those that you are unwilling to compromise on regardless of cost. Desirables, or "value added" features, as some users call them, are those that you would like to have if the cost is right. It usually pays to weight desirable features on a scale of one to five, with five being the most desirable. When the users agree on the importance of the desirables, it is not difficult to select the most effective solution, using a process that is described in the next section. When the cause of the problem is known and the objectives are stated as mandatory and desirable, the next step of developing alternatives is greatly simplified.

Developing Alternatives

Often, when problems and objectives are clearly stated, the solution may be obvious, but usually it is not. With no obvious solution in sight, the next step is to begin developing alternatives. A frequent mistake people make at this stage is to limit the alternatives and thereby deprive themselves of imaginative solutions. Committees have an unfortunate tendency to discard solutions and to suppress alternatives by intimidating members who don't want to appear foolish. One excellent way of developing alternatives is "brainstorming," in which people propose ideas without allowing them to be evaluated until later. The objective of brainstorming or "green light" sessions is to unleash creativity by developing a free-wheeling environment in which people are encouraged to generate ideas. After a list of ideas is developed, a few good ideas are gleaned and expanded, and the rest are discarded.

Brainstorming is often overlooked by people who feel it is an unprofessional way for developing ideas for something as tangible as a telecommunications system, but professional system designers are often stymied by problems that turned out to be so obvious that everyone overlooked them. It is essential that analysts discipline themselves to avoid the easy fix. When the solution is chosen before the problem is understood, it is a sure sign of trouble.

Alternatives are evaluated by comparing them to the objectives. If the alternative fails to meet a mandatory objective, there is no reason to consider it further unless the objective needs to be changed. When the non-conforming alternatives have been discarded, the remaining ones should be weighted by how well they fulfill the desirable features. Usually, the alternative with the best score is selected and becomes the plan. The plan should then be documented to give everyone who has a stake in the outcome a chance to discuss it.

To illustrate the matter of setting objectives and weighting alternatives, let's take a typical case. Assume that an organization has a requirement for data communications between terminals located in the same building. The requirement can be satisfied by a digital PBX, a baseband Local Area Network (LAN), a broadband LAN, or a data PBX. Figure 3.3 shows a sample of objectives weighed against the four alternatives. The objectives are rated on a scale of one to five with five highest. This illustration awards the most points to the data PBX, but in another situation any of the other alternatives could be most effective.

With this view of the overall planning and problem solving process we can turn our attention to specific techniques for preparing and presenting the plan.

LONG RANGE PLANNING

Fundamental or long range telecommunications planning is an essential part of every organization with more than a handful of employees. It asks questions such as these:

- How adequate is our current system for supporting our long range development plans?
- Do we expect to make major organizational or geographical changes in the future?
- Are we losing productivity or effectiveness because of weaknesses in our present telecommunications system?
- Do we have planned changes in data processing, buildings, policies, organization, or vendor relationships that warrant a review or our present telecommunications systems?
- Is dissatisfaction being expressed with present systems, costs, or controls?

Long range telecommunications planning translates the organization's current situation, its strategic plans, and long range business plan into a telecommunications plan extending three to seven years into the future. The extent of the long range telecommunications plan is limited by the scope of the business plan. If the organization has no long range business planning, there is little reason to attempt a long range telecommunications plan because the telecommunications plan is subsidiary and supports rather than drives the organization.

If a business plan exists, it is essential that it is supported by a telecommunications plan that determines such things as these:

- Changes in network configuration such as a change from analog to bulk digital circuits or replacing common carrier circuits with private microwave

FIGURE 3.3
Weighted Alternatives for Local Data Communications

		Digital PBX		*Data PBX*		*Baseband LAN*		*Broadband LAN*	
	Weighting	*Score*	*Points*	*Score*	*Points*	*Score*	*Points*	*Score*	*Points*
Cost	25	2	50	4	100	3	75	2	50
Ease of Connectivity	7	5	35	5	35	4	28	4	28
Ease of Mainframe Access	8	3	24	3	24	5	40	3	24
Ease of Use	35	4	140	5	175	3	105	3	105
Flexibility	10	2	20	2	20	3	30	5	50
Expandability	15	4	60	2	30	4	60	4	60
Total Points	100		329		384		338		317

Note: This figure is for the purpose of illustrating the concept of weighting alternatives and is not intended to be representative of actual differences among alternatives.

- PBX or tandem switch replacement
- Provision of a centralized telecommunications maintenance and control center
- Support of major changes in data processing
- Support of major relocations

The fundamental planning process is illustrated in Figure 3.4. It begins with a complete understanding of business plans and organizational objectives. Problems the organization is experiencing are analyzed, and methods are developed for using the telecommunications system to solve them. Once the problems and objectives are well understood, the next step is to develop alternatives. The alternatives are tested against the objectives to determine how well they meet the mandatory features. Next, each alternative is tested for financial and economic feasibility.

The feasibility study process, which is described in Chapter 4, is used to determine which alternative offers the best economic and technical performance. The plan is documented in detail, approvals are obtained, and the plan is used as a guide for implementers to carry out.

Review of the Business Plan

The importance of reviewing the business plan before starting the telecommunications plan cannot be overemphasized. Such factors as these must be considered:

- Plans for major organizational changes
- Personnel growth and relocation plans
- Sales and earnings forecasts
- Planned acquisitions and divestitures
- Management's stance toward accepting or avoiding risks
- Required rate of return for capital investments

The organization's telecommunications objectives must also be considered. Objectives such as the following should be evaluated:

- Cost vs. service levels
- Features of the telecommunications equipment
- Cost control methods
- Ownership versus lease

FIGURE 3.4
The Fundamental Planning Process

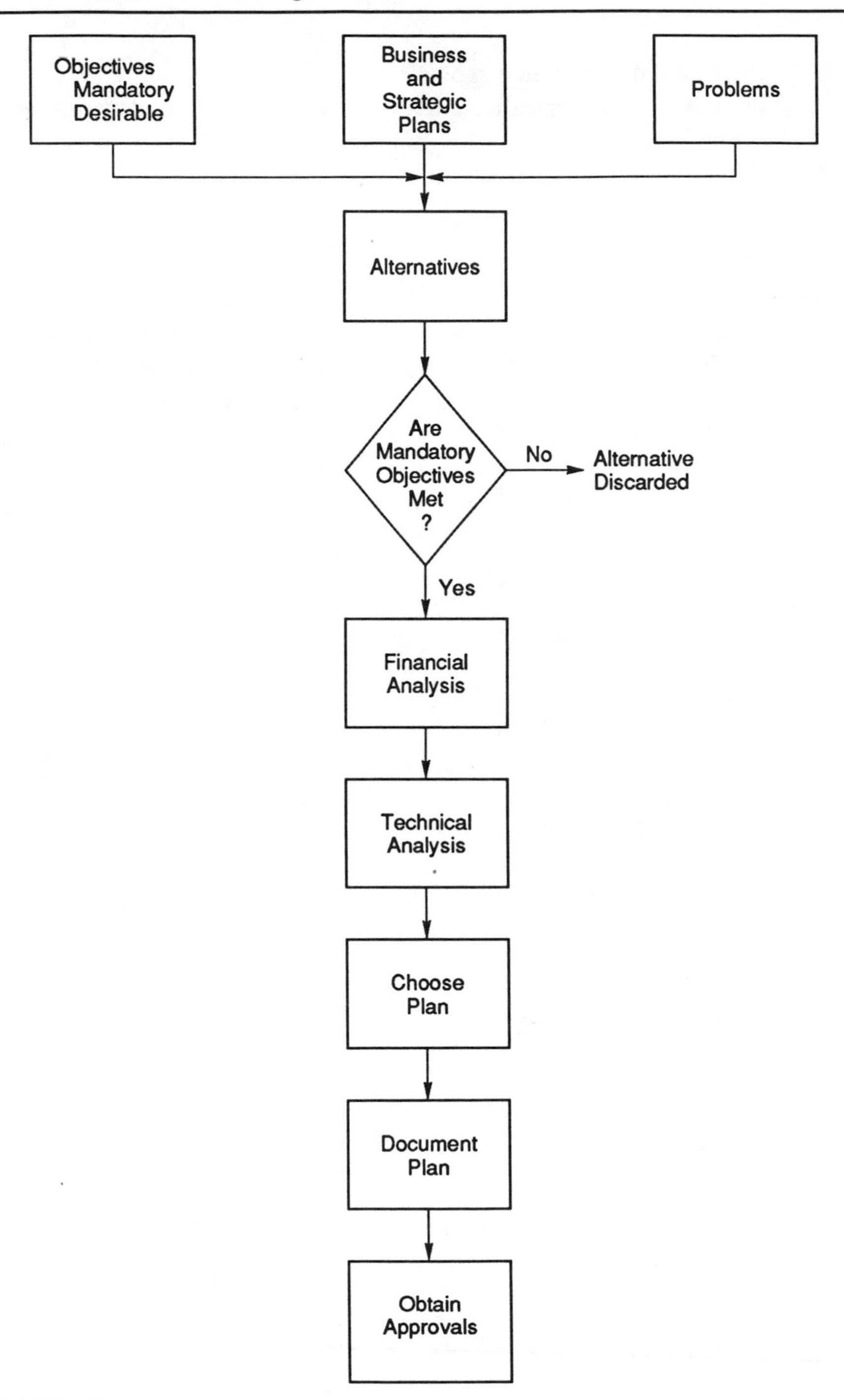

As discussed earlier in this chapter, any problems facing the telecommunications system should also be entered into the equation. When the present situation is thoroughly understood, alternatives are developed and tested for financial and technical feasibility, as discussed in Chapter 4.

DOCUMENTING AND PRESENTING THE PLAN

An effective planning job culminates in a report that higher management can either accept or disapprove. It is always difficult to reach just the right balance between presenting enough information to answer questions before they are asked, and so much detail that busy managers lack the time or interest to absorb it all. The amount of detail to present is a matter of judgment, but a few simple techniques can mark the difference between an acceptable report and one that is sent back for rework. Consider the following points in documenting your plan:

- *Executive Summary:* Prepare a summary of no more than one page that explains the essence of the plan, its alternatives, and its conclusions. This allows managers to spend a minimum amount of time in gaining the understanding you want them to have.
- *Recommendations:* Each recommendation in the body of the report should be clearly identified so a reviewer can scan the report for its conclusions and read the backup detail only if necessary.
- *Section Titles:* The liberal use of section titles enables a reader to scan a report quickly, looking for salient points.
- *Table of Contents:* If the report is more than a few pages long, a table of contents keyed to the major section headings facilitates a quick scan.

The most effective plans are documented for easy reading, following the points outlined above, contain clear and unambiguous conclusions and recommendations, and allow an approving authority to accept the report with no more than a signature. Reports that do little more than present findings, leaving the approving authorities to draw their own conclusions and make the final decision themselves, are the least effective and the least likely to be approved. The report should be written in such a way that it is a recommendation until management accepts it, but after it is accepted, it becomes an implementation plan simply by adding the details.

Alternatives

The plan should describe the alternatives that were studied. Alternatives that were considered but discarded should generally not be mentioned unless they are necessary to explain to management why a favored plan was not feasible. Each alternative should be given some brief identifier that is carried through the plan. For example, the plans can be referred to as "Plan A," "Plan B," etc., or they can be described as "Status Quo," "T-1 Network," or some other such identifier.

A discussion of the main differences, advantages, and disadvantages of each alternative is helpful to aid higher management in understanding the essence of each plan. It is also helpful to include a brief discussion of where the cost and technical data for each plan were obtained.

Financial Analysis

The financial analysis is unquestionably the most important part of a feasibility study because there is little reason to elaborate on a study that isn't cost-justified. Financial indicators such as net present value, internal rate of return, and payback period should be prepared for each alternative studied. These indicators, which are explained in more detail in Chapter 4, should be arrayed into a table such as Figure 3.5 to visually display the financial effectiveness of each alternative.

The financial results should be tested for sensitivity to uncertain factors. Sensitivity tests are readily displayed by reprinting the table showing the financial comparisons and by discussing in narrative form the conclusions you reach from the sensitivity tests. Only the tests that have some significance should be included. The others can be mentioned without cluttering the report with the results of insignificant tests.

FIGURE 3.5
Financial Evaluators of Alternative Plans

	Net Present Value	*Internal Rate of Return*	*Payback Period*
Plan A	$356,553	8.5%	4 Yrs
Plan B	$432,854	9.2%	3.2 Yrs
Plan C	$332,578	7.9%	4.5 Yrs

Technical Evaluation

Occasionally, all of the alternatives being studied use the same technology, but more often the alternatives will have some fundamental technical differences. Where differences exist, the study should discuss them for the benefit of approving authorities who may not be acquainted with the technology. Vast differences among organizations can be found in the degree to which they are willing to apply new technology. Some organizations pride themselves on being on the "cutting edge" of new technology. These companies are willing to pay a premium for fiber optics, VSAT (very small aperture terminal), or whatever tends to be the most forward looking at the time.

Other companies are unwilling to pioneer, and are comfortable only with proven products and thoroughly tested technology. The study should discuss any risks that higher management may be taking, and, where the risks exist, the plan should include a fall-back position and a discussion of the benefits of taking greater than normal risks.

The plan should explain how the recommended alternative fits into the company's future application of other advanced technologies. For example, if an extensive T-1 network is being proposed, the plan might logically explain how T-1 would fit into ISDN when it becomes available. If the plan is implementing an approved strategic plan, reference should be made to that plan.

The plan should also cover other technical issues such as the transition from existing technology to the new technology. The primary conversion steps such as training, new methods of operation, gradual versus instantaneous cutover, and other such issues should be explored.

Testing and troubleshooting methods should be discussed. If the proposed plan is a departure from the existing network, the report should explain how the network will be maintained and what, if any, additional personnel or training will be required.

Implementation

One part of the plan should be devoted to a broad outline of how the plan is to be implemented, including such points as major activities, who is responsible, what resources are required, and a recommended timetable of events.

User Concurrence

In most organizations it is wise to obtain concurrence that the recommended plan is acceptable to the major using organizations. This is usually not difficult if a representative group of users has assisted in preparing the plan initially.

Higher Management Approval

A properly documented plan should encounter little difficulty in gaining higher management approval provided the project is needed, it is technically and economically feasible, sufficient budget is available, and it has the support of using organizations. A well documented plan answers management's questions before they are asked, yet contains no unnecessary information and needless details.

CURRENT PLANNING

Current planning, sometimes called tactical or short range planning, acquires the necessary services and equipment to execute the long range or strategic plans. Contrasted to strategic plans, which are futuristic, goal-oriented, and somewhat theoretical, current plans are result and action oriented. Briefly put, the aim of current planning is to provide the capacity needed to support current operations. Where the tools of the long range planner are discounted cash flow analysis, evaluation of the competitive situation, and strategic positioning several years into the future, the tools of the current planner are requests for proposals (RFPs), budgets, and service orders.

The Demand and Facilities Chart

The demand and facilities or D&F chart is an invaluable current planning tool for visually displaying how capacity compares to present and future demand. As Figure 3.6 shows, capacity is displayed as a stair-stepped solid line that shows graphically the amount of facilities available and planned in the future. The demand line shows facilities used in the past and forecast in the future. If the demand line rises above the capacity line, as it did briefly in 1987, it indicates a capacity shortage. The degree to

FIGURE 3.6
Demand and Facilities Chart for Tracking and Forecasting Capacity

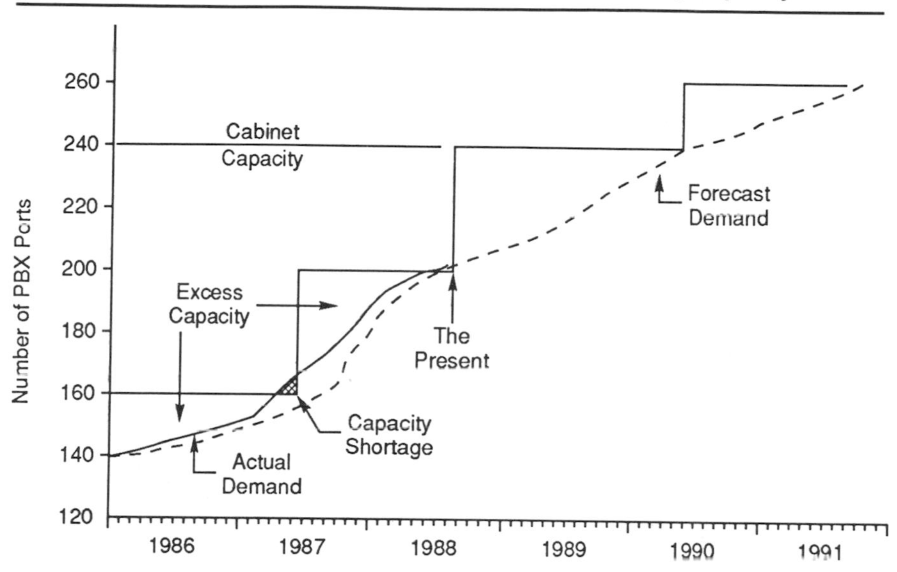

which the capacity line exceeds the demand is an indication of working spare capacity.

The growth increment of the system being tracked determines how much increase in capacity is obtained by each growth job. In the example shown in Figure 3.6, each growth increment adds capacity for 40 ports. When 240 ports are reached, a second cabinet must be added. This is shown by the solid line at the 240 port level.

D&F charts should be maintained on all critical facilities and equipment in the telecommunications system. As actual data become available, it is plotted as a solid line, and serves to validate the accuracy of the forecast. In the example of Figure 3.6, the actual demand consistently exceeds the forecast, which might make the telecommunications manager decide to schedule the 1989-1990 growth job somewhat earlier than normal.

The following are examples of facilities that should be tracked in most systems.

- PBX line and trunk ports
- PBX slot, shelf, and cabinet capacity
- T-1 channel capacity

- PBX trunk group capacity
- Statistical multiplexer capacity
- Multi-drop data communications circuit transaction capacity

To illustrate the use of D&F charts for forecasting local trunks, assume that a group of 22 local trunks is connected to a PBX that serves 150 stations. The service level objective is one percent blockage. (See Chapter 16 for an explanation of trunk group capacity and traffic engineering terms used in this illustration.) From Appendix A2 we determine that 22 trunks at one percent blockage have 13.3 Erlangs of capacity. To allow for future expansion, we create a chart with an upper range of 36 trunks, as shown in Figure 3.7. The horizontal scale allows space for quarterly measurements for the next three years, which requires 12 intervals.

From measurements, we determine that local incoming and outgoing traffic comprises 3.2 hundred call seconds (CCS) per station. From the forecast, we compute the expected demand as shown in the table above the chart. At the point where the demand curve crosses the capacity line, it will be necessary to order additional trunks to keep service at the objective level.

In looking at the chart, the traffic tables, and the cost of adding trunks we determine that it is most economical to add trunks two at a time. The cost of adding capacity is a function of three variables: the non-recurring charge from the telephone company, the cost of installing, translating, and testing the equipment, and the cost of trunk circuit packs. To make the D&F chart a useful tool, it is essential to post actual results to the chart periodically. If your PBX furnishes on-line traffic usage information, the results should be posted weekly or monthly, depending on the amount of effort required to collect and interpret the data. If special studies are required to obtain the data, it should be posted less frequently. The less frequently it is posted, the greater the likelihood of purchasing excess capacity or experiencing service degradation.

Other Current Planning Techniques

A D&F chart is a useful tool for planning and adding future growth, but major changes that reflect the results of reorganizations or acquisitions or

FIGURE 3.7
Using the D & F Chart to Plot Local Trunk Demand

Forecast		*Stations*	*CCS*	*Erlangs*	*Trunks Required*
Year 1	1 QTR	150	480	13.3	22
	2 QTR	165	528	14.7	24
	3 QTR	178	570	15.8	25
	4 QTR	189	605	16.8	26
Year 2	1 QTR	200	640	17.8	27
	2 QTR	212	678	18.8	29
	3 QTR	235	752	20.9	31
	4 QTR	244	781	21.7	32
Year 3	1 QTR	250	800	22.2	33
	2 QTR	255	816	22.7	33
	3 QTR	267	854	23.7	34
	4 QTR	274	877	24.4	35

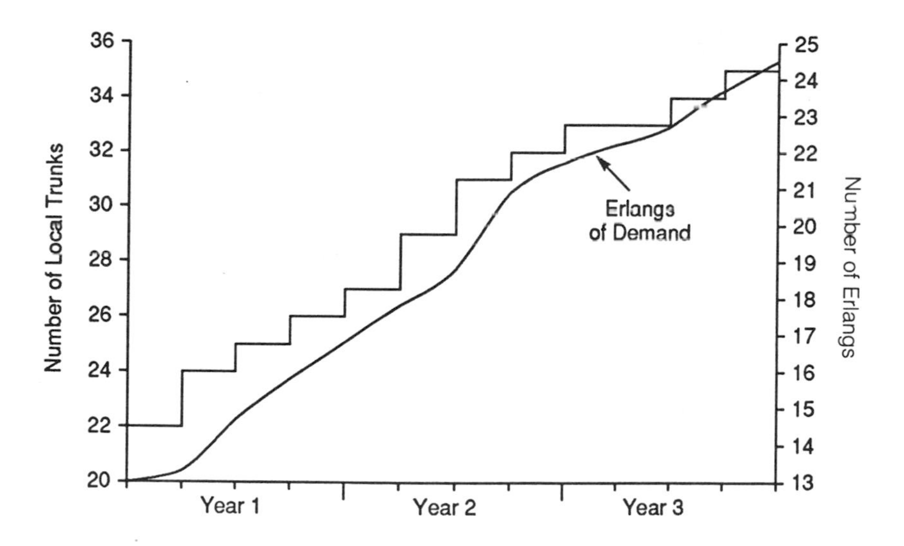

replacements of existing equipment require a more exacting process for acquiring equipment and services. When major growth jobs are engineered, you will undoubtedly use some form of competitive technique to acquire the additional facilities. These techniques are explained in more detail in subsequent chapters.

SUMMARY

If the most frequent mistake made in telecommunications planning is letting the system go with no visible plan, the number two mistake is putting the plan on the shelf and never revising it. A good plan is flexible. It can change as the organization changes, but someone has to review it to be sure it is still pointing in the right direction.

The third planning deficiency is the failure to find out whether the plan delivered its promises. Projects are often justified on the basis of some expected return, but no one goes back to see if it happened according to plan. A good planner verifies actual results to find out whether they were achieved. This activity, illustrated by the feedback loop in Figure 3.1, is an essential part of the planning process. If cost savings or productivity gains aren't realized, you must find out why, and either change your expectations or adjust operations to conform to the plan.

CHAPTER 4

FEASIBILITY ANALYSIS

Feasibility analysis is part of the planning process that is used to evaluate either current or long range plans. The long range planner develops alternatives from knowledge of demand and telecommunications technology. The feasibility analyst (who may be the same person as the planner) uses financial techniques to determine economic feasibility, and technical evaluation techniques to determine technical feasibility.

The two planning chapters just concluded explain the elements of the planning process. This chapter discusses the mechanics, the methods of applying financial study techniques to determine which of several alternatives offers the most favorable result. The purpose of a feasibility study is to examine an alternative from both its financial and technical aspects with the objective of eliminating it from further consideration. If it cannot be eliminated or is the last to go, it is feasible.

This process of screening alternatives through progressively finer sieves finally results in only one alternative surviving. An effective feasibility study process attempts to eliminate alternatives by spending the least amount of effort. It is a waste of time to evaluate an alternative rigorously, only to discover that it has some obvious flaw that should have disqualified it immediately.

In this chapter we will focus primarily on financial evaluation techniques. Although these techniques are somewhat exacting and tedious, they can be simplified by using discounted cash-flow analysis on a personal computer spreadsheet to evaluate the expenditures.

It is important for telecommunications managers to speak the language of financial analysts because investments in PBXs and communication systems, and expenditures for network facilities, must be evaluated along with investments in motor vehicles, computers, office furniture, buildings, and other projects that compete for scarce capital.

Non-profit organizations and government agencies lack the profit motive for evaluating investments, but nevertheless, budgets are always tight, and expenditures for telecommunications products and services must still compete with other claims on the budget.

FINANCIAL EVALUATION TECHNIQUES

There is no single "best" way of evaluating the financial feasibility of an alternative. The techniques used must be geared to the level of sophistication of the individual or group that makes the investment decisions. It is overkill to talk about discounted cash flow to a successful proprietor who bases decisions on the amount of cash in the till, just as it is folly to talk about simple payback to a company with a full time staff of financial analysts.

An investment can be evaluated in many different ways. Each technique tells part of the story, and the accuracy of the conclusion depends on your insight into the meaning of the numbers. Numbers alone don't tell the entire story. One alternative may have an intangible benefit that cannot be expressed numerically. The successful analyst knows how to use a variety of techniques to evaluate alternatives. An old adage says that when your only tool is a hammer, you tend to treat every problem as a nail. Enlarge your professional tool kit so you can approach every analysis from several aspects and choose the one that yields the best answer.

Pay Back Period

The easiest (and least valid) financial evaluator to apply to an investment is the pay back period. Assuming that an alternative reduces expense or increases revenue, the capital investment is divided by the annual savings to calculate how many years it takes to repay the initial cost.

For example, assume a company is leasing telecommunications services from a vendor for $30,000 a year. If it invests $100,000 in a microwave radio, its annual lease expense will be eliminated, but costs of $10,000 per year will be incurred for maintenance and electric power. The annual savings are $20,000. The payback period is:

$$\frac{\$100{,}000 \text{ per year}}{\$20{,}000 \text{ annual savings}} = 5 \text{ years payback}$$

Many investments are evaluated on exactly this basis. The arithmetic is simple, and if the microwave can be used more than five years, the investment may be attractive. This kind of analysis has a serious weakness, however; it ignores the time value of money. Assume that money can be borrowed for 10% or is already invested at 10%. It is evident that to make the $100,000 investment the company is either spending $10,000 per year in interest costs or foregoing $10,000 per year in interest income to save $20,000 a year in operating costs. Suddenly, it appears that the payback period is more like ten years when the cost of money is considered. Even this assumption, however, ignores the fact that $100,000 isn't simply sitting in the bank drawing $10,000 per year of simple interest. The interest is compounded. It draws $10,000 the first year, but the second year it draws interest on the principal plus the interest, or 10% of $110,000, which is $11,000.

The analysis is further complicated by the fact that the savings that result from avoiding the cost of leasing circuits can also be invested, and therefore also help to offset the cost of the original investment. It is obvious that some way must be found to make order from what would otherwise be a hopelessly complex process. Several techniques employing financial tables may be employed. We will consider two, net present value and internal rate of return. Other techniques such as present worth of expenditures and discounted payback period are explained in financial texts and reference books, some of which are listed in the bibliography.

Present Value

The time value of money is an important concept to keep in mind when evaluating alternatives. Often, one alternative requires a large cash payment, but has low recurring costs, while another has a low initial payment but larger recurring costs. Future expenditures are expressed on an equal footing with current expenditures by determining their *present value*. The present value of a future cash flow is determined by discounting it from the future time in which each expenditure occurs to its value at the present. The present value of a future cash flow can be thought of as the amount of money that you would have to invest now to have enough cash in the bank to pay each expenditure when it comes due.

Costs and revenues are moved back and forth in time by applying compounding and discounting factors, which can be obtained from financial tables or calculated with a spreadsheet. Present and future value of a dollar are two complementary concepts that are used to express expenditures in terms of their present worth. The *future value* of an investment is the value to which a dollar will grow if invested at a given rate of interest. The future value is determined by compounding the investment for the required number of periods. Table 4.1 lists the compounding and discounting factors for 10 years at 10 percent interest. Appendix B contains compounding and discounting factors for costs of money from one to 15 percent. Such tables are based on amounts of one dollar. The factor is multiplied by the total dollar amount to arrive at the total figure used in calculations.

The present value of a future cash flow can be visualized as an amount of money which, if invested today, will grow to a given amount in a specified number of years. For example, the present value of $1,000 in five years at 10% interest is $1,000 X .6209, or $620.90. If $620.90 is multiplied by the compounding figure of 1.611, we obtain $1,000. Compounding and discounting factors can be used to move dollar amounts forward or backward in time to express them in terms of their present value. Figure 4.1 shows the complementary relationship between present and future value.

Net Present Value

If cash flows from two plans are arrayed as shown in Figure 4.2, the cash flows can be discounted to their present worth by means of the discount

TABLE 4.1
Compound and Present Value of $1 at 10 Percent

Year	*Compound Factor*	*Discount Factor*
1	1.100	.9091
2	1.210	.8264
3	1.331	.7513
4	1.464	.6830
5	1.611	.6209
6	1.772	.5645
7	1.949	.5132
8	2.144	.4665
9	2.358	.4241
10	2.594	.3855

FIGURE 4.1
Compounding and Discounting: Value of $1 After Year N at 10% Interest

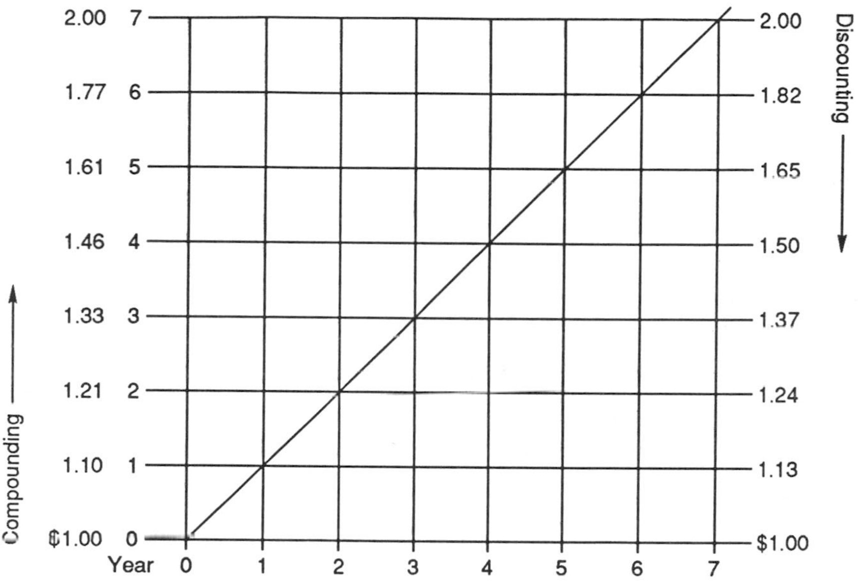

Compounding	Year	Discounting
2.00	7	2.00
1.77	6	1.82
1.61	5	1.65
1.46	4	1.50
1.33	3	1.37
1.21	2	1.24
1.10	1	1.13
$1.00	0	$1.00

Year 0 1 2 3 4 5 6 7

factor. The two alternatives shown demonstrate how to solve the problem of determining whether an initially higher cost is justified by reduced recurring expenses. The factor for making this comparison is known as *net present value* (NPV). NPV is defined as the sum of the discounted cash flows minus the original investment. Since the cash flows of most telecommunications products are negative, meaning they represent a cost to the organization, the NPV of such plans is typically negative. An exception to this rule would occur in an organization that purchases telecommunications products to generate income by reselling them. Examples would be a telephone company, a common carrier, or a company selling PBX services through a shared tenant service.

NPV is usually used to compare two alternative investments. The investment with the highest (or in our case least negative) NPV will consume the least amount of cash during its life. In Figure 4.2, for example, Alternative B offers the best value. Even though its recurring costs are $5,000 per year higher than Alternative A, the higher initial cost of Alternative A is not offset by its lower annual costs.

FIGURE 4.2
Comparative Cash Flows for Two Products at a 10 Percent Discount Rate

Year	Discount Factor at 10% (A)	Plan A Cost (B)	Plan A Present Value (C) (A × B)	Plan B Cost (D)	Plan B Present Value (E) (A × B)
1	0.9091	$(10,000)	$ (9,091)	$ (15,000)	$ (13,636)
2	0.8264	(10,000)	(8,264)	(15,000)	(12,396)
3	0.7513	(10,000)	(7,513)	(15,000)	(11,269)
4	0.6830	(10,000)	(6,830)	(15,000)	(10,245)
5	0.6209	(10,000)	(6,209)	(15,000)	(9,313)
6	0.5645	(10,000)	(5,645)	(15,000)	(8,467)
7	0.5132	(10,000)	(5,132)	(15,000)	(7,698)
Sum of Cash Flows		$(70,000)	(48,684)	$(105,000)	(73,026)
Original Investment			(150,000)		(120,000)
Net Present Value			$(198,684)		$(193,026)

Two plans, A and B are compared in terms of their life cycle costs for seven years at a 10 percent discount rate. Product A costs $150,000, but has annual costs $5,000 less than Product B, which costs $120,000. When costs are discounted to their present worth, we find that Product B is a better value because it has a higher (less negative) NPV.

The feasibility study process consists of the following steps:

1. Determine the cost factors (and revenue factors if applicable) associated with each alternative. Table 4.2 lists typical cost factors for telecommunications projects.
2. Assign values to each factor. Values are obtained from vendor quotations, common carrier tariffs, your organization's cost factors, and other such sources.
3. Determine your organization's cost of money. The finance or accounting department should be able to supply the figure.
4. Discount each cash flow to its present value.
5. Summarize the present values and subtract the initial investment to obtain the NPV.

Obtaining accurate cost information is the most elusive part of a feasibility study. For example, future power and maintenance costs are affected by inflation and other factors that are difficult to predict. In a later section we will discuss how the analysis can be sensitized for

TABLE 4.2
Typical Cost Factors Used in Telecommunications Feasibility Studies

Cost Factor	*KTS*	*PBX*	*Data Comm*	*LAN*
		Capital Costs		
Purchase Price	X	X	X	X
Installation	X	X	X	X
Spares	X	X	X	X
Shipping	X	X	X	X
Sales Tax	X	X	X	X
Equipment Room Preparation		X		
Station Wiring	X	X	X	X
Salvage Value	X	X	X	X
		Recurring Costs		
Electrical Power		X	X	X
Floor Space		X	X	
Maintenance	X	X	X	X
Administration (Changes, etc.)	X	X	X	X
Common Carrier Costs	X	X	X	
Rearrangement Costs	X	X	X	X
Depreciation Tax Effects	X	X	X	X
Software Right-to-use Fees		X		X
Costs of Failures				
Lost production time	X	X	X	X
Repairs	X	X	X	X

The factors marked with "X" above generally apply to that type of equipment.

uncertainty by substituting values within the range of certainty and rerunning the analysis. By computing the difference between the NPVs of the various alternatives, you can rank the products by the amount of cash they are likely to consume throughout their service lives.

Bear in mind that a feasibility study is an incremental study. That is, it compares the *differences* between alternatives, and is normally not concerned with the absolute magnitude of cash flows. Assume, for example, that one alternative has a maintenance cost of $10,000 per year and another costs $15,000. It is equally valid to enter the actual magnitude of each cash flow, or to enter nothing for the lower cost alternative and charge the higher $5,000. If cash flows are equal for all alternatives, they can be omitted without affecting the outcome of the study. Remember, however, that if any cash flows are omitted, it is not valid to compare the percentage of difference among the results.

Internal Rate of Return

Net present value alone is seldom a valid way of comparing alternatives. NPV tells which of several alternatives is the most cost effective, but it does not always answer the question of how effective the alternative really is. Internal rate of return (IRR) is a financial evaluator that expresses the amount of money that an investment returns to the enterprise. It is determined by finding the discount rate at which the NPVs of two alternatives are equal.

For example, Alternative A in Figure 4.2 saves $5,000 per year, but it requires $30,000 more capital than Alternative B. A superficial payback analysis would suggest that the investment is repaid in six years, but as we have seen, the time value of money extends the period longer than that. An analyst would ask how much money is earned on the additional investment.

In Figure 4.3 we have reduced the cost of money to four percent, which increases the discount factor and makes Alternative A more attractive compared to Alternative B. In fact, at four percent the NPVs of the two plans are nearly identical. By varying the discount factor we find that at 4.01 percent the NPVs are identical, so this factor is the IRR. As a practical matter the assumptions used in most financial studies are imprecise enough that rounding off the IRR to the nearest whole number is sufficiently accurate.

IRR isn't always a valid indicator. If difference in investment between Plans A and B is great enough, it would require a negative discount rate to make the NPVs equal. Since a negative discount rate is an undefined quantity, IRR is a meaningless indicator.

Although it is not difficult to compute NPV manually, IRR is an iterative process that is time-consuming by hand. Most personal computer spreadsheets have an IRR function or, as explained later in this chapter, the calculation can be made by varying the discount rate in a spreadsheet used to calculate NPV.

Study Length

The length of the study is often an important factor in determining which product is the most economically attractive. Since Plan A in Figure 4.2

FIGURE 4.3

Year	Discount Factor at 4% (A)	Plan A Cost (B)	Plan A Present Value (C) (A × B)	Plan B Cost (D)	Plan B Present Value (E) (A × B)
1	0.9615	$(10,000)	$ (9,615)	$ (15,000)	$ (14,423)
2	0.9246	(10,000)	(9,246)	(15,000)	(13,868)
3	0.8890	(10,000)	(8,890)	(15,000)	(13,335)
4	0.8548	(10,000)	(8,548)	(15,000)	(12,822)
5	0.8219	(10,000)	(8,219)	(15,000)	(12,329)
6	0.7903	(10,000)	(7,903)	(15,000)	(11,855)
7	0.7599	(10,000)	(7,599)	(15,000)	(11,399)
Sum of Cash Flows		$(70,000)	(60,021)	$(105,000)	(90,030.82)
Original Investment			(150,000)		(120,000)
Net Present Value			$(210,020)		$ (210,030)

The two plans shown in Figure 4.2 are compared, but this time at a discount rate of 4 percent. The NPVs of the two products are nearly identical. The discount rate at which the NPVs are equal is the internal rate of return (IRR).

saves $5,000 per year, it is evident that it will eventually repay its investment. If the discount rate is held constant and the study length is extended until the NPVs of the two plans are equal, we will find the *discounted payback period.* Figure 4.4 shows the same two plans with the study life extended to 10 years, at which time Plan A has the lowest NPV.

The longer the assumed project life, the greater the opportunity for a high initial cost alternative to demonstrate its efficiency through lower recurring costs. Lacking a better figure, the service life allowed by the IRS for depreciating the product is usually a valid factor to use for study life, but many products may have a much longer service life, which will require a longer study life. Any uncertainty can be dealt with by sensitivity analysis, as discussed later.

Cost of Money

In a corporation, the cost of money is the composite cost of capital, which is the weighted average cost of equity, debt, and retained earnings. The accounting department should be able to supply a figure to use in a

FIGURE 4.4

Year	Discount Factor at 10% (A)	Plan A Cost (B)	Plan A Present Value (C) (A × B)	Plan B Cost (D)	Plan B Present Value (E) (A × B)
1	0.9091	$ (10,000)	$ (9,091)	$ (15,000)	$ (13,636)
2	0.8264	(10,000)	(8,264)	(15,000)	(12,396)
3	0.7513	(10,000)	(7,513)	(15,000)	(11,269)
4	0.6830	(10,000)	(6,830)	(15,000)	(10,245)
5	0.6209	(10,000)	(6,209)	(15,000)	(9,313)
6	0.5645	(10,000)	(5,645)	(15,000)	(8,467)
7	0.5132	(10,000)	(5,132)	(15,000)	(7,698)
8	0.4665	(10,000)	(4,665)	(15,000)	(6,997)
9	0.4241	(10,000)	(4,241)	(15,000)	(6,361)
10	0.3855	(10,000)	(3,855)	(15,000)	(5,782)
Sum of Cash Flows		$(100,000)	(61,445)	$(150,000)	(92,167)
Original Investment			(150,000)		(120,000)
Net Present Value			$(211,445)		$(212,167)

The two plans of Figure 4.2 are compared on the basis of their ten-year life cycle costs at 10 percent cost of money. Plan A is now the most cost-effective plan because the longer time frame of the study enables it to recover the additional capital investment.

feasibility study. In proprietorships and partnerships, the cost of new debt capital is usually a valid figure to use. The example in Figure 4.2 uses ten percent as the cost of money or discount factor.

Inflationary Expectations

In an inflating economy, wages and other recurring costs increase with time. Therefore, a capital investment in labor-saving equipment offsets future recurring costs. The higher your expectations of inflation, the more effective a capital investment becomes if it decreases expense. Figure 4.5 shows the same two plans compared with an assumption of five percent annual inflation, which has the effect of raising the annual costs at a compounded rate of five percent. When it is assumed that costs inflate at five percent, Plan A has a net present value nearly equal to B. It is important to include the effects of inflation if you think they will be significant.

FIGURE 4.5

Year	Discount Factor at 10% (A)	Compound Factor (B)	Product A: Cost ($10,000) (inflated) (C)	Product A: Present Value (A × C) (D)	Product B: Cost ($15,000) (inflated) (E)	Product B: Present Value (A × E) (F)
1	0.9091	1.050	$(10,500)	$ (9,545)	$ (15,750)	$ (14,318)
2	0.8264	1.103	(11,025)	(9,111)	(16,537)	(13,666)
3	0.7513	1.158	(11,576)	(8,697)	(17,364)	(13,045)
4	0.6830	1.216	(12,155)	(8,301)	(18,232)	(12,452)
5	0.6209	1.276	(12,762)	(7,924)	(19,144)	(11,886)
6	0.5645	1.340	(13,400)	(7,564)	(20,101)	(11,347)
7	0.5132	1.407	(14,071)	(7,221)	(21,106)	(10,831)
Sum of Cash Flows			$(85,489)	(58,363)	$(128,234)	(87,545)
Original Investment				(150,000)		(120,000)
Net Present Value				$(208,363)		$(207,545)

The same two products are again compared, but this time with an assumed inflation rate of 5 percent per year. When inflation is taken into account, the two plans have close to the same NPV.

Tax Considerations

A capital investment reduces income taxes in two ways. The first effect is through the tax benefits of depreciation. Because depreciation is a non-cash expense, it reduces income taxes without a corresponding cash outflow. Therefore, depreciation has the same effect as a revenue or positive cash flow. The second way taxes are reduced is by deducting from income taxes the annual expenses of operating the system.

To calculate tax effects, you must know the company's incremental tax rate for combined federal, state, and local income taxes. The incremental tax rate is the amount paid on the next dollar of income. It is higher than the composite tax rate, which is the total tax divided by total income. The accounting department should be able to furnish the incremental tax rate.

Conducting The Analysis

A feasibility study is conducted by estimating the cash flow for every factor from the time the project starts until the end of its life. Lacking a computer and software, you can conduct a feasibility study manually by entering every cash flow into a table and multiplying it by the discount factor for the year it occurs as we have done in Figure 4.2. As the calculations are somewhat laborious, and because you will undoubtedly want to play some "what if" games to deal with uncertainty, the use of a computer is recommended.

The outcome of the analysis yields financial indicators to predict the economic performance of the products. Generally, NPV and IRR are the most important. However, the amount of capital that a project requires can be significant to a firm that has difficulty attracting capital. Conversely, a firm with the objective of reducing future expenses by making aggressive present investments may spend more than the results indicate.

Interpreting the Results

Net present value is the most significant indicator of the feasibility of an alternative. Because NPV is the algebraic sum of all discounted cash flows, the product with the highest NPV will extract the least amount of cash from the enterprise. If the alternatives increase revenue, the NPV may be positive. This is sometimes the case when NPV is being used in a feasibility study because the feasibility of a project may be related to the revenue it generates or the costs it saves. In many telecommunications-related projects, however, revenues are equal for all alternatives, and are therefore omitted from the study. When this occurs, the NPV is minus, and the most effective product is the one with the highest or *least negative* NPV. The case study that follows, comparing a PBX with Digital Centrex, is a case in point. A telephone system is essential to a business. Its role is to support the work force, not to generate a revenue. Therefore, it is a cost of business that cannot be avoided, and most of its cash flows will be negative.

As we discussed earlier, however, NPV is not the only financial indicator. Internal rate of return is an indicator of how efficiently the capital is used in the business. If the IRR is not at least as high as the cost of money, the product consumes capital. The amount of capital required is also important if the organization has a shortage of investment capital. Also, you must consider intangibles such as service improvement and the convenience of new features in comparing the value of the alternatives.

Sensitivity Analysis

In any feasibility study there will always be several items that are highly uncertain, but which must be handled objectively. The most effective way of dealing with uncertainty is to rerun the analysis with a range of values. This process is known as *sensitivity analysis*.

Sensitivity analysis allows you to determine which assumptions have the greatest effect on the outcome of the analysis. Those factors that have little effect can be ignored, but those with the greatest effect require further scrutiny before the decision is made. The sensitivity analysis process is another case where a computer is useful in conducting a feasibility study.

USE OF A PERSONAL COMPUTER IN FEASIBILITY STUDIES

As we have seen from the preceding discussion, a feasibility study requires a considerable amount of mathematical manipulation. The arithmetic is not difficult, but it is tedious, especially when you are calculating IRR or running a sensitivity analysis. With the advent of low-cost personal computers, it is not necessary to do the arithmetic by hand. Most spreadsheets include NPV and IRR functions, so it is unnecessary to look up discount factors in financial tables. Also, a spreadsheet can easily inflate a stream of cash flows without the need to look up compounding factors. The case study that follows demonstrates the use of Lotus 1-2-3® for performing a feasibility study.

When you use a spreadsheet to compare alternatives, you must observe some elementary precautions. The first lies in the definition of NPV. As explained above, NPV is the sum of discounted cash flows *minus the original investment*. Spreadsheets are not programmed to subtract the original investment automatically. Therefore, as illustrated in Figure 4.6, the original investment in Cell B1 is subtracted from the NPV of the stream of cash flows shown in Cells B6 to G6 to obtain the actual NPV. The formula in Cell B3 is @NPV (C2, C6..G6) + B1 + B6. B6 is added (algebraically subtracted) from the stream of cash flows because it is a current year's expenditure and not discounted. The model in Figure 4.6 is too limited to be useful in making a feasibility study, but in the next section we will see how it can be expanded to compare two plans.

The second precaution relates to the timing of investments. When the @NPV function of Lotus 1-2-3 is used, it begins discounting with the first cell enclosed in brackets in the formula. If we had begun discounting in Cell B6, which is Year 0 (by convention Year 0 is the current year), it would have been discounted by the first-year factor. Since we do not want Year 0 to be discounted, we must subtract it outside the brackets.

A third precaution relates to the signs of the numbers used in the calculation. Be careful in adding and subtracting negative numbers that you do not obtain unintended results. For example, even though the NPV definition calls for subtracting the original investment, in Figure 4.6 it is added because it is a negative number.

A final precaution relates to calculating IRR. Most spreadsheets include an IRR function that can be used directly, but it is easy when dealing with negative numbers to get an inaccurate result. It is more accurate and just as easy to enter the discount factor in a cell as we have done in Figure 4.6, and to refer to the cell number, in this case B2, in the NPV formula. That way, IRR can be computed by changing the value in the cell that contains the discount factor.

FEASIBILITY STUDY: A CASE HISTORY

To illustrate the feasibility study process, we will follow a case history of the headquarters organization of a large Pacific Northwest corporation

FIGURE 4.6
Net Present Value Model

A	B	C	D	E	F	G
1 Amount of Investment	$(50,000)					
2 Cost of Money	9.0 %					
3 Net Present Value	$(104,757)					
4						
5 Year	0	1	2	3	4	5
6 Cash Flow	$(10,000)	$(10,500)	$(11,025)	$(11,576)	$(12,155)	$(12,763)
7						
8						
9						
10						
11						
12						
13						
14						
15						
16						
17						
18						
19						

through a study that evaluated whether Digital Centrex or a digital PBX was the most feasible alternative for replacing an existing analog Centrex. This organization comprised only 120 stations, but its high visibility to the investment community and to the public made it mandatory that telephone service be superior.

At the start of the study, the company's telephone service was analog Centrex supplied by the local telephone company. Only two alternatives were considered in the feasibility study contracting for Digital Centrex from the local operating telephone company, or purchasing a new digital PBX. For several reasons, the existing Centrex was not satisfactory, and was not considered as an alternative. For example, the largest telephone problem the company had was with handling calls and messages for high-level managers who were absent or on the telephone. This meant that the new system required single-button call forwarding for the managers, and display telephones for the answering positions, features that were not available with the existing Centrex.

This case history illustrates the feasibility analysis process and demonstrates some of the principal differences between a PBX and Centrex.

Determining Objectives

As stated in Chapter 3, it is futile to embark on any feasibility study without a clear idea of your objectives. The need for clearly defined objectives is so obvious that it would seem indisputable, but complex systems are purchased every day by organizations that have only half-formed ideas of what they want to accomplish. The process of setting objectives is often circular—there is no point in setting objectives that are technically or economically unfeasible, but you can't tell whether objectives are feasible until you set them tentatively and test them against reality. It doesn't hurt to dream a bit when setting objectives, because often the objective can be accomplished with only a minor alteration, and the mere act of committing an ideal to paper may cause people to refine it to the point of realism—but don't count on it. Furthermore, there is very little in the realm of telecommunications that can't be done today if you are willing to pay the price, and usually you don't know what the price is until you have received a number of quotations.

The process of setting objectives ideally starts with some knowledge of the products on the market. Vendor presentations are an excellent source of information, and tend to convert partially formed impressions into the beginnings of a list of objectives. A list of features such as the PBX feature list in Appendix D helps to guide your thinking about some of the characteristics a product should have.

In this study, the company's feature requirements were conventional enough that most of them could be met by either a PBX or Centrex with equivalent ease, but a few of the features were so compelling that they became crucial to the decision-making process. The following subsections list some of the principal features that the company found important. Other companies would attach little importance to some of the features, but great importance to features that this company disregarded entirely. Therefore, even though much has been written about the PBX/Centrex issue, it is impossible to generalize about which is the best alternative. The answer is, it all depends on your objectives. The following is how the company made the decision.

Mandatory Objectives

Highest Economic Value: Because the company is regulated by the public utilities commissions in several states, it was essential that the alternative chosen offer the greatest economic value. Otherwise, the public utilities commissions could disallow a portion of the company's investment and expenses from its rate base. The ability to demonstrate the greatest value to regulatory commissions was one of the principal issues in distinguishing among systems. Neither a PBX nor a Centrex was preferred, but the system chosen had to offer the highest value.

Reliability: The company depends heavily on its telecommunications system. Management was unwilling to consider any alternative that did not offer nearly 100 percent system availability, which mandated battery backup and redundancy of the processor and switching network. These features, which are expensive in small PBXs, are inherent features of Digital Centrex.

Improved Messaging System: The company occupies two nonadjacent floors of an office building. The message center is located with the receptionist on one floor. Before this project was completed, paper messages were placed in the receptionist's message box, which illuminated message waiting lamps on the telephones. To retrieve messages, users either had to make a trip to the receptionist, which wasted time, or call on the telephone, which detracted from the receptionist's primary purpose. Neither alternative was satisfactory, so one of the mandatory objectives of the system was to improve message handling capability. Either system was capable of offering this feature, which was unavailable on analog Centrex.

No Number Change for Selected Numbers: Although company management realized that existing numbers would have to be changed since the Digital Centrex was in a different switching machine than the analog Centrex, there were eight key numbers that they insisted would have to remain with the new system. This requirement can be satisfied by the personal line feature, which is supported in a PBX, but is unavailable in Digital Centrex. The telephone company solved this problem by offering call forwarding from the old numbers to the new for the life of the contract.

Availability of Product Support: The company had no intention of managing and maintaining its own telephone system, nor did any of its subsidiaries want to manage an isolated system. Therefore, it was important that support be available to administer and maintain the system chosen. Although both Digital Centrex and PBX require some management in handling adds, moves, and changes, and in maintaining station records, Centrex has the edge because maintenance and translations are handled by the telephone company.

Desirable Objectives

Low Blockage: Because such a large portion of the company's business is conducted over the telephone, it was essential that the system support enough trunks that the probability of blockage would be one percent or less. Either system could support this objective, but not at the same cost. Unlike analog Centrex, in which trunks are automatically provided by the telephone company, Digital Centrex provided by U.S. West Communications requires the customer to contract for trunks at a rate that is somewhat higher than PBX trunks.

Integrated Voice Mail Support: To assist in solving the messaging problems, the company planned to acquire a voice mail system. A highly desirable objective was the ability to turn on message waiting lamps on the telephones from the voice mail system.

Developing Alternatives

Once objectives are thoroughly understood, the next step is to develop alternatives. For some reason, most people tend to reverse the sequence, and therein lies a problem because a great deal of time can be wasted considering alternatives that just won't fit.

The methods of developing alternatives range from brainstorming sessions, as discussed in Chapter 3, to careful planning by one or two experts. One alternative that people often fail to consider is simply doing nothing. It may be that remaining with the status quo is unacceptable, but it should be tested against the objectives along with other alternatives.

In the company's case, there were only two major alternatives available, using Digital Centrex or a PBX. The status quo was unacceptable because of a lack of features in analog Centrex. Neither of the alternatives completely solved messaging problems, however, so secondary alternatives, voice mail and a text messaging system, were considered.

After the alternatives have been identified, the next step is to determine order-of-magnitude prices for them. Usually, you do not have enough information to obtain precise quotations at this stage of the analysis. The best source of information is quotations from vendors. Most vendors have a reasonably accurate estimate of cost per station for telephone systems, and given a few of the key variables, they can run a pricing program that computes prices within a few hundred dollars. The same kind of price information can be obtained from vendors for data communications equipment, local area networks, and major transmission systems.

Comparing Alternatives With Objectives

All mandatory requirements should be listed and each alternative measured against it to determine the degree to which the requirement is met. If an alternative fails to meet a mandatory requirement, it is eliminated from consideration with no further expenditure of effort.

Next, alternatives are tested to determine how well they meet the desirable requirements. An alternative is usually not disqualified for ranking low against a desirable requirement until it is known how economical the alternative is. Figure 4.7 shows the rating of the company's requirements on a scale of one to five against five alternatives: status quo, Centrex with voice mail, Centrex with text messaging, PBX with voice mail, and PBX with text messaging. As Figure 4.7 shows, the only alternative that does not meet the mandatory objectives is the status quo; therefore, it is eliminated.

Comparing Financial Results

As mentioned at the opening of this chapter, the cost performance of an alternative can be evaluated in a number of different ways. The most valid method is to determine the net present value (NPV) and internal rate of return (IRR) of each alternative, considering every cash flow that can be expected over the life of the product. Table 4.2 lists typical cash flows that should be considered in feasibility studies of most telecommunications products. Wherever there are differences among alternatives in costs listed in Table 4.2, you should attempt to identify the differences either in absolute or relative terms. Any factors that are equal for all alternatives can be omitted without affecting the outcome of the study.

The differences between PBX and Centrex illustrate the feasibility study process very well. A PBX has a large initial cost and lower

FIGURE 4.7
Centrex and PBX Alternatives Tested Against Status Quo (Analog Centrex)

		Alternatives			
		Centrex		*PBX*	
	Status Quo	*With Voice Mail*	*No Voice Mail*	*With Voice Mail*	*No Voice Mail*
Economic Value	3	5	5	4	4
Reliability	5	5	5	4	4
Improved Messaging	0	5	3	5	3
No Number Change	5	5	5	5	5
Product Support	5	5	5	4	4
Low Blockage	5	4	4	4	4
Voice Mail	0	5	0	5	0
Total Points	23	34	27	31	24

recurring costs, and Centrex has low initial costs but larger recurring costs. Often, this one difference is compelling enough to induce a company to choose an alternative without even considering other factors. For example, if investment capital is not readily available or if it is needed for other purposes, a Centrex system might be chosen without a detailed study.

It is advisable, however, to consider other factors, many of which are not difficult to quantify. Figure 4.8 is a feasibility study model that calculates the financial evaluators of the two alternatives. To simplify the model, some factors are combined into one entry. For example, the cost of the PBX includes shipping, installation, and sales taxes. Since we are going with order-of-magnitude figures, the refinement of additional detail is not necessary.

FIGURE 4.8
Feasibility Study Model PBX vs. Digital Centrex

Feasibility Study Model

Product Costs	CENTREX	PBX	
Initial Cost	$0	$107,036	INPUT AREA
Installation	$2,000	$8,370	
Station Sets	$39,732	$20,077	
Shipping	$0	$250	
Total Installed Cost	$41,732	$135,733	
One Time Costs			
Trunk Installation	$7,104	$720	
Training	$0	$600	
Total One Time Costs	$7,104	$1,320	
Annual Costs			
Maintenance	$(2,583)	$(9,504)	
Trunk Costs	$(24,209)	$(20,022)	
Line and Card Costs	$(14,669)	$0	
Software Upgrades	$0	$(700)	
Floor Space	$0	$(1,272)	
Power & Air Cond.	$0	$(870)	
NET PRESENT VALUE	$(148,363)	$(187,460)	RESULTS
NPV CENTREX- PBX	$39,097		
Common Costs			COMMON FACTORS
Inflation	5.0 %		
Power $/KWH	$0.05		
Cost of Money	9.0 %		
Svc. Life in Years	7		
Salvage Value	10.0 %		
Floor Space $ per sq. ft.	$15.90		
Incremental Tax Rate	43.0 %		

FIGURE 4.8—*Continued*

AFTER TAX COSTS *YEAR*	*1*	*2*	*3*	*4*	*5*	*6*	*7*
CENTREX							
Maintenance	$ 0	$ (1,546)	$ (1,546)	$ (1,623)	$ (1,704)	$ (1,790)	$ (1,879)
Total One Time Costs	(4,049)	0	0	0	0	0	0
Trunk Costs	(13,799)	(13,799)	(13,799)	(13,799)	(13,799)	(13,799)	(13,799)
Line and Card Costs	(8,361)	(8,361)	(8,361)	(8,361)	(8,361)	(8,361)	(8,361)
Software Upgrades	0	0	0	0	0	0	0
Floor Space	0	0	0	0	0	0	0
Power & Air Cond.	0	0	0	0	0	0	0
Depreciation Tax Effect	3,417	3,417	3,417	3,417	3,417		
Salvage							3,973
Total	(22,793)	(20,290)	(20,290)	(20,367)	(20,448)	(23,950)	(20,066)
PBX							
Maintenance	0	(5,688)	(5,973)	(6,271)	(6,585)	(6,914)	(7,260)
Total One Time Costs	(752)	0	0	0	0	0	0
Trunk Costs	(11,413)	(11,983)	(12,582)	(13,211)	(13,872)	(14,566)	(14,566)
Line and Card Costs	0	0	0	0	0	0	0
Software Upgrades	(700)	(700)	(700)	(700)	(700)	(700)	(700)
Floor Space	(725)	(725)	(725)	(725)	(725)	(725)	(725)
Power & Air Cond.	(496)	(297)	(312)	(327)	(344)	(361)	(379)
Depreciation Tax Effect	10,932	10,932	10,932	10,932	10,932		
Salvage							12,711
Total	$ (3,154)	$ (8,461)	$ (9,360)	$(10,303)	$(11,294)	$(23,265)	$(10,918)

CALCULATION AREA

This model can be used for any type of feasibility study. The model is composed of four parts, the input area, results, the common factors area, and the calculation area. The input area provides space for entering the study factors. Factors are automatically copied to the calculation area of the model. Two alternatives are tested at a time, and the difference in NPV between the two is displayed in the results area.

Either alternative can be entered as Alternative A or B, except that the IRR calculation is valid only if Alternative A is the most attractive. If Alternative B is the most attractive, IRR may be negative, which is an undefined quantity. By convention, all cash outflows are shown as negative and all inflows are shown as positive. If revenues can be attributed to the project, they should be included, but in most cases an investment in telecommunications equipment is considered as an expense with no corresponding revenue. Therefore, with the exception of depreciation, which is discussed later, all cash flows are negative and are shown in parentheses in the model. IRR is calculated by changing the cost of money in the Common Factors section until the difference in NPV is within a few dollars of zero.

In addition to the cost of money in the Common Factors section of the model, other factors can have a significant effect on the outcome of the study. For example, the inflation rate assumption affects how quickly a capital investment pays off in reduced labor costs. The service life period is used to determine depreciation expense by dividing the capital investment by the number of years over which the investment will be recovered. The easiest way to treat depreciation expense is on a straight line basis; however, many spreadsheets have the ability to calculate accelerated depreciation. You may find that accelerated depreciation more accurately reflects your company's accounting practices. Tax exempt organizations can disregard depreciation entirely.

The incremental tax rate determines what portion of the investment comes from company coffers compared to that which comes from the government in the form of reduced taxes. For example, if a company is in the 40 percent bracket for combined state, federal, and local taxes, an expenditure of $10,000 represents a negative cash flow of only $6,000 because tax savings pay the other $4,000.

The calculation area of the model is an extension of the figures entered in the Input Area and the assumptions made in the Common Factors area. The sample in Figure 4.8 considers cash flow over a seven year period. Longer periods can be computed by extending the study period.

Most of the cash flows in this model are self-explanatory, but depreciation deserves special attention. Depreciation reflects the reality that the value of an investment is gradually consumed by obsolescence and wearing out. Since no check is written for depreciation, no cash outflow occurs even though it is shown on the income statement as an expense. As with other expenses, depreciation is deductible from income taxes, and this deduction, therefore, has the same effect as a revenue. The incremental tax rate is used to calculate the effect of depreciation on the investment.

You must select a service life for depreciation purposes. The IRS publishes tables of service lives for various classes of equipment. The service life for depreciation purposes is not necessarily the same service life as used in the study period. You might, for example, assign a five year service life to a PBX for depreciation purposes, but expect it to last for ten years before it is replaced.

Salvage value, which occurs at the end of the project's service life, is another positive cash flow, usually expressed as a percentage of original cost. Ten percent is a frequently used factor, but bear in mind that it may not apply to every item of equipment or to shipping, taxes, cabling, and installation costs. As a practical matter, cabling may have a longer service life than the equipment it supports because it may be possible to use it for the next system. It may, however, have no reusability because it is functionally obsolete. The cost of removal should be deducted from the salvage value of any product if it is potentially an expensive item to remove. Salvage value rarely has a significant effect on a study because the cash flow occurs so far in the future that it is deeply discounted. If alternatives are closely equivalent, salvage can probably be ignored entirely without affecting the study.

Sensitivity Analysis

One of the main advantages of using a computerized model to evaluate investments is the ease with which the results can be tested for sensitivity to different assumptions. Rarely is it possible to estimate all of the factors in a study with a high degree of certainty. More frequently, factors are known within a range. For example, the model allows for inflation of certain variables. By changing the inflation rate, you can determine what happens to the net present value. If the difference in NPV remains fairly constant, the alternatives are insensitive to inflation, but if the difference changes significantly, the alternatives are sensitive. Insensitive factors can be ignored without a great deal of risk, but any factors to which the alternatives are sensitive should be examined carefully. For example, in most cases salvage value can be changed over a wide range without much effect on the decision.

It is often useful to prepare a table such as the one in Figure 4.9 to illustrate the results of sensitivity testing. The difference in net present value is shown for several different tests for sensitivity. For the model used in this case study it can be seen that Digital Centrex is the most economical alternative regardless of what assumptions are made.

SUMMARY

Although the examples in this chapter have illustrated the process of comparing Centrex to PBX, the same process can be used for comparing many other telecommunications alternatives. For example, analog facilities can be compared with T-1. Dial-up data over the telephone network can be compared with a public data network. The use of switched services

FIGURE 4.9
Sensitivity Test Results

SENSITIVITY ANALYSIS

	CENTREX	*PBX*	*CTX-PBX*
NO REDUNDANCY	$(160,037)	$(182,117)	$22,080
FIVE YEAR LIFE	(129,576)	(171,434)	41,858
50% INCREASE IN MAINT. COSTS	(163,739)	(214,725)	50,986
PBX COST REDUCED 10%	(160,037)	(191,275)	31,235
NO POWER COSTS	(160,037)	(198,347)	38,311
PBX COST 10% INCREASE	(160,037)	(209,339)	49,302
FIVE YR. LIFE AND:			
NO REDUNDANCY	(129,577)	(151,701)	22,125
50% INCREASE IN MAINT. COSTS	(132,126)	(181,401)	49,275
2 % INFLATION AND 5 YR. LIFE	(159,508)	(192,166)	32,658
	$(129,329)	$(167,143)	37,814

This table is created by rerunning the analysis of Figure 4.8 making the changes shown in the left column.

can be compared with a private line. The feasibility of private systems such as microwave, fiber optics, or leased facilities can be compared with the cost of using tariffed facilities of the telephone company or common carrier.

Any time that two or more alternatives are available, the feasibility study process is used to determine the economic effects of one versus the others. Of course, economics is not the only determining factor. Other considerations such as security, uncertainty or lack of stability of company plans, attitude of company management toward risk, uncertainty about changing technologies, and dozens of other such factors frequently enter into the decision and may be impossible to quantify in a cost study. If the factors can be quantified, however, the process just described is a fast and easy method of evaluating the cost performance of the alternatives.

CHAPTER 5

FORECASTING TELECOMMUNICATIONS SERVICES

Every manager in every company prepares forecasts, whether they are documented or not. As the telecommunications manager, you can't accurately predict demand for your services without knowing what the other managers intend, unless their demand follows consistently predictable patterns. The forecasting job is complicated by the fact that most forecasts aren't in writing, and those that are seldom align themselves in neat rows of figures that can be plugged into formulas to help you predict trunk and equipment quantities. Some companies are an exception to the rule of haphazard forecasts. These companies have a business research department or a group of statisticians who routinely examine operating statistics, external influences, and business plans to forecast personnel levels, sales, orders, inventory, and other factors that can be used to predict telecommunications demand.

More commonly, especially in smaller companies, the forecasting variables exist as a plan, impression, or dream in the mind of a company officer. As telecommunications manager, you must either develop a predictive facility of your own, or be prepared to react rather quickly when the business changes direction.

Telecommunications forecasts are inherently numerical, even if they don't start out that way. In most businesses, you must be prepared to forecast variables such as the following:

- Numbers of ports in a PBX, front end processor, or multiplexer
- Quantities of trunks, data lines, or cable pairs required

- Volumes of long distance calls or data transactions
- Telecommunications budget requirements

Some services can be obtained on fairly short notice, so forecasting them precisely is not a critical matter. For example, you can obtain additional central office lines or WATS trunks with no more than a week or two of notice unless the telephone company has done a poor job of forecasting and has no cable facilities available. Connecting these trunks to a PBX, however, may be another matter. If sufficient card slots are available, trunks are added rather easily, but if all the slots are full, it will be necessary to add another card carrier. If all carriers are full, it will be necessary to add another cabinet, and if all cabinets are full, the PBX has reached its capacity, and it will be necessary to replace it. Each of these requires progressively more complex and expensive activity. To avoid being caught short, you must forecast requirements far enough in advance to meet the vendor's typical lead times.

This chapter presents an overview of forecasting techniques, and suggests ways to implement them without an extensive background in statistics or forecasting methods. Fortunately, several personal computer software packages can insulate you from the mathematics of forecasting, and reduce it to a managerial task that almost anyone can perform.

WHY FORECAST?

It is no secret that most telecommunication managers do a rather poor job of forecasting usage. In fact, except for large businesses with a full time telecommunications staff, it is unusual to find any kind of telecommunications forecast at all. Most managers tend to follow what professional forecasters politely refer to as "a judgmental forecast," which is like a pilot who trusts the seat of his pants more than his instruments. But as anyone who has flown into a fogbank knows, it is easy to lose your orientation when the horizon isn't in sight, and even a crude artificial horizon, like a half-filled vial of water, is a better indicator of which way is up than your intuition.

There are many valid reasons for not forecasting; you've heard most of these:

- "It's too difficult to collect the information"
- "I don't have time to forecast. I'm too busy running the job"

- ''Something always changes anyway and you have to throw the whole forecast out the window''
- ''This company doesn't know where it will be tomorrow let alone a year from now''

It would be easy to expand the list, but there is one reason for making forecasts that outweighs all the contrary reasons; what you can't predict you can't control. Of course, you can't always control everything you can predict (try controlling the weather) but prediction helps you control the adverse effects. Besides improving control, there are a number of other reasons for making predictions:

To Adjust Circuit Quantities to Meet Demand. Whenever the fixed monthly cost of a temporarily unneeded circuit exceeds the one-time cost of reconnecting it, it is economical to disconnect the circuit temporarily. For example, a wholesale travel broker normally sees a peak in the first quarter of the year as tourists prepare for summer vacations, and then demand wanes through the third and fourth quarters. It may pay to disconnect 800 lines during the lulls and reconnect them during the peaks.

To Predict Telecommunications Budgets. Regardless of whether any other forecasting is being done, nearly every organization forecasts its telecommunications budget. Budgeting can range from a simple projection of expense dollars per station, to a complex forecast of capital expenditures required for major projects. Most businesses prepare budgets for a quarter, a year, or two years or more in advance. To predict the usage-sensitive portions of the telecommunications budget, some kind of forecast of usage is essential. It may also be useful to forecast unit costs as well since costs such as long distance have been declining regularly.

To Forecast Equipment Quantities. PBXs, multiplexers, front end processors, and other such devices are expandable, within limits, to meet demand. Usually, such devices have fairly long lead times between order and delivery for plug-in cards, card carriers, and cabinets. A forecast helps to ensure that capacity is available when the demand occurs.

To Meet Circuit Demands. Some types of circuits can be installed almost immediately, but others may require a lengthy installation interval. This is particularly true of cross-country circuits, T-1 circuit groups, and circuits to isolated locations. The forecast helps you avoid being caught short.

To Predict Building Space Requirements. Additional floor space is one of the least easy commodities to obtain in expanding a telecommunications system, so it is generally necessary to forecast building space further into the future than equipment. Fortunately, the trend is toward smaller sizes for most types of telecommunications equipment, so when equipment reaches the end of its service life it can probably be replaced with a system of greater capacity and smaller size. Auxiliary equipment such as distributing frames and power equipment are not shrinking at the same rate, however, and terminating equipment such as channel banks and multiplexers are being added to consume more floor space.

DETERMINING PATTERNS

The first task in preparing a forecast is to collect repeated observations about the same data and to analyze them to detect the underlying patterns. These patterns are then used to predict the future within limits. The four fundamental patterns that underlie forecasts are:

- Trend
- Seasonality
- Cyclicality
- Randomness

Of the four patterns, all but randomness can be extrapolated from the underlying data. To illustrate the effect of these elements, assume that you have collected data about average usage on long distance trunks for a period of several years, and have plotted it on three charts as shown in Figure 5.1.

If the observations are plotted by hour of day, as shown in Figure 5.1a, a type of *seasonality* can be observed; variation by hour. This shows a traffic pattern typical of many companies—a morning and an afternoon peak, with virtually no traffic between 6:00 PM and 7:00 AM. Different companies, of course, have much different patterns than this. For example, a west coast stockbroker would begin heavy calling at 6:00 AM to coincide with a 9:00 AM opening of stock exchanges in the east. An east coast telemarketing group, on the other hand, would experience the reverse kind of pattern as its representatives work late to accomodate the early evening hours of the west coast.

Figure 5.1b plots usage by day of the week, illustrating another type of seasonality. Most businesses have Monday peaks and many have mid-week lulls in calling volume, with another peak on Friday, dropping off to virtually nothing on the weekend. If calls are plotted by month, as shown in 5.1c, yet another type of seasonality can be observed.

By looking at Figure 5.1c more closely, another pattern is shown by the dotted line. Although the seasonal pattern repeats itself, the overall *trend* is toward an increasingly heavy volume. If the seasonal fluctuations are great enough, they can be used to predict times when WATS lines should be added or removed, or to forecast the long distance budget on a monthly or quarterly basis. The long term trend, however, is upward, and

FIGURE 5.1

Three Different Types of Seasonality (Three different types of seasonality are apparent in these graphs: (a) by hour, (b) by day, (c) by month. Figure 5.1c also shows an upward trend, indicated by the dotted line.)

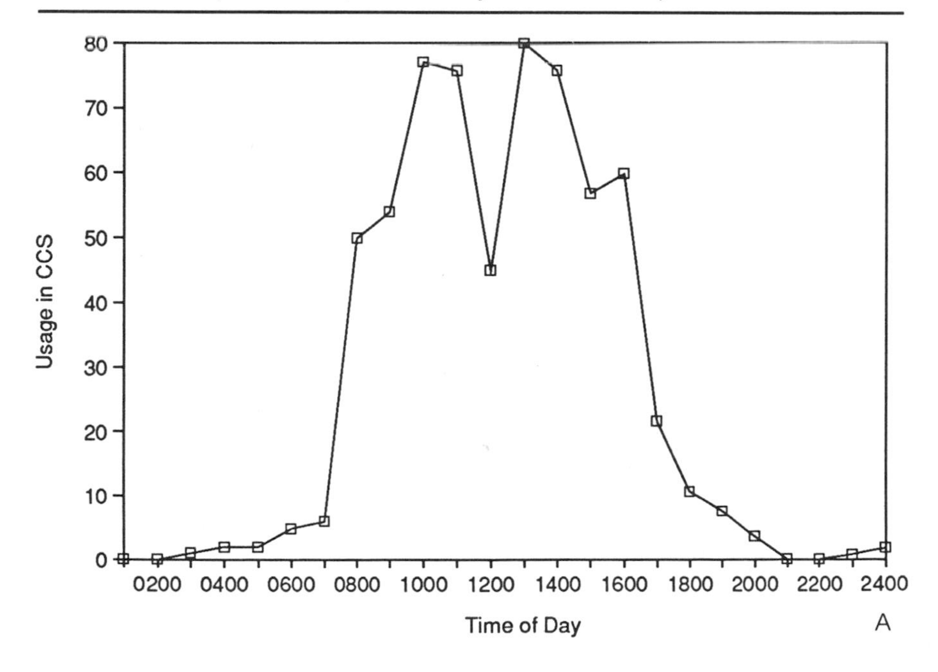

FIGURE 5.1—*Continued*

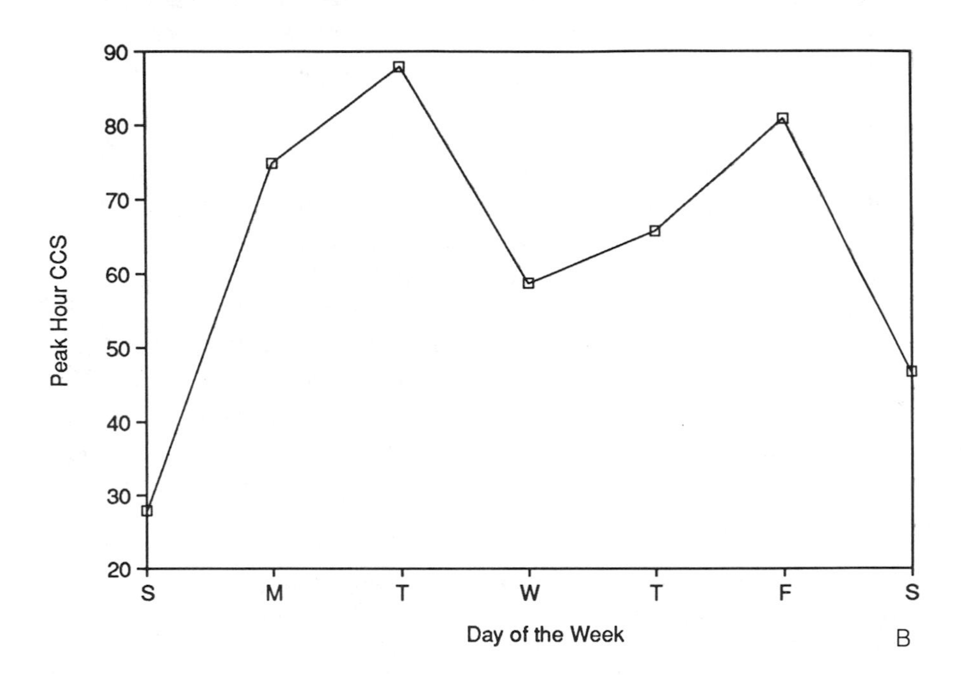

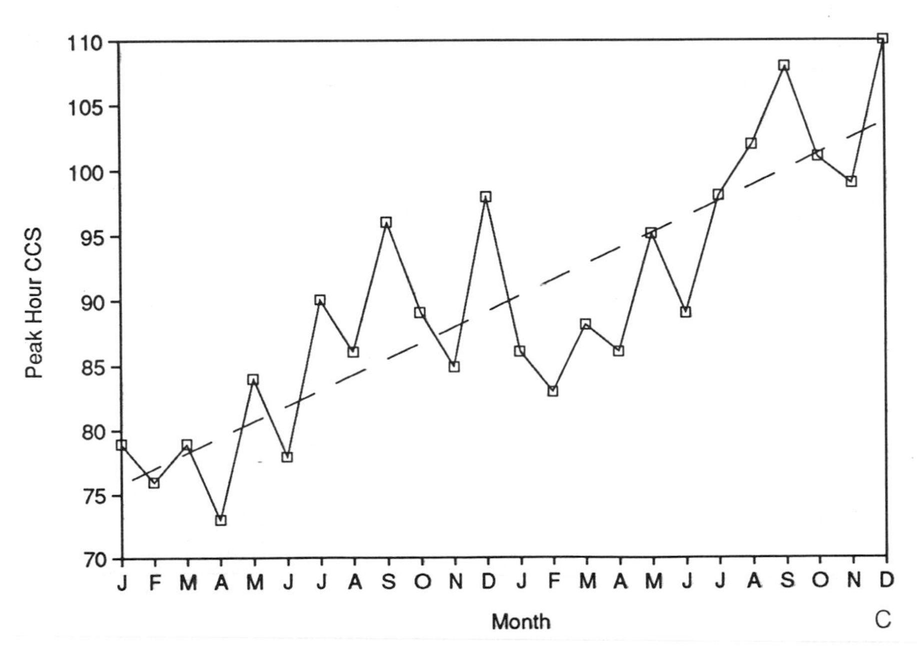

by fitting the pattern to an appropriate forecasting method, you can do a reasonable job of predicting usage.

A *cyclical* pattern is usually observable over a much longer time span. For example, many economists have observed a rising and falling of general business conditions over several decades. While the business cycle may be useful for the company's general business analysts, most telecommunications managers are called on for short term, real time forecasts, and can ignore cyclical effects. Moreover, most companies have too little historical information, and telecommunications usage is influenced by far too many factors that have greater effect than cyclicality to gain much benefit from observing cyclical fluctuations and usage.

If you could ignore cyclical fluctuations and concentrate only on seasonality and the underlying trend, your forecasting job would be simple indeed. You would merely have to collect data, find a forecasting formula that fits, and extrapolate the past and present into the future. The task isn't this easy, however, because of random fluctuations which, by definition, do not follow a predictable pattern. Random fluctuations can be observed, but there is no reason to expect that they will recur in the future. This is not to say that the fluctuations are causeless; on the contrary, the causes often can be discovered after the fact. You can count on them recurring sometime in the future, but the problem is you can't tell when.

A good example of random fluctuations are those caused by the weather. Telephone company traffic managers have always been plagued by the unpredictability of a sudden snow storm that places an immediate overload on circuits. Some of the fluctuations caused by weather are inherently built into seasonality, but the isolated event causes drastic peaks of usage that can't be predicted or controlled. Weather-caused peaks may have little effect on long distance circuits but they definitely can overload local trunks.

Other random fluctuations are caused by news events of national importance; for example, President Kennedy's assassination threw the entire telephone system into a heavy overload. Fortunately, such events are rare, but lesser happenings affect your company's telecommunications system regularly. For example, a special promotion may generate a high volume of voice and data transactions. Although you may know of such promotions in advance, the effect is not built into the forecast because forecasting methods smooth out the peaks that can't be expected to recur. Of course, you can adjust circuits or staff when you know such demands will occur, but you normally have no historical information on which to base a decision.

SOURCES OF FORECAST INFORMATION

A common objection to preparing forecasts is that information is difficult to come by. For a telecommunications manager who is attempting to respond to immediate demands and has little or no historical data on which to base a forecast, the task may appear formidable. The best way to collect data is going forward because it can be massaged with a computer and interpreted while the causes for fluctuations are still fresh in mind. It is effective to collect usage information on a current basis, delaying the implementation of formal forecasting until sufficient historical information is available. The sources of this data are described below.

Switching Systems

PBXs, tandem switches, and other intelligent devices often collect usage information. Traffic usage is usually quoted in Erlangs (number of hours of circuit occupancy) or CCS (number of hundreds of call seconds of circuit usage). These terms are explained in more detail in Chapter 16. PBXs report on circuit usage, either on a recurring or a special study basis. By trending CCS per user for different types of circuits such as local, long distance, tie trunks, etc., the forecaster can predict circuit requirements with a high degree of assurance when a forecast of the number of users is available. Many companies forecast personnel levels since the number of personnel is so directly related to expense levels, and these forecasts can be used to derive telecommunications forecasts.

Station Message Detail Recorder (SMDR)

The SMDR is the most valuable source of information in a PBX. If the data is collected on machine-readable media, it can be sorted and combined in a variety of ways. The SMDR can collect both local and long distance information over central office trunks, private lines, WATS lines, or other common carrier and private services. The data, which is normally collected a month at a time, can be sorted with some auxiliary programs to show the follow kinds of information:

- Distribution of calls by time of day
- Distribution of calls by the day of the week or week of the month
- Calling details by trunk group
- Calling details by destination
- Calling details by originating department

So much data is available from an SMDR that you must be careful not to collect more data than is actually needed. Data collected for its own sake is a waste of time; you must have definite purposes in mind if information is to be of value. Be aware that SMDR data may not express the total demand. For example, uncompleted calls place a demand on the switching system and trunks, but may not be registered by the SMDR.

Toll Statements

The next best source of usage information is the telephone bill. The main problem with the bill is that it comes on paper, which isn't so easy to manipulate. However, inexpensive optic character readers are coming on the market and may be available to convert the toll statement to a machine-readable state. Some long distance vendors make the toll statement available on tape or disc for additional charge, which may be worth the investment. For long distance calling, the telephone bill yields the following kinds of information:

- Time of day and day of week call distribution
- Call destination distribution
- Cost distribution by destination

The data from the toll statement is useful for predicting the requirements for additional WATS lines, determining when foreign exchange or tie lines are economical, checking the validity of the least cost routing pattern in the PBX, and other such similar purposes.

The toll statement has the same limitation as SMDR data of showing only billed traffic and not the total load of uncompleted calls and time used in call setup. Ways of compensating for these limitations are discussed more fully in Chapter 16. The toll statement is also a valuable source of information for forecasting the telecommunications budget.

Front End Processors

The front end processors of many mainframe and minicomputers are a valuable source of statistical information regarding data transactions. Information such as the following can be collected:

- Distribution of message length
- Time of day and day of week message distribution
- Error statistics
- Percent circuit occupancy
- Blockage, delay, or overload statistics

Multiplexers

Time division and statistical multiplexers often collect usage information and error statistics similar to information collected by front end processors. As these are often remotely located, it may be difficult to collect the information, but such data is useful not only for predicting growth, but for measuring circuit performance.

Budgetary Information

The level of expenditures for various types of telecommunications services may be useful for predicting future budgets. Usually, it is more useful to extract actual usage information from the bill, but if no finer detail is available, the raw expenditures for telecommunications services may offer a clue to future expenditures. Certainly, it is useful to observe trends as a way of measuring the effectiveness of cost control programs or circuit reconfigurations that are designed to reduce costs.

FORECASTING METHODS

The basic assumption of forecasting is that some relationship exists between sets of observations that you have collected, and that the relationship can be discovered and used to predict the future within certain limits. The relationship may vary over time, in which case a *time series* forecasting method is appropriate. The three graphs in Figure 5.1 illustrate time series data that can be used to prepare a forecast.

The relationship may not be time related at all, but rather some variable may cause a change in the relationship whenever it occurs. *Causal* or *explanatory* forecasting methods are used to analyze these relationships. Both kinds of relationships are useful for forecasting telecommunications demand. An additional traffic load created by a special promotion is an example of a causal relationship.

Time Series Forecast

Assume that you have a collection of observations summarizing data transactions from sites on a multi-drop circuit and have plotted them by week, as shown in Figure 5.2. It is evident that a pattern exists and that it has a general upward trend, as shown by the dotted line. A time series forecast assumes that history tends to repeat itself, and uses this data to forecast the future. Time series methods do not attempt to explain relationships, only to detect them and use them to forecast the future.

There is an inherent danger in time series forecasting; extrapolating history into the future can result in serious error. The forecast may be

FIGURE 5.2
Time Series Chart Plotting Data Transaction Volumes by Day of the Week

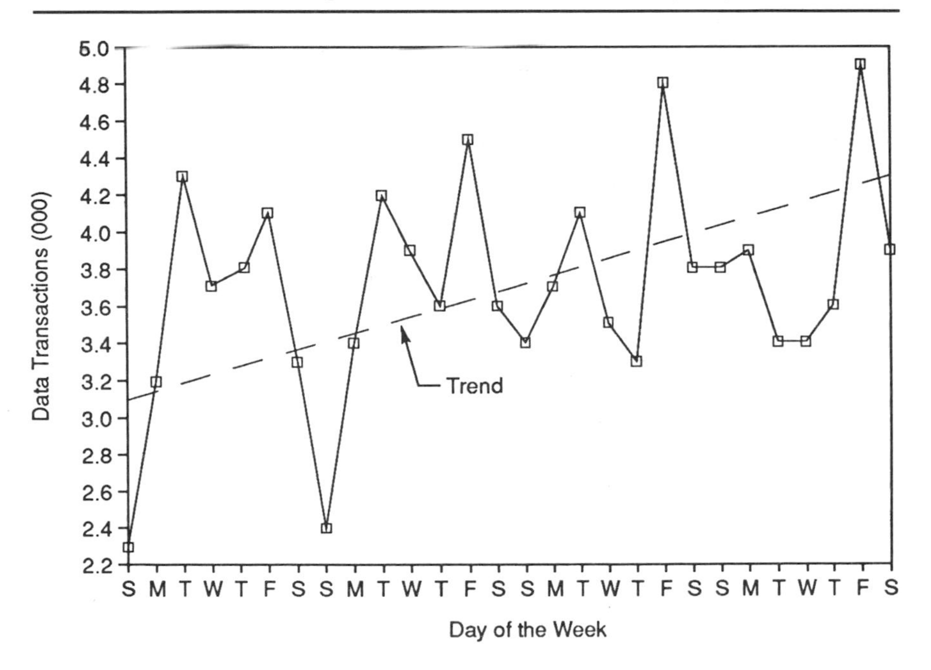

mathematically correct, but still have no relation to what will actually occur. For example, many public utilities made the mistake of over-forecasting future power demand on the basis of historical data that showed an ever-increasing trend, and power-generating capacity was constructed far in excess of actual need. Obviously, one would not attempt to forecast the stock market on the basis of time series forecasting because, although the general trend has always been upward, there have been significant retreats and periods of no growth, and causal factors that are difficult to detect. Blind adherence to a simple time series forecast would have resulted in your making investments at a time when the overall market trend was temporarily down.

With these cautions in mind, however, time series forecasting probably has application in your business. For one thing, time series forecasts are simple to prepare. As we will show, they can easily be prepared with a simple spreadsheet or even projected by eye from a graph such as Figure 5.2. The greater the degree to which the observations vary from one another, however, the less precise is the forecast.

If a collection of observations lies on a straight line, a simple linear formula can be used to extend the observation into the future. For example, Figure 5.3a shows a curve that is described by the formula $X = 2Y/3$. Most data, however, exhibits considerably more variation, as shown in Figure 5.3b. Time series forecasting uses rolling average techniques to smooth out these variations. Forecasters refer to the process of evening out variations as "smoothing."

Moving Averages

Table 5.1 is a spreadsheet model that can be used to develop a forecast using moving averages. The first column represents a collection of observations. Assume, for sake of illustration, that these are data transactions that you want to project into the future. The second column is the average of the previous three observations; the third column is the average of the previous five observations. In a five-month moving average forecast, the forecast is developed after the fifth period by adding the five observations and dividing by five to obtain an average. The sixth period adds an observation and drops the oldest one. Each period drops the oldest figure and adds the newest to develop a new average, which is the meaning of the term "moving average" or "rolling average." The more time periods you include in calculating moving averages, the greater the amount of smoothing that occurs, and within limits, the greater the amount of forecast accuracy.

FIGURE 5.3
Relationships Between Variables (Figure 5.3a shows a linear relationship between two variables, X and Y. The relationship is expressed by the equation X = 2Y/3. Figure 5.3b shows the more usual relationship between variables. Smoothing is used to even out the variations.)

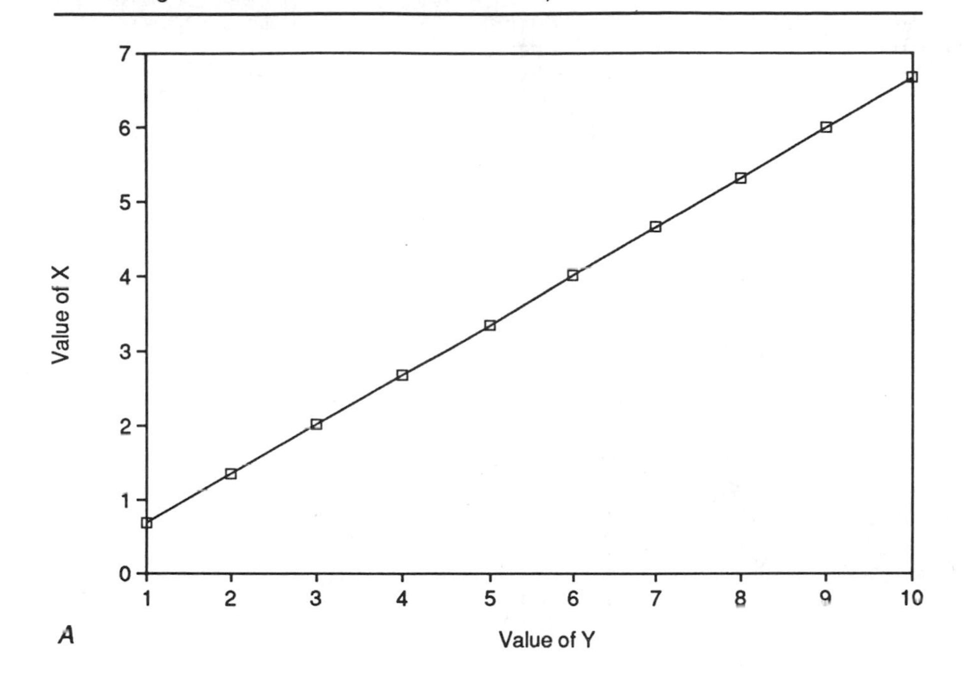

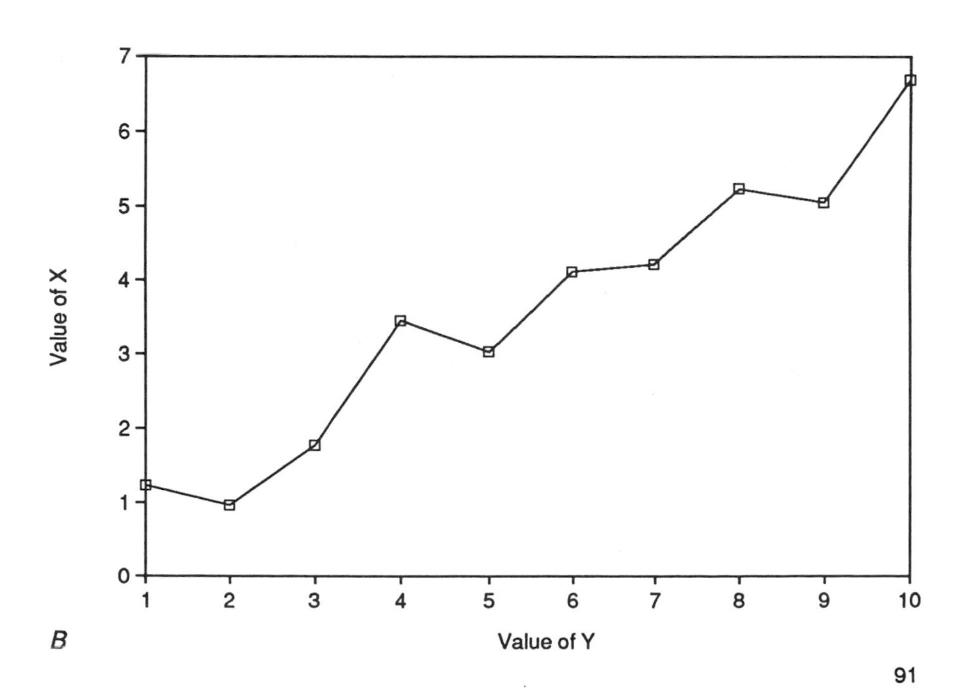

TABLE 5.1
Use of Three- and Five-Period Moving Averages to Forecast Future Demand

Time Period	*Observed Demand*	*Forecast Demand 3-Month Moving Average*	*Forecast Demand 5-Month Moving Average*
1	4336		
2	2927		
3	4228		
4	4282	3830	
5	6721	3812	
6	3794	5077	4499
7	3360	4932	4390
8	2818	4625	4477
9	4770	3324	4195
10	6005	3649	4293
11	5095	4531	4149
12	5587	5290	4410
13	6220	5562	4855
14	3798	5634	5535
15		5202	5341

Exponential Smoothing

The primary disadvantage of the moving average method is that all observations are equally weighted. In actual practice, the recent observations may carry more importance than older ones because they result from more current influences. To give greater weight to recent observations, exponential smoothing is another linear forecasting method that is often used. Table 5.2 is a spreadsheet model that illustrates the exponential smoothing method. The variable *alpha*, shown in the last three columns of the table, is a number between 0 and 1 that expresses the degree of weighting assigned to the most recent observation. *Alpha* is selected by determining empirically which value gives the best results. The higher the value of *alpha*, the less the amount of smoothing in the forecast.

To create a spreadsheet to compute exponential smoothing forecasts, follow these steps:

1. Start with the actual results from the last period. Use this factor for the next month's forecast.

2. When actual results for the next month are in, compute the error. In Table 5.2 the error is 4336 - 2927 = 1409.
3. Adjust the next month's forecast by *alpha* times the error. If *alpha* in Table 5.2 is 0.1, the adjustment is 141.
4. Add or subtract the adjusted error, depending on whether the forecast was high or low, to obtain the next period's forecast.

One of the principle shortcomings of exponential smoothing is the lack of precision in determining the value of *alpha*. A second shortcoming is the inability of this method to account for seasonal changes. To overcome these limitations, variations on exponential smoothing such as Winters linear and seasonal exponential smoothing are often used to accommodate the seasonality of the observations. These methods are more complex and generally require a trained statistician or a computer program to compute the results. One such program is SmartForecasts, a personal computer program that is discussed later in this chapter.

TABLE 5.2
Use of Exponential Smoothing to Forecast Demand

Time Period	*Observed Demand*	*Value of Alpha* 0.1	0.5	0.9
1	4336			
2	2927	4336	4336	4336
3	4228	4195	3632	3068
4	4282	4198	3930	4112
5	6721	4207	4106	4265
6	3794	4458	5413	6475
7	3360	4392	4604	4062
8	2818	4289	3982	3430
9	4770	4142	3400	2879
10	6005	4204	4085	4581
11	5095	4384	5045	5863
12	5587	4455	5070	5172
13	6220	4569	5328	5545
14	3798	4734	5774	6153
15		4640	4786	4033

The forecast is the previous period's forecast plus alpha times the error in the old forecast. The higher the value of alpha, the more smoothing that results.

Regression Analysis

Regression is a method of using one set of known figures or another forecast to predict a dependent variable. Regression analysis makes these assumptions:

- That a linear relationship exists between the independent and dependent variables
- That the errors are constant throughout the range of observations
- That the observed values are independent of one another
- That the values are more or less normally distributed; i.e. they follow a bell-shaped curve when they are plotted

If these conditions are not met, validity of the result is questionable.

Simple regression analysis uses only one independent variable to predict the dependent variable. Frequently, however, the dependent variable responds to more than one factor. For example, data transactions might be a function of sales volume, number of employees, and returned merchandise. If forecasts of all three of these variables are produced by some other part of the organization, you can use multiple regression analysis to predict the dependent variable, data transactions.

Regression analysis is most useful when some other part of the company prepares a forecast that you want to use for developing your own predictions. If a business research group predicts some variable that correlates with your forecast, multiple regression can be used to cause your forecast to track the other one. Figure 5.4 shows an example of a simple regression forecast prepared by SmartForecasts.

From the standpoint of the telecommunications manager, the safest approach is to base the forecast on projections prepared by other parts of the organization. For example, since the demand for telephone service is closely correlated with growth in personnel, it is reasonable to use a personnel projection to drive the voice communication forecast. Parts of the data communications forecast, such as transactions from a payroll system, will also correlate with personnel growth. Demand for transaction-related data transmission will probably be correlated with the sales forecast. If other forecasts can be found, it is not difficult to use regression analysis to develop telecommunications forecasts from them.

The validity of regression analysis depends on how closely the independent and dependent variables track with one another. In statistical terms, this is referred to as the correlation coefficient. Perfect correlation,

FIGURE 5.4
Simple Regression Forecast Prepared by SmartForecasts™

Weekly Data Transactions
Regression forecasts of TRANSACTIONS using: SALES

	APPROXIMATE 90% FORECAST INTERVAL		
Case	*Lower Limit*	*FORECAST*	*Upper Limit*
C13	2345.88	2476.91	2607.94
C14	2429.83	2565.71	2701.59
C15	2357.97	2489.60	2621.23
C16	2547.37	2692.56	2837.75
C17	2605.24	2755.99	2906.73
C18	2309.44	2438.86	2568.28
C19	2333.77	2464.23	2594.69
C20	2582.16	2730.62	2879.07

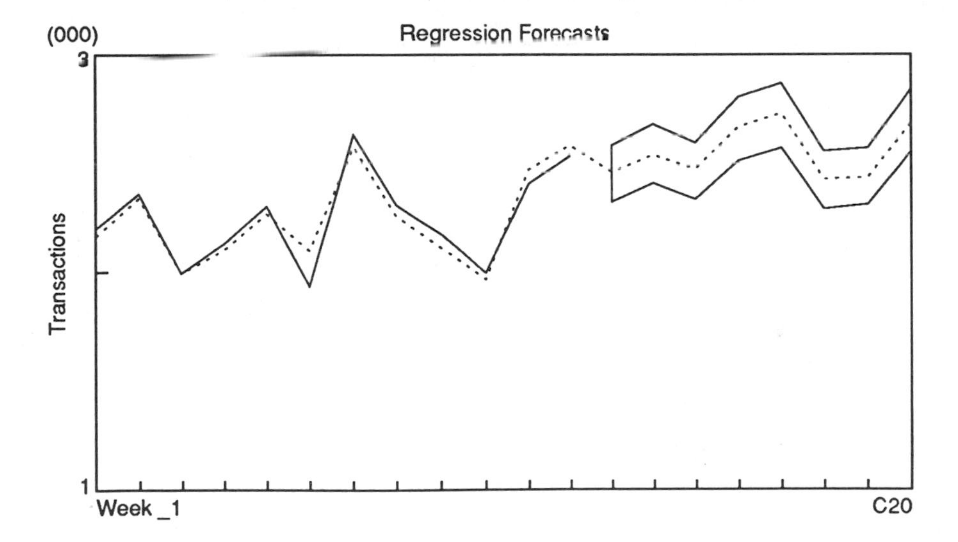

Data transactions are forecast from the company's forecast of sales volume. The sales forecast is entered into an input table. SmartForecasts prepares the table showing the forecast with upper and lower confidence limits. The graph is prepared showing actual data in a solid line with forecast data in a dotted line. The upper and lower confidence limits are graphed with the forecast.

meaning that the amount of change in the dependent variable is exactly proportional to the change in independent variable, would have a 100 percent fit. Figure 5.5 shows a scattergram that plots data transactions against sales volume. To make such a chart, you must collect data about the number of data transactions and sales volume in each of several time periods and plot them on a chart. The more tightly the dots are clustered about a straight line, the higher the correlation coefficient. A correlation coefficient of 1 represents perfect correlation; a coefficient of 0 means there is no relationship between the variables.

Judgmental Forecasts

Until the last few years, forecasting has been a specialized job, the province of experts. Now, with the advent of inexpensive desktop

FIGURE 5.5
Scattergram Showing Correlation Between Sales Volume and Data Transactions

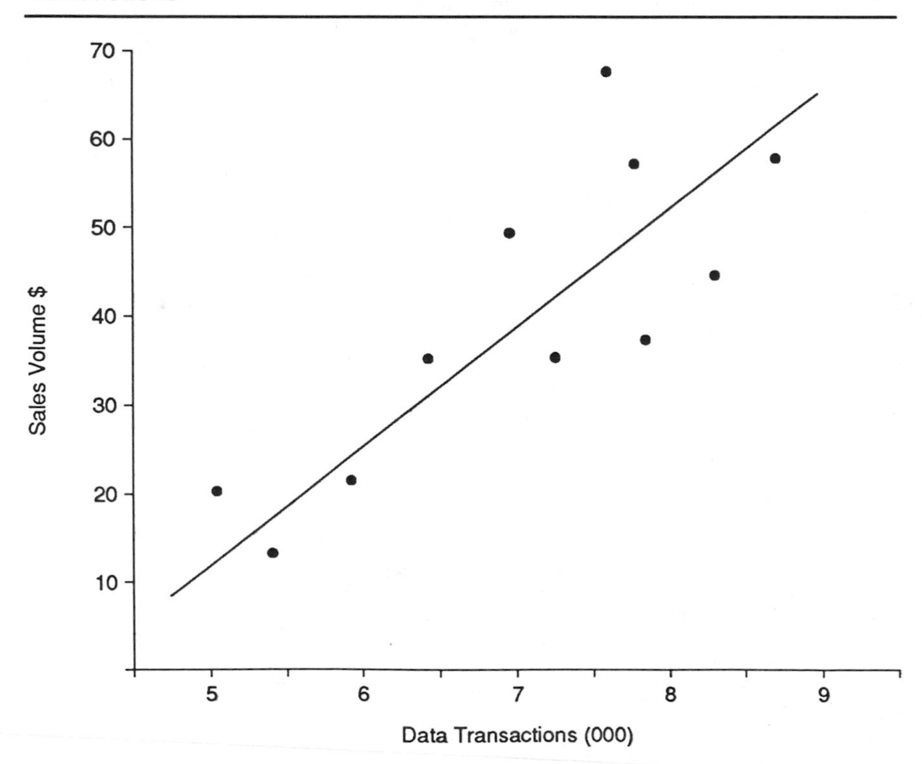

computers and packaged software, telecommunications managers can have the capability of sophisticated forecasting on their desks without seriously damaging the budget. There is always the danger, however, of relying too heavily on the forecast. The neatly aligned tables and charts prepared by forecasting programs are not reality; they are mere projections of what will happen in the future if the past repeats itself, other assumptions do not change, and the same influences continue to drive telecommunications demand. Therefore, you must be not only a producer of forecasts, but also an interpreter.

The numbers are a starting point. They must be carefully watched or they may lead you astray. They must be adjusted to account for future changes that are known to management, but unknown to history, and which cannot be quantified precisely. The process of applying these changes is known as judgmental forecasting. Call it intuition, experience, or feeling, but judgmental forecasting is used by every manager in lieu of blind adherence to the numbers. Some forecasting systems like SmartForecasts give you the ability to tweak the shape of the curves so they reflect your management judgment. Judgmental forecasting is not to be avoided, but rather used as a method of ensuring that results of the forecast and the plan you implement complement each other.

The Delphi Method

The Delphi technique was probably a term dreamed up by a public relations specialist to justify using the time-honored ''seat of the pants'' forecasting technique. In practice, the Delphi method is somewhat more precise than a simple guess, but it still involves considerable risk. In this method, the analyst asks several people, who are in a position to know, to make an estimate of where the organization might be in the next few years. Either a straight average or a weighted average can be used to summarize the results.

To illustrate, assume that data terminal growth is closely correlated to sales volume. If you ask four or five people to estimate sales volume by year, you might receive answers similar to Figure 5.6. Knowing the personalities involved, you might expect the vice president of sales to be overly optimistic and the vice president of engineering to be traditionally pessimistic. The most realistic estimator is the vice president of finance who has a record of being accurate. The vice president of personnel is in a poor position to know and is notoriously inaccurate. Therefore, you weight the estimates as shown in Figure 5.6. This method requires judgment, but is probably sufficiently accurate for most design purposes where no other data is available.

USING A PERSONAL COMPUTER IN FORECASTING

Forecasting techniques often require specialized knowledge that may not be possessed by the average manager. The process of preparing forecasts can be simplified by using either general purpose or specific programs on a personal computer. Examples in this chapter have illustrated the use of a spreadsheet for preparing linear forecasts. Other examples have mentioned SmartForecasts, a special purpose personal computer program that takes most of the work out of preparing forecasts.

SmartForecasts accepts historical data in an input table, and uses a balloting technique to determine which of several forecasting methods yields the best fit. One advantage of a program such as this is the confidence limits that are expressed either in tabular or curve form. Having a reasonable degree of assurance that actual results will fall within certain bounds enables a manager to build alternative plans and determine the service and financial effects of them.

The primary advantage of a forecasting program such as SmartForecasts is that it insulates you from the mathematics of making the forecast.

FIGURE 5.6
Using the Delphi Technique To Forecast Sales

Title	*Estimate $(Mil.) A*	*Weighting B*	*Weighted Average A × B*
V.P. Sales	$6.65	0.4	$ 2.66
V.P. Engineering	4.89	0.5	2.45
V.P. Finance	5.77	0.9	1.24
V.P. Personnel	6.20	0.2	1.24
Total		2	11.54
Weighted Average			$ 5.77

In the Delphi method estimates are obtained from several people and are weighted based on your estimate of their reliability. A weighted average is obtained and divided by the sum of the individual weightings.

For example, the program automatically decides whether Winter's Exponential Smoothing gives a better result than a simple time series forecast. The operator doesn't have to know anything about Winter's method to use it.

SUMMARY

Computer-based tools are gradually moving forecasting from the province of the experts to those whom the forecast is designed to serve. No longer is it necessary to understand the mathematical foundations of the forecast, but it is important that the principles be understood as a guide to using the technique that is appropriate to the situation. In some companies a business research group may regularly prepare operating forecasts. Rarely will these tell the telecommunications managers what they need to know to predict hardware, software, and usage requirements without additional massaging of the numbers. This chapter has presented a brief overview of the principle techniques a telecommunications manager should know to prepare operating forecasts.

CHAPTER 6

DEVELOPING REQUIREMENTS AND SPECIFICATIONS

Anyone who belongs to industry associations or sells telecommunications equipment and services has heard of stories of thousands of dollars wasted and careers damaged because the wrong equipment or product was purchased. Almost invariably, the problem is traced back to a central cause, buying equipment without a complete understanding of requirements, and placing orders with no outline of specifications. As telecommunications managers become more street-wise, the problem is less severe than it was in the past, but you can be certain that every day someone purchases equipment with only a vague idea of what the features do and without understanding whether they are worth the extra cost. At best, the result is an expensive feature that no one uses. At worst, the system will either waste time and money in lost productivity or will be replaced before it reaches the end of its useful service life. The goal of this phase of the equipment selection process is simple: to avoid buying the wrong product.

Except for the simplest equipment and products such as 1200 b/s modems that have achieved a commodity status, it is usually a mistake to rely solely on an equipment vendor to determine your requirements. Many sales representatives do not understand their own equipment well enough, and often do not take the time to analyze your requirements thoroughly enough to give you a comprehensive analysis of your needs. Moreover, if the vendor determines your requirements, you probably won't be aware of features missing from that company's product line.

For the purpose of this discussion, we will define requirements as the non-technical and non-quantifiable features and objectives of a telecommunications system. Specifications, on the other hand, are technical, and usually can be quantified in precise terms. The distinction between the two is not critical. What is important is to express these in unambiguous terms so vendors can match their offerings to your needs as accurately as possible.

DEVELOPING REQUIREMENTS AND SPECIFICATIONS

Requirements relate to delivery dates, environmental limitations, and the features and operating characteristics the equipment must have in order to be acceptable. The danger in feature specification is that vendors may have different terminology for the same feature, or may call different features by the same name. Therefore, it is advisable to write your own feature definitions or to refer to a common set of definitions such as that contained in a widely available publication such as *Datapro Reports*. Appendix D contains feature descriptions that you are permitted to copy and use in your requests for proposals.

Like objectives, feature requirements should be divided into two categories, mandatory and desirable. Mandatory features are those that must be part of the system design and included in the price of any system proposed. Desirables are those that you will buy if the price is right. Table 6.1 lists typical requirements for different types of telecommunications systems.

Specifications are developed by reviewing the environment in which the equipment will operate, and comparing it to specifications of equipment from different vendors. Specifications should be made as tight as necessary without exceeding the limits of equipment on the market. If specifications are made too tight, it may be impossible for many vendors to meet them, and you may find that you have limited your alternatives to the point that you've lost the price benefits that open competition brings. If the specifications are too loose, the equipment may fail to meet your basic needs.

Table 6.2 lists types of specifications that should be considered with different types of telecommunications equipment. Specifications are all expressed as a range of numerical factors or as a specific characteristic the equipment must possess. For example, operating temperature range might be expressed as "0° to 40°C" and an interface might be specified as "must meet EIA RS-232-C."

TABLE 6.1
Sample Telecommunications System Requirements

	KTS	*PBX*	*Data Comm*	*LAN*
		Functional Requirements		
Addressing (Mnemonic addressing, abbreviated dialing, etc.)	X	X	X	X
Attendant Features	X	X		
Call Processing Features (call waiting, speed calling, etc.)	X	X		
Closed User Groups			X	X
Conferencing Features	X	X		
Electronic Mail		X	X	X
Paging	X	X		
		Technical Requirements		
Alarm Reporting Capability	X	X	X	X
Compliance With Codes	X	X	X	X
Compliance With Standards	X	X	X	X
Emergency Backup and Power Outage	X	X	X	X
File Storage Capacity			X	X
Maturity of the Technology	X	X	X	X
Remote Maintenance and Testing	X	X	X	X
Security Provisions		X	X	X
Self Diagnostics	X	X	X	X
Surge and Brownout Protection	X	X	X	X
Traffic Data Collection		X	X	X
Types of Terminals Supported	X	X	X	X
		Support Requirements		
Continuing Availability of Spares	X	X	X	X
Documentation	X	X	X	X
Generic Program Updates	X	X	X	X
Required Service Dates	X	X	X	X
Spare Parts Stocking	X	X	X	X
Training for Users and Technicians	X	X	X	X
Vendor Experience and Stability	X	X	X	X

X-Generally applies to this type of system; may also apply to others
This table is a check list of the kinds of requirments you should consider in planning the acquisition of telecommunications equipment. Appendix D contains a more complete list of features that can be used to review functional requirements.

TABLE 6.2
Sample Telecommunications System Technical Specifications

Feature	*KTS*	*PBX*	*Data Comm*	*LAN*
	Compatibility			
Physical Interface	X	X	X	X
Protocols		X	X	X
Applications Software		X	X	X
Telephone Network	X	X	X	X
Other Networks	X	X	X	X
	Environmental			
Operating Temperature and Humidity Range	X	X	X	X
Air Flow Requirements	X	X		
Floor Space or Footprint		X		
	Technical			
Range of Speeds		X	X	X
Technology (analog vs. digital)	X	X	X	X
Modulation Method			X	X
Switching Method	X	X		
Transmission Medium		X	X	X
Blockage factors	X	X	X	X
Power Consumption		X		
System Redundancy		X	X	X
	Capacities			
No. of Terminations	X	X	X	X
System Total CCS		X		
No. of Transactions per Hour	X	X	X	X
Response time			X	X

DEVELOPING REQUIREMENTS: SOURCES OF INFORMATION

Developing requirements is seldom a straightforward proposition. Until you know what the market has to offer, you can't be sure whether an expressed requirement represents a genuine need or an unattainable ideal. The process of developing requirements, therefore, must begin with a study of the features the market has to offer. To avoid the problem of vendors' using different definitions for features, you should specify exactly the result you want.

For example, one company purchased a PBX with a feature the owner knew he required, least cost routing (LCR). The company used a considerable amount of long distance, and wanted to relieve its employees of the need to decide which long distance service to use. Least cost routing, therefore, became an important item on the company's requirements list. After he purchased a PBX, the owner found to his dismay that the LCR feature translated only on a three-digit basis, which limited it to looking only at the area code.

Company headquarters was located only two miles from an area code boundary. Nearby calls in the other area code were cheaper over Direct Distance Dialing, while distant calls were cheaper over WATS. The PBX, however, could not distinguish between prefixes, so all calls to that area code had to go via either direct dialing or WATS. Either alternative meant that some calls were more expensive than they otherwise needed to be. The problem in this case was the owner's failure to understand the difference between three-digit and six-digit translation until the PBX had been purchased. If the PBX had provided six-digit translation it could have routed calls based on both area code and prefix. This would have enabled it to send nearby calls over regular long distance, and more distant calls over WATS.

This example underscores the need to not only understand your requirements, but also to communicate clearly to the vendors how the requirement must be met. Typically, requirements are written into a request for proposals (RFP), more about which is covered in Chapter 7.

Problems As A Source Of Requirements

One of the best ways of developing requirements for a new system is to find out what problems users are experiencing with the old one. Some factor has triggered management to begin the search for a new telecommunications system, and knowledge of what those factors are is an excellent way of developing requirements for the new system. For example, is the old system out of capacity, costing too much, breakdown-prone, or lacking in essential features? Is a data communications network presenting response time problems, does it lack reroute protection, or does it lack the capacity to add more terminals? As we discussed in Chapter 3, an understanding of existing problems can do a lot to explain the requirements of the new system.

Vendor Presentations

Presentations by an informed representative of an equipment vendor can be an effective way of developing requirements for an investment of some time, but little money. Ask for copies of feature definitions, and try the features on demonstration machines to determine the difference in operation between equipment offered by different manufacturers. Sales people, in subtle ways, can inform you of the issues between their products and those of competitors. Some vendors will even go so far as to write requirements and specifications for you to use in an RFP, a practice that is known as "specing," and which should be avoided.

Obtaining Requirements From Higher Management

It is always good practice to interview higher management to determine company policy on controversial features, and to determine how much management is willing to spend for options. Features such as music-on-hold, trunk queuing, attendant barge-in, and executive override are controversial, and the question of whether they are needed can be answered only by top management in most companies. Higher level managers usually have little interest in minor system details, but a few minutes discussion of what systems can do and what they cost can help you develop requirements that will be accepted by management.

Industry Publications

Industry publications such as *Datapro Reports on Telecommunications* are often an excellent source of requirements and specifications. Such publications describe the features of competing products, explain many of the differences among products, and discuss their limitations. Trade journals also often publish articles discussing the strong and weak points of different products. Appendix H lists the principal publications of this type, and provides information for ordering them.

TYPES OF REQUIREMENTS

In-Service Date. Unless you are satisfied with the vendor's normal installation intervals, an in-service date should always be specified. Frequently, budgetary considerations, organizational changes, facility moves, growth, and other such factors mandate that service be installed by a particular date.

Vendor Support Requirements. Eventually, every manufacturer discontinues support of a product. When this happens, obtaining spares for repair and expansion can be a serious problem. If the system is widely used, third party vendors will continue to supply repair parts. If the market is large enough, parts may be reconditioned; if not, your only recourse may be unreconditioned used parts. As an alternative, you may want to consider requiring the vendor to obligate itself contractually to continuing availability of spares. The contractual aspects of this issue are discussed in Chapter 12.

A vendor's response to this requirement can reveal a lot about its reliability. Most large manufacturers are willing to guarantee availability of spares for at least ten years. If the vendor is reluctant to make a commitment, you should raise a caution flag.

If you plan to have the vendor maintain the system, the details of the maintenance plan should be spelled out. The vendor might be required to submit a maintenance plan that commits its forces to a particular response time on major and minor outages.

Training requirements should also be detailed. Two levels of training may be required, one level for training station users, and another for training your forces on whatever level of maintenance you expect to perform. If you plan to perform station adds, moves, and changes, training will be required, and should be specified in the RFP.

Documentation Requirements. Many systems come with a minimum level of documentation unless you specify what types of information are to be included in the purchase price. You should require the vendor to furnish a complete record of equipment as wired. Whether you plan to maintain the system yourself or contract with the vendor, it is advisable to require that all maintenance documentation be supplied. If the vendor withdraws support in the future, you can usually obtain third party support if you have the necessary documentation.

Administrative Features. One requirement that should be carefully considered is what kind of administrative support is required. This includes such features as the availability of statistical information on trunk usage, data transmission transactions, number of collisions (in a

LAN), and other such information that is essential for sizing trunks and circuit quantities.

Contractual Terms. As discussed in Chapter 12, the vendor may be in a position to dictate terms and conditions with its sales contract. When the RFP is issued, however, you should consider including in the requirements section a list of contractual obligations the vendor must meet. Typical obligations might include a requirement to carry liability insurance, to obtain all necessary permits and licenses, and to demonstrate that subcontractors have been paid before the final bill is paid. It is often advisable to require the vendor to post a performance bond or for you to withhold the final 10 to 20 percent of the contract price until the equipment has been working within the reliability and availability limits specified in the RFP for 30 days.

Features

Virtually all telecommunications equipment significant enough to be purchased with a request for proposals comes with a variety of feature options that may involve extra cost, and may be the deciding factor in selecting equipment. The requirements section of an RFP should list all the features of the desired system, with features listed as mandatory or desirable, as shown in Figure 6.1. The mandatory requirement serves as one of the first screens in choosing equipment. All equipment that lacks the necessary features is rejected. To save time in the evaluation, it pays to think through the mandatory feature requirements carefully. Desirable features are usually weighted on a scale of one to five, indicating the importance of the feature in the equipment selection decision.

TYPES OF SPECIFICATIONS

Specifications differ from requirements in that they are generally technical in nature and refer to some numerical range of performance or generally accepted standard.

Technical Interfaces. To minimize the risk of compatibility problems, the method of connection of telecommunications equipment should be specified. The safest way to specify an interface is by naming a

FIGURE 6.1
Sample Listing of Mandatory and Desirable PBX Features

	Mandatory/ Desirable	*Level*
Automatic Callback-Calling	M	
Automatic Circuit Assurance	D	5
Automatic Recall	M	
Automatic Redial	D	4
Battery Backup	M	
Busy Verification of Trunks	D	5
Call Forwarding All Calls	M	
Call Forwarding Busy Line	M	
Call Forwarding Don't Answer	M	
Call Hold	M	
Call Park	M	
Call Pickup	M	
Call Transfer	M	
Call Waiting Service	M	
Call Warning Tone	D	5
Calling Number Display to Station	D	5
Class of Service Display to Attendant	D	5
Class of Service Intercept	D	5
Code Restrictions	D	5
Conference Calling	M	
Consultation Hold	M	
Cust. Maintenance and Admin. Panel	M	
Deluxe Queuing	D	5
Direct Inward Dialing (DID)	M	

Mandatory requirements are shown as "M" in the above list with desirable requirements weighted on a scale of 1 to 5. When these are entered into an RFP, proposals that fail to meet the mandatory requirements are rejected.

recognized standard, such as RS-232-C or RS-449, published by the Electronics Industry Association (EIA) for data communications equipment. Bell Communications Research and EIA both publish standards for telephone equipment. Some of the EIA standards are adopted and published by the American National Standards Institute (ANSI), as are local area network standards developed by the Institute of Electrical and Electronic Engineers (IEEE). The consultative Committee on International Telephone and Telegraph (CCITT) publishes standards that are accepted in most countries. *The Dow Jones-Irwin Handbook of Telecommunications* lists standards for most telecommunications equipment.

Besides the electrical standards listed in EIA or other standards, you should also list physical standards if these are important. For example, the RS-232-C interface is commonly supported on either DB-9 or DB-25 connectors. The interface between the telephone set may be specified as RJ-11 or other type of connector. The physical interface between the PBX and station wiring should be specified as, for example, horizontal distributing frame with wire wrapped connectors, 66-type punch-on blocks, or other such interface connectors.

Interfaces to outside cable should specify the type of electrical protection along with the type of physical interface. If a PBX or multiplexer must interface T-1 carrier or a digital line, you should specify the digital service unit (DSU) interface.

If you do not specify the demarcation point between existing equipment and that on which you are requesting proposals, you may find that additional equipment must be purchased to make the service work; in the worst case, the equipment may be completely incompatible. Vendors will usually supply the least expensive interconnection hardware unless you give them instructions to the contrary.

Compatibility problems have plagued many telecommunications systems. A carefully written specification is not an air-tight guarantee that you will not encounter compatibility problems, but if the requirements have been explained to the vendor, you have an excellent chance of requiring the vendor to solve the problems.

Protocol Conversion. Data communications systems frequently require support of a specific protocol. PBXs are increasingly capable of supporting data communications protocols. Where a protocol is required, it should be specified.

Environmental Considerations. If space for the equipment room is limited or has certain environmental characteristics, such as ambient temperature, it may be advisable to specify the maximum dimensions of the system and the conditions the equipment must operate within.

System Redundancy. In some systems, processor redundancy or backup power may be required. Where these are specified, you should indicate the degree of protection required; ie., processor only, processor and switching network, power supply, etc. If backup power is required, the number of hours should be specified. In systems without backup power, it still may be advantageous to specify that the system survive a short power interruption without losing translations.

Capacity. System capacity should be expressed in the standard terms for the kind of apparatus being purchased. For example, PBX capacity is typically stated in numbers of port terminations, hundreds of call seconds (CCS) per port, and total CCS for the switching network. (See Chapter 16 for an explanation of CCS). Modem and multiplexer capacity is stated in bits per second. Local area network capacity is loosely stated because it hinges on so many variables, but you should specify the peak traffic throughput and response time requirements in bits per second.

PBXs, key telephone systems, multiplexers, and other expandable devices typically grow by adding circuit cards. Capacity is expressed, as shown in Figure 6.2, as numbers of ports wired, equipped, and ultimate capacity. Wired ports are those for which shelves have been installed and cabled to interface points. Equipped ports are those that are wired and

FIGURE 6.2
Sample listing of Required PBX Line, Trunk, and Station Capacities

	Equipped	*Wired*	*Capacity*
CO Trunks	34	40	48
DID Trunks	4	8	12
800 trunks	0	0	0
Trunks to LD Carrier	8	12	12
Paging Trunks	2	2	4
Recorded Announcement Trunks	2	2	2
Total Trunks	50	64	78
Attendant Consoles	1	1	2
Headsets	2	2	2
Handsets	1	1	1
Off Premise Stations	6	8	12
Single Line Desk Telephones	80	100	120
Single Line Wall Telephones	40	50	50
Electronic Telephones			
6 buttons or fewer	100	120	160
More than 6 buttons	20	25	30
Display	10	15	20
Total Telephones	256	318	392

equipped with plug-in cards. Ultimate capacity is the capacity that can be achieved by adding cabinets, shelves, or other apparatus that is not initially wired. For example, when a PBX reaches its wired and equipped capacity, another piece of apparatus such as a shelf is added to increase the wired capacity. As growth occurs, cards are added to the shelf to increase the equipped capacity. When the last shelf or cabinet has been added to the system, it has reached its ultimate capacity.

VALIDATING THE SYSTEM DESIGN

Requirements and specifications form a vital part of the request for proposals. Before the RFP is published, the data collected to this point should be reviewed with the users. If a users group was formed within the company, they can assist in obtaining agreement that the proposed system meets the organization's needs. If you did not form a users group, it is advisable to circulate the requirements, specifications, and overall system plan to obtain agreement and approval from appropriate authorities.

This step in acquiring a telecommunications system is the most time-consuming part of the entire project, with the possible exception of developing the plan. A consultant or in-house analyst will probably invest at least half of the total project hours in completing these preliminary plans, with few visible signs of progress. Even though this process is time consuming, it is the most effective way of avoiding the mistakes and delays that result from an incomplete analysis. As the old saying goes, "you can pay me now or pay me later." In the long run it usually costs less to pay now.

CHAPTER 7

WRITING REQUESTS FOR PROPOSALS

As all purchasing agents know, no two products are equal. Even a product that is a copy of another has differences in features that may have a significant effect on value. As we will discuss in Chapter 8, you should acquire the product that offers the greatest value in your operations but price alone is a poor indicator of value. Many factors such as rate of failure, vendors' commitment to product support, ultimate capacity, the cost of growth additions, and the ease of feature use affect value. The problem faced by the person who must select from among a bewildering array of products is where to get the information needed to evaluate these factors.

As a purchaser of telecommunications equipment, you are hampered by many factors that make it difficult to assess value. To begin with, not all vendors use the same terminology for the features of their products, or they use a term that has an ambiguous meaning to describe a feature. For example, for a vendor to say that a PBX has least cost routing is not enough. You also need to know how many digits it is capable of screening. Not all specifications are quoted by all vendors, and your specifications may not be expressed in the same language as the vendors' literature. Part of the solution to this dilemma is to use a request for proposals or RFP, a method that is used successfully for competitive purchases by both private industry and public agencies. As telecommunications products and services have become increasingly available through open competition, the RFP has gained acceptance on the part of corporate telecommunications managers, and is recognized as a fact of life by vendors.

In theory, an RFP is a logical way of managing the complexity of acquiring a telecommunications system, but issuing one is costly for a buyer, and responding to them may be even more costly for vendors. Whether you view RFPs from the standpoint of purchaser or vendor, they are rapidly becoming a way of obtaining information for selecting most large telecommunications systems, so you should learn to use them to your advantage.

This chapter outlines some of the most important factors in preparing and evaluating RFPs. Most of the discussion centers around PBX RFPs, but the same considerations apply to RFPs for local area networks (LANs), data communication networks and equipment, key telephone systems, and transmission equipment such as channel banks, microwave radio, and fiber optics.

PURPOSE OF THE RFP

The primary purpose of an RFP is to obtain information you need to make a purchase decision when the product is too complex to purchase on the basis of price alone. Commodities can generally be purchased by issuing a request for price quotation (RFQ), and buying from the low bidder. For our purposes a commodity can be considered as material or equipment that is standardized to the degree that one manufacturer's product can be substituted for another. An RFQ is less detailed than an RFP and usually contains only purchase terms and conditions and a simple specification. Many telecommunications products can be purchased on this basis without much risk. For example, twisted pair cable, mounting hardware, and even standard telephone sets and some modems can be purchased from the lowest bidder who meets the specifications.

A similar method of obtaining price information is the request for information (RFI). The RFI is a series of questions or outline specifications that are used to collect product and price information from vendors. Its primary use is to determine what products are available on the market, vendors' specifications, and a general idea of how the product is applied.

Most telecommunications products are far from being commodities. PBXs, multiplexers, microwave systems, and high speed modems are vastly different in their features and functions, and therefore in their value to the buyer. The buyer can use an RFP to convey the company's requirements and specifications to the seller, and to request the information needed to assess the difference in value among competing products.

An RFP focuses both the buyer and the seller on the important issues. The seller gains from using an RFP because it disciplines the buyer's decision makers to decide what is of real value and to distinguish valuable features from costly ones that are of marginal utility. For the buyer, the alternative to an RFP is to call in several vendors for presentations and quotations. Millions of dollars worth of telecommunications equipment have been purchased on that basis, and countless users have regretted purchasing a product after only a cursory evaluation. This is not to imply that vendors aren't to be trusted; users must understand what they really want, and the discipline of putting this down on paper in the form of objectives, requirements, and specifications is well worth the investment of time and money.

The benefits of an RFP don't accrue only to the buyer--the seller also benefits by gaining a clear understanding of what the buyer wants. Without an RFP, the seller is often faced with a dilemma in deciding what features to quote. If unneeded features are quoted, the equipment may be priced beyond the buyer's reach. If costly but essential features are omitted, the proposal may be rejected. An RFP helps to eliminate uncertainty about the buyer's basis for selecting the successful proposal.

In procuring any product, terms and conditions are frequently more important than price. Lacking an RFP, the buyer is often left with no alternative but to accept the vendor's terms and conditions, discussion of which may be deferred until it is obvious to the seller that the sale is clinched. With terms and conditions clearly delineated in the RFP, and the sellers required to quote on that basis, the buyer is in a better position to compare purchase prices of the alternatives.

Another advantage of an RFP for the buyer is that it causes the vendor to commit to certain standards of performance that might otherwise be treated only verbally, a point that will be enlarged upon in Chapter 12. A final purpose of the RFP is to facilitate the selection process. By choosing an appropriate format for responses, the buyer can organize the replies so that all vendors prepare their quotations on the same basis and in the same format. This eliminates hours of labor in digging through brochures for comparative figures.

On balance, the advantages of the RFP probably weigh in favor of the buyer, but the seller also benefits from gaining knowledge of what is required, what is of greater and what of lesser importance, and what is of no value to the buyer. Furthermore, the RFP separates the genuinely interested customers from the tire kickers and helps the seller concentrate its resources where the business really exists. When buyers undertake the expense of an RFP it is an excellent indication that their intentions are real. Many vendors otherwise waste hours in making presentations to a buyer who may have no intention of making an immediate purchase.

WHEN TO USE AN RFP

RFPs are expensive to prepare, and are not appropriate for all purchases. Expensive apparatus such as PBXs should almost without exception be purchased by RFP. Even if only one vendor is being considered, the RFP offers the advantages of obtaining vendor commitment and a firm price proposal. On the other end of the scale, simple devices such as low speed modems, telephone sets, and LAN interface cards are rarely purchased by RFP except when they are being purchased in quantity or as part of a total telecommunications system. The cost of preparing the RFP and responding to it are too great compared to its benefits when you are making one-at-a-time purchases of low-cost devices.

In between these extremes lie many intermediate-sized purchases such as multiplexers, key telephone systems, and channel banks. Whether to use an RFP for these is a matter of judgment. If you have plenty of time and do not know the principal vendors in the market, the RFP is probably worthwhile. On the other hand, if the deadline is short and you are already familiar with vendors and have established a relationship of trust with them, the RFP is probably unnecessary.

In summary, the RFP should be used when one or more of the following conditions exist:

- The product being purchased is not a commodity
- The purchase is large and complex enough to justify the expense; generally the purchase price should be more than $10,000 to $20,000
- You are not familiar with most vendors on the market
- Prices are elastic and vary with competition
- There are competitive sources for the product
- The penalty of making the wrong decision is severe in terms of cost or service disruption

- More than one type of product can fulfill the requirement (for example, a PBX vs. a LAN vs. a data PBX)
- The purchase is a continuing one rather than a one-time procurement

Most public organizations are required to use competitive procurement for purchases greater than a specified dollar volume, so the question for them is whether to use an RFP or an RFQ. For both public and private organizations, an RFQ is recommended only if the product is a commodity; that is, if it is manufactured to a recognized standard of compatibility and quality, and if initial purchase price is the primary deciding factor. Otherwise, use an RFP, or, if the purchase is simple and risk is low, use a directed purchase from a source of supply known to be reliable. An RFP should not be used when the cost of preparing it is greater than the potential benefits you expect to gain from price competition. Just how complex a product should be to justify the use of an RFP is a matter of judgment and the risk you are willing to assume. The primary risks include these factors:

- *Incompatibility*: By requiring the vendor to warrant compatibility you reduce the risk that the product won't work with your existing system
- *Vendor stability*: By requesting references and financial and experience information, you reduce the risk that the vendor will be unable to perform
- *System reliability*: By requesting reliability information such as mean time between failures and availability of repair forces you reduce the risk of excessive down time
- *Obsolescence*: By requesting information about the status of a product's life cycle, you can differentiate between products that are obsolete and those that are so new as to be unproven

ELEMENTS OF AN RFP

The style and content of RFPs vary widely among the companies that issue them. Some RFPs are restrictive and attempt to foresee every contingency and to place the responsibility on the vendor to correct every fault. By shifting the total responsibility to the vendor, two results often occur; some vendors decline to participate in the competition, and those that do compete increase their bids to take care of unforeseen events. At the other end of the scale, the RFP may be so weak that the buyer is inundated with change orders. It is no secret that many vendors deliberately bid low on RFPs that have obvious deficiencies, with the intent of making the profit on changes.

There are at least as many different RFPs as there are consultants, and each one has a different style and content, but most RFPs have a few elements in common. The sample PBX RFP contained in Appendix C illustrates the points mentioned in this chapter. Although you are free to copy and use this RFP within the restrictions outlined in Appendix C, it is important to understand that preparing an RFP is not a "cookbook" task. It must be customized for the project and the organization. Nevertheless, a great deal of the information contained in RFPs is standard and can be used by almost any type of organization. Organizations that regularly use RFPs will undoubtedly have a format to follow. Other organizations may wish to consider the following elements that should be included in most RFPs.

Project Description

An RFP should typically begin with a narrative description of the organization requesting the product or service and the environment in which it will be used. The existing situation should be described, together with the reasons a change is being sought, and the organization's objectives for a new system. With a clear description of your organization's telecommunications problems and a list of requirements for the new system, vendors can select the best configuration of their products to meet your needs. If permitted by the RFP, they can propose alternative solutions in addition to proposals that conform to the specifications. Such information as the following should be included:

- Addresses of principal locations
- Primary lines of business
- Description of existing systems
- Objectives of the projects described in the RFP
- Outline of the overall system size (numbers of stations, messages, etc.)

General Information

In a few paragraphs, you should explain the principal features of the project so that vendors can quickly determine which, if any, of their product line meets the requirements. Contained in this section should be:

- Type of system being requested (PBX, key telephone system, LAN, etc.)
- Required completion date
- Schedule of major events (date of vendor's conference, response date, order date, etc.)
- RFP amendments: the RFP should state how it will be amended if amendmcnts are required

Boiler Plate

The term ''boiler plate'' refers to standard information, often legal, that describes the buyer's general purchase terms and conditions. This section of the RFP should be reviewed and approved by an attorney, and should be reviewed by higher management to ensure that it conforms to your company's purchasing policies. Public agencies usually have standard language that can be inserted intact into the RFP.

Acceptance of Proposal Content

A controversial subject for some vendors is the requirement that the content of the proposal be incorporated into the purchase contract. In fact, some vendors refuse to accept these terms, and insist instead on using their standard sales contracts. Whether you agree to sacrifice this clause is for your legal advisors to determine, but the inclusion of the response to the RFP in the contract is one of the most important reasons for preparing an RFP. Sales representatives are usually reluctant to admit that their product fails to meet an essential requirement, and, as discussed in Chapter 12, almost all sellers' contracts disclaim any warranty that is not in writing. Therefore, the only evidence you are likely to have that the seller has stated that the product conforms to your requirements is the seller's response to the RFP. Incorporating that response into the contract gives you a measure of protection that is otherwise missing.

Price Increases

Unless there is extreme urgency about a project, it usually takes several weeks from the date responses are received to select the vendor, obtain approval from higher management, negotiate the purchase, and place the order. If prices have increased in the meantime, you may be faced with reviewing the decision before the order is placed. To prevent this, it is advisable to specify the length of time that the vendor must agree to keep the prices valid. Most vendors hold the prices open for thirty days, and this is seldom enough to make a decision. A sixty day clause is reasonable; some companies and governmental bodies may require more.

Performance Bond

A requirement for a performance bond is a matter of company policy. You should consider imposing this requirement if the vendor lacks a strong performance record or if the penalty for non-performance is severe. Withholding part of the payment is often a preferable alternative to requiring a performance bond, and may be less expensive since the seller probably passes the cost of the bond through to the buyer. It is customary to withhold at least 10 percent of the contract price for approximately 30 days to ensure that the system is meeting performance requirements.

Contractor Responsibilities

The contractor should always be required to carry liability insurance, and although almost all contractors will be insured anyway, you should consider making it a matter of contract. The contractor should be held responsible for obtaining all permits and licenses, and should agree in writing that all costs of installing wiring to meet building and electrical code requirements are the contractor's responsibility. You should also require the contractor to submit evidence that subcontractors have been paid before you agree to pay any subcontractor bills or make the final payment to the contractor. The purpose of this requirement is to ensure that subcontractors cannot place a lien on your property because they have not been paid for such work as placing station wiring.

Purchase Terms and Conditions

If you do not specify terms and conditions, the vendor's terms will prevail, and these may not be to your advantage. Examples of terms you may want to consider are these:

- *Warranty:* Indicate whether you require an extended warranty and if you are unwilling to accept the vendor's disclaimer of implied product warranties; see Chapter 12 for further discussion on implied warranties
- *Payment terms:* If you plan to withhold part of the payment until after acceptance, the payment terms should be specified
- *System acceptance:* Benchmark tests that must be met before the product is accepted should be described
- *Purchase contract terms:* If your organization has a standard purchase contract, it should be incorporated by reference
- *Insurance and performance bonding:* The degree of insurance the vendor must carry should be specified. If performance bonds will be required, the amount and terms should be specified
- *Conformance with local ordinances:* You should indicate whether the buyer or seller will be responsible for obtaining permits and licenses, and requirements for meeting fire and safety codes should be spelled out

System Requirements and Specifications

This section is the body of the RFP and explains what the project must accomplish. The requirements and specifications developed in Chapter 6 should be incorporated into this section of the RFP. As discussed in Chapter 6, the RFP should state clearly which requirements are mandatory and which are optional. The RFP should make it clear that the total cost must include the mandatory items, and that the extra cost, if any, of optional items is to be stated. If installation and future maintenance services are required, these should be stated. The general characteristics of the system should also be specified. This includes such features as system architecture, protocol compatibility, interfaces with other systems, traffic volumes, station requirements, required grade of service levels, and other such essential factors.

Of particular concern is your requirement for product documentation. Usually, vendors do not furnish software source code, schematic diagrams of circuit, trouble locating manuals, and other such documentation unless it is specifically requested (and sometimes not even then). The RFP is an appropriate vehicle for clarifying what documentation is required.

Evaluation Procedures

This section is optional for most private companies, and mandatory for most public organizations. Whether it is required or not, it is advisable to explain exactly how responses will be evaluated because this conveys to vendors what factors you deem most important. The process you will use to choose the winning proposal should be developed before the RFP is issued, even if you don't plan to describe the selection process in the document itself.

Information Requests

This section is compiled to obtain information that vendors would not include in their responses as a matter of course. Specific requests should be made for information that you will use to make the decision. This section is valuable to aid you in organizing responses so you can make valid comparisons among different products. Without this section, you will have to search through vendors' literature to find answers to your questions. To reduce the burden on vendors it is sometimes acceptable to give them the latitude of including their published documentation in the response and cross-referencing it by page number.

Proposal Format

To gain the maximum value from the RFP, you should specify exactly how you want the proposals submitted. Response forms, a sample of which is shown in Figure 7.1, should be provided in an appendix. A form such as this ties RFP sections and supporting documentation directly to the response sheet. The M/D column indicates whether you consider the feature to be mandatory or desirable. The Available column requires the vendor to indicate availability information about each feature such as the following:

- I—Included in the price quoted
- O—Optional at extra cost (specify cost)
- N—Not available
- E—Equivalent feature is available (explain the feature)
- D—Under development (indicate when feature will be available)

FIGURE 7.1
Sample Form for Vendors to Use in Responding to RFP

RESPONSE SHEETS

The following pages are included for vendors to enter their responses. Wherever possible, please use these sheets. If insufficient space is available, you may attach additional sheets. The response sheets are in the same order as information requested in the RFP, and generally follow the same format.

Functional Requirements

The following is a reiteration of functional requirements listed the PBX. The M/D column indicates whether the feature is mandatory or desirable. Following each of these requirements, please list these two things:

1. Availability codes as (i) included in the price quoted, (o) optional at extra cost (specify cost), (n) not available, (e) equivalent feature is available (explain the feature) or (d) under development (indicate when feature will be available.
2. Page numbers in proposal or supporting documents where feature is described.

Feature	*M/D*	*Avail.*	*Page*
Add-on Conference	______	______	______
Alarm Display on Attendant Console	______	______	______
Alphanumeric Display for Attendant Console	______	______	______
Area code Restriction	______	______	______
Attendant Busy Lamp Field	______	______	______

This form should be expanded to include all the functional requirements.

You should list what documents are required, and should ideally provide forms for submitting responses. The time you spend in organizing the response format will be more than repaid by the time you save evaluating responses, and will help avoid the necessity of going back to the vendor for additional information.

Pricing information should be requested in a form that allows you to calculate the cost of future expansion. The price of all deferrable plug-in cards should be requested so you can determine the cost of either expanding or shrinking the system. A sample pricing format is shown in Figure 7.2.

Other information that should be included is:

- Deadline for responses
- Person to call with questions

FIGURE 7.2
Sample Pricing Sheet

Pricing Sheet

Please follow the following format and examples in submitting pricing information. Please include enough detail in your response that we can determine the cost of major components such as shelves, cabinets, plug-in cords, power supplies, etc. Subtotal major assemblies and total all items included in your proposal. Submit prices for the following separately:

Hardware
Software
Services such as installation, training, etc.

Summarize your proposal with a grand total price for the intitial cost of all equipment and services proposed. Submit recurring costs such as maintenance contract on a separate sheet.

Component	*Type*	*Quantity*	*Unit Price*	*Total Price*
Line Card	8-port digital	5	1,200	6,000
Trunk Card	4-port analog	4	1,150	4,600
Shelf	Card carrier	2	800	1,600
Cabinet	System	1	225	225
Equipment Total				86,000
Software Total				
Services Total				
Grand Total				

- Number of copies required
- Person to whom responses should be addressed
- Special response requirements such as sealed envelopes and procedures for amending proposals (if permitted)

Appendices

Appendices should contain the vendor qualification forms, floor plans, response sheets, bonding requirements, and other such elements that the project requires. It is advisable to include a feature definition section to ensure that the vendors use your same definitions in responding to features. A sample list of feature definitions is included in Appendix D. Within the restrictions outlined at the beginning of the appendix, you may copy these and include them in your own RFP.

Any information that you require to qualify the vendor should be included in the RFP. For example, if the project requires a substantial commitment of the vendor's funds, you may wish to request a financial statement. You will undoubtedly want to request references from similar projects the vendor has completed, and possibly resumes of the principal employees who will perform or supervise the project. If the vendor does not manufacture the product, you should also consider requesting similar information about the manufacturer, depending on the complexity of the product and the manufacturer's reputation. A sample vendor qualification form is shown in Figure 7.3.

HOW TO PREPARE AN RFP

If you have had no experience in preparing RFPs, get assistance from someone who has or be prepared to spend many hours in research. Casually produced amateur RFPs are rarely effective, usually waste the vendors' time, and may result in costly mistakes on your part.

Assuming you have had some experience with RFPs, the first and most important step is to understand your requirements and specifications in detail. This process has been covered in Chapter 6, and will not be repeated here, but it comprises at least three-fourths of the work of RFP preparation.

Unless you have a burning urge to create an original document, you can probably glean much of the remainder from existing RFPs. Most vendors will give you copies of RFPs they feel are particularly well-done. You can paraphrase and use them as a reminder of points that must not be omitted, or if you have a consultant, he or she will probably have an RFP that can be used with minor modification.

The RFP should be issued with a cover letter that summarizes your request and establishes its return requirements. You should always include a deadline for responses or they will trickle in over several days. Most public and some private organizations will require responses to be sealed and delivered to a specified address by a specified time. Late responses are usually rejected.

VALIDATING THE RFP

Before the RFP is released it should be carefully reviewed for feasibility, accuracy, and completeness. Feasibility is a question of whether the provisions of the RFP can be met by equipment available on the market

FIGURE 7.3
Sample Vendor Qualification Form

Name of Company__________
Address__________ Telephone No.__________
Name of President, CEO or Local Mgr.__________
Name of Primary Marketing Rep__________
Name of Engineering Manager__________
Name of Service Manager__________
Form of Business
- Corporation ______
- Subsidiary ______
- Proprietorship ______
- Partnership ______
- Other (describe) ______

Status of Proposer
- Distributor for Manufacturer ______
- Local office of Manufacturer ______
- Other (describe) ______

How many systems of the same type and model have you installed in the local area?______
How long have you been in business?______
How long has the local office been in existence?______
Number of personnel in local office
- Marketing ______
- Service ______
- Engineering ______
- Administrative/support ______
- Other ______
- Total ______

How many local (within one hour's drive of city) factory trained service representatives do you have on this model of equipment?______
Identification of Manufacturer__________
Address of Corporate Office__________
If you are a subsidiary of another company, identify parent company and state relationship to parent__________

____

To what degree will manufacturer assume obligation for service of the proposed system if you discontinue business or drop the product line?__________

____

List at least three references who have similar systems in this area that have been installed within the past two years. List company name, address, and name and telephone number of contact.__________

____
____
____
____

Please indicate if the following are attacted:
- Annual Report ______
- Dunn and Bradstreet Report ______
- Audited Financial Statement ______

for a price that is within budgetary limits. If your organization's policies permit, an excellent way to test feasibility is to release a draft to vendors for comment. If you have listed mandatory requirements that can't be met by most products, you may want to rethink your approach to gain the benefits of more widespread competition. When the responses are received, if some mandatory features are available at a cost greater than you can afford, you may have to reissue the RFP with a consequent loss of time and increase in cost. It usually pays to state in the RFP that you reserve the right to waive mandatory responses for all vendors (although some public agencies may not permit this).

Validation of completeness and accuracy is an internal function that involves editing the RFP and reviewing it against requirements and specifications. A legal review is usually advisable in addition to the technical review. An attorney should be consulted on any matters relating to warranty, performance clauses, withholding of payments, cancellation terms, and the like.

Vendor Conference

A vendor conference is often a good way of clarifying any ambiguities in the RFP. For public agencies that are required by law to deal openly with all vendors, a vendor conference is highly recommended because everyone has an opportunity to ask questions and hear the same answers. If necessary, the RFP can be amended to correct any errors in the initial issue. Vendors can also be given an opportunity to inspect the site to enable them to better estimate installation costs and respond to environmental concerns. Questions raised at the vendor conference can be answered at the conference or researched after the conference and answered later. In either case you should provide written answers to all questions and send them to attendees.

ALTERNATIVES TO THE RFP

RFPs are not appropriate for all product selections. If the dollar volume is low, there is little reason to invest time and effort in writing an RFP, but some alternative way of disciplining the selection process is needed. This section briefly discusses alternatives to the RFP for selecting products.

Favored Vendor

Many product purchases are made without considering more than one vendor. If an existing vendor gives satisfactory service at competitive prices, purchasing without an RFP can be good business. If your purchase volume is high enough, you may be able to negotiate a pricing agreement for future purchases. The negotiated purchase is particularly feasible when one or more of the following conditions exist:

- There is limited opportunity for competition
- You have a significant investment in existing equipment that requires an addition
- Compatibility between competing products is questionable
- It is expensive to change vendors because of an investment in spares, documentation, and training
- Management policy favors a particular vendor

Informal RFP

With smaller purchases it is often possible to use an informal RFP. Much of the detail of the formal RFP is omitted, and only a few essentials are requested. This method has the advantage of getting the benefits of competition, but reduces the cost of preparing the RFP, preparing responses on the part of the vendors, and evaluating responses on the part of the buyer.

This method is usually the best way of purchasing products that have gained the status of commodities. Telecommunications products such as wire and cable, connectors, distributing frames, 2500-type telephones, and similar items are examples of products where the informal RFP may be advantageous. This method should be considered when one or more of the following conditions prevail:

- First cost is the most important factor
- Comparable products are available from numerous vendors
- The risk of product failure is low
- Numerous products are completely interchangeable

SUMMARY

The RFP is an effective way of purchasing telecommunications products where the scope of the purchase is significant enough to warrant the expense and effort. Considering the time it takes to prepare objectives, requirements and specifications, and issue an RFP, it may sometimes seem that the effort is greater than the benefits, and for many purchases an RFP is not worth the cost and time it takes. Until telecommunications products reach the status of a commodity, which is not to be expected in the near future for most products, the RFP is an excellent way of collecting enough information to make a valid selection decision without an excessive expenditure of effort. If you already know what system you want to buy, or if the system is not complex, there are better ways to purchase than an RFP, but you will save time and expense in the long run by using an RFP for most PBXs, LANs, and complex data communications systems. The time you have spent to this point will be returned during the equipment selection and implementation stages, which we will cover in the next several chapters.

An RFP must be prepared with your own requirements and specifications in mind. Many organizations attempt to issue RFPs by copying what someone else has already done. This is an acceptable practice provided you customize the RFP to reflect your own unique requirements. The primary advantage of the RFP is in gathering enough information to make a selection decision. Several specialized techniques for evaluating the responses are discussed in Chapter 8.

CHAPTER 8

EVALUATING RESPONSES TO REQUESTS FOR PROPOSALS

How much should you pay for quality? Conventional wisdom suggests that quality costs less in the long run, which may be true. But it costs more to begin with and the long run may not be long enough to recover the extra investment in a higher priced product. A typical business problem is how much it is worth spending initially for a product that saves money over time. To select the best product from a group of competing proposals requires you to have a systematic process for evaluating technical and economic performance.

One product may have the lowest price tag, but it may sacrifice quality. A second product may have the highest quality, and yet another the lowest upkeep cost. Whether you are investing in telecommunications equipment, a motor vehicle, a computer, automated manufacturing equipment, or any other type of labor saving device, there are always differences in the values of the products under consideration, and basing the decision on price alone is short-sighted. The question is how to evaluate the tradeoffs between the many variables that are important in assessing value. This chapter suggests a process for evaluating RFP responses with a particular emphasis on life cycle costs.

Life cycle costs are evaluated with a *life cycle analysis* (LCA), which is a systematic process for evaluating both financial and intangible differences between products. LCA is a far more reliable guide than intuition in comparing differences in quality, warranty terms, and original cost of the equipment. It provides a way of comparing differences in recurring costs such as maintenance, and in the timing of major expenditures such as growth additions. Not only does an LCA help to minimize subjectivity, but more important, it offers a way for the purchaser to escape the pitfall of buying an inferior product merely because it has the lowest cost or because the sales person has the most persuasive pitch.

In purchasing a product, only the first cost and other "getting started" costs are invested initially. All other costs will be incurred in the future, and predicting what costs will be in the future is an uncertain process at best. No technique is an infallible guide to the future, but an LCA is a way of documenting assumptions and handling them equitably among all products in a systematic and bias-free manner.

OVERVIEW OF THE PRODUCT SELECTION PROCESS

The objective in reviewing RFP responses is to invest the least amount of effort in screening out proposals that have little or no chance of being selected. Unless there are compelling reasons for following a particular sequence, the evaluation should proceed from the factors easiest to assess to those most difficult to evaluate. For example, technical and operational evaluations usually require detailed discussions with the vendor and visits to inspect working systems. Since these evaluations are more costly than economic evaluations, it makes sense to calculate the economics first to screen out proposals that are too costly to be within the competitive range.

The process suggested in this chapter will enable you to sift progressively through responses until only the winning product remains. The following steps are recommended:

Step 1: Eliminate proposals that fail to meet mandatory requirements. The coarsest screen in the sieve is to review responses for compliance with mandatory features and to be sure that all requested information has been supplied. The value of requiring vendors to follow a prescribed format in their responses shows up quickly at this stage of the evaluation. It is not uncommon for vendors to respond with non-conforming proposals, and depending on your organization's policies, these can either be rejected at the outset or set aside for possible future analysis in case the conforming proposals are all unsatisfactory. In any case, it is impossible to make a valid comparison between products that conform and those that do not conform to the evaluation criteria.

Step 2: Create a response matrix. A matrix chart such as the one in Figure 8.1 should next be created to display the degree of conformity to desirable features and to put costs on an equal footing. If your company policy permits, comparing prices at this stage enables you to reject proposals that are so far out of line with the others that they have little chance of being chosen. You must be aware, however, that first cost is not a reliable indicator of value. You should also compare other cost factors such as maintenance and power consumption before rejecting any proposals out of hand.

The matrix chart makes it obvious that certain proposals such as Product C in Figure 8.1 have little chance of being chosen and can be set aside. The rejection at this stage is tentative because fatal deficiencies may turn up in the other proposals as the analysis continues.

Step 3: Review features in detail. Using the vendors' responses, the next step is to revise the matrix to reflect reality, as opposed to vendors' claims. Your requirements are checked against the response sheets and against the documentation submitted with the proposal. The purpose of this step is to determine how effectively the features are implemented. This stage of the review assesses value. For example, if a vendor proposes a redundant processor, this feature would add value to the overall proposal compared to other vendors who did not.

Step 4: Perform a life cycle analysis. If the system is complex enough to warrant using an RFP, it is usually complex enough to justify the time it takes to use a life cycle analysis for the economic evaluation. Using techniques described later in this chapter, the proposals are subjected to a detailed analysis of costs that will be incurred over the life of the system. At the conclusion of this step, the selection is narrowed to the two or three products that offer the greatest value.

Step 5: Review technical performance of the top proposals. The technical review is often the most demanding part of the entire study, and therefore is reserved for a few proposals. This review is often made on a working system arranged for by the vendor. Feature operation is tested, station and terminal equipment are evaluated, maintenance and administration methods are examined, and the vendor's support procedures are reviewed in detail.

FIGURE 8.1
Sample Decision Matrix for Comparing PBX Proposals

Stations = 117	*Product A*		*Product B*		*Product C*		*Product D*	
	Cost	*Cost/Stn*	*Cost*	*Cost/Stn*	*Cost*	*Cost/Stn*	*Cost*	*Cost/Stn*
Basic PBX	$ 48,543	$ 415	$ 36,454	$ 312	$ 84,323	$ 721	$ 59,717	$ 510
MAT Terminal	4,553	39	3,476	30	2,323	20	3322	28
Station Equipment	26,453	226	34,232	293	43,333	370	34,994	299
Basic PBX Total	79,549	680	74,162	634	129,979	1,111	98,033	838
Auxiliary Equipment								
Call Accounting System	32,113	274	21,323	182	5,443	47	7,653	65
Wiring	2,312	20	10,996	94	22,700	194	12,543	107
Battery Backup & Power	11,200	96	7,193	61	18,764	160	772	7
Miscellaneous	1,508	13	1,043	9	9,225	79		
Aux Equipment Total	47,133	403	40,555	347	56,132	480	20,968	179
Labor	27,743	237	9,210	79	4,538	39	2,225	19
Total System	$154,425	$1,320	$123,927	$1,059	$190,649	$1,629	$121,226	$1036
Features								
Least Cost Routing	YES		YES		YES		YES	
Direct Inward Dialing	YES		YES		NO		YES	
Digital Display Phones	$ 535		$ 476		$ 645		$ 449	
Modem Pool	$ 3,450		INCLUDED		$ 4,455		$ 4,965	

Step 6: Weight desirable features. Not all features are of equal importance, and not all systems implement the features with equal elegance. Some method is needed to evaluate how well the features are implemented. Public agencies often use a weighting scale to arrive at a final score for each feature. Private organizations are less apt to spend the time it takes to make a detailed feature evaluation.

Step 7: Develop a final product evaluation. All of the elements entering into the product evaluation should be summarized and the final evaluation computed. Public organizations almost always use a scoring process that weights each part of the evaluation and assigns a final numerical rating to it. Private organizations have more latitude, but the scoring process tends to eliminate bias.

To develop a final rating, you must decide what factors will enter into the selection process and how important each is. Typical factors are:

- Life cycle cost
- Vendor's qualifications and experience
- Evaluation of mandatory features
- Evaluation of desirable features
- Vendor's support plan (training, maintenance, etc.)

The final score is computed by weighting the factors as shown later in this chapter. This scoring method is not an infallible way of selecting the most effective product, but it does offer a systematic process for evaluating the relative importance of various features. By comparing all proposals on an equal footing, your attention is focused on the strong and weak points of each proposal. The weak points can be strengthened by negotiating corrective measures into the purchase contract.

LIFE CYCLE ANALYSIS

A life cycle analysis is the process of estimating all the cash flows an organization will incur as a result of owning a product, and discounting the cash flows to present worth at the organization's cost of money. An LCA uses the same financial techniques as the feasibility study we discussed in Chapter 4, but it is an altogether different concept. A feasibility study determines which *plan* is the most financially and economically effective. A life cycle analysis takes over where the feasibility study leaves off, and determines which *product* is most effective for executing the plan.

An LCA determines which of several alternatives consumes the least amount of cash throughout the life of an investment. The analysis is accomplished by compiling a financial model of the discounted cash flows expected from acquiring each product alternative. For example, you may complete a feasibility study that shows that replacing your present key telephone system with a new PBX is feasible. The question is which PBX to buy. A life cycle analysis helps to answer this question by identifying all the cash flows that will occur throughout its service life. The present worth of the cash flows is summarized to identify the product with the highest net present value (NPV). The product with the highest NPV will consume the least amount of cash throughout the product's service life.

Steps in Conducting a Life Cycle Analysis

There are five steps in conducting an LCA:

1. Determine your organization's requirements
2. Determine your specifications
3. Identify the life cycle cost factors
4. Obtain life cycle cost information
5. Evaluate life cycle economics

The first two steps are described in Chapter 6. From the requirements and specifications, you must identify the *life cycle cost factors*, which are one-time and recurring cost elements such as purchase price, maintenance, and operating expenses. Life cycle cost information is obtained by compiling your company's costs for the equipment proposed by each vendor and evaluating the costs in a life cycle model.

The most effective way to obtain life cycle information is to request the necessary data in an RFP, as discussed in Chapter 7. This, of course, makes it necessary to determine the life cycle factors before the RFP is written. We have deferred that discussion to this point, however, because of the need to clarify certain terms and concepts.

Determining Life Cycle Factors

An LCA requires that you identify every cash flow resulting from the ownership of a product from the time it is acquired until it is removed from service. Table 8.1 lists typical cash flows for PBXs, key systems, and most other telecommunications products. The cash flows closely follow those that are identified during a feasibility study. A key to conducting a successful LCA is to think out all recurring and non-recurring costs that will be incurred over the product's life. This is the most important part of an LCA because if any factors are overlooked you may come to the wrong conclusion. Remember that the objective of the study is to model as closely as possible the financial effect of choosing one alternative over another.

Many factors can safely be omitted from an LCA. Any factor that is common to all alternatives will have no effect on the outcome whether it

TABLE 8.1
Typical Cash Flows for Telecommunications Products

Intitial or Getting Started Cash Flows
Purchase price of equipment
Shipping, handling and delivery
Sales taxes
Installation and wiring
Training
Instructions and documentation
Initial software and right-to-use fees
Building or floor space initial costs or rearrangements
Non-recurring charges for common carrier services (tie lines, C.O. trunks, etc.)
Recurring or Annual Cash Flows
Maintenance and repairs
Administrative costs
Operator or user labor
Lost production time during outages
Recurring software fees
Operating supplies
Fuel or electric power
Floor space and air conditioning
Tax effects of depreciation and operating expenses
Revenues
Rearrangements (stations, trunks, terminals, etc.)
Common carrier services (trunks, lines, etc.)
Growth additions

is included or not. In the example furnished later in this chapter, we assume that training costs are equal for all products and that the telephone company's charges for local lines and exchange access are equal. Such charges affect the *feasibility* of the system, but if they will be incurred regardless of which alternative is chosen, they can be disregarded without affecting the outcome of the study. When common costs are omitted, however, the total cash flows will not represent the money the company will actually spend on the product. Although the total dollar differences between products will be valid, the percentage differences between alternatives will be meaningless.

Obtaining Life Cycle Cost Information

The most accurate source of life cycle cost information is the vendor, provided the figures can be validated from published documents or experience. The values you will assign to each product in the analysis should be requested in writing, preferably in the RFP. You should specifically request any factors that can be assigned a numerical value, such as purchase price, installation cost, floor space, power consumption, and monthly maintenance costs.

Intangibles should also be requested in writing, with vendors asked to respond to specific questions, such as how many systems are currently in service. Questions should be developed to estimate the cost of training administrators, users, technicians, and operators, and to determine any other factors that are considered important but difficult to quantify. The most effective way to organize information is with a response form that provides space for a response to every tangible and intangible factor that will affect the outcome of the study. Intangible factors are not used in the economic analysis, but may sway the outcome between products that are economically comparable.

Evaluating Life Cycle Economics

The life cycle analysis is conducted by assigning a value to each factor for each vendor and discounting each year's cash flow to its present value. While it is possible to do this manually, the calculations and table manipulations are tedious. Furthermore, if you change any assumptions, more arithmetic is required. A desktop computer with a spreadsheet program offers a convenient way of performing the calculations. Spreadsheet programs such as Lotus 1-2-3®, Microsoft Multiplan®, Sorcim

SuperCalc®, or any other spreadsheet that contains an NPV factor can be used for performing the calculations. The steps in performing the economic analysis are:

1. Determine the length of study
2. Determine the cost of money for your organization
3. Decide on your expectations of future inflation
4. Determine costs for each product
5. Determine the effects on income taxes
6. Run the analysis
7. Interpret the results

Determining the Length of the Study. The length of the study often has a significant effect on which product is the most economically attractive. For example, assume that one product has a higher initial cost, but lower recurring costs than a competitor. The longer the assumed product life, the greater the opportunity for lower recurring costs to offset the effects of higher initial costs. Unless you have a better figure, the asset depreciation life prescribed by the IRS is a reasonable factor to use for study length. Any uncertainty in service life can be dealt with by sensitivity analysis, a process we will discuss later.

Determining the Cost of Money. In a corporation, the cost of money is the composite cost of capital, which is the weighted average cost of equity, debt, and retained earnings. The method of developing this figure is explained in most financial management texts, or your accounting department or Certified Public Accountant firm should be able to supply the figures.

Deciding on Inflationary Expectations. In an inflating economy, wages and other recurring costs increase with time. Therefore, a capital investment in labor-saving equipment offsets future recurring costs more quickly than when labor rates are not inflating. With electric power rates inflating more rapidly than the cost of living in many parts of the country, an investment in energy-efficient equipment will pay for itself in time. The higher your expectations of inflation, the more effective a capital investment becomes if it decreases expense. It is important to include the effects of inflation if you think they will be significant.

Determining Costs For Each Product. Some factors are easy to value and can be taken directly from the RFP response. For example, the power consumption and floor space for each product are a matter of the manufacturer's specifications and can readily be obtained from brochures or the RFP. Other factors such as maintenance cost are more elusive.

The most reliable source of information, providing it can be verified, is the vendor itself. For example, an estimate of the cost of maintaining a product can be obtained from the vendor's quotation for a maintenance contract. Even if you don't intend to use a maintenance contract, its price is a reasonable indicator of comparable costs among vendors.

Another way of calculating maintenance costs is to request mean time between failure (MTBF) and mean time to repair (MTTR) factors from the manufacturer and to apply these to expected labor costs. Here is an example of calculating maintenance costs from MTBF and MTTR:

Assume that a point-to-point data circuit is connected by two multiplexers that each have an MTBF of 10,000 hours and an MTTR of 2.0 hours, including travel time. Assume that labor to repair failures is $60 per hour. Also, assume that the cost of lost productive hours during down time is $600 per hour. What is the expected annual cost of failures?

$$\text{MTBF} = \frac{10{,}000 \text{ Hours}}{2 \text{ Multiplexers}} = 5{,}000 \text{ Hours}$$

$$\frac{8760 \text{ Hours per year}}{5000 \text{ Hours MTBF}} = 1.75 \text{ failures per year}$$

$$1.75 \text{ Failures per year x } 2.0 \text{ hours MTTR} = 3.5 \text{ Hrs/year downtime}$$

Cost of Downtime	=	$ 600	per hour lost productivity
	+	60	Repair labor
		$ 660	per hour total cost
	×	3.5	Hours
		$2,310	Annual Cost of Failures

Calculated repair costs are reasonably accurate if the failure and repair rates are accurate. MTBF can be calculated for most electronic products, but the manufacturer may have no data to validate the calculations. Moreover, failures because of software malfunctions are impossible to calculate and must be derived from experience. With all things considered, the cost of a maintenance contract is the most convenient figure to use and may be a sufficiently reliable guide to future maintenance costs.

All other factors relating to failures, such as lost production time while the system is down, the cost of restoring failures, the cost of spare parts, and the cost of maintenance can be developed from the manufacturer's estimate of system reliability. The first year's repair costs should be reduced to reflect repairs that are performed under warranty.

Salvage value is a positive cash flow at the end of the study. Unless industry figures are available, salvage can be calculated by assigning a percentage of original cost. Often, because the cash flows for salvage occur at the end of the study period, they are so deeply discounted that they have little effect on the outcome of the study, and can be ignored. This is not true, however, for high-cost products that have a substantial resale value. Salvage values can be estimated by reviewing book values of equivalent equipment approaching the end of its service life and comparing them to the original cost. Salvage values should be reduced by the cost of removing the equipment, if any.

Determining Income Tax Effects. A capital investment reduces income taxes in two ways. The first effect is the tax benefit of depreciation. Because depreciation is a non-cash expense, it reduces income taxes without a corresponding cash outflow. Second, the annual expenses of operating the system are deductible from income taxes. In the past, investment tax credit (ITC) has enabled companies to write off part of the initial investment as a direct reduction of income taxes. At the time of this writing ITC is no longer authorized, although it may be reinstated in the future.

Conducting the Analysis. The LCA is conducted by estimating the cash flow for every cost element from the time a product is acquired until it is disposed of. Lacking a computer and software, you can conduct an LCA manually by entering all cash flows into a table and multiplying them by the discount factor for the year in which they occur. Financial tables are included in Appendix B. As the calculations are somewhat laborious, and because you will undoubtedly want to play some "what if" games to deal with uncertainty, the use of a computer is recommended, as discussed in Chapter 4 and shown in the sample LCA later in this chapter.

The outcome of the analysis yields financial indicators to predict the economic performance of the products. Generally, NPV and IRR are the most important indicators; however, the amount of capital that a project requires can be an overriding concern to a firm that has limited funds available for capital investment. Conversely, a firm with the objective of aggressively reducing future expenses by present investments may be prepared to spend more than the results of the analysis would indicate are prudent.

Interpreting the Results. While life cycle results are an aid to decision making, they are by no means the entire decision. The most significant life cycle indicator is NPV. By definition, NPV is the algebraic sum of all discounted cash flows less the cost of the initial investment. Therefore, the product with the highest NPV will extract the least amount of cash from the enterprise.

If the products yield increased revenue, the NPV may be positive. This is often the case when NPV is being used in a feasibility study because the feasibility of a project is most often related to the revenue it generates or the costs it saves. In an LCA, however, revenues and some cost savings may be equal for all products, and therefore omitted from the study. When this occurs, the NPV becomes minus. The most effective product is the one with the highest or *least negative* NPV. When you are evaluating telecommunications equipment, remember that its primary role is to support the work force, not to generate a revenue. Therefore, it is often a cost of business that cannot be avoided, and the most effective system will be the one with the least negative NPV.

NPV is not the only financial indicator. Internal rate of return is an indicator of how efficiently the capital is used in the business. If the IRR is not at least as high as the cost of money, the product consumes capital. IRR can be a deceptive indicator if not used with care. If the extra investment is not repaid through reduced expenses, IRR becomes negative, and is meaningless as an indicator.

LIFE CYCLE ANALYSIS: A CASE STUDY

To illustrate the LCA process, assume that responses from an RFP have been received. Before beginning the financial analysis, the responses should be screened against the mandatory features listed in the RFP.

Nonconforming products can be rejected at the outset or set aside for possible future analysis.

The next step is to develop cost factors for all remaining products. To illustrate, Table 8.2 compares cost factors for two PBXs, called Products A and B. Although PBX A is more expensive, its recurring costs are lower. It is impossible to determine intuitively whether A is worth the extra cost. Furthermore, you can't be entirely certain about the accuracy of some of the estimates. Although the costs of power and floor space can be computed with a high degree of accuracy, labor and maintenance costs are often broad estimates. An LCA model aids you in assessing these tradeoffs by testing a range of assumptions.

The next step is to develop the company's cost factors for owning and operating the equipment, as shown in Table 8.3. These are unique to each company and include such factors as the cost of money, power, and

TABLE 8.2
Cost and Feature Comparisons Between Two PBXs
(Cost Figures For PBX A and B)

Initial Costs	*PBX A*	*PBX B*
Purchase Price (Including shipping, sales tax, and station equipment)	$122,000	$108,000
Spares	2,500	1,600
Installation	8,800	7,500
Software right-to-use fee	10,500	12,000
Total	$143,800	$129,100
Recurring Costs		
Maintenance	12,250	14,600
Rearrangments	1,560	2,435
Software	0	400
Cost-related Specifications		
Floor Space	225 ft^2	320 ft^2
Electric Power	1.2 kw	2.5 kw
Outage time per year	3.2 Hrs.	4.8 Hrs.

Note: These figures are not intended to be representative of actual costs. The figures are chosen to illustrate how differences in recurring and non-recurring costs are handled in a life cycle model such as the one in Figure 8.2

TABLE 8.3
Common Cost Factors For PBX and Key Telephone Equipment

Labor cost per hour	$24.50
Floor space $/ft^2 year	12.00
Power $/kwh	.09
Cost of money	10%
Incremental Tax Rates	
Federal income	35%
State income	12%
Local income	3%
Total	50%
Equipment service life	7 years
Salvage value	10% of purchase price

Note: These factors are not intended to be representative of actual costs. Each organization must derive similar figures for its own operations.

floor space. Because an LCA involves a tradeoff between a present capital investment and future recurring expenses, it may be advisable to estimate inflation rates for each class of expense. The more expenses inflate, the more attractive an investment in cost-reducing equipment becomes.

Tax considerations are always a factor in an equipment purchase, so the study model calculates after-tax cash flow. To calculate tax effects, the company's incremental rate for combined federal, state, and local taxes is needed. We will assume 50% in this example, but any other figure can easily be plugged into the model.

You must also determine the service life of the equipment for both study and tax purposes. In this illustration we assume the equipment will last for seven years, but that it will be depreciated on a straight line basis over five years.

With this information, the life cycle study model shown in Figure 8.2 can be developed in a spreadsheet program. An unlimited number of products can be compared, but to keep the study manageable, they should be compared two at a time. The common cost factors, corresponding to Table 8.3, are stored in separate tables below the data entry section of the spreadsheet so they can be changed readily without altering the formulas in the spreadsheet itself.

This model is identical to the one used in Chapter 4 for a feasibility study, except that the headings have been changed. The cash flows in the bottom of the model are after taxes. Nonprofit organizations can use the same model by reducing the tax rate to zero. Also note that factors such as maintenance costs increase from year to year by the rate of inflation. A discount rate, or cost of money, of 10 percent is assumed. The model can easily be expanded to cover a range of 20 years and to use other forms of depreciation. The length of the study is adjusted by changing the column range in the NPV formulas.

Interpreting the Results

All cash outflows in this model are shown as negative, and cash inflows are positive. Depreciation is a non-cash item, but it has the effect of a revenue because it is a tax saving that increases net income.

FIGURE 8.2
Sample Life Cycle comparison of Two PBXs.

Product Costs	*PBX A*	*PBX B*
Initial Cost	$(122,000)	$(108,000)
Spares	$(2,500)	$(1,600)
Installation	$(8,800)	$(7,500)
Software RTU	$(10,500)	$(12,000)
Total Installed Cost	$(143,800)	$(129,100)
Recurring Costs		
Maintenance	$(12,250)	$(14,600)
Rearrangements	$(1,560)	$(2,435)
Software Upgrades	$0	$(400)
Floor Space sq ft.	225	320
Power KW	1.2	2.5
NET PRESENT VALUE	$(109,745)	$(126,126)
NPV PROD. A - PROD. B	$16,382	
Common Costs		
Labor Cost/Hr.	$24.50	
Maintenance Inflation	5.0 %	
Power Cost/KWH	0.09	
Cost of Money	10.0 %	
Floor Space $/sq. ft.	$12.00	
Incremental Tax Rate	50.0 %	
Svc. Life in Years	7	
Salvage Value	10.0 %	
After Tax Cash Flows		

FIGURE 8.2—*Continued*

YEAR	*1*	*2*	*3*	*4*	*5*	*6*	*7*
After Tax Cash Flows							
Product A							
Maintenance	$(6,125)	$(6,431)	$(6,753)	$(7,090)	$(7,445)	$ (7,817)	$(8,208)
Rearrangements	(780)	(819)	(860)	(903)	(948)	(995)	(1,045)
Software Upgrades	0	0	0	0	0	0	0
Floor Space	(1,350)	(1,418)	(1,488)	(1,563)	(1,641)	(1,712)	(1,809)
Power	(473)	(497)	(522)	(548)	(575)	(604)	(634)
Depreciation Tax Effect	14,380	14,380	14,380	14,380	14,380		
Salvage							12,200
Total	5,652	5,216	4,757	4,276	3,771	(11,139)	504
Product B							
Maintenance	(7,300)	(7,665)	(8,048)	(8,451)	(8,873)	(9,317)	(9,783)
Rearrangements	(1,218)	(1,278)	(1,342)	(1,409)	(1,480)	(1,554)	(1,631)
Software Upgrades	(200)	(210)	(221)	(232)	(243)	(255)	(268)
Floor Space	(1,920)	(2,016)	(2,117)	(2,223)	(2,334)	(2,450)	(2,573)
Power	(986)	(1,035)	(1,087)	(1,141)	(1,198)	(1,258)	(1,321)
Depreciation Tax Effect	10,800	10,800	10,800	10,800	10,800		
Salvage							10,800
Total	$ (823)	$(1,404)	$(2,014)	$(2,655)	$(3,328)	$(14,834)	$(4,776)

Note that cash outflows are shown in parentheses and cash inflows are shown as positive numbers.

The key factor in the comparison is the difference between the NPV of the two products as shown in the last line of the input section of the model. If this figure is positive, an investment in Product A save cash compared to Product B; if the figure is negative, the investment in Product A consumes cash. With the figures shown, Product B appears less attractive than Product A despite its lower first cost.

We can calculate the internal rate of return (IRR) of the extra investment for Product A by increasing the cost of money in the common cost table until NPV is zero. A little experimentation shows this to be 42.2 percent. Most spreadsheets have a built-in IRR function that will produce the same result, but to use it requires paying close attention to the algebraic sign of each year's cash flow. It is easier to obtain IRR by varying the cost of money in the table.

NPV alone is not a reliable indicator of the value of an investment. Frequently, the NPV of one product is higher than one of its alternatives, but its IRR is less than the enterprise's cost of money. In such a case, an investment may be marginally better than an alternative, but the additional investment will earn at less than the cost of money. If the IRR is less than the cost of money, the firm will gain more by choosing the product with highest IRR and investing the additional money elsewhere.

Sensitivity Analysis

Many figures in a study of this kind are of doubtful validity, and should be evaluated by sensitivity analysis, which is accomplished by changing questionable figures and observing the effect on the outcome. For example, suppose you believe that the maintenance cost estimate is accurate within +50%. By multiplying the maintenance cost figures by 1.5 and 0.5, a range of differences between the NPVs from (13,116) to (19,647) is obtained as shown in Table 8.4. In sensitizing these figures further, we find that if the maintenance cost of Product A increases by 22% or if that of B decreases by 66%, B is a better investment. Judgment enters the process at this point. To justify purchasing product B, you would have to estimate high probability of a change in one or more factors.

Table 8.4 shows the effect on NPV and IRR by rerunning the analysis after changing some of the more elusive figures. Sensitivity analysis allows you to determine which assumptions have the greatest effect on the outcome of the analysis. Those factors that have little effect

TABLE 8.4
Sensitivity Analysis of Life Cycle Model in Figure 8.2

	NPV	*IRR*
Initial values	16,382	42.2%
Assume no inflation	14,383	39.9%
Product A 50% increase in mtce.	642	***
Product B 50% increase in mtce.	36,671	72.7%
Both products 50% increase in maintenance	13,116	36.7%
Service life increased to 10 years	28,957	47.5%
Service life decreased to 5 years	12,456	39.4%

***Meaningless figure

can be ignored, but those with the greatest effect require further scrutiny before the decision is made. In this example, the study is sensitive to maintenance costs, which leads you to examine your assumptions about them very carefully. For example, if we have underestimated the maintenance cost of Product A by 50 percent, Product B becomes the most attractive. It is relatively insensitive to inflation, however, which indicates that you don't need to worry too much about your assumptions in that respect.

The model can also be used to determine other interesting facts about the purchase. For example, by changing B's purchase price you can determine the point at which it would break even with A. This information may be an asset in a bargaining session. Changing the service life of the products shows that a longer service life makes A more attractive, but shortening the life to five years narrows A's margin.

Lease Versus Purchase Analysis

The same model can be used to aid in determining whether to lease or purchase equipment. A lease simply substitutes expense for capital costs. It is often useful to obtain lease and purchase proposals from all vendors. To evaluate the economics, all costs can be omitted except for the factors unique to the lease and purchase plans. For example, power consumption can be ignored since it is independent of the acquisition method.

Leasing plans often provide a purchase option. The life cycle model can be used to evaluate the effects of exercising the purchase option at any time in the lease cycle.

TECHNICAL EVALUATION

As mentioned at the start of this chapter, the objective of any evaluation is to avoid spending time on products that have no chance of being selected because of some fatal flaw. Therefore, the first level of technical evaluation, comparing the manufacturer's specifications against the specifications in the RFP, should be performed before the economic evaluation is started. After the initial screening, it is generally easier to perform the economic evaluation as the second step because it requires less effort and expense than the technical evaluation. The goal of the economic evaluation is to narrow the field to two or three alternatives, which are then subjected to more rigorous technical scrutiny.

The technical evaluation usually requires testing features and observing demonstrations of the system's operating characteristics on the vendor's or another user's premises. These demonstrations are time consuming for both seller and buyer, and should be avoided unless the product has an excellent chance of being selected.

Technical evaluations of telecommunications products often strain the ability of the using organization, in which case you may want to rely on evaluations by companies that perform them and publish the results for a fee. Factors that should be considered in a technical evaluation include the following:

- *Status of the Technology*: Does the product use technology that borders on the obsolete or is it new enough that it hasn't been tried in practice? Will replacement parts, maintenance support, and program updates be available throughout the product's service life?
- *Capacity*: Does the product have sufficient capacity to support unexpected growth or to provide margin for activity peaks? How much additional shelf capacity is available before another carrier must be added?
- *Documentation*: Is the product fully documented? Is the documentation logically ordered in keeping with industry standards? Can an

untrained person understand it? Does the vendor refuse to release certain portions of the documentation?

- *Technical Support*: Does the vendor maintain a service force within reasonable driving time? Can the product be remotely tested? Is specialized test equipment required to maintain it locally? If so, how much training does it require? How much does it cost? Does the documentation explain how to perform tests and describe the results to be expected? Does the equipment include self-testing or diagnostic features?
- *Reliability*: What kinds of service protection features does the system include (such as redundancy or backup)? Are reliability figures published? How are they computed? Have they been verified by field experience? What kind of experience have other users had with similar systems?
- *Administrative Functions*: Does the system provide traffic data? What form is it in? Can it be interpreted without extensive training? Can the user easily apply feature changes? How are program updates installed?
- *Transmission Performance*: Are there built-in design problems that may cause dissatisfaction with transmission quality? (Such features as conferencing, remote access, and off premise stations may result in low volume on some types of calls as discussed in Chapter 8).
- *Environmental*: Does the system fit readily into available space? Does it require air flow or air conditioning in excess of that presently available?
- *Wiring Plan*: Does the wiring plan meet present requirements and provide for future terminal additions and expansion? Is the wiring logically terminated and designated for ease in identification? Is the system protected against damage from electrical hazards?
- *Range*: For each type of telephone set or data terminal, what is the maximum distance from the PBX or host that can be supported?

The most effective way to evaluate technical performance is by discussing the experience of other users. The vendor should be able to supply contacts in companies who use similar systems. It may be productive to visit other locations to determine how satisfactory their experience has been. You should pay attention to their trouble history logs and find out if they are satisfied with the vendor's response time.

Dealing with Intangibles

Every product selection study will have certain features that cannot be quantified. In a PBX, for example, ease of feature use, freedom from obsolescence, future compatibility, and an overall experience rating of the vendor are examples of intangible factors that the company should evaluate in deciding whether to select a product.

The best way of handling intangibles is to decide how important they are relative to the numeric factors, and weight them by assigning a score to each product. The product with the highest score is the most effective. The rating tends to be subjective, but when it is combined with objective factors such as life cycle cost, a product may show an advantage that outweighs economic considerations.

Often, the most effective way of dealing with subjective features such as ease of use is to assemble a panel of stakeholders to develop the weightings and evaluate the products. If each person evaluates independently, and if their opinions are weighted, a few products will emerge as clearly the most effective. If this evaluation is performed in several iterations, with the least effective products eliminated, one product will eventually rise to the top.

Ease of Feature Use

Assuming the desirable features are weighted in the RFP as described in Chapter 7, the technical evaluation determines how well the product implements the features. For example, call pickup is a highly desirable feature of both Centrex systems and PBXs. With a Centrex system a user may activate the feature from a standard telephone set by dialing a code. In a PBX equipped with "smart" telephones at a higher cost, the user activates the same feature by pressing a button. A person or team evaluating the system rates how well the system performs the function. By multiplying the value times the degree of importance, a weighted value is obtained. When all of the features have been tested and rated, a table similar to Table 8.5 can be constructed for each vendor and weighted values compared.

The operational features of a system should be observed on a working machine. Verbal descriptions, handbooks, and video tapes are a poor substitute for trying features in actual operation. In many ways, operational interfaces are more important than technical interfaces

TABLE 8.5
Method of Weighting Desirable Features

Optional Feature	*Weight A*	*Rate B*	*Score A × B*
Code call	3	5	15
Conference calling	5	2	10
Discriminating ringing	2	5	10
Do not disturb	4	4	16
Flexible Intercept	3	3	9
Remote traffic measurement	5	5	25
Total			85

A table such as this can be constructed for all optional features. Weights are assigned to the feature and each product is rated according to how well it implements the feature.

because they are used by everyone in the organization, many of whom receive little training. Complex features tend to fall into disuse and generate dissatisfaction. Factors such as the following should be evaluated:

- *Console Operation*: Are console keys placed for maximum efficiency? Is console operation logical? Are aids available for assisting the operator with seldom-accessed features? What facilities are available for training new console operators? Is the console operation fully documented in terms that an untrained person can grasp? Are alarms and signals fully documented with recommended procedures for handling them?
- *Station and Terminal Operation*: Are station features easy to use or do they require dialing access codes? Are terminals and station sets clearly labeled? Is the labeling logical? Are terminal operator's manuals available? Are terminals designed in keeping with current ergonomic principles? Are terminal messages easy to interpret?
- *Training*: What kind of training does the vendor provide? Is there a charge? How will new employees or operators be trained? Are training aids available?
- *Administrative Information*: Does administrative information such as SMDR printouts and usage statistics match your present accounting system? Can departmental or client billing indicators

be entered easily? Can you collect data transaction volumes from the system?

- *User Interfaces*: Does the system supply feedback to let users know all is well? Are prompts logical and easy to understand? Are errors indicated with feedback that offers clues to what is wrong? Are reports formatted for easy interpretation (as opposed to blocks of unidentified register printouts)?

Maintenance Plan

The vendor's strategy for maintaining the system is likely to override many other factors in your evaluation, and deserves careful analysis. The elements of a maintenance plan are discussed in Chapter 21 and are not repeated here. Your evaluation must analyze the maintenance plan for your operating environment. For example, a system with elaborate provisions for owner-performed maintenance may not be valuable if you plan to contract maintenance to the vendor. Remote maintenance provisions will probably be valuable regardless of who performs the maintenance. Of utmost importance to anyone planning to maintain the system is how well it is documented with trouble locating manuals and diagnostic procedures.

Documentation

Documentation is evaluated on the basis of whether it exists at all, and if so, how well it is written. Evaluating it requires an experienced analyst. Whether you plan to maintain the system or contract it, the costs will be a function of how well it is documented. Likewise, costs for self-maintenance depend heavily on documentation because you must depend on either training or documentation to analyze every case of trouble. If you plan to maintain the system, the vendor should be asked to demonstrate diagnostics and trouble locating procedures.

Vendor's Reputation for Product and Support

Vendor evaluation is probably the most intangible factor in the entire selection process. Where the vendor is not the manufacturer, you may want to evaluate the manufacturer as well because a weak vendor can make a strong manufacturer look bad, and vice versa. If the manufacturer is the vendor, a poorly organized field force may fail to support an otherwise top quality product. Lack of tangible information does not, however, mean that buyers are without resources. The vendor's reputation is probably the most reliable factor to use, while its financial stability is the most tangible. The following factors should be considered:

- *Vendor's Reputation*: Does the vendor have a record of satisfactory provision, installation, and support of the product in similar environments? Are the users satisfied? Are field forces cross-trained in enough depth to cover absences and resignations? How long has the vendor supported this type of equipment? Are there others who could support the system in case of difficulty?
- *Financial stability*: Does the strength of the vendor's balance sheet assure you it will be in business at least as long as the product is in service?
- *Results of other users*: Do other users express satisfaction with outage time, response time, capability of the repair force, and other such factors? What has been the repair response time experienced by other users?
- *Size and location of maintenance force*: Does the vendor have sufficient people located within reasonable driving distance to support the product? Consider paying an unannounced visit to the vendor's maintenance location and ask to see evidence that the maintenance force is equipped to perform as they propose.
- *Product stability*: Has the product been in use long enough to be stable, yet with technology new enough to guard against obsolescence?

CHOOSING THE PRODUCT

The product selection decision hinges on a variety of factors that must be weighed to determine which product is the most effective for your application. Public agencies frequently use a weighting table like the one shown in Table 8.5. Private organizations are not bound by the same constraints, but the exercise of building a similar table is a worthwhile method of determining the relative importance of the factors on which you will base the decision.

When the proposals have been evaluated and the successful vendor has tentatively been chosen, it is a good idea to call the vendor in for a conference to clarify any ambiguities in the proposal. A vital question looms at this stage of the process: will you purchase the product under the vendor's contract or your own? Unless your organization is large enough to exert considerable pressure on the vendor, it is often difficult to change many of the terms of the vendor's standard contract. Almost invariably, however, the standard sales contract contains significant differences from the terms and conditions of the buyer's RFP. The time to negotiate these differences is before the contract is awarded. If an agreement cannot be reached, it may be necessary to choose a different product. Contracting considerations are discussed in Chapter 12.

The deciding factor in determining how much pressure to exert on the vendor to change is a matter of the risk in choosing a product. Risk is defined in telecommunications procurement as the potential loss of money, time, and information resulting from a product failure. In a complex system such as a PBX or data communications system, the risk is invariably high because failures affect the entire organization. In less complex products such as telephone sets and modems the risk is relatively low because a failure affects only a small part of the organization and the units are low enough in cost that spare units can be provided.

SUMMARY

It would be easy to misconstrue the intent of the product selection process as a matter of finding numbers to justify the decision you want to make. To be sure, it is possible to use the process to justify your own preferences, but that is not the intent. The purpose of the process is to evaluate systematically all the objective information that can be obtained about a product, and to treat the intangible and uncertain information as a range of values that will augment your decision-making ability. Perhaps the greatest value of the process is the discipline it requires to obtain information that would otherwise be disregarded in the difficult task of choosing a product.

CHAPTER 9

SELECTING CUSTOMER PREMISES SWITCHING EQUIPMENT

Switched services comprise the systems that connect station lines to one another, to trunks, and to special features such as paging systems and dial dictation. This chapter explains many of the important considerations in evaluating three types of switched services, one or more of which are used by most companies; key telephone systems (KTS), private branch exchanges (PBXs), and automatic call distributors (ACDs).

The objective of this chapter is to explain the major distinguishing features among these three types of apparatus. No attempt is made to describe all features; to do so would require a large book in itself. Appendix D contains a short description of the most common features, but be alert to the fact that not all manufacturers mean the same thing when they claim to have a feature, and some manufacturers may have a similar feature, but call it by a different name than Appendix D.

This chapter should guide you to a clearer understanding of your requirements through understanding the alternatives and the major decision points in choosing among various systems. There is a great deal of overlap between switching technologies. The gap between a KTS and a PBX is bridged by a hybrid PBX, which has some of the characteristics of both. Some of the functions of ACDs are found in PBXs and KTSs as well as in stand alone devices. Many features are essentially the same regardless of which type of switching system they are used in.

A PBX or KTS is a major investment, and for most companies one that will last for five to ten years. The high cost and relative permanence make it imperative to do a careful job of selecting among the many alternatives.

DIFFERENCES IN SWITCHING SYSTEM TYPES

The decision is not always clear whether you are in the market for a PBX or KTS. If you have fewer than 25 and more than about 125 stations, the choice is not difficult. PBXs are rarely economical in small line sizes, and large organizations require features and capacity that key systems lack. For companies in between, the market offers a combination PBX and key system that bears the name "hybrid." A hybrid system has some of the characteristics of both the PBX and a KTS. The following are the principle distinguishing features of these three types of systems.

Outgoing Trunk Selection. Users in a key system select outgoing trunks by pushing a button to choose an idle trunk. In PBXs and hybrids, trunks are selected by dialing an access code, usually "9." This feature is often called *pooled access*.

Incoming Trunk Selection. In a KTS, the user or attendant answers an incoming trunk by pressing a button associated with the trunk. Calls are announced from the attendant to the user over an intercom or simply by calling to the person to pick up a specific line. In a hybrid or PBX the attendant answers the call and transfers it to the extension by pressing the direct station select button or dialing the extension number. The trunk select methods are an important consideration in many areas because the telephone company may charge a higher rate for pooled access trunks than for lines that are selected by the user. The difference in rates is based on the theory that PBX trunks are more heavily used than KTS lines.

Intercom Paths. Users communicate between stations in a PBX by dialing the station number through the PBX switching network. In a KTS, users communicate over intercom paths, which may be limited in number. Hybrids have generally more intercom paths than KTSs, but fewer than PBXs. Companies with extensive intercom traffic may find hybrids or KTSs are too limiting for their volume of intercom traffic.

Least Cost Routing. The least cost routing systems in PBXs tend to be more sophisticated than in KTSs or hybrids. Many KTSs have no LCR at all. Hybrids frequently are limited to three- digit screening, which means that all calls to a given area code must go over the same route. This limitation is explained more thoroughly later in this chapter.

Trunk Termination. Most KTS systems and many hybrids are incapable of terminating tie trunks with four wire transmission paths and E&M signaling. Also, direct digital trunk termination is generally available only on PBXs, although not all PBXs offer this feature.

Digital Telephone Features. Most digital PBXs are capable of interfacing digital telephones and switching data without the use of a modem. Most hybrids and KTSs are restricted to analog telephones, and are limited in the display features that often accompany digital telephones.

MAJOR PBX SYSTEM FEATURES

Once the decision is made whether to use a KTS, PBX, or hybrid, an important distinction between competing products is in how well the more important features are handled. A modern PBX, an example of which is shown in Figure 9.1, is compact, fully electronic, and highly reliable. This section explains the principal features that distinguish PBXs. Features that are handled uniformly between competing systems are omitted from this discussion. Appendix D lists the most desirable features and explains briefly what the feature does.

Least Cost Routing

One of the major motivations for obtaining a PBX is to secure true least cost routing capability. Not all systems are alike in their capabilities, so it pays to inquire carefully into how well the system operates.

Full least cost routing can analyze dialed digits and determine the least expensive way of handling the call based on trunk groups available, time of day, and digits dialed. A variation known as *automatic route selection* lacks the time-of-day routing capability of full least cost routing. It is important to know how many digits the system is capable of screening. The simplest systems can handle only three digits, which limits their flexibility. For example, suppose your office is within a few miles of a state border. Neighboring states are usually in WATS Band 1.

FIGURE 9.1
The IBM/Rolm Computerized Branch Exchange

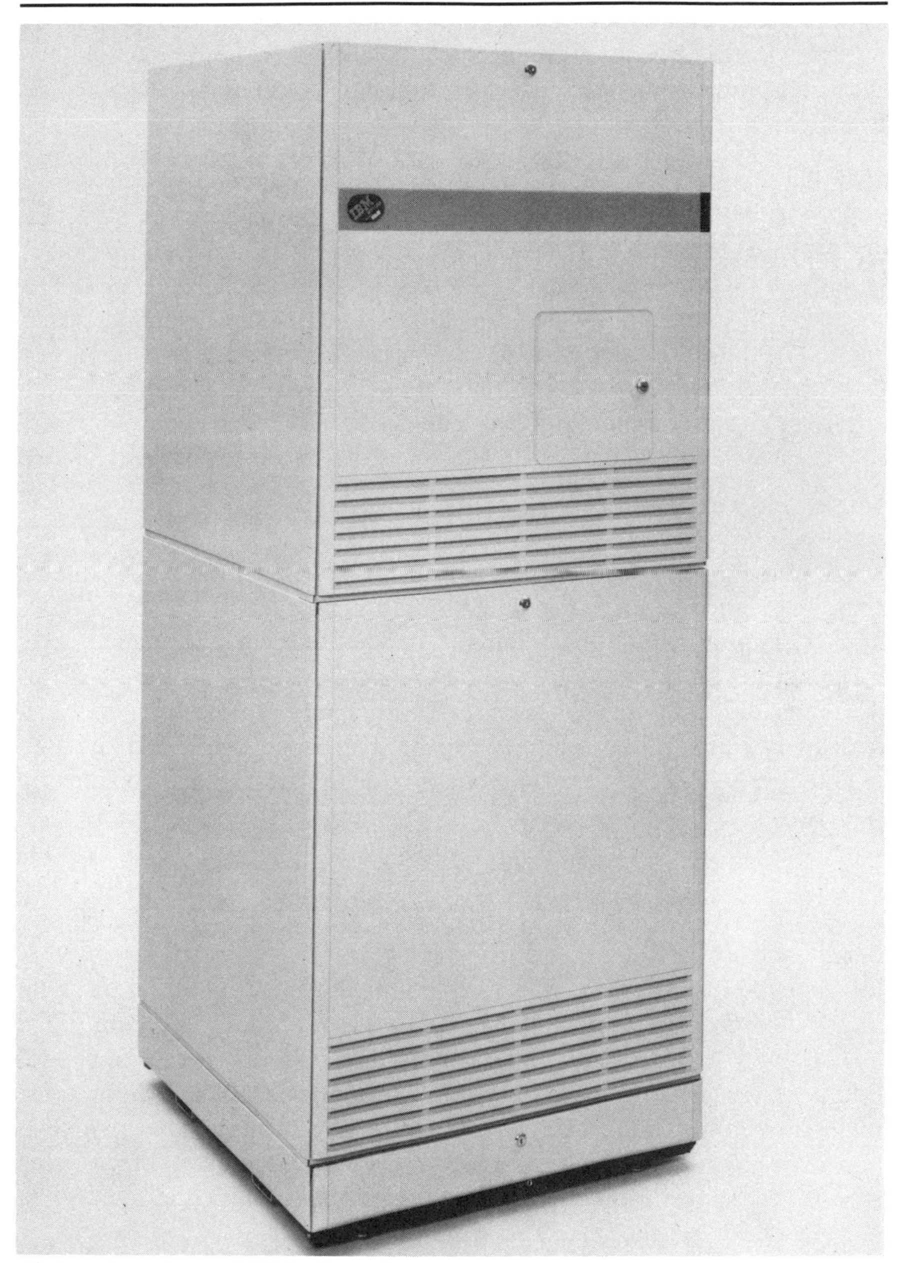

Photograph courtesy of Bell/Rolm Corporation

If your system has only three-digit screening capability, calls to the other area code would all have to route via either WATS or DDD because the system would be incapable of screening on area code and also determining that nearby prefixes were more economical over DDD than WATS. With this limitation, the least cost routing system loses much of its flexibility.

Within a LATA it is often more cost effective to pass long distance traffic to the local telephone company to handle than to an interexchange carrier. InterLATA traffic within the same area code may, however, be more economically handled by a discount IEC, even a different carrier than the one that carries interstate traffic. The only way a PBX can distinguish between intraLATA and interLATA traffic if the LATA crosses an area code boundary is by six-digit screening. A prefix table for the entire area code is built into the system so the PBX can determine which group of trunks should receive the calls.

Messaging Features

Some of the most significant productivity gains with a PBX come from its ability to reduce message-handling problems. Automatic call back allows you to camp on a busy line within the system, and when the called party hangs up, both phones are automatically rung. Message display features are available with many systems. If you place a call to a busy or unanswered line, you can leave a call back message that is displayed on the other telephone. Some systems allow you to return calls automatically without dialing. Determine, however, whether it is possible to return calls other than in the sequence in which they were received. Some systems call back only in the order received, which limits your ability to return calls in priority order. Some systems use a voice synthesizer instead of a display telephone to announce who has called. This method generally allows the use of less expensive telephone instruments, instead of one with a display unit. A typical display telephone is illustrated in Figure 9.2.

Display telephones and voice synthesizers are useful only for internal messages. Messages outside the system must be handled by some other messaging system such as text messaging or voice mail, which is described later. Most PBXs do not provide an integrated text messaging system, but some manufacturers offer integrated voice/data terminals that can be used to display messages left by the receptionist or others on the

FIGURE 9.2
AT&T 7407 Display Telephone

Photograph courtesy of AT&T Corporation

system. In effect, text messaging is an electronic mail system that is implemented on an external processor. The PBX is used for switched access to the computer.

Traffic Usage Measuring Equipment

Of the PBX features that offer a genuine cost saving, few can surpass traffic usage measuring equipment. Paradoxically, this is one of the most under-utilized features of a PBX. With measuring equipment you can monitor usage of trunks and features of all types. The data gathered can aid you in saving cost, either in wasted capacity or ineffective attempts by customers or employees. Most modern PBXs provide this facility in some form, but there is a wide difference in the usability of different systems on the market. The following are some of the principal features to consider in evaluating traffic measuring equipment.

Local or Remote Readout. Some systems can be monitored only from a remote maintenance and testing system (RMATS). RMATS monitoring has the advantage of requiring no premise visit because the measuring feature is turned on from a remote location. Its primary disadvantage, however, is that it requires a trained person to interpret results, and usually carries a substantial charge. The most effective systems can be read locally or dialed from a conventional terminal.

Human Interface. Systems on the market also vary widely in the degree to which they are readable without special training or equipment. Some systems dump unformatted register readings on paper with no decoding as to the meaning of the registers and without identification of the equipment being monitored. The most effective systems read out the information in a formatted form on either screen or paper and identify exactly what each reading means.

Machine Interface. The type and amount of hardware needed by the system to print out traffic reports is another consideration. Some systems print the report on the station message detail register printout, so that reports can be read without auxiliary equipment such as a maintenance and administration panel. Other systems require specialized apparatus that, if not needed for other purposes, adds considerably to the cost.

Flexibility. The most effective systems on the market have the flexibility to be turned on at will by the user and can monitor either hourly traffic or peak-hour traffic. The ability to screen all but the peak hour greatly simplifies the analysis task and reduces the amount of paper required.

Figure 9.3 is an example of traffic register readings from an AT&T System 75 PBX.

Station Message Detail Recorder

One of the most valuable features on a PBX is the Station Message Detail Recorder (SMDR). This device, in conjunction with a call accounting system, is useful both in controlling unauthorized use of the telephone and in distributing costs to cost centers where they can be budgeted and controlled. Chapter 20 discusses the SMDR and call accounting systems in more detail.

FIGURE 9.3
Trunk Group Measurements Report From an AT&T System 75 PBX

TRUNK GROUP MEASUREMENTS
List Measurements Trunk-group Last-hour
Peak Hour For All Trunk Groups: 1200 - 1300

Grp No.	*Grp Size*	*Grp Type*	*Grp Dir*	*Meas Hour*	*Total Usage*	*Total Calls*	*Inc. Call*	*Grp Ovfl*	*Que Size*	*Calls Qued*	*Que Ovf*	*Que Abd*	*Out Serv*	*% ATB*	*%Out BLK*
1	23	CO	two	1300	656	323	170	7	0	0	0	0	0	2	0
2	2	fx	inc	1300	24	70	70	0	0	0	0	0	0	0	0
3	7	fx	out	1300	143	48	0	7	2	6	1	1	1	0	0
4	4	Wats	out	1300	73	81	0	5	1	0	0	0	0	0	0

FIGURE 9.3
Continued

Measurement	*Explanation*
Incoming Calls	The total incoming calls via the trunk group to the system. These calls may be incoming from a 1-way trunk group or the incoming portion of a 2-way trunk group. This measurement is zero (0) for outgoing trunk groups.
Trunk Group Overflow	The number of calls that reach a trunk group and find all the trunks in the group busy. Calls directed to queue are considered as trunk group overflow calls.
Trunk Queue Size	A number between 0 and 100 that identifies the slots assigned to respective trunk group queue. The maximum queue slots that can be assigned to the system is 100.
Call Queued	The number of calls that enter the trunk group queue after finding all trunks busy.
Queue Overflow	The number of trunks that were not queued because all slots in the queue were occupied.
Queue Abandoned (Abd)	The number of calls that are removed from the queue either by the system for staying in the queue for the maximum allowed time (currently 30 minutes) or when the user cancels the auto callback.
Out-of-Service	The number of trunks in the trunk group (listed as maintenance busy) that are out of service at the time data is collected. This measurement is obtained by sampling each trunk in the system during the hourly interval.
Percent (%) All Trunks Busy	The percentage of time within the polling interval that all trunks in the trunk group were unavailable for use.
Percentage (%) Outgoing Blocking	The percentage of outgoing calls not carried on a trunk group to the outgoing calls offered. For trunk groups with no queue, the calls not carried are those calls that find all trunks busy. For trunk groups with queues, the calls not carried are those calls that find all trunks busy and cannot be queued because the queue is full. An "*" is displayed for incoming trunks because this column pertains to outgoing or two-way trunks.

Figure Courtesy AT&T.

Number of Wires Per Station

Most PBXs can accommodate ordinary 2500-type sets with single pair station wire. Some machines can also handle analog or digital feature sets on single pair wire, whereas other systems require more than one pair. If all stations are in the same building, it makes little difference if the PBX requires more than one pair because the cost of wiring is largely concentrated in labor instead of materials. When a PBX is installed in a campus environment, cabling to other buildings makes the cost of multiple pair station wire a disadvantage.

Station Range

Telephone station range is specified by most manufacturers in either feet or ohms of line resistance. The limiting factor is not identical for all systems, and varies according to the type of telephone set. In most cases, digital telephones have the least amount of station range; the limitation is usually 3,000 feet or less. The range for hybrid or 2500-type sets is usually more. By using electronic range extension, 2500 sets have a virtually unlimited range. Additional range may be obtained with off-premise extension (OPX) line cards in some systems or by using external range extenders.

Remote Maintenance and Diagnostics

Most PBXs are capable of being monitored either from a colocated maintenance console or from a remote maintenance center. Remote maintenance and diagnostics can more than pay for itself in reduced service calls. A remote maintenance and testing system (RMATS) dials the maintenance port on the machine over a separate central office line. With the proper password access, a technician can duplicate most of the functions of the local maintenance console. Often, trouble can be cleared from the RMATS, and if not, the right kind of technician with the right kind of equipment and spares can be dispatched.

Data Transmission Capability

The latest generation of digital PBXs was widely touted for its capabilities of directly interfacing digital trunks and data lines. PBX vendors promote their systems as the integrators of the automated office because of their ability to switch 64 kb/s data and to interface a variety of data communications protocols. A digital PBX is capable of directly interfacing data terminals and telephones alike, and can switch voice and data with equal ease. Integrated voice and data has become an industry buzzword, but its capabilities are not needed by a majority of users. Most voice/data PBXs that are installed are used exclusively for voice, and for many of those that do use data, interconnection of data devices is insignificant compared to voice.

Before you invest in digital capability, you should investigate the alternatives to determine whether voice/data integration through the PBX is the most economical method of gaining the capacity. The problem with data communications through a PBX is cost. Digital instruments with data communications capability are more expensive than analog instruments in most PBXs, and because the number of telephones is high, their use can rapidly inflate the cost of the PBX.

For switching data communications over a narrow range, users generally have two alternatives besides the PBX; a local area network (LAN) or a data PBX. The primary advantage of the digital PBX over the other two is that additional wiring and an additional switching system are not needed. If a building is being wired for both voice and data, it costs little more to pull in the additional circuits needed for data devices, and this is often less than the cost of purchasing digital telephones and line cards for the PBX.

Another advantage of a PBX for data communications is its accessibility to long-haul communications circuits. If you have substantial dial-up data requirements, a digital PBX can eliminate the need for individual modems. Modems are connected to a terminal in a modem pool, and are readily connected to the public switched telephone network (PSTN). This capability is more difficult with LANs and data PBXs, and if access to the PSTN is required from one of those systems it is often advisable to hook them through a voice PBX.

A third advantage of the PBX for data switching is the ability to access management cost control features. With features such as call accounting, least cost routing, and station restrictions, it is possible to control cost more easily than with LANS and data PBXs, which often lack restriction features.

Beyond the three advantages mentioned, there are few good reasons for integrating voice and data in the PBX. It is a costly option, and is one that should be carefully weighed against the other alternatives. Data communications alternatives are discussed in more detail in Chapter 11.

Direct Digital Trunk Interface

Where large numbers of circuits are required between PBXs or between a PBX and an IEC, a digital trunk interface is an essential feature. A T-1 line, which is described in Chapter 11, can be connected directly to the PBX where all of its 24 channels are available for either voice or data. The digital trunk interface will also be a requirement of the Integrated Services Digital Network (ISDN) in the future when the primary channel interface to the local central office is introduced. See Chapter 2 for further discussion on ISDN. Some telephone companies are offering T-1 connections for local trunks at a rate competitive with individual analog trunks. This feature offers two advantages to users. First, the cost of the PBX may be reduced. The per-port cost of a digital trunk interface card is usually less than the cost of analog trunk cards in PBXs large enough to require at least 24 trunks. The second advantage is the improved transmission performance that T-1 offers.

A related feature enables PBXs to connect directly to a mainframe computer, as shown in Figure 9.4. This feature, known by various proprietary names such as Digital Multiplex Interface (DMI), is specific to the PBX and computer pair. It enables the PBX to operate as a port concentrator, connecting a large number of terminals through a smaller number of computer ports.

Some PBXs are inherently incapable of interfacing a digital line because of the type of switching technology they employ. In the pulse code modulation (PCM) system, which is used by T-1, the voice signal is encoded into a 64 kb/s digital signal. Twenty-four of these 64 kb/s signals are multiplexed to form a 1.544 mb/s signal. The PCM modulation system is compatible with T-1, but other technologies such as delta modulation and pulse amplitude modulation, which are employed in some PBXs, are not compatible.

Direct Inward System Access (DISA)

Direct Inward System Access allows callers from the outside to access features within the PBX. It is used most frequently to allow station users to gain access to features such as long distance services and dial dictation.

FIGURE 9.4
Digital Multiplex Interface

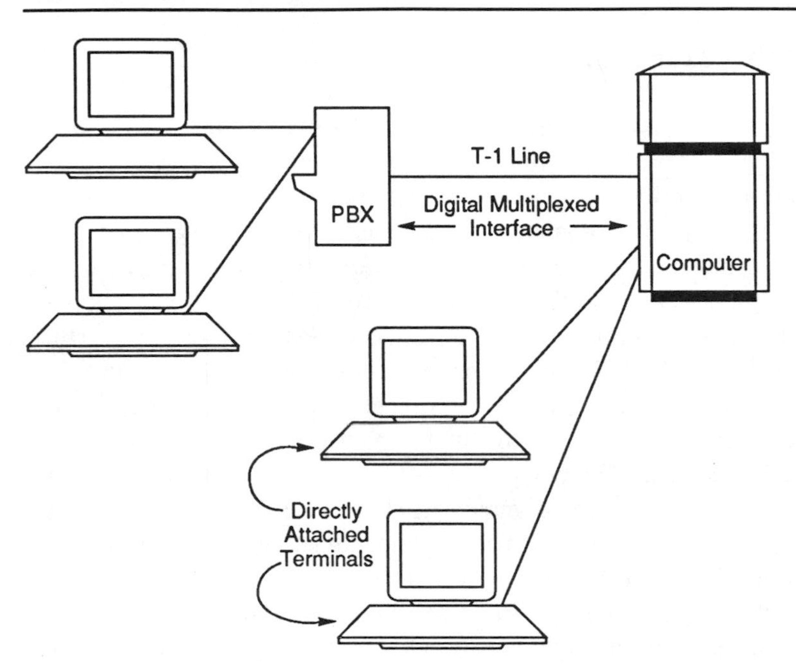

Terminals are either directly attached to the computer or switched via DMI.

The danger of such a feature is obvious; if an unauthorized person obtains the access code, it is possible to run up large long distance bills before it is discovered. Two different methods of limiting access are employed, a password and unlisted telephone number.

In some older DISA systems that lacked password protection, the only limit on DISA access was allowing only a restricted group of trunks to gain access to the features. The telephone number of the trunks was guarded. In later systems password access is also required. DISA can be a valuable feature in some systems, but it must be controlled to prevent abuse.

Paging

Most PBXs offer access to a paging system. Total system paging allows the attendant to address all zones in a system simultaneously. In a zone paging system, the speakers in various regions of a building or campus are actuated by addressing different zones from the console. The purpose of zone paging is to prevent the disruption caused by paging throughout the entire building. Many key telephone systems allow the operator to page through speakers in the telephone instruments.

A major disadvantage of voice paging is the annoyance of continuous announcements over the loud speaker system. Recently, several companies have begun to produce silent paging systems. These operate on the same principle as powerful radio paging systems, but operate over a narrow range, usually no more than two or three blocks. The paging receiver either vibrates or produces a tone to alert the wearer. A message or the number to be called is displayed on the receiver.

When integrated with a PBX, a silent paging system offers interesting possibilities. Users who are in the building but away from the telephone can forward their calls to a paging number. Anyone dialing that extension reaches a recorded announcement with instructions to dial the last two digits of the extension number, which is also the pager number. When the pocket receiver is signaled, the extension to be called is displayed. If the operator receives a trunk call for the absent person, the call is parked with the park number displayed on the receiver. This method, although expensive, offers a significant advantage over voice paging, where silent paging is preferred.

Dial Dictation Equipment

Dial dictation systems are available for most PBXs and key telephone systems. A caller from outside the system accesses the dial dictation unit over a central office trunk. Users inside the system dial an extension number or pick up a dial dictation trunk by pushing a feature button. The recording medium is usually tape, although in systems such as the one shown in figure 9.5 there is a growing tendency to use digitized voice and fixed disks. The advantage of a digitized system is in improved management information, increased storage capacity, and the random access capability afforded by a fixed disk. Digital systems however, tend to be more expensive than tape.

The normal functions of a dictating machine—record, review, fast forward, and rewind—are actuated by the telephone DTMF dial. Before you purchase dial dictation equipment, you should consider the potential drawbacks of control from the telephone. Interruptions may be inconve-

FIGURE 9.5
Harris Lanier VoiceWriter 4800 Dial Dictation System

Photograph courtesy of Lanier Voice Products, Division of Harris Group

nient to handle. With a separate dictating unit, you can set dictation aside for visitors or telephone interruptions, pick it up later to review what you have written, and resume dictating. Not all dial dictation units can be put on hold, so you may find it inconvenient or impossible to break the connection, review your work, and resume at a later time. Users who are frequently interrupted will almost invariably prefer a separate dictating machine.

The lack of direct machine control may be disconcerting to some users. People who have become accustomed to the direct control of a dictating machine may find the indirect control of dial dictation difficult to accept. If the dictation unit is located in a central pool it is less convenient to give special instructions to the transcriber or to get priority treatment for a document that is embedded in a shared tape.

Offsetting the disadvantage of dial dictation is the major advantage of dictating from a remote location. Doctors frequently find it convenient to dial their offices from a hospital to dictate patient reports. Business executives find it convenient to call the dictation unit from a distant location, even across time zones, and have the completed document waiting on their return. As a general rule, dial dictation is most effective when callers are outside the switching system.

Direct Inward Dialing

Direct inward dialing (DID) offers the ability to bypass the attendant and reach a PBX extension directly, which is an important feature in many organizations. The telephone central office must be equipped for this feature, in which the central office relays the extension identification to the PBX. DID has several benefits, the most obvious of which is deloading the attendant. It is also a valuable feature when the office is occupied outside hours when the console is covered. Without DID it is necessary to assign night transfer numbers to the trunks so people can be called from outside the PBX. It also saves on long distance calls for the calling party, who may be put on hold when the attendant is busy on other calls.

Attendant Console Features

The attendant's functions can be highly supported by mechanization in a modern PBX. The principal features are described briefly in Appendix D.

The attendant can "camp" a caller on a busy line so the call can be completed when the called party hangs up. The called party hears a brief tone, and can either choose to hang up or ignore the camped-on call. Most systems automatically provide timed reminders to alert the attendant to reenter a line that is not answered or that has been camped-on for a predetermined time.

Digital display features tell the operator a great deal about calls that are returning to the console for further attention. For example, the photograph in Figure 9.6 shows the status of a call in progress to a particular extension. When restricted telephones attempt to place long distance calls they are routed to the console, together with the class of service of the line. If a caller signals the attendant, the identity of the line is displayed. The digital display enables the attendant to do a better job of personalizing call-handling.

FIGURE 9.6
The IBM/Rolm 9755 Attendant Console

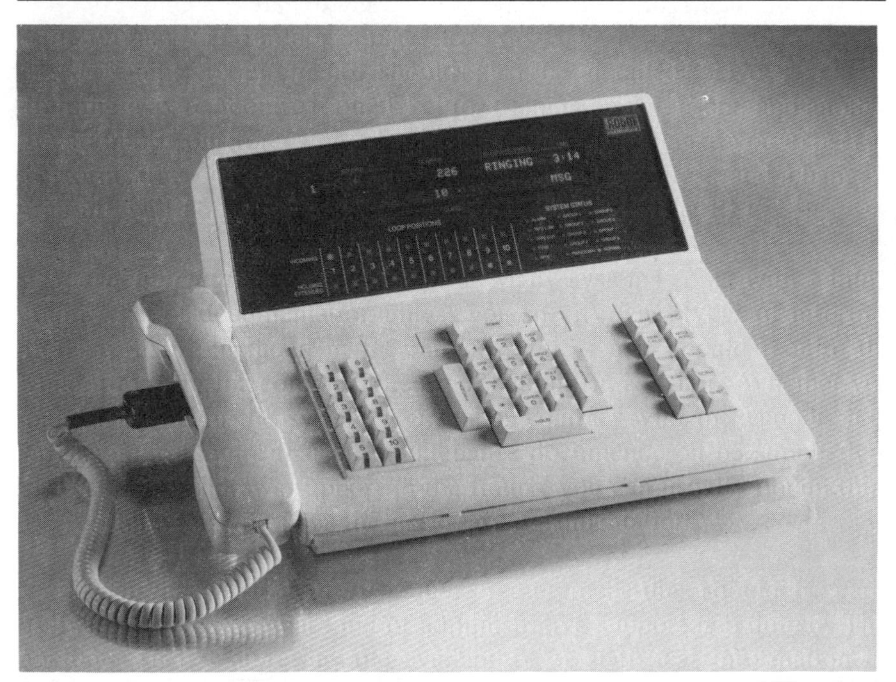

Photograph courtesy of Bell/Rolm Corporation

Call Forwarding

Although call forwarding is a standard feature of most PBXs, it is not used to its maximum capability in most systems. Part of the difficulty is the need to remember codes, but this problem can be alleviated by using feature buttons on the telephone. A well-designed call forwarding plan can do a great deal to eliminate the annoyance of message handling.

All users should have a primary destination to which calls are forwarded when they are absent. This is either an answering position or the console operator, and is actuated with a "forward all calls" button. If the button is not activated, calls are forwarded after two or three rings, but this unduly delays the calling party. When the forward all calls button is pushed, calls forward immediately without waiting for rings. When the answering position is unoccupied, the forward all calls button on that telephone can send calls to a secondary position.

Call forwarding can be also be used in a number of effective ways. If the system is equipped with off-system call forwarding, calls can be routed to home, a vacation cottage, or any place the recipient happens to be. Calls can also be forwarded to the paging system, voice mail, or other facility for dealing quickly and effectively with incoming calls. The way calls are handled in an organization has an important effect on customer perception of that organization's efficiency, and call forwarding is one of the most effective ways of handling calls for mobile users.

VOICE MESSAGING SYSTEMS

Voice messaging systems, also known as voice store-and-forward and voice mail, were first introduced about 1980, and have grown rapidly into a multi-million dollar per year business. Most people are not indifferent to voice mail; they either love it, they hate it, or at best, barely tolerate it. A frequent complaint about voice mail is that is little better than a telephone answering machine, a device that most people hate to talk to but love for its convenience. Voice mail is an external messaging system that is accessed directly or through the PBX.

Voice mail systems are generally one of two types, integrated or stand alone. An integrated voice mail system turns on the message waiting light on the user's telephone when the voice mail message is received. With a stand alone system, users must dial their mailbox numbers to determine whether they have messages. The ability to drive the message waiting lights greatly adds to the value of voice mail.

Another voice mail integration feature is "return to operator." With this feature, a user can dial a code, usually "O," to return to the PBX attendant. Without this feature the user must hang up and call again to reach the attendant. Stand alone voice mail cannot offer this feature, which is one of its greatest disadvantages.

Voice mail is much more versatile than an answering machine. Aside from the option of dialing a number to reach an operator, the caller who reaches an answering machine has little choice but to leave a message or hang up. With voice mail the caller can signal an attendant, replay the message, or be routed to other information sources. To the called party, voice mail has a number of significant advantages. Messages can be retrieved from any DTMF telephone by dialing the voice mail code. Other codes allow you to listen to the message more than once, delete it, file it, or forward it to someone else to handle either with or without annotation.

Because many callers dislike the impersonal aspects of voice mail, some companies route calls to an answering position where the attendant offers to transfer the caller to the voice mail. If the caller objects, the message can be taken manually and later the attendant can read the message into the voice mailbox.

Selecting Voice Messaging Systems

In selecting a voice mail system, there are several important factors to consider. First, as mentioned earlier, is the question of integrated versus stand alone systems. If an integrated system is available, its features will greatly enhance the value of voice mail for people who return to their telephones frequently. If users are out of the office a great deal and call in for messages from outside the system, there is little advantage to the message waiting light. Stand alone units are available for private use or through a service bureau. Often, trying voice mail on a service bureau is the most effective way to find out if it will be accepted.

The next issue is administration. Voice mail systems digitize the voice and store it on magnetic discs. Unless the system is regularly policed, storage can be used up with dead messages. Look for a system that offers the option of automatic purging, or at least provides management information to allow the administrator to purge overage messages.

A third issue is how calls will be forwarded to voice mail. Will callers be transferred directly to the mailbox or will an attendant intervene? Another issue is who will be given mailboxes. Users who receive a large number of calls from customers may be better off not using voice mail for fear of customer dissatisfaction. Users who are frequently away from the office and receive frequent intracompany calls can profit from voice mail.

Voice Mail Applications

If you are considering voice mail for your office, you should consider which of its several applications fit your company. Voice mail has several distinct applications, not all of which are available in the same machine:

- Answering telephone calls while users are unavailable
- Disseminating information to callers
- Collecting information from callers
- Call routing

Most people tend to think of voice mail as an expensive telephone answering machine, and thereby limit its usefulness. This is often the case with a new technology; a product is introduced to fill a void or enhance an existing product, only to find that the real market was something not envisioned at first. For example, movies were at first perceived as a way of disseminating stage plays to a wider audience. It took a few years for people to realize that the true value of the technology was taking the audience to locations that couldn't possibly be shown on stage, filming action outside its normal sequence, and assembling it into a finished product.

Although most people prefer to have their telephones answered by a human being, once the initial resistance is overcome voice mail provides a way to transmit information across time zones, to collect information from those outside the immediate organization, and to transmit one-way information as well as collecting messages when the user is away from the telephone. Information can be disseminated with voice mail by sending information from a DTMF telephone on a one-to-one or a one-to-many basis. It can also be collected one-to-one or many-to-one. Caller and called party do not have to be on the same connection simultaneously. Calls tend to be much briefer on voice mail, and 100 percent of them are completed.

How Voice Mail is Connected

Figure 9.7 shows a typical voice mail connection. Calls reach the voice mail system through direct inward dialing, transfer by the attendant, or calls are forwarded to voice mail when the user is away from the telephone. Many service bureaus also offer voice mail, in which users forward calls to a mailbox number or they have a telephone line terminating directly on the voice mail system.

Telephone Answering Applications

The most common application for voice mail is simple telephone answering. Users can forward their calls directly to voice mail or give others their mail box number. To give people a choice, it may be preferable to forward to a secretary who takes the message manually and then forwards it to voice mail. Depending on whether the system is integrated or not, the user dials the mailbox at intervals or when the message waiting light is illuminated.

FIGURE 9.7
Voice Mail Access Methods

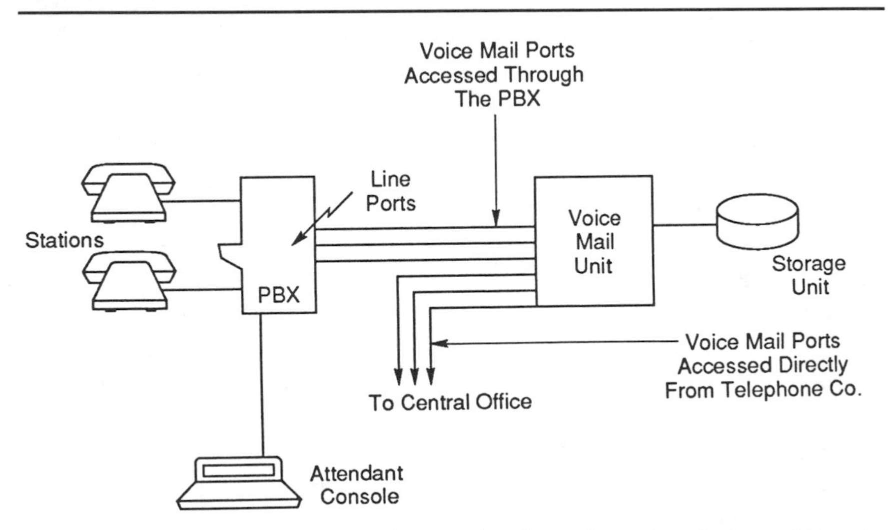

Voice mail is accessed by stations or the attendant forwarding or transferring calls to voice mail. Users can be assigned mail boxes that are accessed from outside the system over ordinary telephone lines.

Mailboxes are protected by passwords. A properly authenticated caller can listen to messages, scan the mailbox, reply to messages from someone within the system, save messages, discard them, or forward them to a single user or a group of users. Messages can also be time and date stamped just like the familiar pink telephone slips.

Dissemination of Information

A less familiar application of voice mail is to disseminate information within an organization. A manager can receive a message over voice mail and route it to the mailboxes of other people in the organization or messages can be created, and by selecting a group address, disseminated to a group of users. In this application, voice mail substitutes for written memorandums. Where the announcement does not have to be in writing, it is an extremely fast and efficient way of spreading information among a group of people.

Information can also be disseminated to the general public. For example, a highway department could disseminate road information over a voice mail system. Callers could be given a choice of routes such as "press one for north, two for south, three for east, four for west, or stay on the line for an operator." The operator alternative is important when you are dealing with the general public because more than one-fourth of the telephones in the United States are still rotary dial and cannot activate a voice mail system.

Information Collection

An application of voice that has been exploited little to date is its use to collect information. For example, the park department of one state requires that payroll information be transmitted to the payroll department within one day following the end of the pay period. Mail is too slow, and a data transmission system is too expensive to collect information from more than 100 locations, so the department uses electronic mail to gather information. The caller enters a password and then a one-or two-digit employee identification number. The machine responds with the employee name and prompts for regular and then extra or premium hours. The system is fast, efficient, and inexpensive.

Call Routing

One of the more controversial uses of voice mail is routing incoming telephone calls, sometimes known as an "automated attendant." The caller is given a short menu of alternatives and asked to dial a digit. If the desired extension number is known to the caller, it can be dialed to give the equivalent of direct inward dialing. Although such a system is effective in some applications, you should be cautious in asking the general public or customers to use an automated attendant. Many people simply hang up. For internal callers the automated attendant has some of the benefits of direct inward dialing in systems that are not otherwise capable of it. It may be feasible to provide separate trunks groups, one answered by an attendant for outsiders and one for insiders that is answered by an automated attendant. The automated attendant is not universally applicable, but in some situations callers can save time by dialing directly to the extension they want.

Voice Mail Summary

The telephone answering machine stigma still hangs over voice mail, and causes many people to reject it. Its acceptance is a cultural matter, and when you are dealing with an organizational culture and introducing change, you must be aware of the pitfalls. The same precautions should be observed as you would in any major organizational change:

- Study the application carefully
- Understand how work is performed, how information is handled, and how communications take place within the organization
- Analyze the effects of the change on the company culture
- Prepare an implementation plan
- Educate and train users
- Follow up to be sure that your expectations are being realized

SELECTING KEY TELEPHONE SYSTEMS

Many PBX features are also available in key telephone systems. The question of which type of equipment to choose is primarily a matter of size and features. If you require features that are not available in a key telephone systems (KTS), you must acquire either a PBX or a hybrid. In small line sizes, neither a PBX nor a hybrid is economical; there are clearly applications where only a KTS will fit.

Most vendors quote KTSs by a number that designates the maximum capacity in central office lines and stations. For example, a 412 system could accommodate a maximum of four lines and 12 stations. Each line is terminated on a button on a telephone instrument. The number of buttons that can be physically terminated on the telephone tends to put an upper limit on the feasible size of a KTS. At some point it is awkward to select lines manually for outgoing calls and to use the intercom for announcing incoming calls. A hybrid, in which the attendant transfers calls and users dial an outgoing call access code, is usually justified.

Telephone Sets

Most KTSs use proprietary stations; 2500-type sets are supported only with an off-premise extension (OPX) line card. Nonproprietary stations have no provisions for picking up outside lines. An older technology, usually called 1A2 key after the original AT&T designation, can support multiple button nonproprietary 2500 sets. The primary disadvantage of 1A2 key equipment is the requirement for 25-pair station wire. An attendant position requires even more pairs. Some manufacturers still produce 1A2 compatible equipment. It lacks most of the features described in this section, and has essentially none of the PBX features described earlier. It is, however, useful for additions to existing systems, and may be feasible for new installations where 25-pair wire is already in place.

The proprietary telephone sets in some systems are upward compatible with larger KTS or PBXs in the manufacturer's same line. If you expect to grow into a larger system or PBX before the service life of a KTS has been reached, this upward compatibility can be an important feature.

Expansion Capability

Most modern KTSs have electronic line cards that allow the station to communicate with the central unit over a thin cable, usually two to four pairs. Stations are terminated in line cards, which usually accommodate from four to 16 stations each. Up to the maximum capacity of the card carrier, adding stations is a matter of adding more cards. When the carrier is full, the system is at its maximum or must be replaced. Some systems have the capability of adding an auxiliary carrier, but most do not. It is important when purchasing a KTS to determine the maximum capacity and growth increment.

Hands-free Answer and Intercom

Because KTS attendants must announce each call to the called party, most modern systems have an internal paging feature that allows the attendant to hold a two-way conversation over a speaker/microphone built into the telephone. The same speakers can be used for on-hook dialing, in which the caller can monitor call progress over the speaker, and does not have to lift the handset until the distant party has answered. Some systems also support internal paging, a feature that allows anyone who can access the paging feature to page over the built-in speakers.

EVALUATING AUTOMATIC CALL DISTRIBUTORS

An automatic call distributor (ACD) is required in many organizations to apportion incoming calls to service positions. Telemarketing organizations with large banks of local trunks and 800 numbers, service organizations such as utilities, classified advertising departments of newspapers, information service providers, and emergency services agencies such as 911 bureaus can improve their effectiveness with ACDs.

Numerous alternatives are offered by telecommunications equipment vendors. ACDs are offered as stand alone devices, or as options in PBXs. A full ACD switches incoming calls to answering positions based on instructions programmed by the administrator. A less complex device is the call sequencer that holds calls until an available agent answers. Calls are organized by *splits*, which correspond to a work group with unique duties. For example, in the airline industry it is common to assign one split to agents who handle calls from travel agents and another to agents who deal with the general public. The general public split could be further subdivided to give special attention to frequent fliers.

Whatever the device, the objectives are common. Agents are assigned to answering positions. Incoming calls are directed to idle positions based on some characteristic such as the least busy or first available position. If all positions are occupied, the system answers the call, places the caller in a holding queue, provides periodic announcements and perhaps music, and directs the call to the first available position. The system also collects statistics so the administrator can prepare work schedules and evaluate service.

This section explains typical features of an ACD, discusses the kinds of organizations that can profitably apply one, and explains how ACD statistics are used to manage the customer service job. In this section, the person who answers incoming calls is referred to as an agent.

ACD Features and Functions

Although not all ACDs and call sequencers provide a complete selection of features, the following discussion explains the benefits of the most common features and indicates the benefits the organization will derive from using them.

Call Distribution

The basic function of the ACD is to distribute calls to agents based on variables programmed into the system. If not all agents are occupied, calls can be distributed to the agent who has handled the least amount of work during the shift. If agents are occupied, the system can distribute the calls based on first come first served or some form of preferential treatment. For example, in the airline case cited earlier, a person who had dialed a frequent flier number might move ahead of other customers in the queue.

Statistical Information

The classical job of balancing cost and service is difficult without valid statistical information about callers' habits and how well the work group handles incoming calls. The office administrator needs the following kinds of information to manage a service function:

- Distribution of incoming calls by time of day and day of the week
- Distribution of the length of service time by type of call and by agent
- Frequency and average time in which callers are held in queue
- Number of calls abandoned in queue
- Average revenue per incoming call summarized by agent
- Number of calls handled by each agent

The statistics collected from the ACD are used by the administrator for shift and break scheduling, monitoring customer service, and measuring agent productivity. By analyzing statistics, the administrator can evaluate ways of reducing the waiting time and processing time of the average transaction.

Provide Delay Announcements

A fundamental requirement of the ACD is to answer incoming calls with a recorded announcement if no agent is available. In some systems, music is provided while the customer holds. The ACD should return periodically with an announcement to reassure waiting customers. Some systems are capable of varying the announcement depending on how long the caller has been in queue.

Machine/Customer Interaction

When customers are placed in queue, the ACD may be able to transact part of the business automatically, therefore reducing the total transaction time. For example, callers to a department store billing group could be asked to dial in their account number. The first free agent could greet the customer by name and have the full details of the account displayed on a screen to start the transaction.

Automatic Cut Through

The agents' transaction time can be reduced if the ACD automatically connects an incoming call instead of requiring the agent to select the call with a button.

Cost Reduction

Cost reduction comes from at least two sources with an ACD. Agent productivity is improved by features such as automatic cut through, machine interaction with the customer, and the ability to measure agent productivity. Productivity is also increased by not requiring agents to interrupt calls in progress to answer an incoming line and place it on hold. A second source of saving is in reducing telephone costs. Placing 800 numbers on hold causes costs to mount at 25 to 30 cents per minute, and may result in lost business when calls are abandoned.

Call Monitoring

Some ACDs provide for service observing, which is to say, monitoring calls to determine how effectively the agents handle them. Anyone using this feature should be aware of privacy issues that have been raised in the courts, although the right of an employer to observe calls for evaluating service has generally been upheld.

Agent Displays

The effectiveness of agents can usually be improved by providing them information on their displays as to the number of calls waiting and how long the callers have been on hold. If agents have this information when they answer the calls, they can be more responsive to the customer's actual situation. By expressing regret for the length of hold, the customer may be somewhat mollified.

Types of Call Distributors

The market offers three types of devices to distribute calls; automatic call distributors (ACD), automatic call sequencers, and uniform call distributors (UCD). An ACD is the most sophisticated and expensive of the three devices. A UCD is designed to function as part of a specific PBX. It distributes calls to an agent group in a predetermined sequence such as "top down," or in a circular hunt. Unlike an ACD, which allows customers to direct calls, a UCD routes calls on a first-in- first-out basis. It has little, if any, real time statistical information, and usually has no management console for making dynamic changes to call routing patterns.

A call sequencer is usually designed to work with a key telephone system, either behind a PBX or on a group of incoming lines. It does not distribute calls, but merely alerts agents to incoming calls by lighting a button on the telephone set. The sequencer holds calls until they are answered, usually with an announcement message. Its statistical information is more basic than that of a true ACD.

Statistical Information

The primary issue in managing any service function is how to determine how many positions should be provided, and how many agents should be occupied at any time. The arithmetic for this decision is not complicated, but it requires the administrator to have some reliable data for making the decision. If variables such as numbers of calls during peak seasons and the average holding time of each call are known, the number of agents can be calculated with the queuing concepts presented in Chapter 16. Of course, you must have a service level objective to complete the calculation. If the average revenue per call is high, it pays to have more agents, but if it is relatively low, you may want to have customers wait longer during peak periods. Chapter 16 explains how to use ACD data to calculate the effectiveness of adding agents. Figure 9.8 is a sample of an on-line status report from an ACD.

Management Software

Many ACDs also provide software for managing the work group and the ACD system. The software is proprietary to the system, and varies greatly among manufacturers, but generally falls into the categories listed in this section.

Scheduling

Scheduling software designs work schedules based on call volumes. The program should determine the number of people needed for each shift, and should be able to predict when to schedule shift starts, part time help, and break periods.

Personnel Tracking

Tracking software keeps track of absences of all kinds, including, in some systems, a real time system that predicts the remainder of the day when an unexpected absence occurs. The program calculates how many people are required, and should recommend changes in staffing levels.

Forecasting

This function forecasts call volumes based on historical information. Some forecasting software also monitors the dollar volume of transactions and forecasts revenue.

Trunking

Trunking software calculates the number of local trunks and 800 lines based on call statistics.

Planning and Budgeting

Planning and budgeting software is available for planning staff levels, expenses, and revenues. A desirable feature is the ability to make "what if" calculations to see the effect of changes in staffing or load.

FIGURE 9.8
Example of an On-Line Status Report From a Siemens ACD

ACD GROUP	STNS TOTAL	STNS REMOVED	STNS ACTIVE	STNS BUSY	STNS RING-ING	STNS IDLE	TRKS ACDQ	TRKS ACDRA	TRKS WAIT-ING
0	20	2	18	12	2	6	2	0	2
1	10	0	10	1	1	8	1	0	1
2	8	0	8	5	2	1	2	1	3
3	7	7	0	0	0	0	0	0	0
4	15	1	14	9	4	1	4	0	4

ACD GROUP	STNS TOTAL	STNS REMOVED	STNS ACTIVE	STNS BUSY	STNS RING-ING	STNS IDLE	TRKS ACDQ	TRKS ACDRA	TRKS WAIT-ING
0	20	2	18	10	3	5	3	0	3
1	10	0	10	3	2	5	2	0	2
2	8	0	8	3	2	3	2	1	3
3	7	7	0	0	0	0	0	0	0
4	15	1	14	6	2	6	2	1	4

ACD GROUP	STNS TOTAL	STNS REMOVED	STNS ACTIVE	STNS BUSY	STNS RING-ING	STNS IDLE	TRKS ACDQ	TRKS ACDRA	TRKS WAIT-ING
0	20	2	18	17	1	0	1	2	3
1	10	0	10	5	3	2	3	0	3
2	8	0	8	3	2	3	2	0	2
3	7	0	7	0	1	6	1	0	1
4	15	1	14	10	3	1	3	0	3

ACD GROUP	STNS TOTAL	STNS REMOVED	STNS ACTIVE	STNS BUSY	STNS RING-ING	STNS IDLE	TRKS ACDQ	TRKS ACDRA	TRKS WAIT-ING
0	20	1	19	10	4	5	4	0	4
1	10	0	10	3	3	4	3	0	3
2	8	0	8	7	1	0	1	2	3
3	7	0	7	3	1	3	1	0	1
4	15	1	14	12	2	0	2	2	4

ACD GROUP	STNS TOTAL	STNS REMOVED	STNS ACTIVE	STNS BUSY	STNS RING-ING	STNS IDLE	TRKS ACDQ	TRKS ACDRA	TRKS WAIT-ING
0	20	1	15	9	2	4	2	0	2
1	10	0	10	8	2	0	2	1	3
2	8	0	8	5	0	0	0	0	0
3	7	0	7	6	1	0	1	2	3
4	15	1	14	10	0	4	0	0	0

FIGURE 9.8—*Continued*
The meanings of the status information are:

(1)	ACD GROUP	- ACD group number (0-63).
(2)	STNS TOTAL	- Total number of stations assigned to the ACD group.
(3)	STNS REMOVED	- Total number of stations in the ACD group temporarily not accessible by ACD hunting (i.e., stations in the out-of-service or line lockout state, or stations that have one of the following features activated: Stop Hunt, Do Not Disturb, Call Forwarding - Busy Lines, Call Forwarding - No Answer, Call Forwarding - All Calls, Call Forwarding - Secretarial, and Call Forwarding to Public Network).
(4)	STNS ACTIVE	- Total number of stations in the ACD group available for hunting.
(5)	STNS BUSY	- Total number of stations that are busy (i.e., in any state other than ringing).
(6)	STNS RINGING	- Total number of stations that are ringing.
(7)	STNS IDLE	- Total number of stations that are idle (i.e., not ringing or busy).
(8)	TRKS ACDQ	- Total number of trunks in associated ACD group having calls in a silent or music state (waiting to be routed to an idle ACD station) or connected to EPABX ringback tone (waiting for an idle station to answer.)
(9)	TRKS ACDRA	- Total number of trunks having calls listening to either the first or second recorded announcement.
(10)	TRKS WAITING	- Total number of trunks in associated ACD group having calls waiting to be answered.

Courtesy of Siemens Corporation

SUMMARY

The advent of low-cost electronics has resulted in significant improvements and cost reductions in customer premises switching equipment. Equipment has become much smaller, uses less power, and takes fewer wire pairs to operate stations. Reliability has increased dramatically compared to electromechanical switching apparatus of earlier years. Not only do systems fail less frequently, remote maintenance capabilities enable repair forces to cut outage time as well. More and better information is available to help you determine how many trunks and tie lines to provide to optimize cost and service. Restrictions and call accounting help cut down the amount of unauthorized calling that occurs, and when it does occur, information sources help you track it down to the source quickly.

Although the equipment and software have become markedly more effective in the past few years, users still experience problems getting service after the equipment is installed. The critical link in the chain between the manufacturer and user is the vendor that installs and supports the system. The importance of a reliable source of continuing support for the product cannot be overemphasized.

CHAPTER 10

SELECTING LONG DISTANCE SERVICES

In the highly competitive long distance market that has developed since divestiture, a bewildering array of services have emerged. The choices were once simple: Direct Distance Dialing (DDD) or Wide Area Telephone Service (WATS) from AT&T, Bell Operating Companies (BOCs), and independent telephone companies. Now, MCI, U.S. Sprint, ITT, and Allnet serve most metropolitan areas, and countless regional long distance carriers offer long distance services within certain geographical boundaries. The BOCs, which were once the exclusive long distance interface for all but the largest customers, are permitted to offer long distance service only within their Local Access Transport Areas (LATAs).

The economics of various long distance alternatives have changed dramatically in the past few years. Many companies haven't examined the new services, and may be paying more than necessary. Even if you prefer not to change carriers, it still pays to evaluate other service offerings to see if they can save money. The evaluation process requires knowledge of your calling patterns and costs, and tariffs or charges the interexchange carriers (IECs) apply to handle your traffic.

This chapter describes the different types of long distance service, discusses when each type tends to be most effective, and offers a process for evaluating the various services. Long distance services are evaluated on the basis of three factors:

- Economics, normally judged on the basis of cost per minute
- Transmission grade of service, normally judged on three variables; loss or volume, noise, and echo

- Switching grade of service, normally judged on the basis of percentage of blocked calls

In this chapter we will discuss how to make objective evaluations of the cost of various services using spreadsheet programs. We will also discuss how to make subjective evaluations of transmission and switching grade of service.

CHARACTERISTICS OF LONG DISTANCE SERVICES

An understanding of the characteristics of different long distance alternatives is essential. In this section, the different types of service are explained. Table 10.1 lists the generic types of long distance service available on the market today, and the trade name of the service offered

TABLE 10.1
Comparison of Long Distance Service Offerings for Major Nationwide Interexchange Carriers

	AT&T	*MCI*	*US Sprint*	*Allnet*	*ITT*
Message Telephone Service	Direct Distance Dialing	MCI Prime Calling Option	US Sprint DIAL 1	Dial Up	Custom Call 100 and 200
Wide Area Telephone Service	WATS	MCI WATS	Advanced WATS		Smart WATS
Virtual WATS	---	Prism II Prism III	Advanced WATS Plus	Maxcess II and III	Custom WATS 200
Dial-one WATS	Pro America II and III	Prism Plus	Dial 1 WATS	Maxcess III Plus	Custom WATS 100
Bulk Rated WATS	Megacom	Prism 1	Ultra WATS	Maxcess I	Custom WATS 300 None
INWATS	800 Service Readyline	MCI 800 Service	Sprint 800 Service	None	

Note: These services were current at the time of this writing, but may no longer be current.

by the major IECs. Most regional carriers offer similar services under other trade names.

Message Telephone Service (MTS)

The most widely used and generally most expensive long distance alternative is directly dialed calls over the selected common carrier's network. AT&T's term for this service is the well-known Direct Distance Dialing (DDD). Many users refer to it as "dial-one" service. In this chapter we will generally use the more generic term MTS. All IECs provide services that are identical to MTS with some differences in features and rates.

As most major metropolitan areas in the United States are now converted to equal access, you have a choice of any of the long distance carriers serving your area. MTS is accessed in one of two ways: by dialing 1 plus the seven- or ten-digit called telephone number, or 10XXX plus the called number (where XXX is a three-digit code assigned to the common carrier). Dial-one access routes your call to the facilities of the IEC you have selected to handle your interLATA traffic. Other IECs are selected by dialing the additional four digit number, 0XXX. Table 10.2 lists the dialing codes for the major nation-wide IECs.

Note that in most states dial-one intraLATA MTS traffic is assigned to the local telephone company even though you have chosen another carrier for your long distance service. If you want an IEC to carry your intraLATA traffic you or the least cost routing system in your PBX must dial the 10XXX code for each call.

TABLE 10.2
Special Access Codes for the Major Long Distance Carriers

Carrier	Code
AT&T	10288
Allnet	10444
MCI	10222
US Sprint	10333
ITT	10488

These codes may be used to access the carrier even if another carrier has been chosen as the "dial-one" carrier of choice.

MTS has the following characteristics and features:

- Each call is individually detail-billed
- A minimum billing period, usually one minute, applies to each call
- Calls are billed in whole-minute increments; after a grace period, typically six seconds, an additional minute is billed
- The initial minute is billed at a higher rate than subsequent minutes
- Calls are billed in airline mileage bands; see Table 10.3 for AT&T's mileage bands, which are typical of those used by all IECs
- Discounts are allowed for evening calls between 5:00 PM and 11:00 PM, and night calls between 11:00 PM and 8:00 AM
- All calls are carried over local trunks by the local telephone company and switched to the IEC

Intrastate calls are regulated by state utility commissions, and have rate structures that may differ significantly from interstate calls. Within a state, the local telephone company can carry traffic only within its LATA. A LATA is a metropolitan area with surrounding territories that the Department of Justice and AT&T agreed would be the boundaries over which the telephone company would be permitted to provide long distance service. For example, Figure 10.1 shows the LATA boundaries for New Jersey.

LATA boundaries do not necessarily follow area code boundaries, which offers untold opportunity for confusion. In some regions, the boundaries cross state lines, which makes it difficult to determine which carrier is authorized to handle your traffic. For example the Portland,

TABLE 10.3
AT&T Mileage Bands for Rating Interstate Calls

Mileage Bands	
1–10	431–925
11–22	926–1910
23–55	1911–3000
56–124	3001–4250
125–292	4251–5750
293–430	

FIGURE 10.1
Local Access Transport Area (LATA) Map of New Jersey

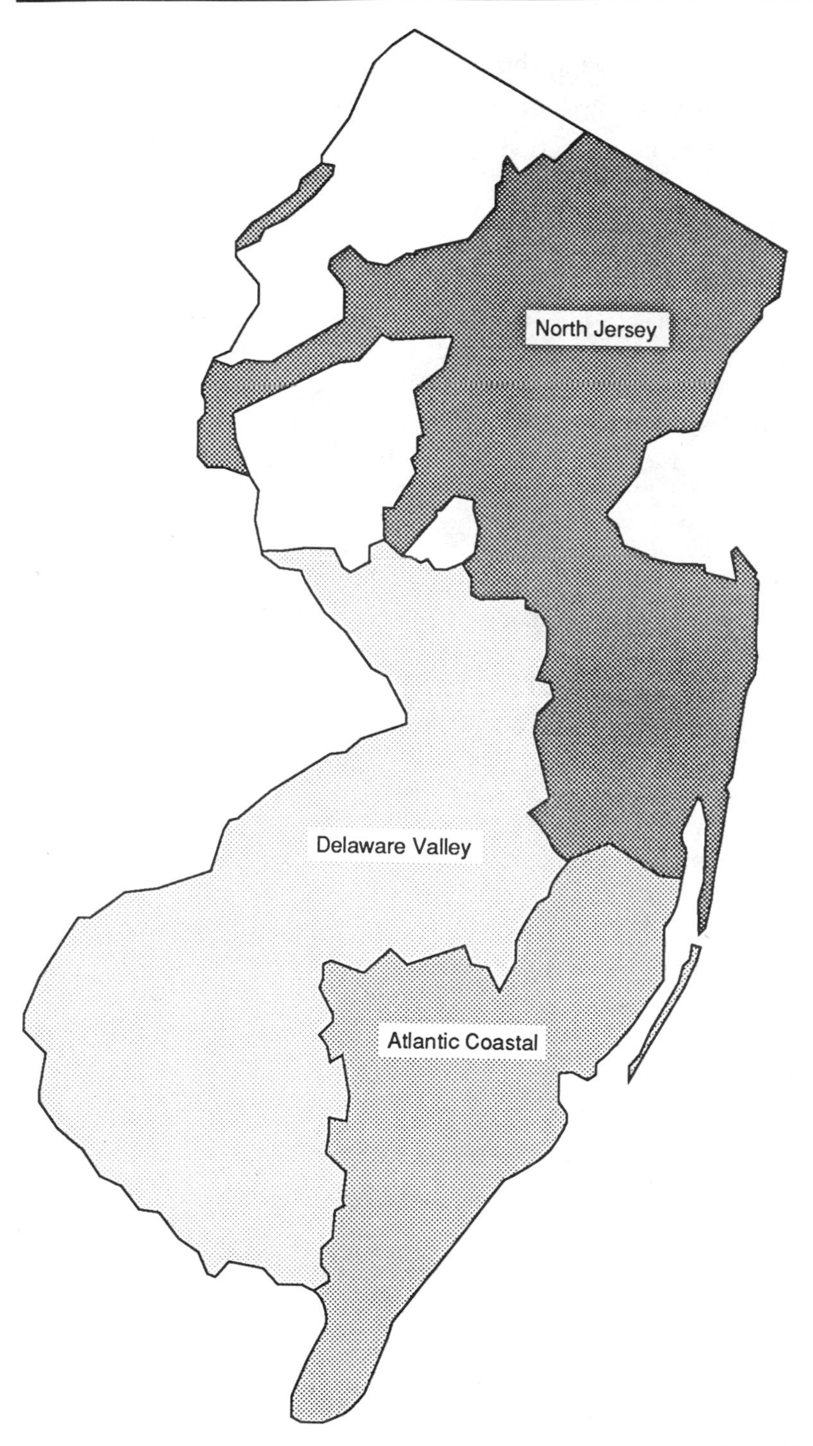

New Jersey is divided into three LATAs with boundaries as shown. The white area is independent telephone company territory.

Oregon, LATA extends into southwestern Washington. MTS calls from Portland to Southwestern Washington are handled by U.S. West Communications, but WATS calls are handled by the IECs.

Airline Mileage Calculations

The most effective way to compare costs among the various long distance alternatives is to calculate the cost based on your actual usage profile. Because DDD is rated on mileage bands that do not coincide with NPA or state boundaries, it is difficult to make valid cost comparisons because most users do not have the distance tables that the long distance carriers use. Distance calculations are based on a system of vertical and horizontal (V & H) coordinates that are contained in a table for every city in the United States and Canada. These tables are used for computing airline miles between points. Appendix I lists V & H coordinates for the Class 1 rating centers in the United States as defined in FCC Tariff No. 264. The formula for computing airline mileage is:

$$\frac{\sqrt{(V_1 - V_2)^2 + (H_1 - H_2)^2}}{10}$$

For example, to compute the distance between Chicago and New Orleans, follow these steps:

Step 1: Obtain the V & H coordinates from Appendix I.

	V	H
Chicago	5986	3426
New Orleans	8483	2638

Step 2: Subtract the V coordinates and the H coordinates and square the difference.

Difference	2497	788
Difference squared	6,235,009	620,944

Step 3: Add the squares of the differences and divide by 10.

Sum of squares	6,855,953
Sum divided by 10	685,595

Step 4: Take the square root of the result obtained in Step 3. This factor is the airline distance between the two points.

$$\text{Distance} = \sqrt{685{,}575} = 828 \text{ miles}$$

Most call accounting systems include V & H coordinates for calculating rates on every call. The tables must be kept updated to reflect the addition of new central offices or changes in NPA boundaries.

Call Timing Considerations

Call timing becomes an important consideration in selecting IECs and MTS service. Because of its historical connection with the Bell Operating Companies, AT&T uses a call timing method that relies on an answer supervision signal from the local central office. The signal is produced when the called party lifts the telephone receiver. IECs often refer to this as "hardware" answer supervision.

Most other IECs do not use answer supervision on all calls because some of their circuits are connected to the local telephone company over channels that do not signal when the called party answers. Therefore, these IECs use alternate methods of detecting call answer. One method is to begin timing after a certain number of seconds has elapsed. Some IECs wait for, say, 45 seconds, and then automatically begin call timing. Others use a longer interval, but begin timing earlier if their equipment detects a voice signal from the distant end or after ringing ceases for one or two cycles. These timing methods are often referred to as "software" answer supervision. Software timing methods are inherently less accurate than hardware methods.

With IECs that use software answer supervision, you may find many one-minute calls being billed that actually have not been completed. If you use an IEC that employs software answer supervision, users should be cautioned to hang up if the call isn't answered after a number of rings that corresponds to the carrier's rating procedures. You should also have users scrutinize bills for one-minute calls that were not completed. It is worth noting that PBX call accounting systems also use the same criteria for call timing because they are likewise unable to detect the actual answer time.

The greatest advantage of MTS is that all costs are strictly usage based. All other alternatives substitute a fixed monthly cost for a lower rate per minute. MTS is a satisfactory alternative for long distance under these conditions:

- When the calling volume is insufficient to justify lower cost facilities

- For overflow when lower cost facilities are blocked
- To carry traffic to locations that are not served by lower cost facilities; for example, overseas locations are rarely served by alternative services

Wide Area Telephone Service (WATS)

WATS is offered by most of the major IECs with some differences in rates, serving area, and timing intervals. At the time this book was written, the market offered three different types of WATS service: banded WATS, virtual WATS, and dial-one WATS. A fourth type of WATS is offered to users with a large enough call volume to justify a T-1 line to the IEC. This service is billed like virtual WATS, and is discussed separately below. All three WATS services are discounted long distance, with the following differences between them.

Banded WATS. A dedicated line is run from the IEC's central office to the user's premises. The cost of this line is included in the fixed monthly service charge. With banded WATS you must determine the most economical combination of the six WATS bands. A particular WATS band can offer calls within that band and all bands below it. Calls are billed, however, at the rate of the band over which the call was placed.

Virtual Banded WATS. Dedicated lines are also used with virtual WATS, and the same six WATS bands apply for rating calls. The main difference is that the dedicated lines to the IEC carry calls for all WATS bands. The IEC rates the calls by destination, but at a somewhat higher rate than for banded WATS.

Dial-One WATS. This service is essentially the same as MTS except that a fixed monthly fee is charged in exchange for a deeply discounted rate structure. Dedicated lines are not provided. Calls are routed over local central office trunks to the IEC.

With some minor differences, most carriers use the same format for their WATS service offerings. Rates are divided into six bands, which are based on distance from the state of origin. Appendix E lists AT&T's WATS bands for the 50 states. In AT&T's tariffs, WATS rates are classified into 18 different rate steps which further reflect distance between points. For example, Band 5 (shown as Service Area 5 in Appendix E) from New York is in Rate Step 18, and from Illinois it is in Rate Step 15, which carries a lower rate. Rates are not published in this book because they change too frequently to be reliable. You can obtain current WATS rates from any of the common carriers.

Note also in the Rate Step tables in Appendix E that SA (Band) 5 and 6 have the same rate steps. This means that you can place calls to Alaska which is the only state in Band 6, for the same cost per hour as calls to Band 5 locations. This change occurred in 1987, and many companies have not changed their WATS banding and least cost routing tables to take advantage of it. Telecommunications managers must monitor tariffs constantly to remain current with changes such as this.

Banded WATS

WATS bands generally are based on state boundaries, but some of the more populous and larger states such as New York, Illinois, Texas, and California are subdivided into two or more parts. Most IECs use the same six WATS bands as AT&T.

The only differences between banded and virtual WATS are in rates and sharing of access lines. In a banded WATS rate structure, a WATS line can carry traffic to destinations covered by lower bands, but not to higher ones. For example, assume you have chosen Band 1 and Band 5 WATS in a banded rate structure. The carrier would provide two separate groups of trunks. If the Band 1 lines were full, calls could overflow to Band 5, and would be carried at Band 5 rates. If the reverse occurred, however, and all Band 5 trunks were full, calls to Band 5 destinations could not be carried over idle Band 1 lines.

Banded WATS rates taper according to the amount of usage during the billing month. AT&T has three taper points: first 25 hours, next 75 hours, and over 100 hours. Other vendors may have more or fewer taper points, but all offer an incentive for heavy users to place as much traffic as possible on the lines. Before January 1, 1987, AT&T used average line usage to calculate total cost for a month. For example if a total of 200 hours of traffic was placed over two lines, the average usage per line was 100 hours and that was the basis on which usage would have been rated, and the result multiplied by two. On January 1, 1987, AT&T eliminated average line usage and now rates calls based on total usage. Other vendors may have a different rating process.

Compared to other forms of long distance, banded WATS has these advantages:

- For heavy users it is less expensive than other forms of long distance; it is advantageous where traffic is heavy to a single terminating band
- The dedicated line handles traffic that otherwise might flow over central office trunks, therefore reducing the number of local trunks required
- The fixed cost per month tends to be lower than many other discounted services

Offsetting these advantages are several significant disadvantages:

- Banded WATS is inflexible; if your calling pattern is seasonal you must keep readjusting bands to keep the most economical balance
- If usage is low enough to require only one or two trunks, you will encounter the inefficiencies of small trunk group size
- If higher banded WATS lines are used to gain the advantage of the larger trunk group size, calls are placed at higher rates

To understand the issue of trunk group efficiency, assume that during your peak load period you have 1.2 hours of traffic to Band 1 destinations and 1.2 hours to Band 5 destinations. The traffic tables in Appendix A1 show that four trunks are required for each of the two bands, for a total of eight trunks, assuming five percent overflow to MTS. See Chapter 16 for an explanation of how to use the traffic tables. If the traffic is combined into one group handling the total 2.4 hours of peak-period traffic, only six trunks are required. The smaller number of trunks reduces the fixed monthly cost of the service.

In a banded WATS service, it is usually most economical to allow at least five percent of the traffic to overflow to MTS. Here is how to determine how much traffic should overflow:

> Assume that you use approximately 100 hours of traffic per month, primarily to Band 5 locations, and that it averages $14.40 per hour at your usage level. MTS traffic to the same locations averages 36 cents for the first minute and 28 cents per minute thereafter. The average call length is four minutes.

The average cost per hour for MTS would be

1 minute at \$0.36	=	\$0.36
3 minutes at \$0.28	=	0.84
4 Minute call	=	\$1.20

1 Hour = 60/4 x \$1.20 = \$18.00

WATS is \$18.00 - \$14.40 = \$3.60 per hour less than MTS.
If the fixed cost of a WATS line is \$40.00 you can afford to have \$40/\$3.60 = 1.11 hours of traffic overflow before the fixed cost of another WATS line is justified. Since the usage level is 100 hours, this means that 11.1 percent of the WATS traffic should overflow before another line is added.

Intrastate services cannot be handled over the same trunks as interstate WATS by some carriers. This means that separate WATS trunks are required for intrastate calls. Large states such as California divide the state into two sections for intrastate calls, and the carriers provide either full-state or half-state WATS.

Calls within the LATA over intrastate WATS are carried by the local telephone company. Calls outside the LATA are carried by an IEC; usually AT&T and the local telephone company share the WATS line. Usage is billed in two sections, intraLATA and interLATA. The amount of usage must be combined to determine the correct quantity of WATS lines.

The principal characteristics of WATS service are:

- Calls are billed by hours or minutes of usage; the higher the amount of usage per month, the lower the cost per minute; it is therefore advantageous to pack as much traffic as possible into WATS trunks
- Some carriers provide call detail, but others bill only the number of hours of usage in bulk
- The common carrier is accessed over private lines, not the local switched telephone network
- Rates are based on distance between states, which makes it easy to compare rates, since area codes do not cross state lines
- When all WATS lines to a band are full, traffic is blocked or overflows to a higher-rated band or to MTS
- Banded WATS is most advantageous when relatively large volumes of traffic are going to a narrow range of terminating locations.

At the time of this writing the future of banded WATS is in doubt. Many industry observers feel that AT&T is positioning itself to drop WATS service in favor of services that use the switched network. Very large users will still find it economical to access the IEC's switch with T-1 facilities as discussed later, but the next few years may see the demise of traditional WATS services.

Virtual WATS

Virtual WATS service overcomes some of the disadvantages of banded WATS. A single group of dedicated lines allows you to access any WATS band. Each call is rated for the band it used. Gone is the necessity of choosing the most ideal combination of bands. You can obtain the efficiencies of larger trunk groups while still retaining the advantage of lower rates. Dedicated lines are still run to the IEC's central office. When the capacity of lines of the group is exceeded, calls are either blocked or they overflow to another service such as MTS.

Unlike banded WATS, the cost of virtual WATS offered by most IECs does not vary with load. Taper points that reduce the cost of long distance as the usage increases each month are not employed in most virtual WATS tariffs. With banded WATS, the incentive is high to pack the lines with as much traffic as possible to gain the lower cost as usage increases. Except for the fixed monthly fee for access lines, there is little incentive for a company to pack virtual WATS with traffic.

Virtual WATS bands are not necessarily the same as conventional WATS bands. For example, MCI's banded WATS service uses bands identical to AT&T's, but MCI's virtual WATS services, Prism I and II^{SM}, use a different banding structure. AT&T does not offer a virtual WATS service except for Megacom,^{SM} which is intended for companies with enough long distance to justify a T-1 line to the AT&T central office. Megacom and other high-volume services are discussed later.

Unlike banded WATS, many IECs carry intrastate traffic on virtual WATS. Generally, intrastate rates are the same as for service Band 1, and may or may not be economical compared to MTS or WATS provided by the telephone company.

The dedicated lines of virtual WATS remove the traffic load from local trunks and may reduce the requirement for expensive local trunks. The primary disadvantage of virtual WATS compared to dial-one WATS, which is described in the next section, is its lack of flexibility. A fixed number of lines is installed, and during usage peaks traffic is blocked or overflows to MTS at a higher cost.

The method of evaluating the economic effectiveness of virtual WATS service is nearly identical to banded WATS, except that the calculation is much simpler because costs do not vary with usage. The process, shown in Figure 10.2, is to obtain usage by WATS band and enter it into a spreadsheet that calculates the total cost.

The principal characteristics of virtual WATS are:

- Calls are billed in bands that coincide with state or area code boundaries, but the bands are not necessarily the same as WATS bands
- Calls to all locations share a common group of dedicated lines to the IEC
- Calls are billed at a flat rate per unit of time to all locations in a band regardless of the amount of usage
- When all WATS lines are full, the traffic is blocked or overflows to MTS

Dial One WATS

A relatively recent service, "dial-one" WATS, is, in effect, a throwback to the earlier practices of many companies of discounting calls from

FIGURE 10.2
Worksheet for Calculating Virtual WATS Costs

Range		*Minutes Used*	*Prism Plus Rates*	*Prism Plus Usage Cost*	*Prism 3 Rates*	*Prism 3 Usage Cost*
	1	2347	$ 0.1962	$ 460.48	$ 0.2107	$ 494.51
	2	4431	0.2097	492.17	0.2185	512.82
	3	6543	0.2241	525.96	0.2308	541.69
	4	336	0.2313	542.86	0.2455	576.19
	5	8765	0.2448	574.55	0.2559	600.60
Total		22422		2,596.02		2,725.81
Busy Hour Erlangs		2.9				
Dedicated Trunks		6			$35.00	210.00
Addl.C.O. Trunks		3	$52.00	156.00		
Total Cost				$2,752.02		$2,935.81

This spreadsheet model calculates the total cost of long distance over two alternative services, both of which use mileage-based pricing. The model has an Erlang B table out of the visible range. The program looks up the number of trunks required to handle the entire load. For dial-one services such as Prism Plus it calculates the number of central office trunks that must be added to accommodate the added long distance load.

AT&T's rates for a fixed monthly fee plus usage charges. This service differs from banded WATS and virtual WATS in three significant ways: the switched local telephone network is used for access instead of dedicated lines, the fixed monthly cost is not a per-line cost, but is a per-billing number cost, and the rates are based on mileage bands instead of state boundaries. The use of mileage bands makes it difficult to compare services. Unless you have a V and H table you cannot easily determine the mileage between points.

Dial-one WATS distance bands, shown in Table 10.4, are generally the same as MTS bands except that MTS has more mileage bands at the low end. From your MTS bill it is relatively easy to determine which mileage band a call was billed in. Given the number of minutes and the cost, the call can be looked up in the dial-one WATS rate table.

Bear in mind that MTS calls are usually billed in one-minute increments and WATS calls are billed in six-second increments. Therefore, if you want to compare MTS and dial-one WATS, remember that MTS bills do not show the actual call duration. Since there are ten six-second intervals in a minute, and calls are rounded up after the first interval, each call will have, on the average, approximately 0.5 minutes of time that was billed for but not used. To make a comparison between MTS and dial-one WATS on a sample of calls, you should assume that MTS calls are 0.5 minutes longer on the average than dial-one WATS calls.

Since dial-one WATS is indistinguishable from MTS except in the billing process, it is not used for carrying intraLATA traffic. As stated earlier, most telephone companies automatically carry dial-one traffic within the LATA unless the user dials the 10XXX access code.

TABLE 10.4
AT&T Mileage Bands for Dial-One WATS Service (Pro America II and III)

Mileage Bands	
0–292	1911–3000
293–430	3001–4250
431–925	4251-5750
926-1910	

Dial-one WATS, offered by most IECs, is well worth investigating if you have enough traffic to offset the fixed charge, which ranges from about $60 to $80 per month. In many cases, dial-one WATS will be more economical than either banded or virtual WATS, particularly when administrative costs are considered. One large advantage of dial-one WATS is ease of administration. It is no longer necessary to monitor usage to determine whether the WATS configuration is correct. It is, however, necessary to monitor local trunk usage since this is the medium used to originate the call.

The principal characteristics of dial-one WATS are:

- Calls are completed over local trunks rather than dedicated lines to the IEC
- Calls are billed in mileage bands that do not coincide with state or area code boundaries
- The service is virtually identical to MTS except for six-second billing and a fixed monthly fee
- Calls are detail billed
- All calls are carried on the service; no calls overflow to MTS or other service

Bulk-Rated Access WATS

For very large users of long distance service, most IECs offer access over T-1 lines directly to the IEC's switch. The rates for these services are generally among the lowest in the industry because of the large amount of usage that is required to justify the access costs. The IECs charge a fixed cost per month for access, and may pass their costs from the local telephone company for the T-1 line directly to the user. Some IECs allow large users to install their own microwave or fiber optic links to bypass the local telephone company entirely.

The principal characteristics of bulk-rated services are:

- Usage must be high to justify the fixed access cost; generally, at least 600-1,000 hours per month of usage is needed
- T-1 lines are terminated directly on the IEC's switch; the user must terminate directly on a digital PBX or provide a channel bank to split out individual channels
- Usage is billed on the same basis as virtual WATS

From the pricing of these services, the strategy of most long distance companies appears to be toward dial-one and virtual WATS. Any organization that presently has banded WATS should evaluate its long distance usage to determine if a change to virtual or dial-one WATS is advantageous. In most cases it will be.

INWATS (800) Services

INWATS or 800 service is usually thought of as a competitive tool to allow customers to call at your expense. Although this is definitely a major reason for using 800 service, it can also be a cost-saving tool to reduce the cost of long distance, collect, and credit card calls to a home office. Depending on the length of the call, collect calls may cost as much as $3.00 per minute. By contrast, 800 service costs are in the order of 25 cents per minute. If the load is heavy, a considerable amount can be saved by using 800 service for intra-company calling. It isn't even necessary for callers to be on the road. Companies with several branch offices, many of which are too small to justify WATS, can use 800 numbers to reduce the cost of calls to headquarters by 10 to 15 cents per minute.

The switching systems of the local telephone companies were not configured to allow 800 number calling from multiple IECs prior to 1987. At the time of this writing, changes are in place to enable other carriers to provide 800 service by assigning a block of 800 numbers to each IEC. This means that users can switch between IECs only by changing the 800 number, which deters many companies that advertise their 800 numbers.

Dedicated trunks are used for ordinary 800 service. A fixed cost per month covers the cost of the trunk, and usage costs vary with the amount of usage. The bands and usage taper points are generally the same as WATS, but the usage cost is lower.

Unlike WATS service, in which intrastate and interstate trunks are separate, a single number 800 service is available. Effective January 1, 1987, AT&T added a single number 800 service to allow customers to receive both intrastate and interstate calls on the same WATS line. Before 1987, single number service was available, but at a cost that made it unattractive for most companies. The primary advantage of single number service is that it eliminates the need to publish separate numbers for interstate and intrastate 800 service. Most IECs also offer a hunting arrangement so that when all trunks in a band are busy, a call can be completed over a higher band at a higher rate. Calls cannot be completed over a lower rated band, however.

In 1987, AT&T introduced a specialized 800 service with the trade name Readyline. Whereas 800 service uses dedicated trunks, Readyline calls are routed to the user's local telephone number. The fixed cost is lower, but the usage cost is higher than the cost of 800 service. This type of service is advantageous to the small company that has no space on its telephone system to terminate a dedicated 800 trunk or too little volume to justify the higher fixed cost. It is also a good service to use for a facsimile network because it is possible to reach a single machine by dialing either an 800 number or a ten-digit telephone number.

The process of evaluating 800 service is much the same as evaluating WATS. When toll statements have been converted for spreadsheet manipulation, it is a simple matter to sort the calls to look for patterns of usage to intra-company locations. Using the techniques described earlier, calls are summarized into minutes of use and cost by WATS band. The results are entered into an 800 service rating model to determine the potential savings. Collect and credit card calls, which are normally excluded from a conventional WATS study, should also be evaluated.

OTHER LONG DISTANCE ALTERNATIVES

The switched services described in the previous sections are justified for carrying the bulk of long distance traffic. Many organizations, however, have large amounts of traffic to a few locations, either to a customer, supplier, or another company site. Where the circumstances warrant, other types of long distance service may be more economical. This section describes the alternatives and the situations in which they are usually justified.

Foreign Exchange Service

Foreign exchange service (FEX) is a combination of switched and dedicated line service. A dedicated line is used to bring dial tone from one

FIGURE 10.3
Foreign Exchange Service

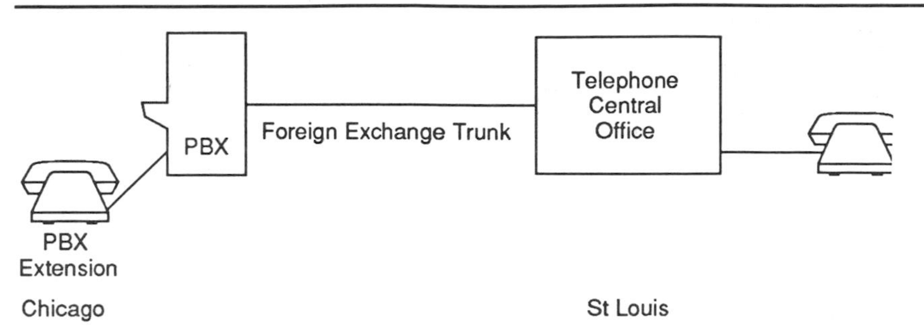

A PBX station user in Chicago can call the St. Louis local calling area over a dedicated trunk. Callers from St. Louis can reach the Chicago PBX by dialing a local call.

city to another, as shown in Figure 10.3. FEX has two uses. The first is to allow people in a distant calling area to reach you by dialing a local telephone number. The second use is to enable you to access the local calling area in a distant city without paying a long distance charge. For example, as shown in Figure 10.3, users in a PBX located in Chicago could dial the St. Louis local calling area over a Chicago-to-St. Louis dedicated line. In this configuration, the FEX line is accessed by pushing a button on a key telephone, dialing a special access code, or selected by a least cost routing system in a PBX.

In its original implementation, FEX was not usage sensitive, but many telephone companies are now measuring FEX service. Calls are generally measured in both the originating and terminating directions. FEX is an effective service when you want to be reached from the entire local calling area in a distant city. It is particularly advantageous to give the public access to your company. It is also useful for enabling users on your PBX to reach the local calling area of a distant city.

The cost of FEX consists of two components, usage cost and the fixed cost of a dedicated line between two cities. The cost of a dedicated line is usually high enough that FEX is not economical for long distances unless you have enough trunks and traffic to keep the line utilization high. For long distances the cost of FEX is usually more sensitive to distance than MTS or 800 service. Therefore, FEX is usually more expensive than other services for long distances.

Remote Call Forwarding

For giving the public access to your company, remote call forwarding (RCF) is an effective alternative to FEX if the call volume is too low to justify a dedicated line. RCF provides a local telephone number that can be reached from the calling area. Translations in the telephone company's switching machine simply redial the call to another number. For this service, you are billed a fixed charge for the telephone number plus long distance charges.

Tie Lines

When a great deal of traffic flows between two locations, dedicated tie lines between PBXs may be an attractive alternative. As shown in Figure 10.4, the tie line is connected to a trunk port in the PBX, which gives users the ability to dial station numbers directly without going through the attendant.

A tie line connects to the trunk side of the switching machine so that incoming and outgoing calls are treated much like local trunk calls; ie., they can access the attendant, directly dial other stations, and if permitted, access special features such as dial dictation and long distance trunks in the distant machine. Tie lines are accessed from within a PBX the same as FEX, by dialing a special access code (codes starting with "8" are common), or the least cost routing system in the PBX chooses the trunk without involving the user. Two or more PBXs can be tied together with a uniform dialing plan that effectively links the machines as if they were one.

FIGURE 10.4
PBX Tie Trunk

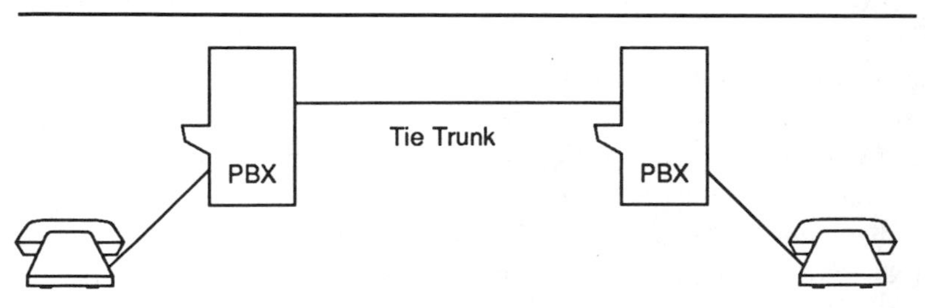

Station users in either PBX have access to the local calling areas of the other PBX and to many of the features of the other system.

Off Premise Extensions

An off premise extension (OPX) is similar to a tie line. An OPX extends a station line in a PBX or KTS to a station in a distant location over a dedicated private line as shown in Figure 10.5. Users in the PBX can dial the extension number to reach the distant station, and the distant station user has access to most of the features of the PBX.

For the purpose of evaluating the cost effectiveness of tie lines and OPXs, they can be considered as identical. To illustrate the process, assume that a company has two PBXs located seventy-five miles apart. A tie-line between the machines costs $175 per month. An evaluation of long distance bills shows that 265 hours of traffic flows between the two locations at an average cost of $.24 per minute. The process for calculating the cost effectiveness of a tie line is:

1. Determine the peak hour traffic volume. If actual statistics are not available, divide the total hours by 22 to get the average daily traffic and multiply by 0.17 to get the peak hour's volume. In this example, the peak traffic would be 265/22 x 0.17 = 2.05 hours.
2. Determine from the traffic tables in Appendix A1 how many tie-lines are required to handle peak hour traffic. Assume that it is economical to allow 5 percent of the traffic to overflow. In this case five trunks are required.
3. Calculate monthly cost for trunks. 5 X $175 = $875

FIGURE 10.5
Off Premise Extension

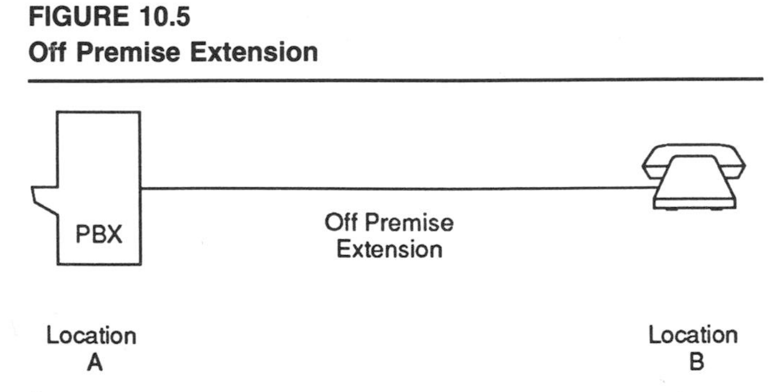

A user at location B has access to many of the same features as any local extension in the PBX including access to the local calling area (if permitted by the class of service).

4. Divide the trunk cost by the number of minutes of traffic and compare to the present cost. $875/(265 X 60) = $875/15,900 = 5.5 cents per minute.

In the above example, the cost saving from establishing a tie line network compared to using MTS is significant, and the tie lines should be installed. It may also be possible to remove local trunks corresponding to the reduction in long distance load. Remember, however, that the tie lines represent a fixed cost, where MTS is a variable cost. If the load drops off, tie lines must be removed to keep costs from getting out of hand.

OPTIMIZING LONG DISTANCE SERVICES

This section discusses how to make a valid comparison among long distance services. The first step in optimizing long distance is to determine traffic quantities to various terminating locations. The best source of this information is output from a call accounting system if one exists. If not, the second best source is long distance bills. If possible, the information should be obtained in a form that can be read by a spreadsheet, data base management system, word processor, or some other program that can sort the data. It is not necessary to obtain a 100 percent sample of calls; depending on the size of the bill, a 10 to 50 percent sample should be enough.

If there is a great deal of intracompany traffic, the full telephone number for each call should be obtained. If calls are going to random destinations, the area code is enough. It is essential that the number of minutes and cost of each call be detailed. If the time of day is available, it is a useful way of determining the peak traffic volume; if not, it is usually safe to divide the total number of minutes by the number of working days, and multiply by 17 percent to approximate the peak volume.

With the sample of calls in machine readable form, it is easy to sort and combine them by area code. A spreadsheet is a useful tool for performing this sort. The WATS band can be looked up manually, or by using a table-lookup function if your spreadsheet contains one. Remember that not all IECs use the same WATS bands for all services.

If calls were initially sampled, they should be expanded to a 100 percent sample and processed against the rating tables. The model in

Figure 10.2 is used for virtual WATS. Virtual WATS tables are much easier to construct than banded WATS tables because separate trunks are not required for each band, and the usage costs do not taper with volume. A reasonably close approximation of mileage band-based service costs can be made by assuming that all of an area code fits in a particular mileage band. This is not entirely accurate because mileage bands split some area codes, but it is usually close enough for you to compare dial-one WATS with banded and virtual WATS.

Traffic between locations in the same company should be examined separately from traffic to random destinations. As discussed earlier, 800 service is an economical way of carrying intracompany traffic. It is less expensive than WATS, and is an excellent alternative when one location, such as a headquarters, receives many calls from satellite locations that are too small to justify discounted long distance. If the traffic volume is high enough to justify dedicated lines such as tie lines, FEX, or off premise extensions, these should be evaluated as well.

Comparing Long Distance Services

Not all long distance services are created equally, and one of the most difficult jobs a telecommunications manager has is determining which service offers the greatest value. This section discusses differences among services and suggests methods you can use to evaluate services on bases other than cost, which has previously been discussed.

Transmission Quality

Transmission quality on voice circuits generally consists of three elements: loss, noise, and echo. These variables are easy to measure with the right kind of test equipment, but most companies do not have the ability to make extensive tests.

If a transmission measuring set is available, loss and noise measurements can be made quite easily and serve to distinguish between services. Most central offices have assigned the 1000 line number to a transmission measuring circuit so that all you need to do is dial the area code, central office prefix, and 1000. A one milliwatt tone (0dBm) alternates with terminated or quiet periods. The amount of loss in dB and noise in dBrnc can be measured from your PBX or key system to any spot in the country.

Loss and noise vary greatly with the type of circuit the IEC uses. Different methods of accessing the exchange telephone network and different circuit design standards may cause one carrier's circuit to have much greater loss than another. The amount of noise on a circuit depends on design standards and, in the case of analog circuits, depends on how well the equipment is maintained. Several IECs advertise the use of quiet fiber optic transmission facilities, but the important question is the extent to which they actually deploy them. A few dollars spent in making actual transmission measurements can enable you to measure quality more accurately than subjective judgment.

Without the use of transmission equipment, subjective transmission tests may be next best. A subjective measurement is probably the most practical way of determining if echo is a problem. If the IEC uses satellite circuits extensively, echo may result, and users may find the delay objectionable. You should be aware, however, that it is very difficult to assess the difference between carriers on loss and noise by subjective measurements. A few dB of difference in either variable is scarcely detectable to the ear, but when several circuits are connected in tandem or when a location is accessed that has frequent transmission problems, the difference of a few dB may make the difference between a satisfied and dissatisfied user.

Companies with extensive private networks should obtain the test equipment to make transmission measurements. Tests should regularly be made to all of your terminating locations to approximate the service that users actually receive. Chapter 18 contains further information on transmission concepts.

Billing Detail

A major difference between services and IECs is the degree of billing detail that is provided. If you have an effective call accounting system, call detail from the long distance carrier may not be needed. Without this billing detail from a call accounting system, however, a long distance statement may be the only way of obtaining usage information to enable you to optimize the service. The availability of billing detail may be a significant advantage, particularly if someone within the organization monitors usage on an ongoing basis.

Vendor Stability

In the past few years there has been a large shake-out in long distance carriers and many observers believe there is more to come. IECs that specialize in reselling AT&T WATS will find it difficult to compete unless they can build their own networks of dedicated facilities. Changing long distance vendors may not be a significant problem to small users, but the disruption to large users on finding that their IEC has gone out of business or sold its account to another IEC may cause serious problems.

Coverage Area

Some long distance carriers offer attractive prices to some parts of the country, but are not competitive to others, particularly the small towns. Some vendors use multi-tier pricing in which calls to large metropolitan areas are offered at lower rate than to smaller areas. Besides being almost impossible to evaluate, multi-tier pricing is not cost competitive for companies which have a great deal of calling to rural areas.

Billing Accuracy

As discussed earlier, some carriers are unable to measure actual call answer time, and use alternate ways of determining when a call is answered. Some studies have suggested that as much as two to three percent of unanswered calls are actually billed in error by the IEC. On the other side of the coin, the first few seconds of some answered calls are not billed by the IEC, so the inaccuracies may have a tendency to even out. Nevertheless, billing accuracy should be carefully monitored with all IECs.

ACCESS CHARGES

A long distance telephone call usually consists of three elements, as shown in Figure 10.6. The originating link consists of a local loop, a switched connection in the telephone central office, and a trunk to the IEC. An interexchange circuit connects two exchanges, and the connection terminates in a link that is identical to the originating link. Before divestiture, the originating and terminating connections were owned by the operating telephone companies, and most interstate links were owned by AT&T.

The local telephone companies, independent companies, and AT&T went through a division of revenue process that allocated part of the cost of every call to each owner. The FCC prescribed a rate of return on investment for AT&T, and all parties were allowed to earn the same return on the portion of their plant that was allocated to interstate traffic.

FIGURE 10.6
Elements of a Long Distance Telephone Call.

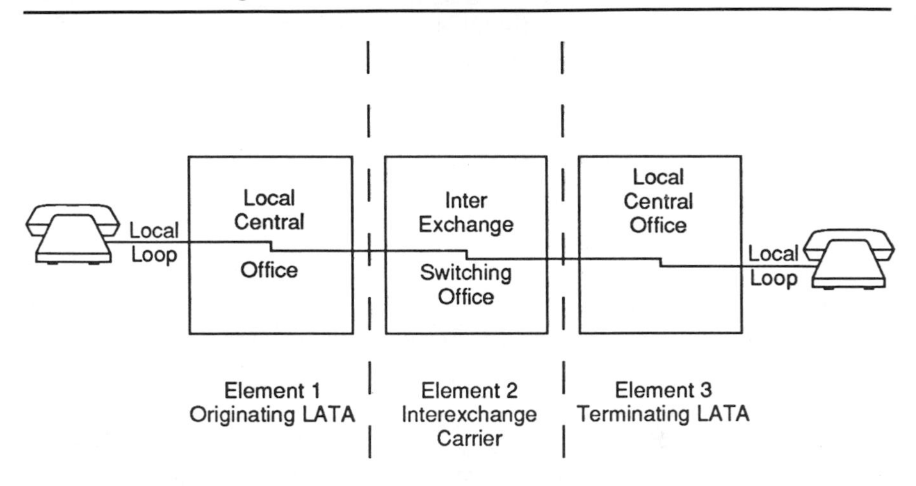

The question was what percentage of investment in a local switching machine and local loop were used for long distance. The utilities commissions wanted a large allocation of investment to long distance because it helped hold down local rates. One of the principal issues of debate between AT&T and its competitors before divestiture was the subsidy that local rates received through the division of AT&T's revenue. This process was changed by divestiture.

After divestiture the FCC decided that exchange access costs would be borne by both the end users and the IECs. These costs are recovered by fixed and variable usage charges on exchange access trunks. The users' portions are billed directly as access charges. The FCC intends to shift more of the access charge from the IECs to the end users. At the time of this writing, access charges are $6 per month per local line for multi-line business users and $3.20 per line for residence and single line users.

FCC rules also require IECs to bill a special access surcharge on private lines that bypass the local telephone network. This surcharge is sometimes referred to as a "leaky PBX surcharge." To illustrate its application, consider the diagram in Figure 10.4. Callers in either PBX can bypass the local exchange network to reach the local calling area that serves the other PBX. The IEC, therefore, is required to assess a $35 per month surcharge on each trunk.

The special access surcharge is a source of great confusion to users and IEC employees alike. It is frequently applied incorrectly, and if you are being billed such a charge, you should be certain it is warranted. The IEC will remove the surcharge upon receiving written certification that the charge is not appropriate.

SUMMARY

Telecommunications managers today face an excess of alternatives for obtaining long distance service. To stay abreast of all the services is impractical; there are simply too many of them and they change too frequently. AT&T dominates the long distance industry, and is therefore the bellwether. Regardless of who your IEC is, you should be acquainted with AT&T's services and rate structure because the rest of the carriers tend to follow AT&T. When a major restructuring of long distance rates and services occurs, you should study its implications and reexamine your service to determine whether a change is warranted. Many companies have failed to change, and have consequently spent more money than necessary on obsolete services.

CHAPTER 11

SELECTING DATA COMMUNICATIONS EQUIPMENT AND SERVICES

The range of products for communicating between computers and remotely attached data devices is so extensive that we can discuss only the most general considerations for choosing them. Three distinct areas of application are discussed in this chapter; wide area, metropolitan, and local area networks. There is some overlap in the products used for these three types of network, but for the most part the applications are different. The choice of the type of computer or existing equipment may drive the entire selection process, but where latitude exists, the guidelines in this chapter should assist you in selecting products. This chapter discusses the principal alternatives for obtaining data communications circuits and explains how to evaluate them.

DATA CIRCUIT QUALITY

Quality of a voice circuit can be judged on somewhat subjective criteria, but data applications are usually far less tolerant of impairments. To make a valid comparison among alternatives, you should get the vendor's published specifications for the criteria discussed in this section wherever they apply. Note that not all criteria are published for every kind of circuit, but those that are should aid you in assessing value.

Amplitude Distortion. This variable describes the amount of variation in amplitude of different frequencies within the passband of an analog circuit. Modems that use a combination of phase and amplitude modulation techniques such as quadrature amplitude modulation (QAM) are affected by amplitude distortion.

Availability. Circuit availability is the ratio of circuit up time to total time. For example, if a vendor quotes 99.96 percent availability, the outage time in one year would be:

8760 hours per year X .9996 = 8756.5 available hours or 3.5 hours of outage.

Bit Error Rate (BER). This figure specifies the average number of errored bits that will be encountered as compared to total transmitted bits. The BER is usually specified as an exponential factor. For example, a BER of 10^{-5} means that one errored bit can be expected for each 100,000 transmitted bits.

Block Error Rate. The block error rate is more indicative of circuit performance than bit error rate in a synchronous application. When an error occurs, a block is rejected regardless of how many errors there are, so if noise comes in bursts, it may obliterate several consecutive bits in the same block. The block error rate expresses how frequently blocks must be retransmitted because of error, and gives a valid insight into the effect of errors on throughput. Block error rate is normally not quoted by vendors, but is measured by test equipment or data communications equipment such as multiplexers.

Envelope Delay. Envelope delay expresses the difference in propagation speed of different frequencies within the passband of a circuit. An excessive amount of envelope delay will be detrimental to modems that rely on phase shift or trellis coding techniques to encode high speed data.

Error Free Seconds (EFS). The EFS variable is quoted for most digital circuits. EFS is the number of seconds per unit of time that the circuit vendor guarantees the circuit will be free of errors.

Harmonic Distortion. Data signals are composed of many harmonics, which are frequencies that are a multiple of the fundamental or basic frequency of the signal. For reliable data communications, the second harmonic must be at least 25 dB lower than the fundamental frequency and the third harmonic at least 28 dB below the fundamental.

Reliability. This variable is quoted in terms of mean time between failures (MTBF). Reliability figures are usually available for equipment rather than services.

Throughput. Mathematically, throughput is the ratio:

$$\frac{\text{Information Bits Correctly Received}}{\text{Second}}$$

Throughput is not a variable quoted by circuit vendors because it is affected by the number of overhead bits transmitted, which is a function of the protocol. Throughput is, however, an essential variable in assessing the performance of a data communications network. The method of calculating throughput is explained in Chapter 17.

Improving Circuit Performance

Assuming a data circuit is operating according to specifications, circuit performance can be improved in two ways, by obtaining digital circuits or by line conditioning. Digital circuits, available from most telephone companies and interexchange carriers, accept digital data signals directly from the source without need for a modem. Types of digital circuits are discussed later in this chapter.

Line conditioning can be applied to an analog circuit by the telephone company to minimize the effects of amplitude, harmonic, and envelope delay distortion. Type C conditioning, which comes in five different levels from C1 to C5, tightens circuit requirements to reduce envelope delay and to equalize amplitude distortion. Type D Conditioning controls the amount of harmonic distortion. Figure 11.1 illustrates these three types of distortion. Note that circuit conditioning is applied only to dedicated circuits. Dial-up circuits can take so many different routes to the destination that conditioning is not possible. Modems often, however, include adaptive equalization circuitry that compensates for distortion and makes high-speed communication over nonequalized circuits practical.

SELECTING DATA COMMUNICATIONS SERVICES

Five architectural alternatives are generally available for carrying data communications:

FIGURE 11.1
Types of Distortion in a Telecommunications Circuit

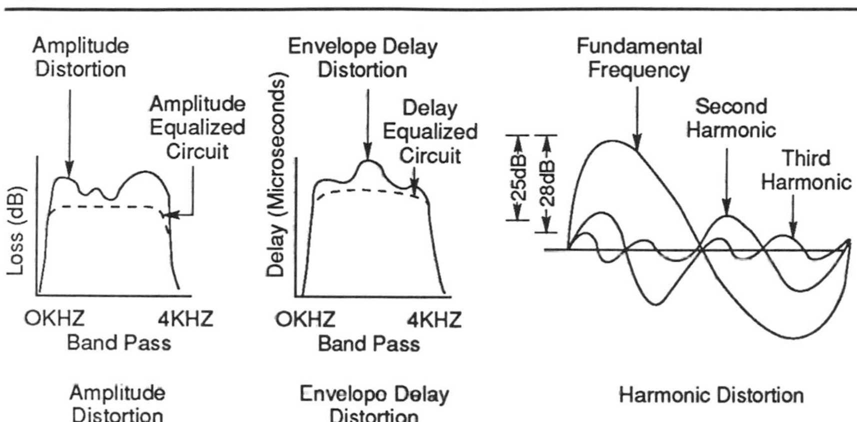

- Dial-up
- Polled multi-drop
- Point to point
- Hierarchical
- Packet-switched

This section discusses the characteristics of these alternatives and explains when each is most likely to be effective from both a cost and service standpoint.

Dial–Up

The public switched telephone network (PSTN) is used for a great deal of data communications. Where usage is not measured, dial-up can be a very attractive alternative to dedicated line service. Some companies dial a connection at the opening of business and leave it connected throughout the business day. If the cost of two business lines (one at each end of the connection) is less than the price of a dedicated line and service is not measured, this alternative is attractive.

When service is measured, calculating the effectiveness of a dial-up line is simple if you know the usage volume. Assume, for example, that a company places ten data calls per day to a location 200 miles away. The DDD cost is \$.28 per minute with a one minute minimum. A dedicated line costs \$230 per month. The method of calculating which alternative is most attractive is:

1. Determine the total number of calls per month (22 days X 10 calls = 220 calls per month).
2. Determine the long distance cost per month. Include the cost of the local telephone number if applicable. If the telephone line is used for other purposes, its cost can be ignored. In this example assume that each call requires one minute. The total usage would be 220 minutes at $0.28 per minute, or $61.60.
3. Compare the costs of the two services.

In the above example, dial-up is obviously the way to go, but if data calls are longer, a dedicated circuit is least expensive. The break even point is:

$230 per month / $.28 per minute = 821 minutes of usage.

If the telephone line is used only for the dial-up service, its cost must be added. If the line costs $50 per month, the break even point drops to 643 minutes of usage.

An important consideration in evaluating dial-up service is the billing increment charged by the interexchange carrier. As explained in Chapter 10, some classes of long distance are billed in six-second increments with a six-second initial period. Other services bill a thirty-second or one-minute minimum period and may round up to the nearest minute for fractional periods of six seconds or more. These billing methods have a substantial effect on the feasibility of dial-up service.

Call set-up time is a major drawback to dial-up data communications. It often takes longer to dial a data call than it does to complete the transaction. A further disadvantage is that dial-up may not be compatible with some forms of protocol. Some systems are designed to be connected only over a dedicated network.

Dial-up circuits have limited bandwidth, which, for many years, restricted the practical maximum full-duplex data communications speed to 1,200 b/s. Within the past few years CCITT has recommended V.22bis and V.32 standards, which support full duplex communications at 2,400 and 9,600 b/s respectively. The 9,600 b/s speed approaches the theoretical maximum for the bandwidth of a dial-up circuit, and is not likely to be exceeded.

Security is inherently low with dial-up circuits. Elaborate precautions must be taken at the host to prevent unauthorized persons from gaining access to the data base and obtaining confidential material or damaging the data. One method of overcoming this problem is the use of dial-back modems in which the remote terminal calls the host, identifies itself, and the host dials it back. Security in this arrangement is good, but the double telephone call is costly. If all traffic is from the host to the terminals, security is not as great a problem as when terminals are accessing a single host.

One company that makes extensive use of dial-up data is a truck servicing company. The company matches shippers who have cargo to transport with truckers who are seeking a load. Video display terminals, similar to those announcing airline schedules in airports, are located in truck stops throughout the country. Shippers wishing to post loads call the company on an 800 number with the details of the cargo. The company's computer dials the appropriate truck stops and sends a short data message that is automatically displayed on the screen. To link all the truck stops on the network with dedicated lines would be prohibitively expensive. Dial-up service is an ideal alternative for this application.

Dial-up data communication is advantageous under these conditions:

Where traffic volumes are light and usage is occasional

Where usage is not measured, or is measured in short increments

Where multiple locations must be accessed by a single host, but not simultaneously

Where a location must access multiple hosts

Where call set-up time is not critical

Polled Multi-Drop

In a multi-drop circuit the terminals share a common transmission medium as shown in Figure 11.2. The host computer or front-end processor (FEP) sends polling messages to each controller to determine whether it has traffic to send. The host or FEP also selects terminals to download host-to-terminal traffic. Terminals are connected to the network through controllers attached to analog bridges, which are devices that match the impedance of the circuits and connect all controllers to the common transmission medium. When a controller is polled, it sends traffic if it has any; if not, the controller responds negatively and the host polls the next controller in turn.

The primary advantage of a multi-drop circuit is the reduction in circuit costs that results from sharing the transmission medium. Where

FIGURE 11.2
Polled Multidrop Network

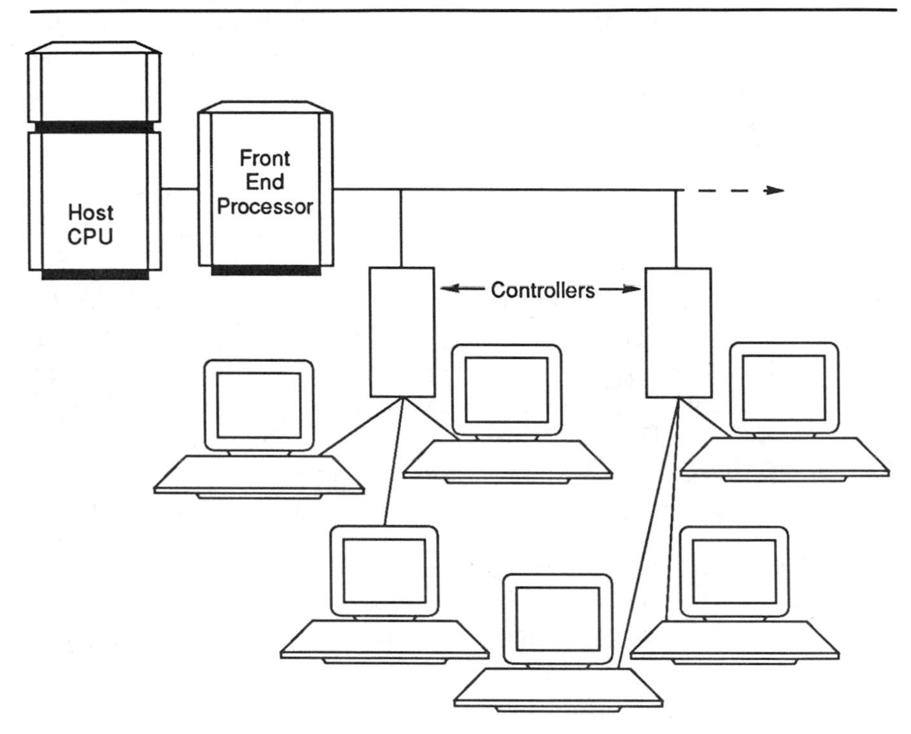

terminals are idle a substantial portion of the time, the circuit capacity would be wasted if it was not shared with other terminals. The primary disadvantage of a multi-drop circuit is the amount of overhead or non-information bits that occupy the circuit capacity. A substantial amount of overhead is inherent in multi-drop circuits, and is proportionate to the number of devices on the circuit, the length of the polling message, and circuit equipment variables that are described in Chapter 17.

Even if none of the terminals has traffic to send, the circuit has a built-in minimum delay known as the *poll cycle time*. Traffic adds to the poll cycle time, and if traffic is excessive, terminal response time deteriorates to an unacceptable level. As explained in Chapter 17, it is possible to model mathematically a multi-drop circuit and approximate the terminal response time within reasonable bounds.

Security is moderately low in a multi-drop network. Because the circuits are dedicated, the general public can be kept from accessing the network, but anyone who has access to the transmission medium can easily monitor any messages. The only effective way of ensuring security on a multi-drop network is encryption.

Multi-drop circuits tend to be effective under the following conditions:

- Circuits are long with clusters of terminals at distant points
- Traffic volumes are light
- The host computer architecture requires multi-drop

Point-to-Point Circuits

Point-to-point circuits, an example of which is illustrated in Figure 11.3, are an effective way of connecting devices with a large amount of traffic to send. The full bandwidth of the circuit is dedicated to the connection between terminal and host, and throughput can easily be calculated as described in Chapter 17.

If the full bandwidth of the circuit is not needed by a single device, the circuit can be subdivided with a multiplexer to form multiple circuits over a single backbone point-to-point circuit. Multiplexers are of two types; *time division* and *statistical*. A time division multiplexer (TDM), sometimes referred to as a synchronous multiplexer, gives each data device a full-time path to the host. During times when the device is idle, its circuit time is wasted. A statistical multiplexer is able to assign channel time, up to the capacity of the circuit, to devices when they have traffic to send. Therefore, a statistical multiplexer can support more terminals than a TDM.

FIGURE 11.3
Point-to-Point Circuit

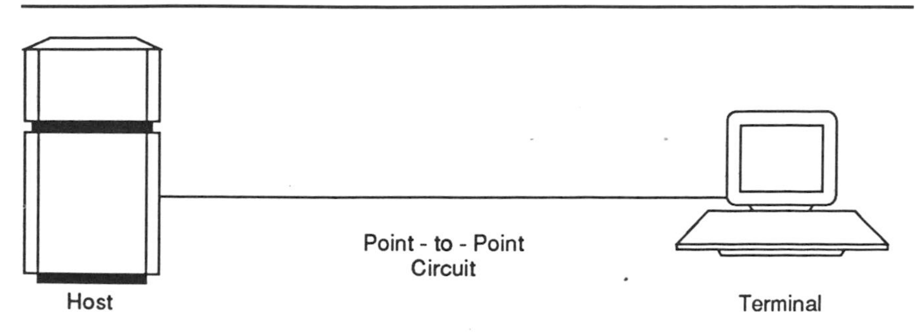

Security is good with a point-to-point circuit. The only way a transmission can be intercepted or interrupted is by gaining physical access to the circuit itself.

The primary disadvantage of point-to-point circuits is the cost. Unless the traffic volume is enough to keep the circuit occupied a high percentage of the business day, some method of circuit sharing is required.

The conditions under which point-to-point circuits are justified are:

- High traffic volume or high speed transfers between processors
- High security requirement

With multiplexing, point-to-point circuits are justified whenever circuit costs exceed multiplexer costs. Multiplexing is particularly effective in asynchronous networks because the multiplexers provide error detection and correction, which are not otherwise available with asynchronous transmission.

Hierarchical Circuits

A hierarchical network uses some form of switching device such as a circuit switch, a switching multiplexer, or a hierarchy of data communications devices such as FEPs and controllers to connect devices to the host. Figure 11.4 illustrates two types of hierarchical networks. The first is a fixed network similar to IBM's Systems Network Architecture (SNA). In this type of network, every session passes through and is controlled by the host. With IBM's support of Logic Unit 6.2, communication between peers is supported, which eliminates the need for all communications to go through the host.

With IBM's dominance in the market, the majority of data communications takes place over this architecture. From the standpoint of the telecommunications manager, SNA simplifies the design and selection task because it is relatively inflexible and offers few choices.

A second type of hierarchical network uses an intelligent switch to either route a bit stream to one destination or another, or to switch an entire circuit. Most of the major manufacturers produce switching multiplexers, which have the ability to route data to multiple destinations. Other devices such as data PBXs, digital PBXs, or digital tandem switches can also route data on a circuit-switched basis.

FIGURE 11.4
Hierarchical Networks

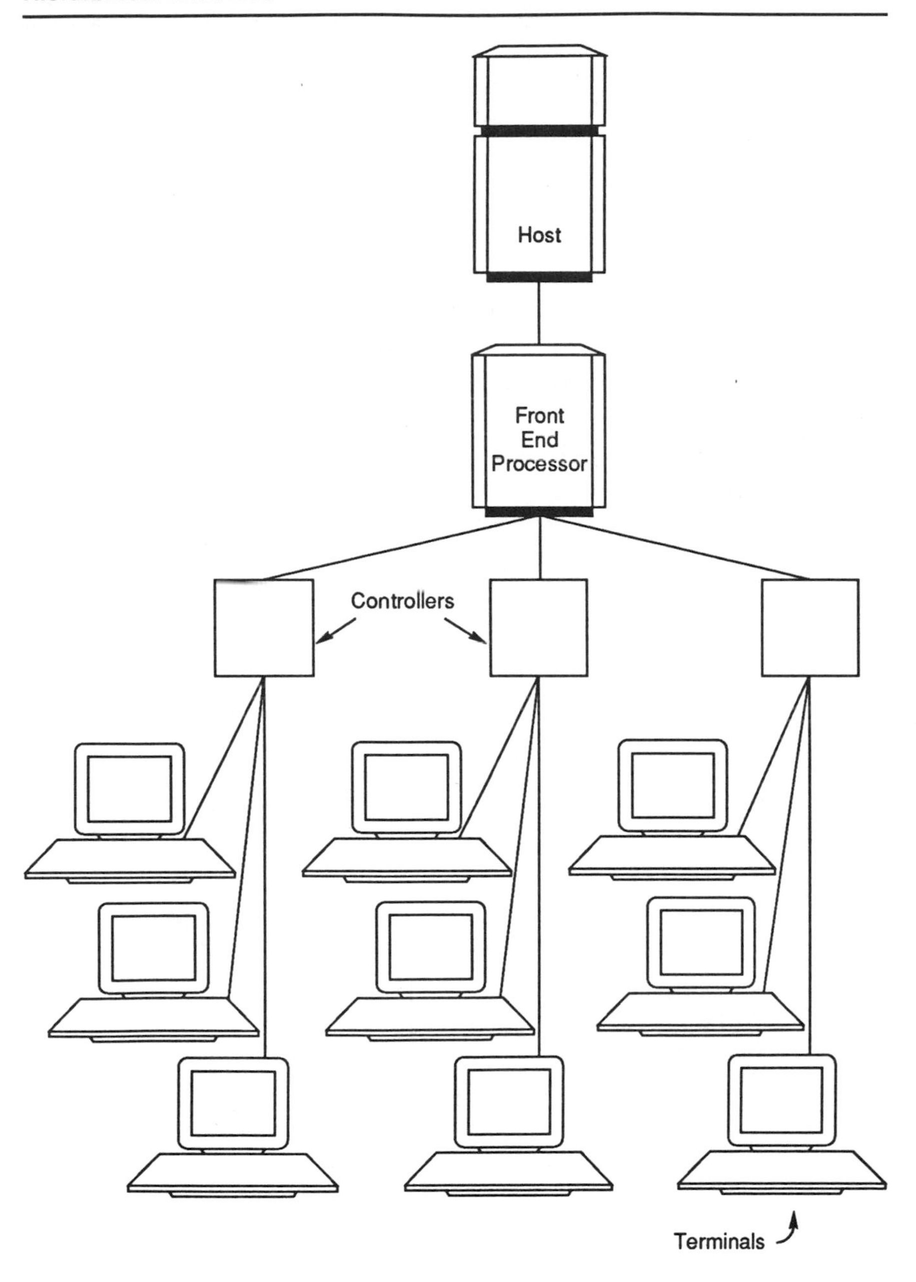

FIGURE 11.4—*Continued*
Hierarchical Networks

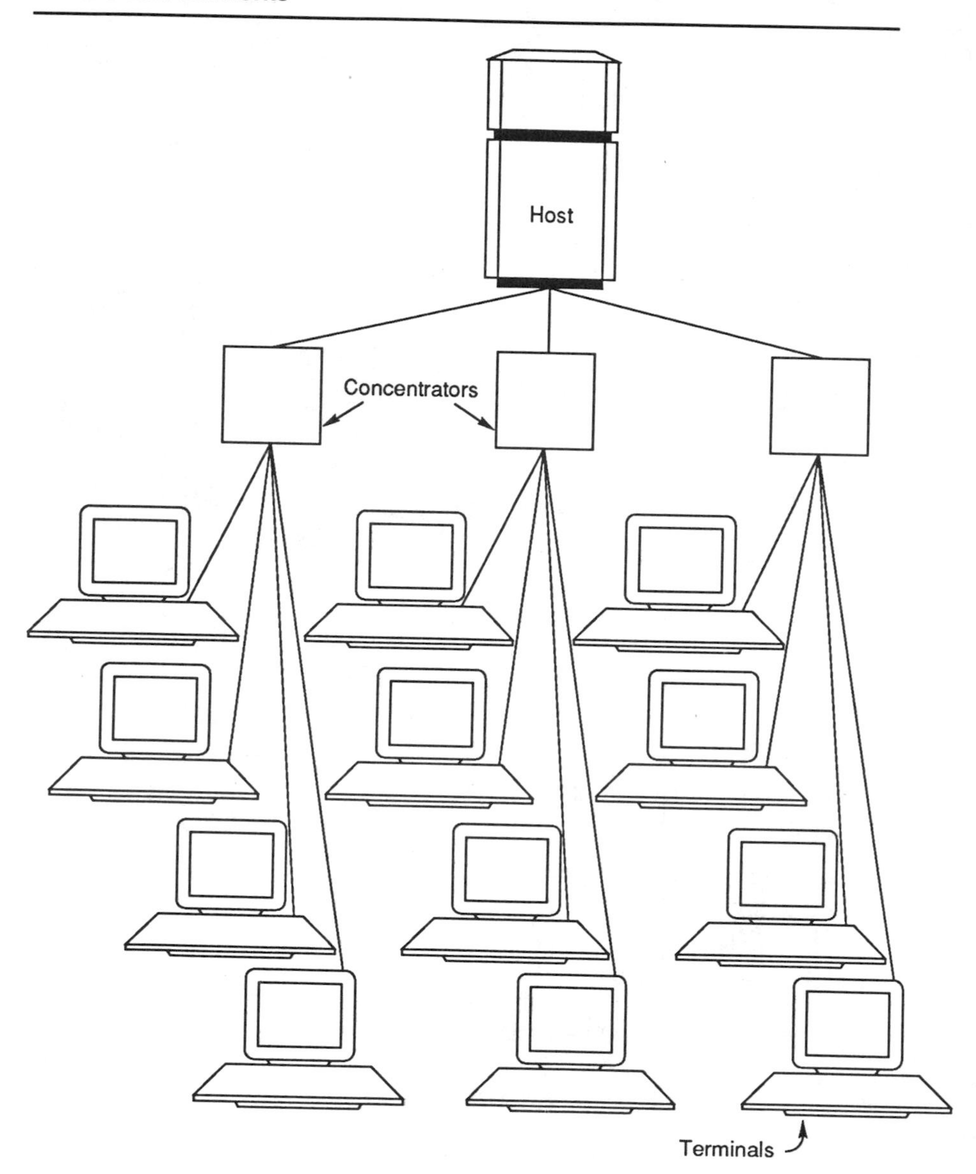

A hierarchical network is often a good compromise when point-to-point circuits are required, but traffic volume is not enough to justify the cost. The trade-off is relatively simple; the cost of multiplexer or concentrator is balanced against the cost of the circuit by using a topological optimization algorithm. Such algorithms are described in most data network design books, several of which are listed in the bibliography.

The primary disadvantage of a hierarchical network is its complexity and inflexibility. Circuits and devices cannot easily be rearranged, but the rigid structure makes it easier to test and isolate trouble.

The primary applications for a hierarchical network are:

- When it is required by the computer vendor's architecture
- When devices need to access multiple hosts or other destinations
- When traffic volume is high

Packet Switching Services

Packet switching is available over both public value added networks (VANs) and private networks. Devices are connected to the network through a packet assembler/disassembler (PAD). The PAD takes data streams from its attached devices and slices them into packets. A packet header containing addressing, control, and routing information is appended, a trailer is added with error correcting information, and the packet is handed off to the node for delivery to the destination. At the receiving end, the PAD reassembles the original data stream from the arriving packets and delivers it to the receiving device. Packet switching nodes are equipped with specialized computers that receive packets and route them to other nodes based on the address contained in the packet and the routing algorithm of the network.

The primary advantages of packet networks are economy (if the traffic volume is appropriate) and robustness. A packet network inherently has more than one path to carry data, so the network is relatively immune to overloads. The fact that data can take multiple paths to the destination also provides a measure of security. If the network does not have a fixed routing scheme, it is possible to capture all the elements of a session only at the source and the destination. In between, only isolated packets could be intercepted.

Companies with a large amount of data traffic can often justify the cost of a private packet switching network. Companies with smaller amounts of traffic can use a public packet network through services offered by nationwide carriers such as Tymnet and Telenet or local packet networks such as those offered by the Bell operating companies. Access to the networks is usually one of three methods as shown in Figure 11.5:

- Dedicated X.25 connection from a host computer to the node
- Dedicated X.25 connection from a user-supplied PAD to a network node
- Dial up or dedicated circuit to a carrier-supplied PAD

While dial-up can be used for originating messages, some packet networks do not deliver traffic over a dial-up connection. The recipient must have a dedicated connection to the provider's location.

Packet switching networks tend to be advantageous under the following conditions:

- Short, bursty traffic
- Low traffic volumes to widely dispersed points

SELECTING LOCAL AREA NETWORKS

In recent years demand has developed for some kind of network to support data communications over a narrow range, usually within a building or campus. The network is required for four primary reasons:

- To facilitate sharing of files including gaining access to files on mainframes and minicomputers
- To enable sharing high-cost peripherals such as printers and file servers
- To enable sharing high-cost software such as word processing and spreadsheets
- To gain independence from the products of a single vendor
- To enable workers to communicate with one another over electronic mail and messaging services

The above objectives can be accomplished with virtually any combination of architecture, access method, transmission medium, and

FIGURE 11.5
Packet Switching Network Showing Access Alternatives

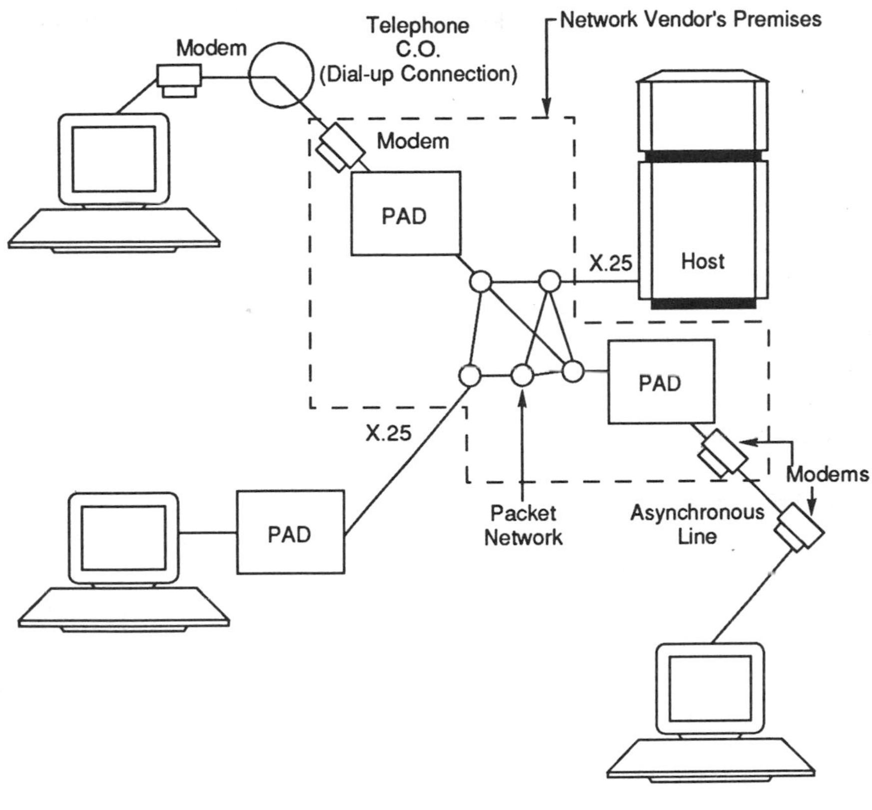

modulation method, although each combination of technologies has characteristics that make it more or less suitable, depending on the application. Selecting a LAN is a matter of compromise. No technology is universally applicable.

As the demand has developed, products have proliferated, offering many choices of method. For example, PBX vendors contend that a digital PBX is all the LAN most users need. Computer vendors provide most of the functions of a LAN through multiuser computers. Data PBXs can fulfill many LAN requirements, and on top of these alternatives are families of true local area networks. There are so many ways to fulfill local data communications requirements that you must have some kind of screening process to narrow down the options. This section suggests some general principles that you can follow to choose among LAN alternatives. The principal criteria most users seek when evaluating a LAN are:

- Reliability
- Performance under heavy system traffic
- Service and support from the vendor
- Expandability
- Multi-vendor connectivity
- User friendliness

Standards

The standards problem is more acute with LANs than with wide area data networks. Although there are numerous noncompatible wide area data products on the market, at least there are a few dominant suppliers and architectures that narrow your choices. With LANs the choices are almost infinite.

The matter of standards is particularly confusing. The IEEE 802 standards have been widely accepted, but there are three basic variations: a contention bus, a token bus, a token ring, and variations within each standard. For example, 802.3, which is the contention bus standard, recognizes thick RG 6/U cable, thin RG 59/U cable, and twisted pair wire. Further complicating the standards picture is the fact that 802 standards cover only the first two layers of the International Standards Organization's Open Systems Interconnect (OSI) model. These layers, the physical and data link layers, form only a transport mechanism. Higher order protocols are required to form a complete network, and unfortunately, standards for the higher protocols are not part of the 802 standard.

Some widely used protocols such as the Department of Defense's Transmission Control Protocol/Internet Protocol (TCP/IP) and Microsoft's Network Basic Input/Output System (NETBIOS) are gaining the status of *ad hoc* standards. Not all products support these protocols, however, so the objective of vendor independence remains an unattainable ideal.

Access Protocol

All LANs provide a method for multiple terminals to share a common transmission medium. Three principal methods are used in LANs; carrier sense multiple access with collision detection (CSMA/CD), token passing, and switching. Each method has characteristics that make it acceptable or unsuitable for certain applications.

Carrier Sense Multiple Access With Collision Detection

CSMA/CD is a protocol with no master control; all stations are peers. This protocol is often called Ethernet, which is a trade name of Xerox, Digital Equipment Company, and Intel. Stations listen to the network before transmitting, and if it is idle, are free to transmit a packet. Sometimes two or more stations transmit simultaneously, in which case the packets collide and are mutilated. When collisions occur, the stations back off for a random interval, and attempt to retransmit. Packets are sent to all stations simultaneously; the addressee copies the message and all others discard it.

The main advantage of CSMA/CD is its simplicity. Controllers are built into boards that plug into personal computer expansion slots or into transceivers or access boxes arranged for RS-232-C interface. Because there is no central controller, a CSMA/CD network is robust. A terminal or cable failure can disrupt the network, but failure of a terminal does not affect more than one device. Because the network is simple it is relatively inexpensive and easy to implement.

The primary disadvantage of CSMA/CD is its lack of assurance of timely data delivery. Because stations contend for one another for network access, it is possible that in heavily loaded networks data delivery may be delayed. One or two stations with heavy transmission loads can dominate the network.

The primary applications for CSMA/CD have these characteristics:

- Tolerance for some delay in data delivery
- Personal computer networks where ease of application is important

- Interface with mainframe network protocols such as DECNET
- Messages are short and bursty
- Widely distributed asynchronous terminals where wiring is difficult

Token Passing

Unlike the statistical access method of CSMA/CD, token passing is a deterministic access method that ensures that every station obtains its fair share of network time. A software mark, or token, circulates throughout the network in a predetermined pattern. Only the station in possession of the token is permitted to transmit, so collisions are eliminated.

Token networks are more complex than CSMA/CD. Provisions must be made for regenerating the token if it is lost or mutilated, such as would happen if a station failed while it possessed the token. One station must be designated as network control, but to prevent disruption of the entire network if the control station fails, every station is equipped to assume control responsibilities.

Token passing networks are implemented in either a ring or bus topology. In a ring topology stations are connected in series through a closed loop. The transmitting station removes the token and replaces it with a message. All stations regenerate the message bit-by-bit and place it on the ring; only the addressee copies the message. The originating station removes the message from the ring, replaces the token, and passes it to the next station in turn.

In a token bus network, all stations are bridged in parallel across a common transmission medium, usually a broadband coaxial cable. Packets are broadcast to all stations, but like a CSMA/CD network, only the addressee copies the message. Possession of the token denotes permission to transmit. After a message is transmitted the token is passed to the next station in a logical ring.

The primary advantage of token passing is its assurance of data delivery. Each station is given a proportionate share of network resources if it needs them. In exchange for this capacity, token networks are more complex and somewhat more expensive and difficult to implement. Applications using token passing protocol will usually have these characteristics:

- Assurance of data delivery is mandatory
- Messages are more uniform in length and volume

Circuit Switching

Although data PBXs and digital PBXs are often used for local network applications, they are not, strictly speaking, a local area network. Nevertheless, their applications with LANs overlap to some degree, and a switched network has advantages that may make it ideal in some cases.

In LANs, the terminals share a common transmission medium. In a switching system the transmission medium, usually twisted pair wire, is run to each device from a central location, and the connecting circuits are switched together on command from the terminal to the central controller.

The primary advantage of circuit switching is that the transmission medium is not shared. The parties to a session have the exclusive use of a connection through a common switching network for the duration of the session. When the session is completed the network is released for use by other devices. In some cases this is less expensive than other forms of local network.

The primary disadvantage of circuit switching is the lack of interactivity between the network and its attached devices. In a LAN a personal computer can save and retrieve files to and from a file server as if the server was directly attached to the personal computers. Unless a circuit switch has an option for directly attaching to a LAN, it is unable to provide this kind of functionality. It can only transfer data, which means the personal computer's user must prepare a file for transmission through the switch and send it to the file server.

A second disadvantage of circuit switching is that the network is controlled by a central device which, if it fails, disables the entire network. Another disadvantage is the need to run twisted pair wire from every device to the central location. In widely distributed networks, switches are available with distributed switching units, which limit the length of the wire runs.

The primary applications for a circuit switched network tend to have the following characteristics:

- Devices have no need to share a central file server
- Privacy of the connection is an important consideration

- Twisted pair wiring is already in place or easy to install
- The network can tolerate the loss of a central control station

Modulation Method

LANs either impose data directly on the transmission medium (baseband modulation) or superimpose the data on a radio frequency carrier (broadband modulation). Broadband LANs are generally more expensive than baseband, but they are capable of supporting more than one LAN on the same cable, or of carrying other services such as video, point-to-point data, or in some cases, voice. Aside from the capacity differences, there is little difference to the end user between broadband and baseband LANs. The same access protocols are used with minor differences.

The primary deciding factor between broadband and baseband is what other services in addition to data need to be transported. If a cable is in place for closed circuit television, security monitoring, or other such service, it can often be used to support a LAN for relatively low cost.

The choice of modulation method might also be made because of preference for a particular manufacturer. Some manufacturers specialize only in broadband or baseband, and a particular product or technology might be the only choice offered. This could be particularly important if the hardware you plan to use supports only one of the two methods.

Transmission Medium

The preferred transmission medium is another important consideration in selecting a LAN. The three media, twisted pair, coaxial cable, and fiber optics, have distinct ranges of applications. Although there is some overlap among the three, each fits more readily into certain environments than the others.

Twisted Pair

The primary advantages of twisted pair are its ease of installation and the fact that a considerable amount of excess wiring already exists in many buildings. Although twisted pair was rarely used in early local area networks, it has rapidly gained acceptance with the introduction of products that use existing telephone wiring under the right conditions. Twisted pair wiring is more susceptible to interference from external sources than either coaxial cable or fiber optics. A properly balanced twisted pair circuit is relatively immune to interference, but balance is easily disrupted by improper installation, taps, or unbalanced terminal equipment.

In any data network there is a tradeoff between transmission speed and network diameter. Twisted pair networks support lower transmission speeds and narrower range than other media. Twisted pair LANs generally support transmission speed of 1 mb/s or less over ranges less than 800 feet. At the time of this writing, several vendors have announced LANs that operate at 10 mb/s over twisted pair. The technology has been introduced, and the range that can be achieved is controversial, but is generally conceded to be a maximum of 300 feet.

The attractive feature of using twisted pair for a LAN is the fact that many buildings are already wired, or if they must be rewired, they can be wired with twisted pair for a lower cost than coaxial cable or fiber optics. Many telephone systems were wired with 25 pair cable, some of which is usable for a LAN. If the cable has been properly terminated, it will support many types of baseband LANs using either CSMA/CD or token passing protocol. Proper installation means that the wire is tightly twisted, it contains no bridging conductors, and any cross connections at terminals are carefully balanced and neatly dressed. Splices of all kinds are generally to be avoided.

If new wiring is being installed for the telephone system, as covered in Chapter 13, it pays to consider installing it for a LAN at the same time. Many LANs use twisted pair wiring from a wiring closet to the terminal, and some specify twisted pair between wiring closets.

Coaxial Cable

Coaxial cable is somewhat more difficult to install than twisted pair wire and is considerably more expensive per running foot. Compared to wire, however, coaxial cable offers an increase in capacity that far exceeds its greater cost. Coax is an unbalanced medium. A single center conductor is surrounded by an insulating medium and a shield. Special taps connect the terminal to the coax through a transceiver.

Coaxial LANs are implemented with either "thick" or "thin" cables. Thick cable, one-half inch or more in diameter, is relatively inflexible, and in some types of coaxial systems is rigid. Thin cable, about the diameter of a pencil, is flexible and easier to install, but it has less bandwidth and supports fewer devices over a narrower range than thick cable. Thin cable is looped past every device on the network, which is acceptable in open work spaces, but difficult to achieve when devices are located in enclosed offices.

Besides having greater bandwidth, coaxial cable has much greater noise immunity than twisted pair. A properly installed cable can survive in most high-noise environments. Care must be taken with baseband systems to ensure that they are grounded only once per cable. Multiple grounds can cause undesirable currents to flow, resulting in data errors and hazards to personnel and equipment.

Fiber Optics

The transmission medium with the greatest bandwidth and noise immunity is fiber optics. It suffers from the disadvantage, however, of higher cost and greater difficulty of installation. It is also very difficult to tap, which tends to limit its usefulness to a star topology. It can also be used in a ring, but it is not practical for a conventional bus with taps at each node.

Fiber optics is likely to become the preferred transmission medium for LANs in the future, but at present it is impractical for most LANs except those requiring high speed, long range, and noise immunity. It is also used for linking segments of CSMA/CD networks.

Topology

Topology is the least important consideration in selecting a network. It tends to be inherent with the network selected, and is not a driving force in the network itself. Four topologies, illustrated in Figure 11.6, are commonly used in local area networks. The star topology is used with all circuit switched networks. It is also used in fiber optic and twisted pair CSMA/CD networks.

The bus topology is used with both broadband and baseband Ethernets and in token bus networks. The ring topology is used with token ring networks linked by twisted pair wire. The branching tree network is used with broadband coaxial systems; each branch is connected to the trunk through an impedance matching device.

Connecting Asynchronous Terminals

If a host computer is connected to the LAN, the issue will arise of what to do with existing terminals. The typical problem facing most companies is the high cost of long coaxial cable or RS-232-C cable runs, congested

FIGURE 11.6
LAN Topologies

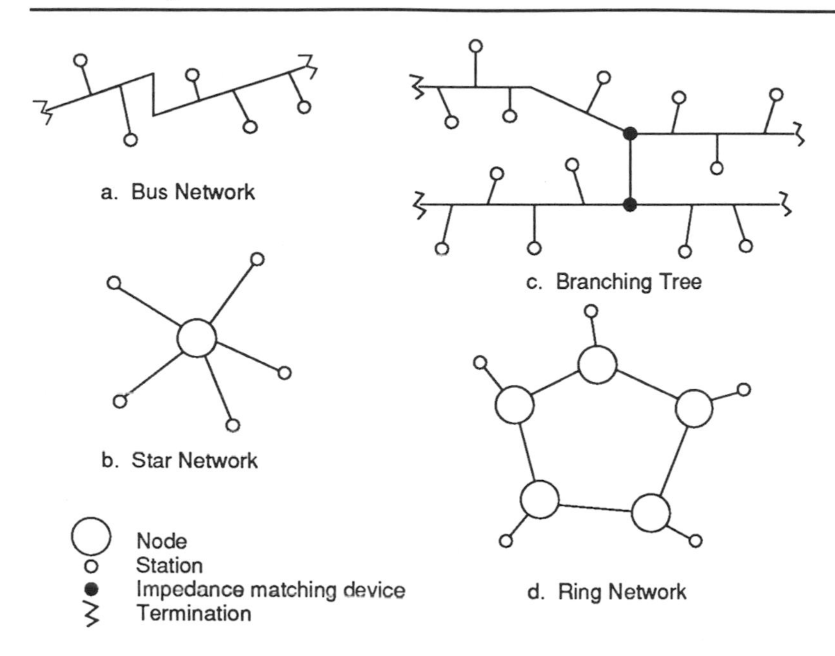

raceways and conduits, and the difficulty of moving workstations. The LAN is often seen as a solution, but it is not always feasible or desirable to connect terminals to a LAN.

Part of the problem stems from the nature of asynchronous communications. Data is transmitted one character at a time, which, in some LANs, results in a packet being transmitted for each character. The problem is particularly acute if the host computer uses echoplexing so that each character results in two packets being placed on the network. The most extreme case is with Ethernet or IEEE 802.3. The minimum data block size is 40 characters. If the data block isn't that long, the protocol pads it to the required length. In conjunction with normal packet headers, this means that the network could be required to carry as many as 75 characters of overhead for each information character. A lightly loaded network can tolerate the overhead of asynchronous terminals, but an excessive number of terminals will affect service.

The problem is aggravated with a protocol such as Ethernet that cannot tolerate overload. As load increases, throughput diminishes with Ethernet, and eventually reaches a point where it drops to zero. Token passing networks are less susceptible to degeneration from an asynchronous terminal load because the packet overhead is less, and because the protocol always retains some throughput, even under heavy load.

A solution to the asynchronous terminal problem is to add a terminal server to the network. A terminal server accepts and buffers asynchronous data, packetizes it, and releases it to the network in larger packets. If your network uses echoplexing, however, be certain that the terminal server echoes the characters or keyboard operations may be adversely affected by the time delay of the server.

SELECTING DATA COMMUNICATIONS TRANSMISSION SERVICES

A variety of different media are available for carrying both voice and data communications. The characteristics of the dial-up network have already been discussed. Where dedicated channels are required a number of options are available. This section discusses what the market offers and when each service is generally feasible.

Analog Private Lines

The vast majority of data communications circuits are carried over analog private lines furnished by the local telephone company and an interexchange carrier. Figure 11.7 shows the layout of a typical circuit, known by most common carriers as a "3002" circuit. An analog private line has a nominal bandwidth of 0 Hz to 4000 Hz, with a practical bandwidth of approximately 300 Hz to 3300 Hz. With these bandwidth limits, complex modulation methods are needed to carry high-speed data. A 3002 circuit can support half-duplex data communications up to 19.2 kb/s. This speed is approaching the theoretical maximum of a 3002 circuit.
Besides speed limitations, analog circuits have a higher error rate than their digital counterparts; the longer the circuit the higher the probability of error. The twisted pair cable, microwave radio, and coaxial cable that support analog circuits are amplified at regular intervals, and each amplification increases both the signal and noise. Noise, which causes most data errors, is cumulative through amplifiers.

Despite the noise, most data is sent over analog circuits because they are so readily available. Data circuits can be provided at a reasonable price in small quantities at virtually any location that has telephone service. Services are offered by all local telephone companies and most interexchange carriers. The rates are based on five elements shown in Figure 11.7, two local channel charges, two central office termination charges, and an interexchange mileage charge. The cost is proportionate to distance.

Digital Private Lines

Some of the bandwidth and error-rate limitations of analog private lines can be overcome by using digital circuits. These circuits are derived over T-1 carrier by the local telephone companies and connected to digital radio or fiber optic circuits of the interexchange carrier. The signal is

FIGURE 11.7
3002 Data Circuit

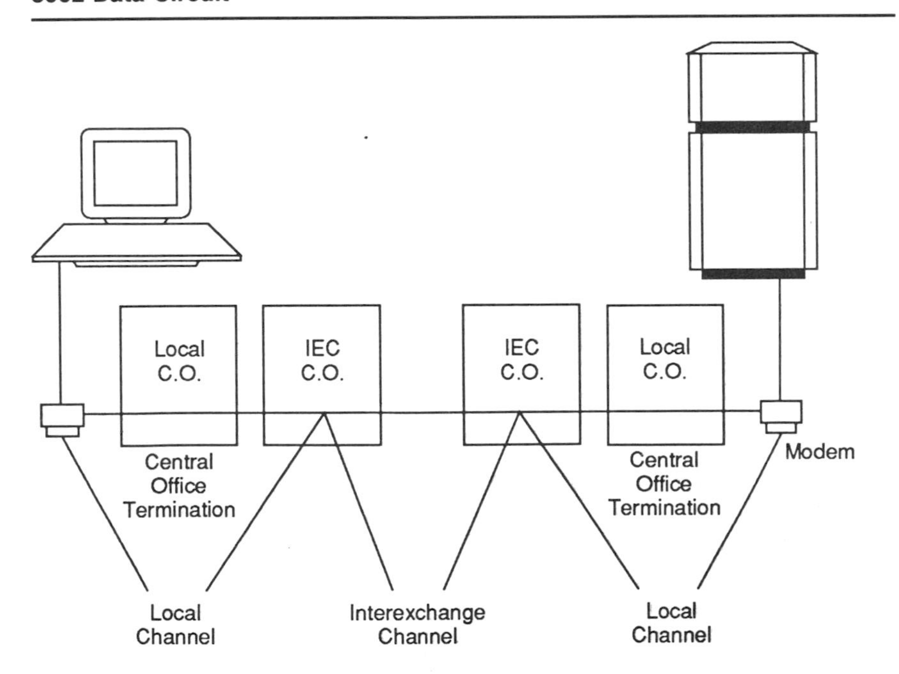

regenerated at each repeater point so that a digital circuit does not suffer from the cumulative noise effects of an analog circuit. The error rate is dependent on the transmission medium, but is normally at least one order of magnitude better than an analog circuit.

The original digital circuit configuration routed all circuits through a series of central hubs that provided testing access. Hubbing is not economical for local digital circuits, so many telephone companies offer a hubless service that connects digital circuits between the user's end points. A local loop carries the digital signal from the telephone central office to the user's premises. A data and channel service unit (DSU/CSU) furnished by the user interfaces the digital circuit and provides signal conversion and testing access. Digital services are available at speeds of 2.4, 4.8, 9.6, and 56 kb/s.

EVALUATING T-1 SERVICE

One of the hottest topics in telecommunications today is the use of bulk T-l transmission services to replace individual private lines for both voice and data communications. A T-1 channel, which offers 1.544 mb/s of bandwidth, provides capacity for 24 64 kb/s channels. This section describes the methods of terminating and subdividing a T-l channel and explains how to use it for combinations of voice, data, and video transmission services.

Architecture of a T-1 Channel

The elements of T-1 channels are shown in Figure 11.8. The line, furnished by the telephone company or other common carrier, is connected over twisted pair cable, fiber optics, or microwave radio. T-1 services are available on both terrestrial and satellite services. The line is terminated in a Channel Service Unit (CSU) that changes the T-1 line coding to a format used by the terminating equipment. A CSU, a picture of which is shown in Figure 11.9, also provides testing and line-powering functions. The T-1 bit stream is demultiplexed into voice and data channels by a channel bank or a T-1 multiplexer.

If multiple digital lines are terminated on the same premises, they are usually connected through an access panel known as a digital system cross-connect (DSX) panel. The purpose of the DSX is to provide a

FIGURE 11.8
Elements of a Private T-1 System

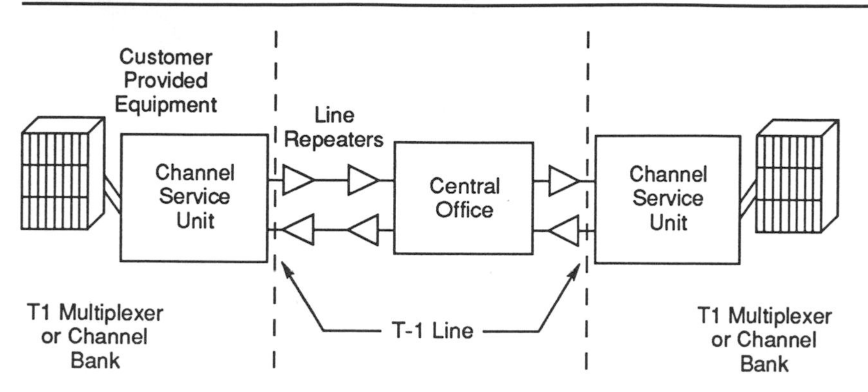

testing and patching point. The DSX is a convenient place for patching to spare lines or spare digital terminating equipment.

Digital Channel Banks

A digital channel bank is a non-intelligent device that separates the bit stream into 24 channels. Each channel has a bandwidth of 64 kb/s, but only 56 kb/s is usable for data transmission because of an in-band signaling system that robs the least significant bit in every sixth frame. Channel banks are packaged in 24 or 48 channel increments. The most common bank is the D-4 bank which packages two 24-channel digroups in a single shelf.

The channel bank can be equipped with a variety of different plug-in units to accommodate different signaling systems and to directly interface digital data. The voice interfaces are analog in all channel unit types except for the dataport units. The talking and signaling paths can be connected to analog ports on PBXs, private line extensions, or any other standard 2- or 4-wire telephone interface.

Dataport channel units have digital interfaces that come in standard speeds of 2.4 kb/s, 4.8 kb/s, 9.6 kb/s, or 56 kb/s. Dataport channel units occupy the entire bandwidth of the voice channel. For example, a 2.4 kb/s dataport unit uses the entire 64 kb/s of the voice channel; the unused bits are not available for information transfer, and are therefore wasted.

If additional speeds are required either in a dataport or a conventional channel unit, they can be derived with a multiplexer. For example,

FIGURE 11.9
Kentrox T Serv™ Channel Service Unit

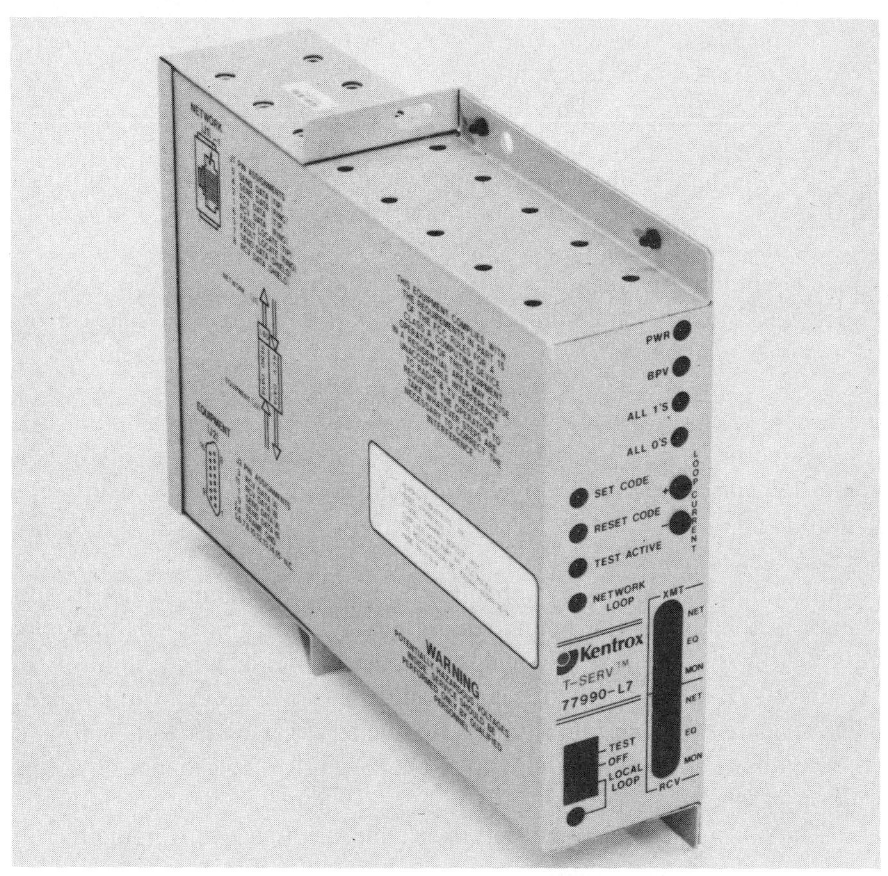

Photograph courtesy of Kentrox Industries, Inc.

a 9.6 kb/s dataport unit could be subdivided into four 2.4 kb/s channels with a time division multiplexer.

T-1 Multiplexers

A T-1 multiplexer performs many of the same functions as a channel bank, but has more intelligence and versatility. Plug-in cards for T-1 multiplexers can support lower speed channels and voice channels using different modulation forms to impose more voice channels in the 1.544 mb/s bit stream.

T-1 multiplexers also have drop-and-insert capability which allows you to split off one or more channels, to extend them to a distant location. When channel banks are used on a T-1 line, the drop and insert capability can be provided only by using two channel banks connected back-to-back. T-1 multiplexers also have service-protection capabilities built into most models. For example, redundancy is frequently built into common equipment and power supplies. Multiplexers may also have switching capability so an attached terminal can choose alternate routes through the device. They also may have the ability to collect statistical information for measuring load and traffic flow. T-1 multiplexers generally are about twice as expensive as channel banks, but in many applications the versatility is worth the additional cost.

Low Bit Rate Voice Equipment

The voice-channel capacity of the T-1 system can be multiplied by using low bit rate voice equipment. The most common method is adaptive differential pulse code modulation (ADPCM), which is the low bit rate voice method that has gained CCITT approval. ADPCM doubles the capacity of a T-1; depending on the signaling method used, either 44 or 48 channels are achieved on a single T-1 line. The gain in capacity is offset somewhat by a reduction in data handling capability. Analog data in excess of 4800 b/s cannot be carried over an ADPCM channel. For voice purposes, however, the difference between ADPCM and a normal T-1 channel cannot be detected by ear.

Other proprietary methods of low bit rate voice are produced by other manufacturers. Continuously variable sloop detection (CVSD) and delta modulation are the most widely used, and yield about the same gain in voice channel capacity as ADPCM.

Digital Access Cross-Connect System

When a network of T-1 channels is used, it is often necessary to split the bit stream to route individual voice channels to alternate destinations. If a digital access cross-connect system (DACS) is placed at the junction of several T-1 circuits under direct or software control, it is possible to redirect voice channels either permanently or temporarily. A DACS is a specialized type of digital switch. As shown in Figure 11.10, a bit stream coming from one source can be split without the use of back-to-back channel banks.

FIGURE 11.10
Digital Access Crossconnect System (DACS)

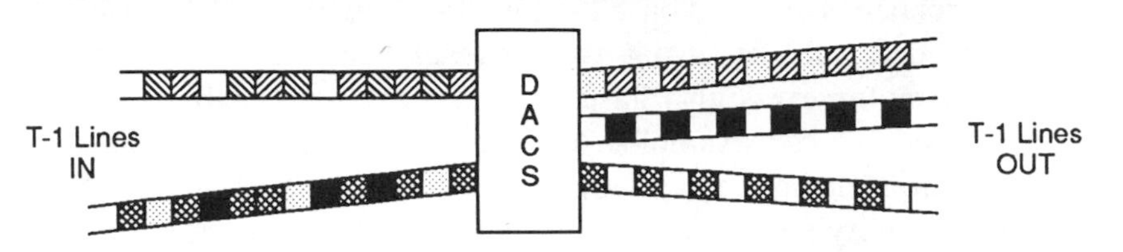

DACS has the capability of separating channels in an incoming bit stream and putting them out on other channels. (Only one direction of transmission is shown).

DACS takes the place of a wired cross-connect panel. Its primary advantages are the ability to reconfigure the network easily from a remote location, and the elimination of channel banks at the hubbing location. The T-1 circuits interface the switch directly, and the switch has access to the individual 64 kb/s bit streams. A DACS enables you to use a T-1 network for voice communication during the day and rearrange it for data communication outside normal working hours. If temporary overloads occur, circuits can be re-routed around the source of blockage or they can supply additional capacity to alleviate congestion.

DACS is widely used by common carriers to configure their T-1 networks. It is also usable in many complex private networks. A company can either purchase its own DACS or lease capacity from a common carrier. Many local telephone companies sell digital cross-connect capability, and since the T-1 lines usually terminate in their central offices, this can be an effective way of obtaining capacity without the capital investment. The important question to ask is the degree to which you are given independent control over the facility. To make the most effective use of DACS, you should have the ability to control your own T-1 circuits without intervention of the common carrier.

Comparing T-1 Services to Analog Private Lines

T-1 services offer flexibility and versatility compared to individual analog private lines, but there are still many instances where analog lines are more effective. In small concentrations of service, T-1 services are too expensive. The monthly cost of a T-1 line is not the only consideration in choosing services. To the cost of the circuits must be added termination equipment such as channel banks and CSUs. Analog circuits require modems, and individual digital circuits require channel service units to terminate the lines.

SWITCHED 56 KB/S SERVICE

High speed switched data has not been widely applied, generally because services are not readily available to most users. AT&T offers a service called Accunet™ Switched 56 that is available in locations served by a 4ESS toll switching office. The user must run a 56 kb/s line to AT&T's central office; the inability to switch through a local central office and use the local loop is one of the principal drawbacks of the service. Applications are expected to develop as the service becomes more widely accepted. Switched 56 kb/s service supports services such as high speed Group 4 facsimile and slow-scan video conferencing in addition to high speed data transmission.

CABLE TELEVISION SYSTEMS

If a cable television system (CATV) is designed for data transmission, it can be an effective transmission medium. It is particularly effective when multiple locations are involved because the signal is broadcast throughout the area served by a single headend. The signal is modulated to a specific frequency, broadcast throughout the network, and is picked off at any location by a receiver tuned to the channel frequency.

The cable system must be designed for two-way operation to support data transmission. Only a single cable is usually used; the frequency band is split into the upstream direction (from terminal to the headend) and downstream (from the headend to the terminal). Figure 11.11 shows the layout and principal components of a CATV system equipped for data transmission.

The data signal is applied to the cable through a radio frequency (rf) modem. The modem converts the digital signal to an analog rf signal, and couples it to the cable through a matching device. A full-duplex channel is derived by transmitting on one frequency and receiving on another. The headend equipment receives a signal from a terminal, shifts its frequency to the downstream channel, and retransmits it. All stations on the network

FIGURE 11.11
Diagram of Cable Television System Equipped for Data Transmission

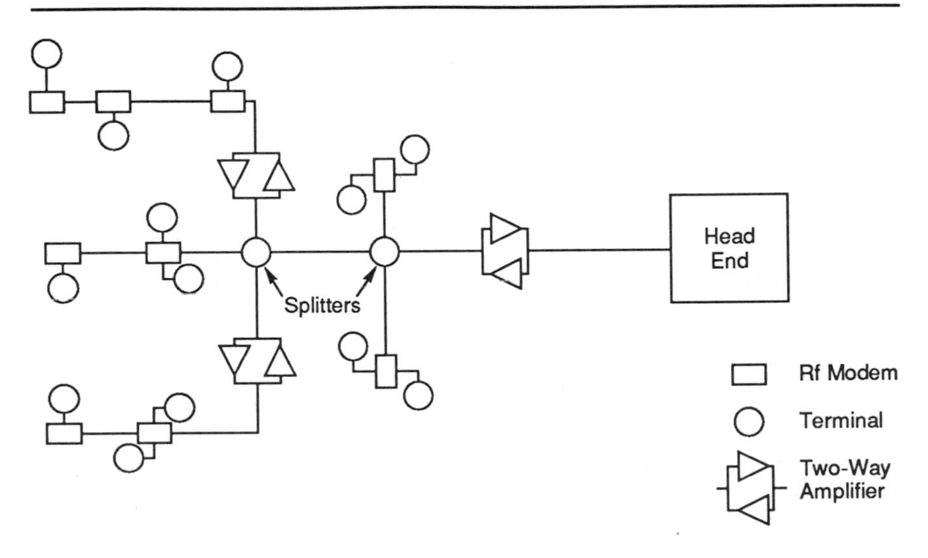

are able to receive the transmission, which may have security implications.

CATV systems are capable of supporting four types of transmission; point-to-point data and voice, packet switched data, T-1 service, and analog video. The applications for these types of services are identical with the applications described earlier under long haul circuits. Two examples should illustrate the application of different types of CATV services.

CATV Applications

A suburban library system serves eight libraries with a central computer that is used for patron registry, book checkout, and access to a data base of library materials. Public terminals can be used by patrons to search for author, title, book number, or subject. All the sites within the library system are served by a two-way CATV cable. Each library is connected to the central with a 19.2 kb/s channel. Terminals and printers are connected through a multiplexer that is attached to a radio frequency (rf) modem. Patron registration is handled over asynchronous terminals. Bar code readers are used for checking books in and out.

The same CATV system serves a suburban fire department over a packet switched network using a CSMA/CD-like protocol. Terminals and printers at each fire station are attached to the network through rf modems. The 911 dispatch center, equipped with a data base that relates addresses to the nearest fire station and hydrant, is also attached to the network. When an alarm is received, the dispatcher types in the address and other details of the alarm. Pertinent information is extracted from the data base and relayed to the fire station. The network has a considerable amount of excess capacity, enough that the fire stations use it for administrative purposes including word processing.

CATV Summary

CATV can be an economical alternative for data communications under the right conditions:

- *Medium speeds:* CATV supports up to 19.2 kb/s over wider distances and for less cost than analog services obtained from the telephone company
- *Properly Equipped CATV System:* The cable must be equipped for two-way communication
- *Multiple Points:* CATV is generally a less expensive multi-point medium than the local telephone network because the signal is broadcast throughout the cable coverage area
- *Narrow Ranges:* The CATV network normally covers a franchised area, and unless arrangements are made to connect to another system, the service area is restricted compared to the area served by the telephone company

A CATV system may be more vulnerable to failure than other types of services. Amplifiers are connected in series, and when an amplifier fails, service may be disrupted or impaired. Depending on whether the cable has a status and alarm reporting system, restoral time may take longer than other types of services. Before you contract to use a CATV network, you should find out from the cable operator what their reliability and availability figures are.

SUMMARY

This chapter has presented an overview of the principal choices available for implementing data communications services. More comprehensive discussions can be found in several books listed in the bibliography. A key point to remember in evaluating data communications services is the great amount of commonalty that is developing between voice and data, particularly in the transmission facility. Any network that is being constructed today should at least consider merging voice and data on the transmission side. The trend is toward bulk circuit provision, and by combining voice and data requirements, you will often find that a single T-1 channel is less expensive than fewer than half the number of analog channels. By acquiring circuits in bulk, you can obtain a great deal of extra capacity for little increase in cost.

CHAPTER 12

CONTRACTING CONSIDERATIONS

All telecommunications managers, whether their job descriptions say so or not, are involved to some degree in contracting. Not even a company that is totally self sufficient in manufacturing, installing, maintaining, and administering its telecommunications system can avoid contracting for telecommunications services. In companies that have their own contract specialists, telecommunications managers are still involved in specifying the technical content of contracts, which means that they must understand some of the elements, benefits, and pitfalls of contracting.

This chapter provides an overview of some of the principal issues involved in telecommunications contracting. If there is a single message you should gain from this chapter it is this: contracting is no place for amateurs. Properly administered, your effectiveness can be multiplied many times by judicious use of outsiders to furnish goods and perform services that are beyond your capabilities. Unwisely administered, contracts can tie you up in litigation and can cost many times the value of the service. You won't gain all the knowledge you need about contracting from this or any other book, but you should gain the knowledge of how to use an expert to help you pick your way through the minefield.

The contracting role of the telecommunications manager is a natural consequence of the procurement function we have discussed in earlier chapters. The culmination of a procurement action of any kind is a contract, whether it is in writing or not. Except for certain contracts covered by the Statute of Frauds, oral contracts are fully as enforceable as written ones, even though it may be more difficult to prove the intent of the parties when the agreement is unwritten.

Contract law bestows certain rights and imposes certain obligations on all parties to the agreement. The law does not, however, protect you from making unwise business decisions, and you may find yourself or your company the uncomfortable defendant in a suit that has nothing to do with fairness, but is concerned only with whether the language or intent of your contract is enforceable. In this chapter we will cite instances of how unwary managers have cost their companies many times the amount they saved by not obtaining expert assistance when they should have. These cases are disguised to protect the identities of the companies, but are nonetheless real.

As a telecommunications manager you will undoubtedly find yourself involved in one or more of the following contracting situations:

- Purchasing telecommunications equipment
- Contracting for consulting services
- Contracting for installation and repair services
- Purchasing software or contracting for its development
- Contracting to perform or supply services for others

Besides the above list, you will undoubtedly venture into many types of agreements such as employment contracts that have nothing to do with telecommunications. To limit the scope of this discussion, we will omit the mention of such contracts, but you should be aware that they exist.

EQUIPMENT PURCHASES

Any time you purchase new or used equipment, you inherit the obligations and protections of a body of law documented under the Uniform Commercial Code (UCC). The UCC has been adopted by all states except Louisiana, which has adopted portions of the code. This means that most transactions are carried out under its provisions. Most commercial transactions result in the satisfaction of both parties. Goods are exchanged for money, and ideally each party feels that it has received more value than it gave up. When satisfaction does not result, a dispute arises, and the rights of the parties are subjected to scrutiny under the UCC, either with or without legal assistance. In this section we will touch on only a few of the UCC provisions that most frequently affect telecommunications contracts. Note that the UCC applies to purchases of goods, but not of services.

Warranties: Expressed and Implied

When goods change hands, they are covered by either a written (expressed) or unwritten (implied) warranty. The expressed warranty is documented by the manufacturer or vendor, and states the terms and conditions under which defects will be corrected and what, if any, charge will accrue to the buyer. You should always request that a copy of the warranty statement be included with the vendor's response to RFPs. Before equipment is purchased on a directed purchase (one made without an RFP), you should read the warranty before you buy. Most used equipment is sold "as is," without warranty. If the seller has made any promises about the equipment, these should be in writing because verbal warranties on used equipment are difficult or impossible to enforce.

The UCC provides for two classes of implied warranty on new equipment. The UCC language refers to warranties of *merchantability* and *fitness for a particular purpose*. The warranty of merchantability means that the product meets the generally accepted industry standards to which that product is manufactured. Fitness for a particular purpose means that the product is warranted to fulfill the purpose for which it was manufactured, but not for other purposes that were not intended.

The purchase of an ordinary telephone instrument illustrates the principles of implied warranty. A standard push button telephone is expected to perform certain functions to meet the standard of merchantability. It should dial without error, ring reliably, transmit and receive voice signals with satisfactory volume and intelligibility, and be sufficiently durable to survive under conditions of ordinary usage.

The telephone is warranted to be fit for a particular purpose, connection to a single party line in the telephone network for the country in which it is marketed. It is not warranted to operate in other countries, and may not operate with some multi-party ringing systems. It also is probably not warranted for use as a remote data entry terminal unless it is sold as such or becomes so by customary usage.

The UCC assumes that the implied warranty provisions apply to an equipment purchase unless the parties agree to waive them. For the waiver to be enforceable, it must be written, generally in bold print so it cannot easily be overlooked. In consumer transactions where it can be shown that the buyer did not understand the effects of accepting the waiver, the courts may not release the seller from its implied warranty obligations even though the contract language is clear. In business transactions, however, the buyer is assumed to be sophisticated enough to understand the effects of waiving the implied warranty.

In all contracts you should be wary of language in the seller's warranty that reads "THIS WARRANTY IS IN LIEU OF ALL OTHER

WARRANTIES, EXPRESSED AND IMPLIED, INCLUDING WARRANTIES OF MERCHANTABILITY OR FITNESS FOR A PARTICULAR PURPOSE." Language such as this is used to negate the implied warranty. You may have no choice, however, but to accept it or shop for a competing product that does not require you to waive your UCC warranty rights.

Software warranties almost invariably include an implied warranty waiver. In the case of a system such as a PBX, where software is not separable from the system, the waiver provisions may be particularly onerous. One way of dealing with the problem of warranty waivers is to detail your requirements in the RFP and incorporate both the RFP and response as part of the purchase contract. That way, it is clear that you and the vendor intended a particular result from the equipment.

The case of a residential hotel illustrates the hazard of purchasing solely under the vendor's contract. The hotel purchased a PBX with a specialized call accounting system manufactured by another company. The call accounting system purportedly included a feature package that allowed the hotel to mark up long distance calls and bill them to the tenants each month or when they checked out.

After a few months of operation, and after the warranty had expired, the hotel discovered that some calls were being omitted and others were being under-rated. The vendor, which was a distributor for both the PBX and the call accounting system, made numerous ineffective attempts to correct the problem, and, because the warranty had expired, the hotel was billed for the vendor's work. The PBX manufacturer disclaimed any responsibility for the problem and offered to solve it by installing its own call accounting package, which was significantly more expensive. The call accounting manufacturer disclaimed responsibility because it said the package was being used in a manner not intended.

The vendor in this case had made no written representation about compatibility of the call accounting system and the PBX. Hotel management had purchased the PBX without an RFP, and with no contract other than the seller's standard agreement, which limited their liability and disclaimed the implied warranty. Although the call accounting system was merchantable, it clearly was not intended for the purpose for which it was sold. The lack of language to document the understanding of the buyer and seller limited the hotel's ability to obtain satisfaction. After losing revenue, it replaced the call accounting system at its own expense.

The Battle of the Forms

The vast majority of commercial sales are consummated under vendors' contracts. The reason for this is understandable; if vendors had to renegotiate a contract for every sale, their costs would be prohibitive. Most sales involving a small amount of money are carried out with no contract at all beyond the manufacturer's warranty and the provisions of the UCC. In contracts for larger items of equipment, the vendor, seller, or both, attempt to dictate the terms of the transaction by the use of standard forms.

On the vendor's part the standard form is printed on the back of the sales agreement; on the buyer's part it usually consists of preprinted terms on the back of a purchase order. Whether you are a buyer or seller, you are undoubtedly aware that contracts are designed to protect the interests of the party who prepared the agreement. The question of whose language prevails is for the courts to decide, based on the facts of the case. You should, however, be alert to conflicts between the forms, and if the language makes a difference, be prepared to negotiate the conflict. The time to resolve any disagreements over form language is before the contract is signed, not after.

Negotiating and Writing an Equipment Contract

Unless your company is large and possesses purchasing power, you will probably have little choice but to purchase telecommunications equipment under the seller's contract. The contract will give the seller all the protection it can reasonably attain, and will probably give the buyer little more protection than it is automatically entitled to under the UCC. As we discussed previously, if the RFP is worded to require the vendor to automatically adopt the affirmations it makes under its RFP response, you will obtain a measure of protection that may supersede the more restrictive terms of the sales contract. In addition, the sales contract is apt to be silent with respect to provisions that may be vital to the success of your project.

Completion Dates

The contract will probably provide blank space for entering the completion date, and will excuse the vendor for nonperformance because of unforeseeable events such as work stoppages and *force majeure*, which is an unexpected event or "act of God" that prevents contract performance.

If the consequences of delay are severe, you should attempt to negotiate a liquidated damages clause in the agreement. A liquidated damages clause is similar to penalty clauses that contracts often included in years past. The courts, however, have frequently failed to support penalty clauses, so purchasers have tended to insert wording such as the following into their contracts:

> "In case of delay in meeting dates in this contract, buyer and seller agree that buyer will suffer damage, the amount of which will be difficult to ascertain. The parties agree that an amount of $____ per day shall be paid by the seller as liquidated damages."

Your attorney can choose the proper wording to use.

Insurance

Most vendors maintain enough liability insurance to satisfy claims for damage that occurs on your premises. The contract should require the vendor to maintain liability insurance, ensure that its subcontractors are covered, and possibly name your company as an additional insured.

Payments to Subcontractors

More than one buyer has been found liable for payments to subcontractors when the primary contractor has defaulted. The consequences are not severe unless you have already paid the prime contractor. If the prime contractor defaults on its obligations to subcontractors, you may be held liable for payment even though the prime contractor has already been paid. The remedy is to withhold payment until you have evidence that subcontractors' claims are satisfied. The contract should be worded accordingly.

Payment Schedule

It is infinitely easier to negotiate performance claims before final payment has been made instead of after. The vendor will typically require a percentage of the contract amount when the agreement is signed, another portion when equipment is delivered and installed, and a final payment on acceptance. These terms are reasonable if you have a clear understanding of what constitutes acceptance. If acceptance terms are written into the contract, you should be entitled to withhold final payment until the vendor has demonstrated that it has fulfilled its part of the agreement.

Acceptance Terms

No general rules can be written about what constitutes acceptable performance for telecommunications equipment because acceptability depends on what you as the buyer want and what the vendor has promised to deliver. For a PBX or key telephone system you would typically require the vendor to demonstrate operability of all features and functions, after which a failure-free period of specified length should ensue. A local area network or data communications system might be accepted on the basis of throughput or response time under controlled load conditions. A digital microwave radio system might be judged on its technical performance such as frequency stability and bit error rate. A T-1 network should deliver the vendor's claimed reliability and error free second performance. With any type of equipment you should consider what variables constitute acceptable performance, determine how they are measured, and write these into the contract.

Documentation

The standard documentation that comes with telecommunications equipment will, in most cases, be insufficient for you to install and maintain the system. As discussed later, even if you have no plans to perform this work yourself, the lack of such documentation may preclude the future use of a third party maintenance contractor. Therefore, the agreement should specify exactly what kinds of documentation you are entitled to receive. Most vendors charge for this material, so it may not necessarily be in your best interests to receive it before it is needed. Wording in the contract entitling you to documentation on request is excellent insurance if it is needed in the future.

Continuing Availability of Product Support

Many companies have purchased telecommunications equipment, only to find that parts for repairs and growth are no longer available, therefore rendering the equipment functionally obsolete. Every agreement should require the vendor to support the product with spares for a specified time. The vendor will be unable to provide any better service than the manufacturer guarantees unless it maintains its own spare stock after the manufacturer has discontinued the product line. This wording in a contract is no better than the stability of the manufacturer, but may earn you support you would not otherwise receive.

CONTRACTING FOR CONSULTING SERVICES

Most businesses eventually engage a telecommunications consultant to perform a special study, substitute for temporarily absent staff, or furnish advice and assistance outside their own capabilities. In this section we will discuss how consulting services are customarily performed and how to go about selecting a consultant. It may seem obvious that the first step is to know clearly what services you want a consultant to perform, but it is not unusual for a manager who has only a superficial knowledge of telecommunications to be unaware of the services that are available. Telecommunications consultants generally provide the following kinds of services:

- Perform needs assessments
- Prepare telecommunications plans
- Evaluate requirements and specifications for equipment and services
- Prepare requests for proposals for equipment and services
- Perform economic and technical evaluation of responses to RFPs
- Design telecommunications networks
- Perform long distance analysis to choose the optimum service configuration
- Manage users' telecommunications networks
- Implement new equipment and services

Of course, not all consultants provide all the services listed above. Many are specialists that offer a narrow range of services and have little interest or capability of providing a full range. Services usually range from offering advice and assistance on an hourly basis to taking full responsibility for implementation of a project. Consultants customarily deal under one or more of the following types of agreement:

- Fixed price
- Hourly rate
- Hourly rate with price ceiling

- Retainer fee
- Contingency or "share the saving" basis

In this section we will discuss the factors you should consider in determining what kind of contract to enter with a consultant. As we discussed earlier, the contract does not have to be in writing to be valid and enforceable. Whether you should insist that it be in writing depends on your company's policies and the method by which you secure consultants' services.

Initial Consultant Interview

The most effective way to obtain the services of a consultant is to hold a series of interviews with firms that appear to have the qualifications you are seeking. Qualified firms can be located through ads in the telephone book, lists published by services such as *Datapro*, or by word of mouth referrals. The latter method is preferable, and can often be obtained through associations of telecommunications managers or other professionals. You can also obtain lists of consultants that belong to the Society of Telecommunications Consultants (STC), a professional organization that is discussed later. The address of the STC is included in Appendix J.

In the initial interview you should expect to obtain the following information from the consultant:

- Description of range of services offered by the firm
- Brief resumes of the principal members or employees of the firm
- A list of client references and brief description of the nature of the project
- An explanation of consultant's method of billing and rates
- Any conflicts of interest such as vendor relationships
- Consultant's code of ethics
- Professional associations to which the firm belongs

During the interview you should be prepared to express to the best of your ability what you want the consultant to perform. To obtain a reasonable quotation from the consultant it is essential that you convey your expectations and the result you want.

In the initial interview you should pay particular attention to ways the consultant can use standard processes to fulfill your requirements. Most firms have standard processes for preparing RFPs, performing economic analyses, developing plans, evaluating long distance services, and other such aspects of the consultant's job. The degree to which your expectations mesh with a standard process will determine the cost and complexity of the project. Whether payment is by the job or the hour, consultants base their quotations on their estimate of the time the project will take. If it is necessary to develop new processes, either because your requirements are unique or because the consultant has not performed this type of project before, the process will be developed at your expense. Pay close attention to standards and ask to see examples of previous work. Do not fall into the trap of assuming your requirements are so unique that a new procedure must be developed. To some degree your business is unique, but there are probably many similarities with previous projects a qualified consultant has performed.

Most users want an estimate of the project cost during the initial interview. If the project is simple, you can probably get a reasonable approximation, but if it is a complex project, you should not rely on an estimate during the interview because too many uncontrollable factors are at work for the consultant to give more than a rough approximation. Instead, ask for a written proposal, and be prepared to furnish information such as telephone bills, project plans, and problem statements to assist the consultant in determining the scope of the project.

Consultants' Proposals

It is always advisable to request a written proposal from consultants before determining who to award the project to. You may choose to specify what the proposal is to contain, or leave form and content to the consultant's discretion. In either case, the proposal should contain the following:

- A reiteration of information about the firm obtained during the interview
- Statement of consultant's understanding of the scope of the project
- A description of the work to be done
- A description of the results to be delivered
- Schedule of major events or milestones
- Cost of the project
- A statement that the consultant will accept no fees or valuable consideration in conjunction with the project from anyone other than the client

In addition, the response may include a statement of the methods the consultant expects to employ, the extent of involvement of the user's staff, expectations of progress payments, and details about why that firm feels it should be selected.

It is often advisable to break the project into phases and to request separate quotations for each. This is advantageous to both the user and the consultant. It gives the user the option of stopping the project at the end of any phase if it is not advantageous to proceed or if the wrong consultant has been chosen. From the consultant's standpoint, it is often difficult to provide an accurate estimate of the cost of later phases until earlier phases are complete. For example, many projects begin with a feasibility study or the development of a preliminary plan. The scope of the total project is often impossible to predict until these planning phases are complete.

Your request should state the basis under which you want the consultant to estimate costs. As listed earlier, a variety of costing methods are available, and the choice should be appropriate to the nature of the project.

Fixed Cost

A fixed cost agreement is logical when the scope of the project is clear to both you and the consultant. Under the fixed cost method, you gain the benefits of competition, and frequently consultants will reduce their bids to get the award. In contracting for consulting services, however, one rule is always valid: *never* award a project solely on the basis of lowest cost. Award it on the basis of best value.

Consulting firms vary greatly in their abilities to handle certain types of projects, and unless you are sophisticated enough to do the project yourself, you will find it almost impossible to determine whether you got the optimum result. Also, don't make the mistake of attempting to shave the project so closely that the consultant has little opportunity for profit. The probable result is that the consultant will take shortcuts that you won't discover until later.

The fixed price method has other disadvantages that you should consider before applying it. The most obvious is that if the project turns out to be less complex than it appears, you have no opportunity to share the savings. Also, the fixed price method requires you to have a clearly documented understanding with the consultant. When unforeseen circumstances arise, they will be covered by a change order. It is no secret that contractors of all kinds, not just consultants, bid low on projects that have obvious omissions and expect to make much of the profit with change orders. As a general rule, you have less control over a fixed price agreement than other types. You specify a result, the consultant agrees to deliver, and any deviations will result in additional expense. Nevertheless, when you can clearly express the result you want, and the project has measurable standards of performance, the fixed price agreement is a reasonable way to contract for consultant services.

Hourly Rate

The hourly rate consulting contract is prized by consultants, just as the cost plus contract is sought by builders. The hazards are obvious: the consultant has little incentive to economize, and to control costs and preserve the result, you must expect to participate to the maximum degree. Despite the hazards, some projects have an indefinite scope and do not lend themselves to other forms of agreement. When you want to purchase the expert assistance of a consultant to lend advice or to perform a technical investigation of some kind, the hourly rate is probably the only feasible method. In effect, you are entering a labor contract.

If the project is lengthy, you may be able to negotiate an hourly rate that is much lower than the consultant normally charges for more speculative projects. Even though the firm publishes a rate schedule, the prices are probably negotiable if the magnitude of the project is great enough.

Hourly Rate with Price Ceiling

This approach has most of the characteristics of the hourly rate and the fixed price agreement combined, and may, under the right circumstances, overcome the deficiencies of both. The buyer is able to create a contract with less detail than the fixed price agreement. Although the consultant has no incentive to economize up to the limits of the agreement, the uncertain nature of the work in most contracts will lead consultants to be as economical as possible with the expectation of profiting up to the limit of the agreement. This method does not solve the change order problem. Any work outside the agreed-upon scope is still subject to additional payment if the total cost of the project exceeds the limit. This method offers the buyer more control than the straight fixed cost agreement because it is possible to specify tasks rather than results. This can be either a benefit or a disadvantage, depending on the type of contract.

Retainer Fee

Engaging a consultant on a retainer for conducting a long term project or continuing work such as administration of a telecommunications system is often advantageous. Both the consultant and the client can find a retainer to be an advantage. In exchange for an assured source of income, the consultant usually performs work for a lower cost than on a project-by-project basis. As the client, you are assured of the consultant's availability, and gain the continuity that such an agreement offers.

Before entering a long term agreement with a consultant, you should be sure that the consultant has the capabilities you require and sufficient staff to meet your needs. You should also be sure you have enough consulting work to justify the expense, and that the nature of your project is continuing rather than one-time. If these conditions are met, engaging a consultant on a retainer may be an excellent alternative. The retainer contract or agreement should call for specific services to be delivered. The agreement may also require a minimum level of activity per month, and require the consultant to deliver certain results each month, such as review of long distance bills, evaluation of service on telecommunications systems, review of a subcontractor's work, or other such service.

Contingency or Shared Savings

Some consultants offer to evaluate your telecommunications systems on a contingency fee basis. Their fees are based on a percentage (usually half) of savings they discover after reviewing your telephone bills and services. Savings typically result from overbilling by common carriers and from reconfiguring local and long distance services to reduce the number of trunks or use of lower cost services.

You should be aware that most consultants do not deal on a contingency basis for reasons we will explain. To the client, the contingency offer has a superficial appeal; if no savings are found, you learn that your system is configured properly and pay nothing for the service. If savings are found, you receive half the benefit for no monetary outlay.

This method has several disadvantages, however, that usually outweigh the advantages to the client. First, if you engage the consultant on an hourly or fixed cost basis, you receive all the savings, not just half. If the savings are significant, the consultant's half will be a substantial reward for relatively little effort. A second problem relates to service. As discussed in Chapter 1, the proper configuration of trunks is always a balance between cost and service. If you are willing to sacrifice service, your telecommunications costs will decrease while business costs increase. The consultant has little incentive to monitor service, which puts his or her interests directly in conflict with the client's.

A third problem relates to the method of calculating savings. Savings that accrue from overbilling are non-controversial; the consultant is paid from common carrier rebates. For savings resulting from reconfiguration of service, the amount of saving is much more elusive. The case of a nation-wide transportation company illustrates the pitfalls of contingency billing.

The company, with offices in three major cities, uses 800 service extensively for access by its customers. The company entered a contract with a consultant to share savings on overbilling and service reconfiguration. The consultant found several cases of overbilling, and was successful in obtaining refunds from the common carriers. She also recommended numerous changes in 800 service bands and quantities of trunks.

The company followed most, but not all of the consultant's recommendations. The contract specified that the consultant would share 50 percent of the savings for five years after implementation. The company was required to submit copies of the telephone bill monthly so the consultant could calculate savings. When several thousand dollars worth of consultant bills were submitted, the company became concerned about the magnitude of the bills and stopped paying them. Eventually, the consultant filed suit for more than $200,000 for her share of presumed savings.

Another consultant, hired by the company to review the billings, found the following problems:

- Shortly after the consultant's changes were recommended, blockage, as measured by overflows registered on the 800 service bills, increased by 100 times compared to the previous years' bills
- Actual savings on changes recommended by the consultant were offset by increases in other aspects of telephone service
- In calculating the base from which savings were measured, the consultant had omitted numerous costs, including the costs billed by the carriers for circuit rearrangement
- Changes made by the company after the consultant's recommendations were implemented masked the consultant's recommendations, so it was no longer possible to compute savings

Although it was clear that the consultant had performed some valuable services for the client and actually achieved savings, the real savings were less than 10 percent of the amount billed by the consultant. Moreover, the cost of service degradation, which undoubtedly resulted in lost revenue, could not be computed. The case demonstrates the serious flaw in the contingency method of billing; the consultant's interests are often directly opposed to the client's interests.

Before you enter this kind of contract, have it reviewed by someone who understands the technical aspects of telecommunications in addition to the usual legal reviews. Pay particular attention to the method of calculating savings, and do not agree to any changes unless the consultant presents documented evidence that service will not be adversely affected. Also, do not agree to calculate savings on only the changes recommended by the consultant without considering costs that may appear elsewhere. For example, if you reduce the number of WATS lines, the usage per line will increase and WATS cost will drop. Calls that overflow to an alternative service such as message telephone service, will raise that cost, and possibly more than offset WATS savings.

Contracting With Telecommunications Consultants

As with other contracts for professional services, contracts with consultants usually result in a fair exchange of service for money received. In the complex world of telecommunications, it is frequently necessary to seek the help of a specialist to perform a service that cannot be furnished by equipment vendors.

The telecommunications consulting business is unregulated, and anyone can enter without demonstrating qualifications. One way of obtaining some protection is by hiring consultants who belong to a professional association. The Society of Telecommunications Consultants (STC) is an association that establishes a code of ethics and certain minimum standards that consultant members must adhere to. Only a minority of consultants belong to STC, however. Some choose not to belong for a variety of reasons, but some are excluded by the Society's insistence that consultants have no ties to vendors. Although lack of STC membership is not a reason to disqualify a consultant, you should at least insist that the firm adhere to its code of ethics, which is reproduced in Appendix J.

PROCURING INSTALLATION, MAINTENANCE, AND ADMINISTRATIVE SERVICES

Most companies contract with outside firms for technical services that are beyond the capabilities of their own staffs. Installation, maintenance, and administrative services are provided by most equipment vendors, and in most metropolitan areas, are also available from third parties. This section discusses contracting for these services versus performing them with company staff.

Contract Installation

Complex equipment such as a PBX is almost invariably installed by the vendor or the vendor's subcontractor. Many vendors subcontract portions of the work of installing a new PBX such as wiring, shipping, drayage, and storage. With the possible exception of station wiring, it is almost never worthwhile to perform these services with your own staff because of the special training, tools, and information resources required.

Station and terminal wiring and installation can often be handled by company staff. The facilities needed to do the job are inexpensive, and limited knowledge is needed to move and change terminal apparatus.

The installation job requires certain information that the manufacturer may not supply as a matter of course. At the least, you must know pin designations on connectors, testing procedures, wiring limitations, and other such information. Unless you obtain this information from the vendor, you will find it difficult or impossible to perform the work yourself, and you may be unable to use the services of a third party contractor. Therefore, the purchase contract should always specify the kind of support information you are entitled to.

Maintenance Service Contracts

With complex telecommunications systems the question will invariably arise whether to maintain the system with your own forces or by contract, and if by contract, whether to obtain the service on call or with a maintenance contract. The question of using company forces or contract forces is essentially an economic matter. Company forces will usually be able to respond to outages more rapidly than contractors, but if your system is widely distributed, contractors may give better response.

In-house maintenance is usually economically feasible only if the staff is available for other purposes or if the system is large enough to justify the costs of equipping and training repair forces. If you choose to maintain the system yourself, the same precautions about documentation should be taken as discussed under installation contracting.

For smaller operations, contract maintenance has several significant advantages. Because they are working on the equipment for several different customers, contract forces are more apt to be currently equipped and trained, and will have superior experience compared to internal staff. The contractor's force must be trained somehow, however, and it may be at your expense. The contract should, therefore, require a minimum standard of competence.

The question of whether to use a maintenance contract or obtain services on call is also primarily an economic one. Most vendors favor contract customers when conflicting priorities arise. Aside from the question of priorities, you will generally pay more for a maintenance contract than paying for services on an hourly rate. A good way to make the decision whether to remain on a maintenance contract is to monitor performance during the initial year while the product is still under warranty. If numerous failures are encountered it may pay to purchase a maintenance contract.

Like sales contracts, most maintenance contracts are written by the seller, and excuse it from nonperformance under some conditions. Review these carefully. The contract should provide you with a guaranteed response time maximum. Generally, you will be given a choice of normal business day or 24 hour-per-day service, with the latter at a premium rate, often about 65% above normal business hour rates. Response times should be specified for emergency and normal maintenance conditions. You should be certain that you have a meeting of the minds on what constitutes an emergency. Vendors' definitions range from 20 percent of the lines and stations down to total system failure. Be sure the definition is spelled out in the agreement. Also, be certain the contract specifies what, if any, parts are included in the cost, and what exclusions apply. If the contract covers parts, it will probably exclude failures caused by other than normal wear and tear.

Third party maintenance agreements, which are performed by contractors other than the vendor, are advantageous under some conditions. If the vendor has withdrawn support of the product, you will have little choice but to obtain services either under contract or on call from third parties. In other cases a third party contractor may provide better service, shorter response time, or be located closer than a manufacturer's distributor.

The same precautions should be observed whether contracting with third parties or the equipment vendor. Here are some questions to ask:

- How many qualified maintenance people does the contractor have on its staff? Are there enough people to cover for vacations, illness, and turnover?
- How many years of experience do they have in total and with the specific equipment?
- Where are spares obtained, the contractor's own stock or from a warehouse? If the latter, where is the warehouse? How long does it take to obtain emergency parts?
- What is the price schedule for normal business day and out-of-hours coverage?
- What is the response time for emergency and nonemergency maintenance? What kind of guarantee is offered? What is the penalty (if any) if the time is exceeded? What have the contractor's actual results been for the past 12 months?
- What does the contract cover in addition to labor?
- What penalty is assessed if no trouble is found or if it is found in another vendor's equipment or services?

In addition to asking the above questions, ask for a list of references and call them to determine how well the contractor actually performs.

SUMMARY

For all but the largest companies, judicious use of contracting is more effective than attempting to perform telecommunications work with internal staff. There are several principles, however, that should be followed to control the contractor's work:

- Define exactly what you want the contractor to do
- Put the agreement in writing
- Specify a measurable result
- Obtain a clear understanding of how the service will be billed
- Know how to evaluate the contractor's work

The final precaution to bear in mind is that contracting law is complex and requires special training to administer. The law offers many protections to one who buys under contract, but it does not protect buyers from entering unwise agreements.

CHAPTER 13

WIRING PLANS AND EQUIPMENT ROOMS

Considerable attention has been paid to the expensive "last mile" of a telecommunications system, the connection between the serving central office and the customer's premises. Even more difficult and expensive in many buildings is the last hundred feet, the building wiring that connects the telephone entrance or PBX to the station. There are no ideal wiring methods. All involve the use of brute force, quantities of copper, and expensive compromises.

In bygone days of electromechanical key telephone equipment, the solution was clear-cut and expensive: pull enormous quantities of 25-pair station wire to connect every telephone to a central location or to distributed wiring closets. Nearly every building built before 1980 is still wired with this cable, which ranges in quality from neatly terminated new wire to unsightly jumbles of undocumented spaghetti whose major value is its scrap content. More recently, electronic key telephone equipment has reduced the number of pairs required to a more manageable size. Most current installations are wired with cable of three to six pairs.

If the telephone wiring problem wasn't bad enough, terminal wiring came along to complicate matters. Conduits and raceways became choked with twisted pair RS-232-C wiring, and the weight of coaxial cable threatened to collapse the suspended ceilings that supported tons of it. About the same time station wires grew skinny, local area networks appeared on the scene to further complicate the station wiring problem. LAN designers decreed that coaxial cables, some pencil-thin, but some stiff and unwieldy, would serve the ubiquitous terminals and the personal computers that were just beginning to capture the attention of building designers.

This chapter discusses the compromises you must make to install a building wiring system. We will discuss wiring plans, which are the methods of connecting wires from source to destination; terminating plans, which are the methods of terminating the wires at both ends; and distribution methods, which are the methods of getting the wire from the suspended ceiling or raceways to the desk. We will also discuss problems of retrofitting older buildings, which present an entirely different problem than new construction.

WIRING MEDIA

Three types of wiring media are in common use: twisted pair, coaxial cable, and fiber optics. The three media have some overlap in their applications, but for the most part the uses are distinct and are an inherent function of the design of the system they support.

Twisted Pair

Twisted pair is the preferred medium for all voice applications. It is also rapidly becoming significant in LANs and is replacing the coaxial cable that connects many synchronous terminals to controllers. A cable consists of multiple wire pairs that are twisted to ensure that each wire has an equal exposure to every other wire in the cable and to external interference. This twist is essential for reducing noise and crosstalk.

Compared to other transmission media, twisted pair is inexpensive, compact, and easy to install. Its primary disadvantages are its narrow bandwidth, which limits the data transmission speed it can support, and its susceptibility to interference. The effects of interference are minimized by shielding the wire or using it in a balanced configuration. When twisted pair is connected to balanced terminating equipment, interfering voltages are equal on both wires of the pair, and balance each other out. Shielding intercepts interfering signals and drains them off to ground.

Some devices, such as synchronous terminals, are intended to be connected to coaxial cable, which is inherently an unbalanced medium. By use of *balun* coils, twisted pair wire can replace coaxial cable for up to 1,000 feet. LAN terminating devices can also be constructed in a balanced configuration so they can be supported over twisted pair wiring.

Coaxial Cable

Coaxial cable is extensively used for connecting terminals to host computers and controllers. It is also used for LANs and video transmission, but rarely for voice except by common carriers in intercity networks. Coax is manufactured in a variety of configurations specific to the application. Thin coax consists of a single center conductor surrounded by an insulating medium, and covered with a braided shield. Such cable, as shown in Figure 13.1, is inexpensive, flexible, and easy to install. Rigid coax is constructed with discs of insulating material spaced at six inch intervals and enclosed in a rigid copper or aluminum tube. This type of cable is more difficult to install and terminate, but has lower loss and is suitable for transmission over longer distances.

FIGURE 13.1
Coaxial Cables

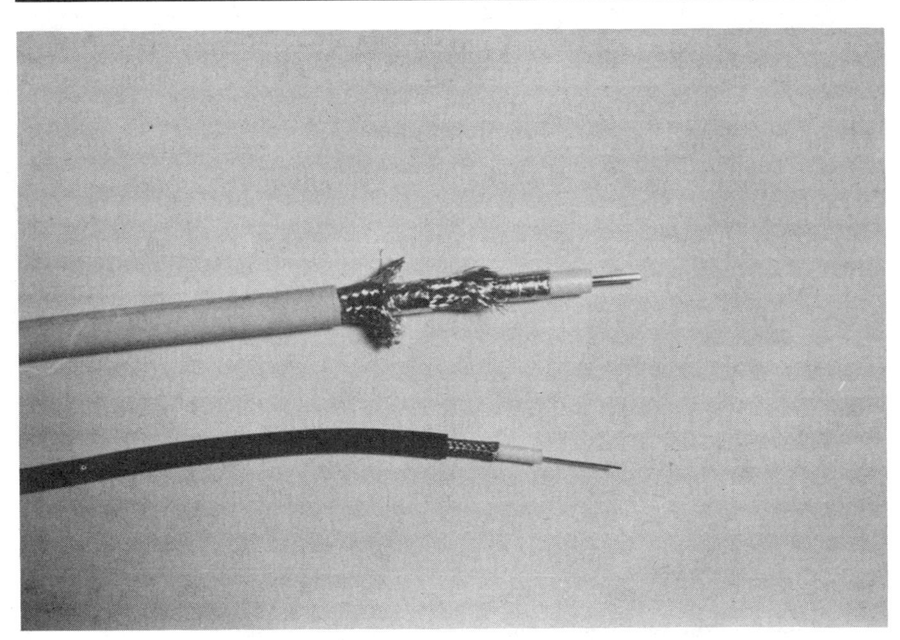

Photograph courtesy of author

The larger of the two cables is a 50 ohm cable designed for Ethernet. It has two braids and two servings of aluminum foil around the Teflon® insulator and inner cooper wire. The smaller cable is 75 ohm RG 59/U, which is used in cable television and thin Ethernet.

Coaxial cable ranges from low to high in cost and difficulty of design and installation. The cost and difficulty are a function of the application. Simple terminal-to-controller connections and packaged LANs are installed by almost anyone who knows how to use simple hand tools. Rigid coax used for broadband LANs and cable television requires skilled designers and installers.

Fiber Optics

Fiber optic cable has found limited application in premise wiring to date, but its use is expected to increase in the future. Tiny strands of highly transparent glass are surrounded by a cladding material that guides light pulses from one end of the medium to the other. Compared to other media, fiber optics has extremely high bandwidth, and is immune to electromagnetic induction. It is more expensive and difficult to install than other media because special skills and apparatus are required to align and splice the fibers.

Fiber optics cannot be tapped easily, which is advantageous for security, but limits its application to point-to-point use, and in some LANs, to a star topology. In a star configuration, conductors radiate from a central bridging point to peripheral equipment. The cable is light weight and easy to support in risers and suspended ceiling. For its bandwidth, it occupies far less space than any other transmission medium.

BUILDING ENTRANCE WIRING

The building entrance must be closely coordinated with the telephone company. Most telephone companies have building industry consultant services that coordinate among the architect, builder, and the telephone company's outside plant engineers. The building entrance will be one of three types: conduit, direct burial, or aerial.

Conduit Entrances

Conduit entrance is the preferred method for both the builder and telephone company because of its flexibility. If an additional entrance cable must be placed, or if a new service such as fiber optics is required after the building is completed, the availability of an empty conduit greatly facilitates the expansion. Besides the aesthetic advantage of conduit, it also offers the best mechanical protection of any entrance method.

As a general rule, conduit should be four inches in diameter and made of a non-corrosive material such as concrete or fiberglass. It should be buried to a minimum depth of 18 inches, with two to three feet the preferred depth. As a rule of thumb, consider providing enough conduit capacity for at least one cable pair per 100 square feet of floor space. A four-inch conduit can accommodate as many as 3600 pairs, depending on the wire gauge and number of bends in the conduit run. Most buildings will require fewer outside pairs than this, but the conduit cost is negligible compared to the cost of the trench.

Even if conduit is not placed to the point of intersection with the telephone company's facilities, short stubs or sleeves should be placed through the building wall to facilitate rearrangements. When a building is being constructed or remodeled, the cost of placing four sleeves is little more than the cost of one.

Conduits should not be shared between communication and power cables, but there is little reason for not placing twisted pair, coaxial, and fiber optic cable in the same conduit if their physical dimensions permit. Sub-ducts are usually recommended to divide the capacity of the larger conduits to accommodate smaller cables. The conduits should terminate at a telephone company manhole, pole, or distribution terminal at one end, and near the equipment room at the subscriber's end of the connection.

Direct Burial

Direct burial is usually less expensive than conduit, and if future additions and rearrangements are not expected, it is an adequate method. Cable is either placed in a trench or plowed to a depth of about two feet. Communication cables can share trench space, but the National Electrical Code requires a separation of at least 12 inches from power cable. The cable is routed through a sleeve in the building wall, and is terminated in the same manner as with conduit.

Aerial

Aerial cable entrances have all but disappeared except in rural and suburban areas. Aerial entrance is much less expensive than conduit or direct burial, but it has two disadvantages; it is unsightly and vulnerable to damage. Nevertheless, aerial cables are often advantageous within plants to interconnect buildings. The primary considerations in its use are to be certain that enough ground clearance is provided, cable is routed away from potential sources of damage, and that the cable is adequately supported. Support is provided by a strand (called a *messenger* in the industry) that is either fastened to the poles or embedded in the cable sheath during manufacture.

Entrance Cable Termination

In most cases, the telephone company will own the entrance cable and will carry it to a point of demarcation with the building owner. For maximum flexibility, the cable should be terminated on a distributing frame. As illustrated in Figure 13.2, the distributing frame provides a convenient spot to crossconnect user services to the entrance cable and to assign equipment to local distribution pairs. It also is a convenient place to install electrical protection equipment.

Electrical Protection

To prevent damage to equipment and hazards to personnel, all entrance cables must be protected from electrical hazards. Cable protection consists of two types, high current and overvoltage. If a cable accidentally comes in contact with a downed power line, is crossed with a circuit carrying high voltage, or is struck by lightning, the protector prevents damage to equipment and protects users from shock.

Overvoltage protection usually consists of carbon blocks connected from each side of the cable pair to face a matching block that is connected to a ground. If high voltage strikes the circuit, it arcs across the block and permanently grounds the circuit. Some types of electronic equipment also require protection in the form of a gas tube to conduct excess current to the ground while the carbon is operating.

Current protection is usually in the form of devices called *heat coils*. A heat coil, placed in series with the circuit, surrounds a spring loaded plunger that is held away from ground by a low-melting point substance. When excessive current flows, the heating effect melts the substance and grounds the circuit to prevent further damage. Figure 13.3 shows a diagram of the most common types of circuit protection.

FIGURE 13.2
PBX Distributing Frame

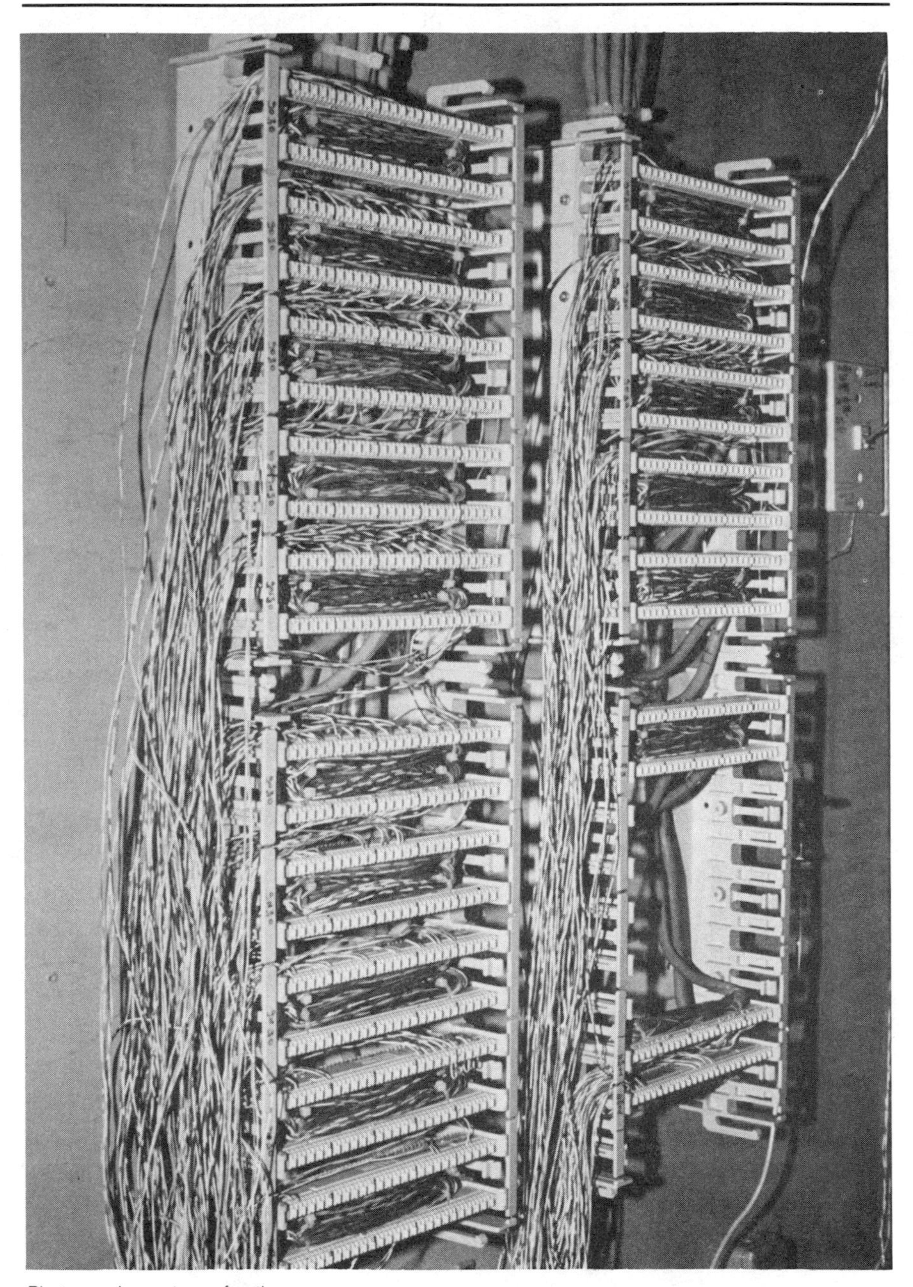

Photograph courtesy of author

FIGURE 13.3
Station Protection Equipment

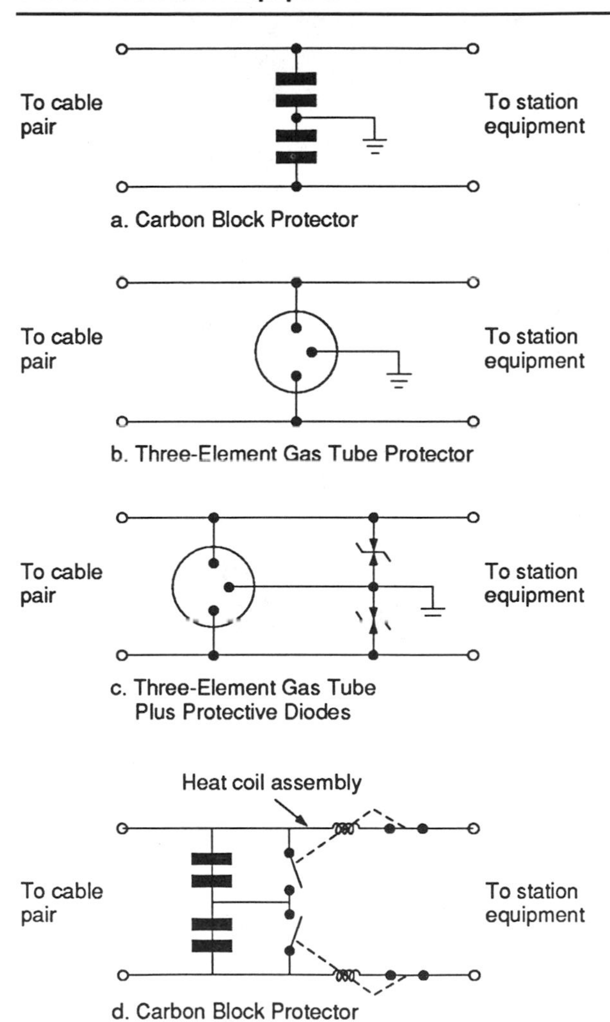

The telephone company is responsible for placing protection on any circuits that are potentially exposed to hazardous voltage. Users are responsible for their own protection in interbuilding cables.

INSIDE WIRING

Inside wiring consists of five elements, each of which is discussed in this section:

- The main distributing frame (MDF)
- The riser cable system
- Intermediate distributing frames (IDF)
- The horizontal distribution system
- Terminating jacks

Not all buildings use all five components of the wiring system. Riser cables are used in large and multi-story buildings and in campus environments for wiring runs from the MDF to intermediate wiring closets. In smaller buildings wiring is run from the MDF directly to the terminating jacks. This wiring system is often called the point-to-point or "home run" system, and has the advantage of simplified record-keeping compared to using intermediate wiring closets. It is also easier to locate trouble in a point-to-point wiring system because there is no intermediate crossconnect point. In larger buildings, however, the use of intermediate distributing frames located in wiring closets is more flexible and less expensive at the price of more complex records and more difficult trouble-shooting.

Main Distributing Frames

Some type of main distributing frame should be employed in every telecommunications system. The first point of crossconnection is in the equipment room. Here are terminated outside cables from the telephone company, riser cables to other floors in the same building, distribution cables to other buildings, and cables to equipment located in the same room. Termination methods range from floor-mounted main distributing frames to wall mounted terminals such as those shown in Figure 13.2.

The following principles should be observed in selecting equipment and terminating services:

- Different types of services should be segregated and interconnected with crossconnections or *jumpers*
- All terminations should be clearly designated and pairs numbered by cable and pair number
- Cables should be neatly dressed, securely fastened to a supporting structure, and unused pairs either terminated on blocks or wrapped and taped to the cable sheath
- Crossconnections should be dressed through rings or structures designed to keep them sufficiently taut and free from movement

As explained in Chapter 21, records of all crossconnections should be kept. In small systems it may be acceptable to tag the cables and not keep separate paper records, but in larger systems, complete cable records should be maintained on paper or in a data base management system.

The Riser Cable System

Risers connect the equipment room to wiring closets located on each floor. Figure 13.4 shows the layout of cables in a typical multi-story building. As discussed in the next section, wiring closets are placed to minimize the length of terminal wiring runs. Riser cables are placed either with a separate cable to each floor, with multiple cables to each floor, or with cables serving multiple floors. The choice depends on the size and layout of the building. The multiple cable arrangement is somewhat less expensive than serving each floor with a separate cable. Some buildings may require multiple cables to each floor to maintain separation between services that cannot coexist in the same cable sheath because of potential interference.

Riser cables are routed from the equipment room to building floors by one of two means, enclosed in conduit or surface mounted in a utility shaft or raceway. Surface mounting has the advantage of low cost. Sleeves are installed through the floor, with unused sleeves and space around cables plugged with fireproof material. Conduit is preferred if its cost can be justified. It is easier to pull new cables between floors through conduit, particularly if the wiring closets are not vertically aligned. Conduit offers better protection from damage, but it may restrict the number of cables that can be accommodated between floors.

In both wiring methods, cables are terminated on distributing frames in the wiring closets. Considerations for selecting and placing wiring

FIGURE 13.4
Multi-Story Building Telephone Equipment Layout

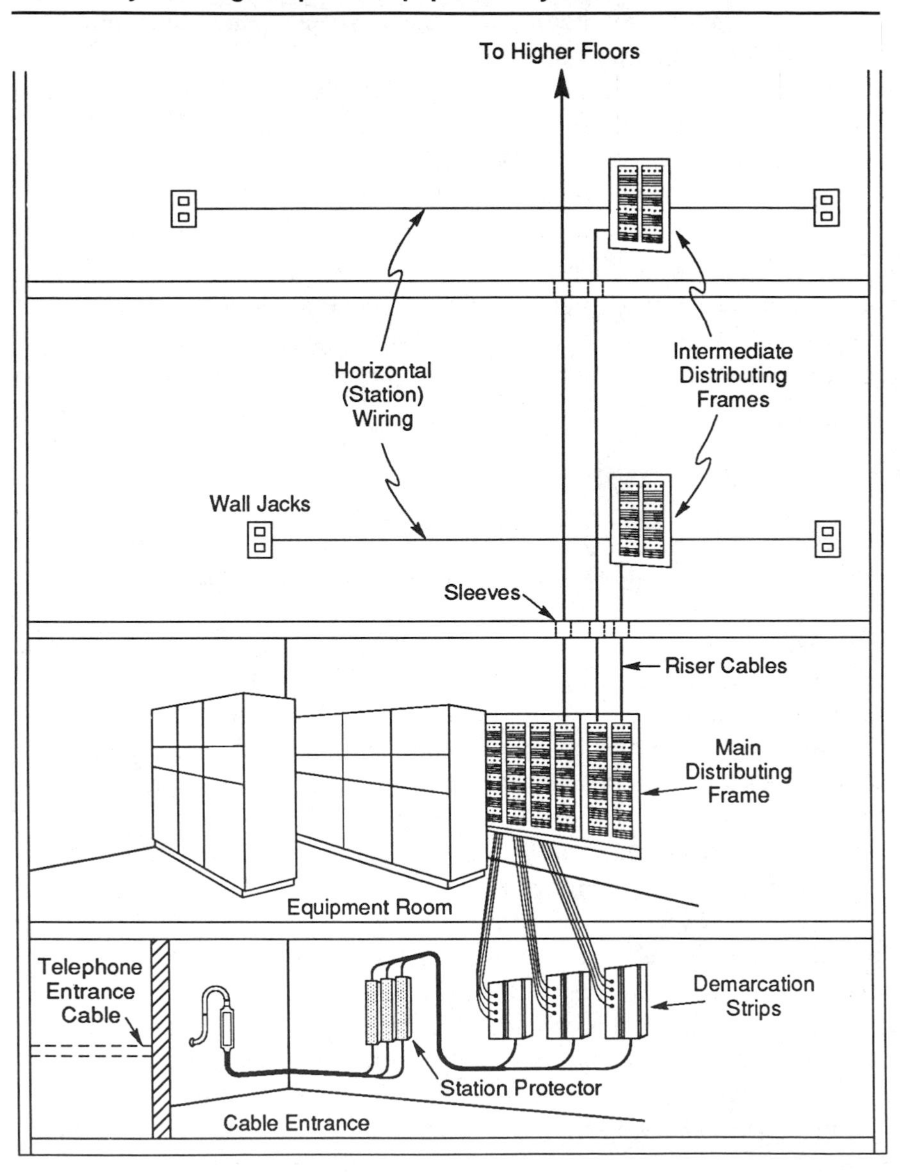

closets are discussed later in this chapter. Riser cables should be provided more liberally than entrance cables because of the need to distribute communication facilities within the building. As a general rule, from 1.5 to 2.0 riser cable pairs per 100 square feet of space should be provided for telephones. If wiring is being placed for terminals also, these figures should be doubled.

Outlying buildings are serviced by distribution cables that are installed and terminated like riser cables. Installation methods are the same as those discussed for entrance cables; encased in conduit, plowed, placed in a trench, or suspended between poles or buildings. The primary considerations are prevention of damage and provision of enough capacity to avoid expensive rearrangements in the future.

Intermediate Distributing Frames (IDF)

The IDF usually consists of 66-type blocks such as the ones shown in Figure 13.5. One set of blocks terminates the riser cable and the other set terminates the horizontal distribution wires to the stations. Terminals are connected to riser cables by running crossconnects or jumpers between the blocks. This method offers a flexible wiring plan. More station wiring than is initially needed can be placed because the cost is relatively low. Jacks can be wired on all four walls of a room, terminated in the wiring closet, and connected only as required. The need for accurate records cannot be overstressed, however, because it can be extremely time-consuming to clear trouble in an undocumented wiring closet.

The Horizontal Wiring System

The most complex part of the wiring design is cabling between the wiring closet and the terminals. The perfect system has yet to be designed, so choosing the wiring method is a matter of making compromises. Generally, the older the building the more compromises you will have to make between cost, aesthetics, flexibility, safety, security, and disruption of the work force during wiring changes.

Studies have shown that the typical office has up to 50 percent rearrangements per year, and from 13 to 20 percent annual growth in office equipment requiring communications support. This means that the initial cost of cabling is but a small part of the total cost of the wiring plan. If you anticipate frequent rearrangements, it pays to design for

FIGURE 13.5
66-type Terminating Blocks

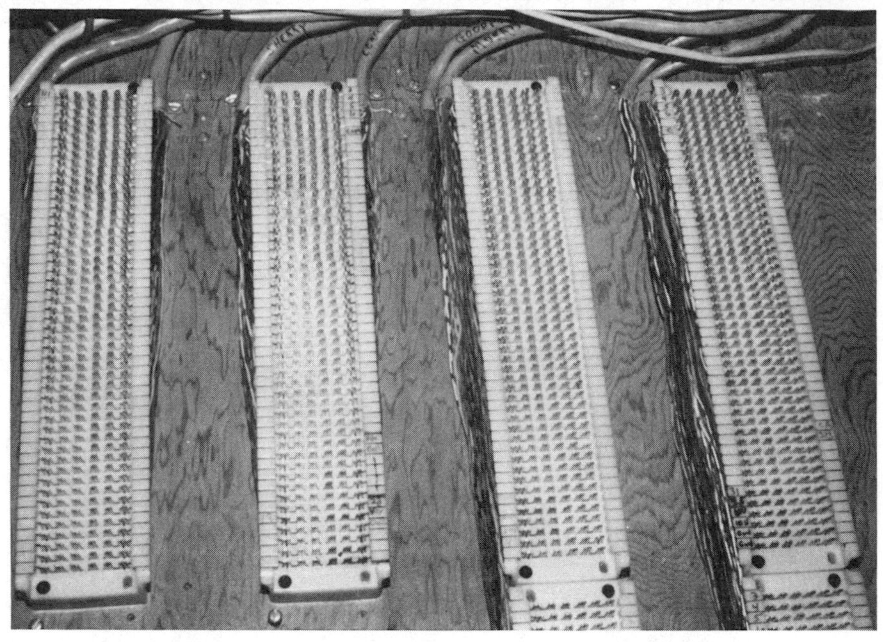

Photograph courtesy of author

The lugs in these blocks are split so they pierce the insulation of the terminated wire. Wire is placed on top of the lug and inserted with a special tool that punches the wire down and cuts off the excess.

maximum flexibility. The cost of the wiring system should be evaluated by the life cycle analysis method discussed in Chapter 8.

Productivity can also be adversely affected by the wiring system. Workers fishing cables through walls and suspended ceilings and drilling floors for access to ducts can disrupt the work force and even cause safety hazards in the office.

Another important consideration is the degree to which wiring costs are deferrable. If the method is flexible enough to allow deferring the placement of wire until space is occupied, a considerable reduction in total cost may result. The considerations in choosing the wiring method are discussed in the remainder of this section.

Exposed Wiring

Many older buildings have been wired with exposed cabling simply because no other alternative was available at a reasonable cost. Cabling is fastened to baseboards or wainscoting and run around windows, doors, and other openings. The exposed method is probably the least costly and most unsightly way of cabling a building, but the method is perfectly acceptable in warehouses, factories, and other locations where cables are unnoticed among exposed pipes and electrical wiring.

In most office areas, exposed wiring is an unacceptable alternative. Surface raceways, which are metallic ducts fastened to the wall, may be an acceptable alternative in some office buildings. Raceways offer better security and protection from damage than exposed wiring, and are somewhat more attractive than exposed cable.

Conduit

Conduit is an excellent, although costly, method of cabling. It is practical only on new construction or when buildings are extensively remodeled. Ordinary electrical conduit can be used for telephone wiring provided you have no more than two 90 degree bends, or have pull-box access at bends to facilitate the job of wire pulling. Conduit is often used in conjunction with the suspended ceiling wiring method, which is discussed later. Conduit stubs are placed in the walls and terminate in the ceiling. Wire is run through the suspended ceiling area, but is routed through conduit to the terminals.

The main advantages of conduit are ease of wiring installation and deferral of wiring until it is actually needed. It also meets electrical safety codes without using specially coated cable.

The primary disadvantages of conduit are its cost and lack of flexibility. Once placed, conduit cannot easily be rearranged, so desks and terminal locations are fixed or attached to the ends of long extension cords. Unlike electrical wiring, which is multipled through outlet boxes, telephone wiring must be cabled individually back to the wiring closet or equipment room.

Poke-Through

The poke-through method is widely used in residences and small frame business buildings where it is easy to drill into basements, crawl spaces, or attics. Terminations are mounted on the floor, wall, or baseboard. Wiring is concealed by boring into the cap or sill plate of a wall and fishing the wire into the attic or basement.

The poke-through method is also frequently used in multi-story concrete structures with suspended ceilings as illustrated in Figure 13.6. A hole is cut in the floor with a core drill and a floor fixture or monument

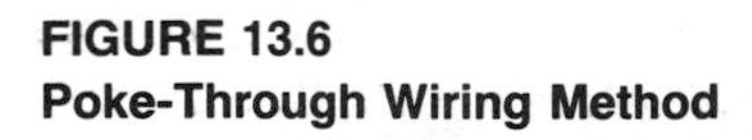

FIGURE 13.6
Poke-Through Wiring Method

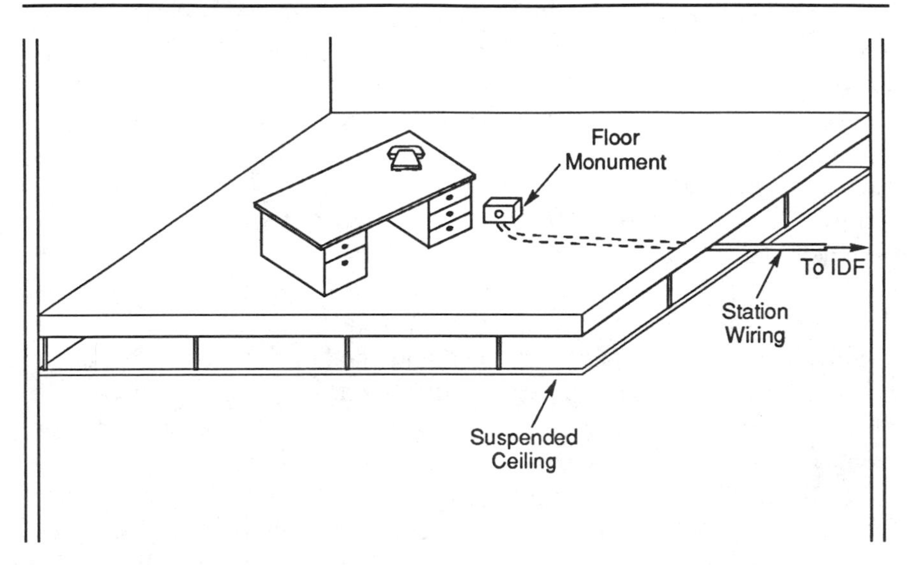

The floor is drilled through to the suspended ceiling in the room below. Station wiring is run above the suspended ceiling.

houses the termination. Wire is poked through the floor and routed to the wiring closet or MDF through the suspended ceiling of the floor below. Fire codes permit no more than one fitting per 65 square feet of floor space.

The poke-through method is reasonably flexible. Wiring can be placed anywhere except where an obstruction such as a column or beam prevents drilling. In an open space, the floor fixtures create a trip hazard unless they are protected by furniture. Once the fixture is installed, the location of the desk tends to be inflexible, but it is possible to cap the hole and drill another, provided the building structure is not weakened or the fire code restrictions exceeded.

An important drawback to this method is disruption of the work force. Core drilling is noisy and expensive and affects the workers on the floors above and below. If the space is occupied by different organizations, it may be difficult to get permission from the tenant below to remove the ceiling and tolerate the noise to accommodate the tenant above. Security is poorer than with many other methods because cables are accessible in work space that may be occupied by another organization. The possibility of electromagnetic interference from other devices such as fluorescent fixtures should also be considered with this method.

When the ceiling space forms an air return plenum, the National Electrical Code requires that cable sheaths have adequate fire resistance and low smoke-producing characteristics. Fluoropolymer materials such as Du Pont's Teflon® have been approved for air plenum use, as discussed later in this chapter.

Cellular Floors

The cellular method is similar to the poke-through method except that steel or concrete cells are embedded in the floor. Alternate cells contain electrical and communication wiring as shown in Figure 13.7. Core drilling is used to cut into the cell. Cells are typically placed on one-foot centers, which allows flexible placement of the floor fixture. The same disadvantages of noise and floor-mounted fixtures are characteristic of this method, as they are with poke-through. Also, the initial cost of the cells represents a significant unused investment until the floor space is occupied.

FIGURE 13.7
Cellular Floor Ducts

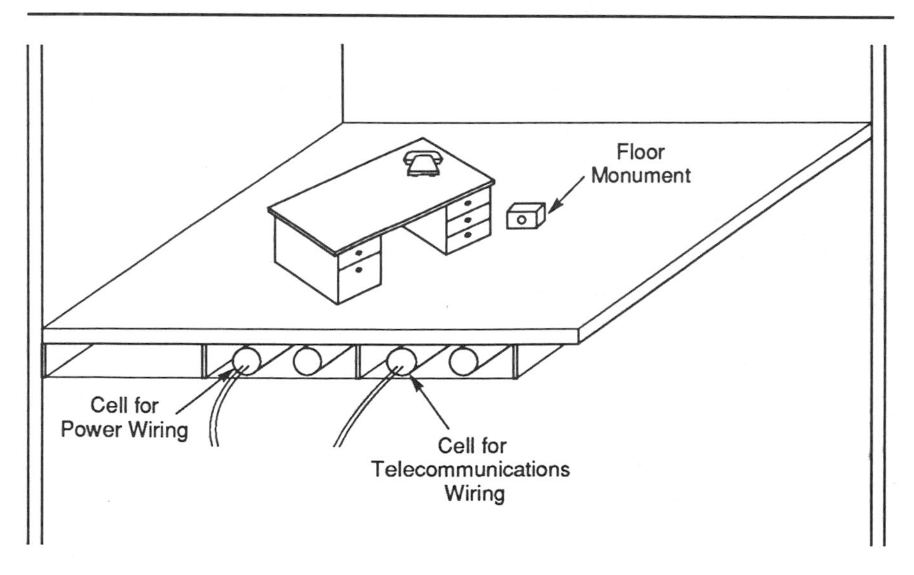

The primary advantage of the cellular structure is that it is unnecessary to gain access to offices below to place wiring. A second advantage is the contribution to structural strength of the building resulting from cells. The cellular structure offers high security and protection from electromagnetic interference. It is necessary, with this method, to ensure that dead wiring is removed because congested cells prevent placement of additional wiring. This is particularly hazardous with cells containing multiple coaxial cables.

Suspended Ceilings

In many buildings wiring is placed in a suspended ceiling and routed through fixed or portable walls or through metal "telepower" poles that contain chambers for both power and telephone wiring. Although the poles are more aesthetically pleasing than exposed wiring, many people find them unattractive, and they are costly to rearrange. Wiring ducts contained in portable partitions are a satisfactory method of enclosing wiring, but they are also expensive. Other advantages and disadvantages of the use of the ceiling area are the same as discussed earlier with the poke-through method.

Raised Flooring

Raised flooring is widely used in computer rooms and other locations that require a great deal of cabling and frequent rearrangement. Floor modules are typically two feet square, and consist of carpet or tile squares over a metal or concrete supporting structure that provides six to 12 inches of space for cabling. Raised flooring is probably the most flexible method of concealing wiring, but it is also one of the most expensive. A portion of the cost may be justified, however, by the tax saving that may result from the use of a shorter depreciation schedule than the IRS permits for the building. Other disadvantages of raised flooring are the ramps it requires to adjoining levels, and the potential safety hazard that results when sections are removed and left unprotected.

Undercarpet Cable

Both flat telecommunications and electrical cabling can be installed under specially designed carpet squares. Some manufacturers even provide flat coaxial cable for under-carpet installation. The cable is taped directly to the floor by a machine that applies and sticks the cable in a single operation. Terminations are contained in floor fixtures resembling those used for poke-through and cellular floors.

The layout of under carpet cable must be carefully planned to avoid cross-overs, which tend to cause noticeable lumps under the carpet. To avoid crossovers it is necessary to design telephone and power wiring as part of a single plan.

The undercarpet method is one of the most flexible within the limits of quantities of cable pairs supported and the necessity to avoid crossovers. It is possible to place terminations almost anywhere, but after the carpet is installed, rearrangement is costly. Another drawback of this system is that carpeting is more expensive than would otherwise be required.

Terminating Jacks

Telephone wiring is normally terminated in four-conductor RJ-11 or eight-conductor RJ-45 jacks. Where data and telephone terminals are both installed in the same wall outlet, they are often terminated in dual jacks. Care must be taken to terminate the wiring correctly and to label the jacks so the wrong type of terminal is not plugged into the outlet.

WIRING PLANS

Wiring plans have several objectives, most of which are mutually exclusive. Cost is always an important consideration, but unlike electronic apparatus, a well-designed wiring plan should outlive several generations of telecommunications equipment. PBXs and local area networks will likely have a life of only five to 10 years, but if station wiring is well designed, it should last about as long as the building's interior. When the building is functionally obsolete and ready for a complete renovation, it makes sense to replace the station wiring. The cost of a high class wiring job can be recovered in reduced cost of station rearrangement. With the frequency of movement of most offices, this obviously requires better planning than most organizations have done in the past.

Before the end of the lifetime of wiring plans being installed today, it is safe to predict that most offices will be equipped with a data terminal in addition to a full-featured telephone. It pays, therefore, to install wiring for a future LAN or terminal in each work space at the same time it is being wired for a telephone.

Cable Materials

The principal issues in selecting cable material are the wire gauge and number of pairs to use, composition of the sheath material, and special armoring or shielding around all or some of the pairs.

Cable Gauge

Selecting the right cable gauge requires a balance between physical size, transmission quality, and ease of handling. Gauges for communications cable range from 19 to 26 AWG (the higher the number the finer the wire). For inside wiring 22 and 24 gauge are the most common. At most termination points you must select the wire to accommodate the physical method of termination. For example, wire wrapping tools may not be effective if the wrong gauge is used, and punch block terminals that operate on the principle of penetrating the insulation generally require 22 gauge wire.

Finer gauge cable is advantageous wherever it must be pulled through restricted space such as conduits. The use of fine-gauge cable permits many more pairs to be pulled through a conduit. The disadvantage of finer gauges is their effect on transmission and signaling range. The finer the gauge, the more loss cabling has, and the shorter the range from a PBX to its off-premise stations.

Number of Pairs

Determining the quantity of pairs to place for riser and distribution cable is not complicated. Place as many pairs as you will need in about five years unless conduit space is too limited to place multiple cables. In case of limited conduit space, place the ultimate sized cable at the outset.

The size of station wiring should be two or three pairs plus the pair requirements for data terminals. The current generation of PBXs and key systems requires from one to three pairs. The most popular twisted-pair local area networks require two pairs. Asynchronous terminals using an RS 232-C interface can get by with as few as two and may require as many as four pairs. Your choices are to use a commercial wiring plan such as IBM's or AT&T's, which are described later, or to design your own using standard twisted pair and an interface jack such as the RJ-11 or RJ-45.

Sheath Composition

The composition of the cable sheath is an important decision that depends on the wiring environment. The least expensive coating is polyvinylchloride (PVC). PVC is acceptable for routing through floor ducts and conduit, but violates the National Electrical Code in air plenums. An air plenum, which is the space above ceilings, is often used for air return in the heating system. The National Electrical Code requires that cabling in plenum space either be enclosed in conduit, or be of a fire-resistant and low smoke-generating composition. Fluoropolymer materials such as Du Pont Teflon® have been approved for use in air plenums, but their cost is higher than PVC.

Shielding

A properly constructed shield connected to a ground protects against interference from electromagnetic induction. All outside cabling should be constructed with shield continuity. Shielding is not required for telephone cables, and may or may not be required for terminal and local area network cabling. Shielding is advisable for RS-232-C cabling more than 50 feet long, although in environments without electrical interference, many companies successfully run unshielded terminal wiring over much longer distances.

Two of the most prominent LANs using twisted pair, the AT&T Starlan and the IBM Token Ring, can operate over unshielded twisted pair. IBM recommends shielded wiring when the distance from terminal to wiring closet exceeds 600 feet or the number of attached devices exceeds 72. Within those limits the network will work with unshielded wire if the wire quality is sufficiently high.

Baluns

A balun is a device that converts the unbalanced wiring of a coaxial terminal system to a balanced twisted pair system. The name is derived from the words "BALanced/UNbalanced." A balun supports a transmission rate as high as the coaxial cable it replaces over a distance of approximately 1,000 feet. The use of twisted pair wire is less expensive and takes much less room than coaxial cable.

The AT&T Premise Distribution System (PDS)

The AT&T PDS system is considered by many experts to be the most effective wiring system on the market. It consists of an integrated set of components, as shown in Figure 13.8, that support all the elements of a wiring system. The 110-type connector is a particularly effective means of crossconnecting devices in either the wiring closet or at the MDF. The system has these advantages:

- Equipment is available for all elements of the wiring system
- It is easy to install and requires little space
- A standard color-coded backboard and labels make the system easy to administer
- It is relatively inexpensive, particularly because it supports non-shielded twisted pair

The IBM Wiring Plan

IBM's building wiring plan is designed to support its token ring network as well as voice and other forms of data. Its topology is similar to the AT&T plan except, as shown in Figure 13.9, it links wiring closets in a ring. The overall topology of the plan is a star-connected ring. The wiring media are:

- *Type 1*: Two 22 AWG shielded twisted pairs
- *Type 2*: Two 22 AWG shielded and four non-shielded twisted pairs
- *Type 3*: Four pairs of 22 AWG non-shielded twisted pairs
- *Type 5*: Two fiber pairs
- *Type 6*: Two 24 AWG shielded twisted pair patch cables

Type 2 is the most common wiring system, but it is much more expensive than Type 3 cable. Where noise performance is important, the Type 2 cable should be used. The token ring network can operate over standard telephone wiring if it is properly installed. IBM will test existing wiring for a fee to determine if shielding is required.

The Digital Equipment (DEC) Wiring Plan

The DEC wiring plan uses four cables: four-pair non-shielded twisted pair for voice, four-pair non-shielded twisted pair for data, a baseband

FIGURE 13.8
Wiring Plan of AT&T's Premise Distribution System

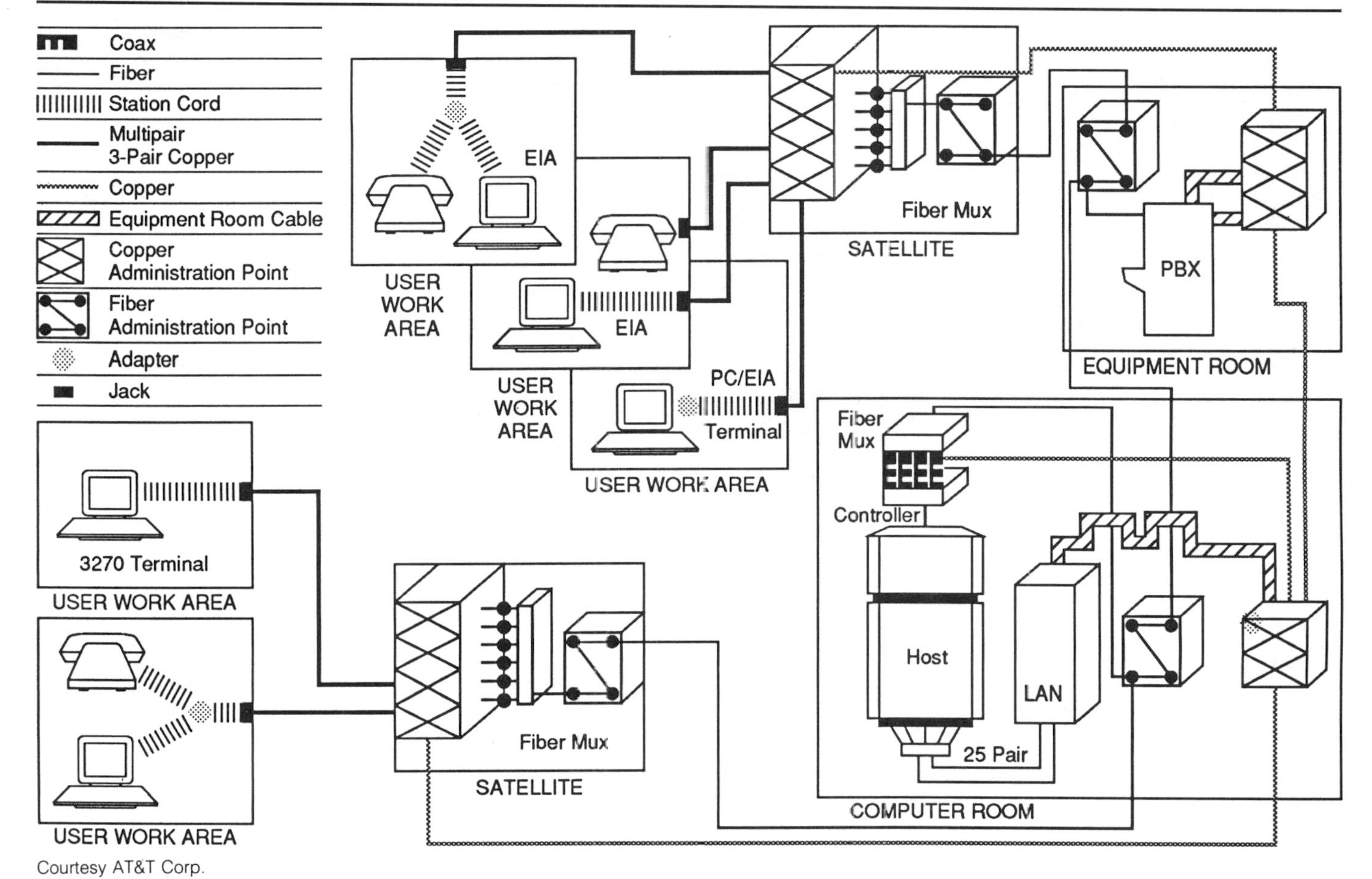

Courtesy AT&T Corp.

FIGURE 13.9
IBM Star-Wired Ring Wiring Plan

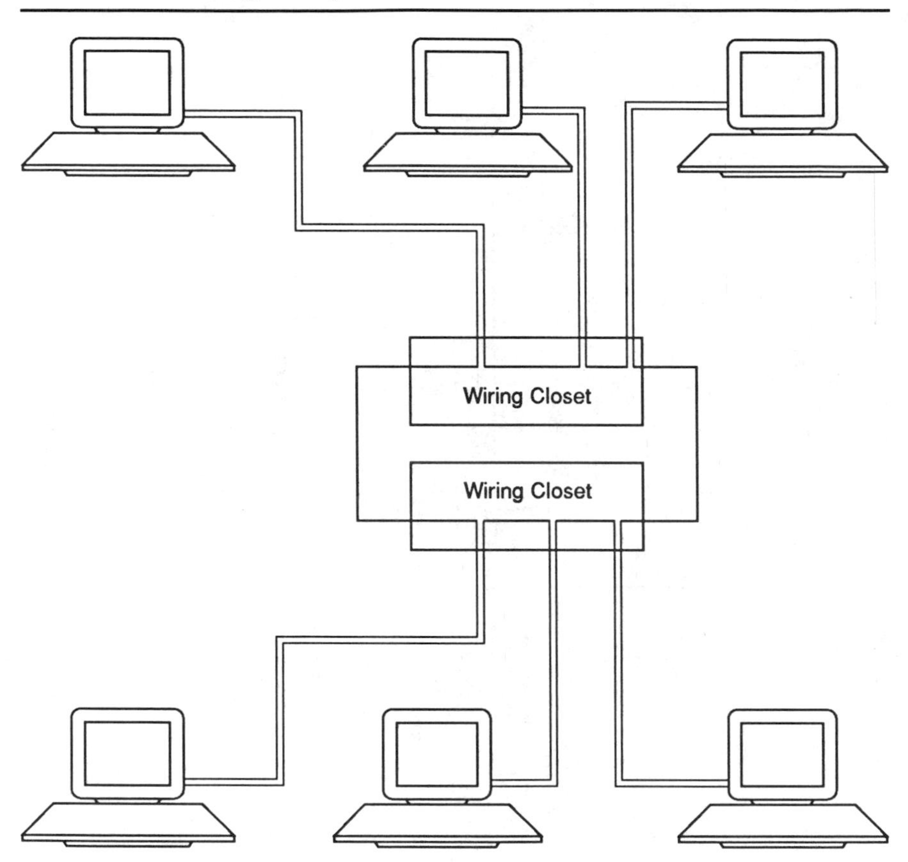

Ethernet cable, and a broadband television cable. The cable will support any kind of telecommunications (except for those requiring fiber optics), but it is expensive compared to other wiring plans.

GROUNDING

It is essential that manufacturer's grounding recommendations be followed rigorously. PBXs and key telephone equipment must be securely grounded with at least a No. 6 AWG ground wire to a ground of known integrity. The best grounds are provided by the electrical power entrance on the power company side of the meter, a metallic water pipe, or a grounding plane embedded in the building foundation. Be certain that none of these sources is isolated from absolute ground with an insulating joint such as a section of non-metallic water pipe. The metal framework of distributing frames must also be connected to the central ground.

Grounding precautions must also be observed in cabling. Metallic sheath of multi-pair cables must be securely bonded to ground to prevent electromagnetic induction. Coaxial cables used for Ethernet LANs must be grounded at only one point to prevent ground loops that may cause errors in data transmission or hazardous voltages on the cable sheath.

INSIDE WIRING REGULATORY ISSUES

Before the Federal Communication Commission 1980 Computer Inquiry II, which detariffed customer premise equipment, all inside wiring was placed under tariff by the local telephone company if the company owned the terminal equipment. Since Computer Inquiry II, inside wire ownership has been a jumble of conflicting opinions and misinformation. As of January 1, 1987, an FCC order detariffs the installation and maintenance of inside wiring everywhere except New York. The issue of who owns previously installed inside wire is not settled, and is subject to different interpretations by telephone companies and state regulatory agencies. This section outlines some of the FCC orders that affect inside wire ownership.

In March, 1981, the FCC issued its order 79-105, which required the telephone companies to expense, rather than capitalize, inside wire. The telephone companies were permitted to sell existing wire if state regulators authorized it. Previously installed wire was to be amortized over a ten-year period. In November, 1983, the FCC issued Order 83-457, which deregulated all installations of complex inside wiring. (Complex wiring is generally that used for PBX and key telephone systems).

In subsequent decisions, the FCC has decreed that simple inside wiring would be deregulated and that customers would be given ownership of wire beyond the point of demarcation (usually the station protector). This means that unless state rules conflict, the following is the case with respect to inside wire ownership:

- If wire was placed by the telephone company after 1981 it was expensed and the customer probably owns it
- If the wire was placed by the customer's contractor at the customer's expense, the telephone company has no jurisdiction over it
- If the wiring was placed prior to 1981, the telephone company owns it, but has limited jurisdiction in restricting what the customer may do with it; the customer and telephone company are generally prohibited from removing, rearranging, or damaging it without the other's consent

If there are any doubts in your mind about inside wire ownership, you should first ask the telephone company. They may relinquish ownership to you. If the telephone company's ruling is not satisfactory, you should seek clarification through the state regulatory agency and, perhaps, through any industry associations in your area. Although the FCC orders limit the amount of control telephone companies can take over inside wire, it does not require them to divest themselves of wire before it is fully amortized.

Interbuilding wiring was usually placed under separate accounting rules by the telephone company, and does not fall under the purview of the FCC's inside wire rulings. You may be invited to purchase such cable if you wish to acquire jurisdiction over it, although some telephone companies turn it over to the customer on request.

Some telephone companies also sell their block cable maps for a nominal fee. These maps show the size and type of cable, cable counts, and routing. If you have acquired ownership of the cable, it will be well worth the expense to obtain the telephone company's records.

SELECTING EQUIPMENT SPACE

Requirements for telecommunications equipment rooms depend on the type of equipment. Simple key telephone systems are usually mounted on the wall and require little space. Large PBXs require separate rooms with special attention paid to heating, ventilation, lighting, and air conditioning. Security is an important consideration in all telecommunications systems. Not only is it vital that information be guarded, but the equipment itself must be protected from unauthorized access. In this section we will discuss considerations in selecting equipment rooms and wiring closet space.

Floor Space Considerations

If telecommunications equipment has been selected before floor space planning takes place, the manufacturer's recommendations should be followed. Add space for such auxiliary equipment as power, uninterruptible power supplies, channel banks, maintenance terminal, network channel terminating equipment, and distributing frames. The amount of space required for this equipment varies. The trend is toward more compact equipment, but the trend is also toward mounting more equipment such as T-1 on the user's premises, and this equipment requires floor space. Distributing frames are not shrinking to the same degree as the electronic equipment they support, so it is advisable to provide space for the ultimate equipment configuration to avoid the need for future building expansion. To calculate equipment space, consider providing space for the following:

- The initial or future installation of battery backup or an uninterruptible power supply (UPS)
- Protection and distributing frames
- Channel banks or multiplexers
- Data communications equipment such as modems and multiplexers
- Analog network channel terminating equipment such as repeaters and signaling equipment
- Aisle and working space around equipment cabinets
- Working spares and equipment held for growth
- Administrative space for desks, files, workbenches, office supply storage, etc.

Power Equipment

The equipment room must be wired with enough power capacity to support the PBX and termination equipment. If battery backup is included, it may be necessary to place the batteries in a separate room to prevent corrosion. Electrical circuits should be equipped with surge suppressors that prevent damage to equipment during power surges. If voltage fluctuation is a problem, you should consider installing voltage regulators to keep power line voltage within limits required for telecommunications equipment.

Equipment Room Location

The following factors should be considered in locating the equipment room:

Centralization: The room should be located near the center of the served area to reduce the amount of cabling and to avoid long wire runs that may result in transmission trouble or require heavy gauge wire.

Security: The room should be kept locked at all times to prevent unauthorized access. The key should be kept by someone who is in a good position to monitor access.

Accessibility: The room should be easy to reach from elevators for moving equipment, convenient to the telephone company entrance cable, and located close to major users. The room must also have an entrance large enough to accommodate equipment deliveries. It may be necessary to locate the PBX close to the computer room if the two systems share common equipment such as UPS supply, air conditioning, or are interfaced by a direct computer to PBX connection.

Environmental Considerations: Telecommunications equipment differs significantly in its requirements for heating, ventilation, and air conditioning. Manufacturers' recommendations should be observed with respect to temperature limits and air flow. Most equipment operates within the temperature limits of office work space, but some PBXs require a closely controlled air conditioned environment. If the specifications require air conditioning, you should inquire closely into the effects of loss of air conditioning. If satisfactory operation depends on temperature control, it may be necessary to provide backup air conditioning in case the primary source fails. Most equipment also specifies temperature minimums. As a general rule, however, any space that is comfortable for office workers will meet the temperature minimums.

Wall Space: Sufficient wall space is needed for backboards that mount terminals.

Storage Space: Spare equipment, including circuit boards, telephones, and terminals may be stored in the equipment room. If so, space must be provided for them. Space may also be required for retaining records of wiring and equipment.

Lighting: The equipment room must be well lighted so service technicians can see to maintain the equipment.

SUMMARY

The key to an effective equipment room and wiring arrangement is planning. Once the overall plan is developed, future rearrangements are usually expensive, disruptive, and difficult to implement. You must plan not only for the arrangement of the facilities, but for the records that support their administration.

A wiring plan is a long-term investment that should outlive the equipment it supports. A properly planned and designed system will repay its cost in administrative cost savings over the years it is in operation. It is usually short-sighted to consider only telephone wiring; data communications requirements should be considered at the same time, and the wiring plan designed to support both voice and data.

CHAPTER 14

CUTOVER PLANNING AND MANAGEMENT

The transition from an old to a new system is called a *cutover* in telecommunications vernacular. Cutover planning consists of determining what activities will be required to complete the actual transfer to the new system. These activities comprise most of the elements of the project plan that should be compiled by the project team. Project management techniques will be discussed in Chapter 15.

A telecommunications system is the nerve center of most organizations, the means by which information is moved to customers and internal work groups. A cutover, by its nature, disrupts the organization; the objective of the transition plan is to minimize disruption. This is accomplished by preparing not only the equipment, but more important, the people so they adapt to the new systems as quickly and effectively as possible.

In pre-divestiture days, when the telephone company furnished essentially all the circuits and hardware in a PBX, data communications system, or major corporate private network, they ran the cutover. The user's role was relatively simple; print a new directory and stationery, decide where the equipment would be mounted and who was entitled to receive what types of station equipment, assign features such as call pickup groups and hunting groups, and be the point of contact while the telephone company's forces installed the equipment, trained the users, and performed the cutover. Major systems with more than about 1,000 stations required substantial involvement of a user's representatives, but for the most part, the telephone company ran the show.

The post-divestiture world requires a great deal more user involvement. It is risky to turn the whole project over to the major equipment vendor because a great deal of inter-vendor cooperation is required for most modern cutovers, and the user is in the best position to cause it all to happen smoothly. A cutover is no trivial matter; it involves major surgery, and unless it is performed skillfully, the patient could be in for a long stint in the recovery room. Someone who understands the issues should orchestrate the cutover.

In this chapter we will consider the following cutover requirements:

- Planning the cutover
- Conducting trunk and station reviews
- Establishing a fall-back plan
- Testing the new system
- Conducting post cutover tests and follow up

PLANNING THE CUTOVER

The key to a successful cutover is to anticipate all of the tasks that must be accomplished. Tasks are sequenced, assigned to someone to perform, and executed. Figure 14.1 lists the major tasks of a telecommunications cutover in a form you can use for managing a project. Not all of these functions will be required in every project, but from this list you can prepare a custom list for your project by deleting the tasks that are not required and adding tasks unique to your cutover. Some of these tasks are required only for data communications projects, some only for telephone systems, and some for both. Each task must have one individual assigned as responsible, with a required due date. The larger the organization, the greater the degree of coordination that is required because of the difficulty of placing and testing facilities.

Reuse of Existing Facilities

In managing a cutover, several considerations warrant special attention by the project manager. An important question is whether existing facilities such as station wiring and central office trunks are to be reused or replaced. When an existing system is being replaced, it is often advantageous to reuse existing station wire. Unless sufficient spare pairs are available, the reuse of existing wiring makes it more difficult to test in advance of the cutover, but to minimize the risk you should perform as many tests as possible.

FIGURE 14.1
Telecommunications System Cutover Requirements

	Responsible	*Due*
Establish Cutover Committee		
Establish Key Cutover Dates		
Determine Feature Requirements		
Identify Equipment Locations		
Conduct Station Reviews		
Determine Trunking Requirements		
Determine Protection Requirements		
Determine Call Accounting Requirements		
Determine Wiring Plan		
Place Equipment Order		
Prepare In-House Publicity		
Prepare Station Translations		
Prepare Trunk Translations		
Develop Least Cost Routing Plan		
Assign Station Numbers		
Assign Addresses, ID codes and Passwords		
Assign Station Features		
Determine and Publicize "Freeze" Dates		
Prepare Equipment Room		
Determine Station Termination Arrangement		
Place Orders for Telephone and IEC Services		
Develop Contingency Plans		
Develop Trouble Report Plans		
Develop Intercept Plans		
Develop User Training Plans		
Develop Testing & Acceptance Procedures		
Develop Call Accounting System Plans		
Place Station Wiring		
Obtain System Documentation		
Install and Test Equipment		
Perform Pre-cutover Station Tests		
Perform Pre-cutover Trunk Tests		
Train Users		
Perform Acceptance Tests		
Cutover System		
Remove Old Equipment and Wiring		
Clean Up Work Areas		
Perform Post-cutover Tests		
Prepare Equipment Location Records		
Prepare Wiring Records		
Prepare Feature Assignment Records		
Conduct Follow up Training		
Review Trunk and Circuit Usage Data		
Adjust Trunk Quantities as Required		
Verify Billing From All Vendors		
Inventory Equipment		
Accept System		

If existing telephone instruments are being reused, testing is often facilitated by the use of cutover switches, as shown in Figure 14.2. Cutover switches are also useful for making Centrex-to-PBX cutover go smoothly. Stations can be tested by throwing the switch to the new system and restoring it when the test is complete. If instruments aren't being reused, it will be necessary to change the instrument long enough to make the test, or to test stations by temporarily connecting them at the PBX. The method used will depend on how large the system is and how much disruption can be tolerated. If the system can be turned down for an evening or a weekend, and if the vendor has enough people to place and test stations out of service, there is little risk in not testing stations in advance. If the system is large, or if it operates 24 hours per day, it will be necessary to test each station for operation of all features and restrictions ahead of the cutover.

Similar decisions must be made for transferring trunks. The most desirable method from a technical standpoint is to duplicate trunks so the entire system can be tested prior to the actual cutover date. To duplicate the trunks, however, means paying installation charges and at least one month's extra service charge. Whether to duplicate trunks or to transfer existing trunks to the new system depends on costs, the amount of work to be done, the nature of the project, and the amount of risk you are willing to assume in transferring untested trunks. In any case, it is a question that must be addressed on every cutover.

FIGURE 14.2
Cutover Switch for Transferring Stations Between and Old and a New PBX

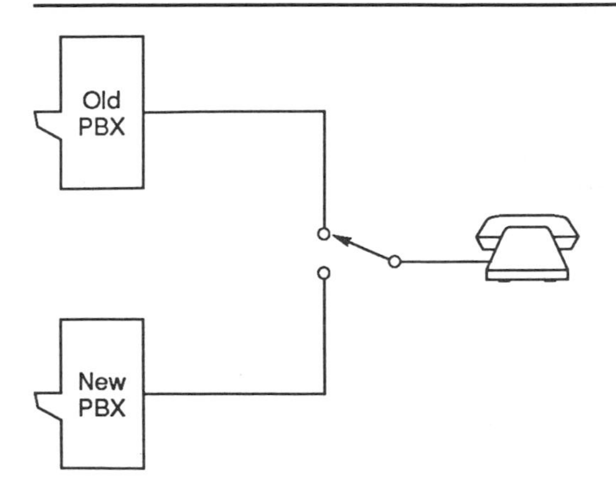

The switch can be operated for testing before cutover and removed after cutover.

User Training

User training is often given too little attention in cutover management. Unless training is carefully administered, the risk is high that expensive features will be used improperly or not at all. To ensure the success of the cutover, you should insist that everyone from top management down receives user training. Training is usually conducted on the new system after it is installed, but before it is placed in service, by temporarily installing training telephones in a conference room.

Cutover Task Lists

The following list of cutover tasks is segregated by major category and includes additional detail to explain what the tasks consist of. This list covers most activities that are needed for a major project, and includes activities that have been discussed in previous chapters as well as some we have not yet covered. The activities listed below can be sequenced and included in a project plan of the type that is discussed in Chapter 15.

System Procurement

Determine System Requirements: Review users' needs and prepare a detailed summary of requirements and specifications.

Develop Request for Proposals: Compile system requirements and specifications into an RFP and issue it to equipment and service vendors.

Select The System : Review RFP responses and select the successful vendor.

Negotiate Contract : Negotiate a purchase contract with vendors. Agree to terms, conditions, and delivery dates. The final contract should specify the cutover date and major events such as the date the order is placed, equipment is shipped, installation is complete, and other project milestones. It should also specify acceptance criteria.

Cutover Planning

Note: If multiple equipment vendors are involved in this project, it will be necessary to develop separate task lists for each of them.

Determine Trunk Transition Methods: Determine whether existing trunks will be reused or new trunks will be placed to support parallel testing.

Determine Trunk Quantities: Determine quantity and types of all trunks such as local trunks, WATS, tie lines, private lines, etc.

Plan Station Requirements: Determine station types and quantities, develop restriction policies, and select wiring method. A policy on what types of instruments will be assigned to users should be developed as discussed later in this chapter.

Develop Training Plan: Determine who will present user training, and what methods will be used. Select a training room and develop a training schedule.

Develop Station Numbering Plan: Determine how extension numbers will be assigned. If DID is being introduced, coordinate with available DID numbers, which are obtained from the telephone company.

Determine Protection Requirements: Determine whether electrical protection will be required on any stations. Protection is generally required for cables between buildings, particularly if cable is above ground.

Develop Cutover Methods: Determine whether the cutover will be progressive or "slash" cut (all lines and trunks cutover on the same day or evening). Determine whether stations will be rewired or transferred on cutover switches.

Develop Cutover Schedule: Develop the overall schedule and determine project milestones. Establish primary responsibilities and develop resource requirements such as additional people, special equipment or facilities, etc.

Develop Testing Methods: Determine how trunks, stations, and auxiliary equipment will be tested before and after the cutover.

Develop Interim Change Procedures: Develop a policy and process for handling changes and rearrangements during the transition period. Develop a "freeze" date beyond which no further changes will be accepted until after cutover. This cutoff date is the date beyond which changes can no longer be entered in the data base.

Station Work

Conduct Station Reviews: Complete reviews of all users to determine terminal equipment type, features, call coverage location, and to prepare information for station translations.

Determine Station Terminating Arrangements: Determine what types of station and terminal jacks will be installed (eg. single or dual RJ-11, RS-232-C, BNC, etc.).

Develop Wiring Plan: Determine what wiring method will be used (see Chapter 13). Determine whether point-to-point or wiring closet plan will be employed. If necessary, locate wiring closets.

Compile Station Records: Enter station type, features, cable, and crossconnect assignments into a station record list or data base.

Assign Station Numbers: Assign extension numbers to telephone stations or password and user ID numbers for data equipment.

Prepare Station Designations: Prepare designations to identify functions of telephone buttons and to place the station number on telephone instruments.

Design Distributing Frame: Determine the block size and location for each major class of equipment on distributing frames and in wiring closets.

Assign Station Connections: Assign distributing frame, riser pair, wiring terminal, and termination wiring to the station equipment.

Place Station Wiring: Reuse old station wire or replace with new wire and plan for removal of old wire. Drill floors if required. Place station jacks. Terminate wire on distribution frames. Designate all station jacks.

Place Station Equipment: Place new station equipment and connect to the station jacks.

Equipment Room

Select Location: Select equipment room and wiring closet locations and coordinate with the building manager.

Prepare Equipment Room: Finish the interior walls, ceiling, and floor of the equipment room. Install raised floor if required. Install the fire alarm and sprinkler system, and install security equipment.

Prepare Electrical Wiring: Arrange for electrical circuits of the proper size, type, and wire gauge for the PBX and data communications equipment. Obtain uninterruptible or backup power if required. Arrange for surge suppression and voltage regulation equipment.

Prepare Heating: Arrange for the heating source in the equipment room.

Prepare Air Conditioning and Ventilation: Arrange for the air conditioning and/or ventilation source.

Equipment Installation

Place Equipment Order: Place the equipment order with the manufacturer. Orders must be placed for all types of additional equipment such as distributing frames, modems, etc., that the primary vendor is not supplying.

Place Trunk Orders: Place orders for trunks with the telephone company, interexchange carriers, etc.

Ship Equipment: Equipment manufacturers will arrange for equipment manufacture and shipping. This date is often carried on the work schedule because it is a critical item that may cause the project to slip if it missed.

Install Equipment: Place framing, relay racks, and electronic equipment in the equipment room space. Cable and power up the system.

Test Equipment: Perform the manufacturer's recommended operational tests on all equipment.

Conduct Acceptance Testing: Develop methods for determining when the system is functioning properly. Conduct or observe overall system tests. Establish corrective action plans to clear any deficiencies. Develop agreement on performance levels and criteria for releasing the vendor from further responsibility and for the buyer's making the final payment.

Initialize Software

Prepare Station Translations: Prepare system translations for station port assignments, hunt groups, restriction levels, and other major features.

Prepare Trunk Translations: Prepare system translations for trunk signaling type, port location, and routing preference. Assign trunk numbers.

Program Least Cost Routing (LCR): Program LCR to conform to levels of station restrictions and intended usage of long distance carriers' facilities.

Operational Functions

Develop Contingency Plans: Develop plans for how the cutover can be postponed or advanced if major problems develop or if an opportunity to advance the transition occurs.

Develop Intercept Plans: Determine whether any telephone company intercept of changed telephone numbers is required. (Usually required for a major change from Centrex).

Reprint Telephone Directory: If extension numbers are changing, arrange to reprint the internal telephone directory. Update the console operator's employee locator guide.

Reprint Stationery: If telephone numbers are changing or if DID is being introduced, reprint business cards, stationery, promotional material, etc. Consider printing stickers that can be placed on such material to announce impending changes several months in advance.

Develop Trouble Reporting Procedures: Develop and publish trouble reporting procedures, indicating where each station reports trouble and what information is required in the trouble report. Be certain all station users know this before the first day of operation of the new system.

Train Users: Prepare the training schedule and administer training. This is usually done the week before cutover using actual equipment in specially prepared training rooms.

Post Cutover Activities

Follow Up on Training Effectiveness: Within a month of cutover, visit a sample of station users to determine whether training has been effective. Indicators of lack of training are telephones not answered within three rings, calls not forwarded to answering positions, or station features not used.

Follow Up on Least Cost Routing: Check usage of long distance service and overflow to DDD to verify that the least cost routing has been correctly programmed.

Follow Up on Trunk or Circuit Quantities: Review trunk usage or data communications circuit usage to ensure that the proper balance between cost and service has been attained.

Check Billing: Check the first month's billing for telephone company, long distance carriers, and other vendors to be sure that charges are reasonable and that charges for new services are justified. Check long distance bills to be sure traffic is routed to the right carrier. Check equipment vendor's bills against an inventory of equipment proposed in the response to the RFP. Verify that change orders are in conformance with the contract.

CONDUCTING TRUNK AND STATION REVIEWS

The transition from an existing system into a new one is a task that most telecommunications managers encounter only a few times in their careers. It is also a task that you shouldn't entrust to the vendor without a great deal of your own participation because it impinges so deeply on your company's internal operations and can have such a significant effect on costs.

Although the term "station reviews" suggests a PBX cutover, it applies equally to conducting terminal reviews for office automation and data communications systems. During the station review, the needs of each user for features and specific types of terminal or telephone instruments are evaluated. The total cost of any system is greatly affected by terminal costs, and the effectiveness of any system is determined by how well the features meet the needs of the users. The station review process is, therefore, of utmost importance in managing a transition.

A review should be made of the different types of jobs within the organization to determine the degree to which they can profit from the use of extra-cost features. The features the telephone is equipped with become a matter of prestige, and controlling them may not be easy. On the other hand, if the features are provided and not used, several hundred dollars per station can be wasted. An effective way of evaluating station features is to appoint a user's committee representing different classes of users, and explain to them the differences in features and costs among telephone instrument types. Once a standard has been established for a particular class of user, deviations should be made only with the approval of appropriate authorities.

Consideration should also be given to the type of instrument placed in common areas such as conference rooms, cafeterias, and reception areas. Often, these areas can be equipped with the most basic type of telephone, and special requirements such as a speaker telephone in a conference room can be added as an outboard device.

While on the subject of telephone instrument types, it also pays to establish restriction levels for the different classes of users. For example, telephones in public areas should be restricted to local calls with long distance calls placed only through the attendant. Some classes of users may be permitted to place long distance calls over tie lines or WATS lines, but not permitted to overflow to direct distance dialing. Most modern systems provide a great deal of flexibility in establishing restriction levels as a way of both controlling usage and ensuring that essential users aren't delayed by blocked trunks.

PBX Station Reviews

Most modern digital PBXs offer at least four types of terminals, listed in ascending order of complexity and cost:

- Single line telephone set
- Multiple line set with feature buttons
- Multiple line set with digital display
- Integrated voice/data terminal

The cost of these telephone sets varies from less than $50 for a single line set to more than $1,000 for an integrated voice/data terminal. Figure 14.3 shows one manufacturer's line of telephone instruments. Since the typical digital PBX supports one hundred or more stations, it is obvious that expensive terminals should be restricted to those who can justify their use. You should consider the criteria in the following sections when selecting terminal types.

Single Line Telephone Sets

Single line telephone sets are generally of the plain "2500" type of DTMF set, or their equivalent. They are generally compatible with any type of telephone system, and are therefore inexpensive compared to proprietary telephones. The disadvantage of this type of set is that special PBX features such as call pickup and call forwarding must be activated by switch hook flashes and dialing special codes. This suggests their use for locations such as clerical positions that primarily use the telephone for incoming calls, and for positions that have no responsibility for answering for someone else.

FIGURE 14.3
Telephone Instruments for AT&T System 75 and 85 PBXs

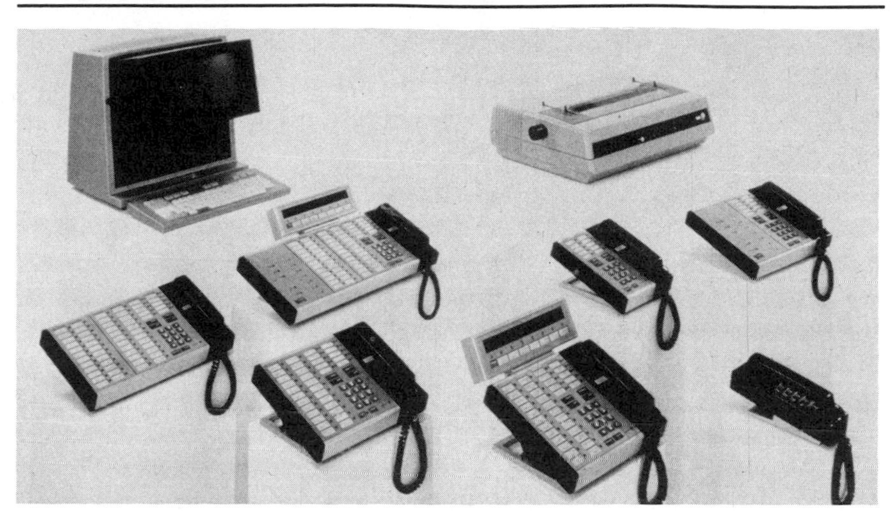

Photograph courtesy of AT&T Corporation

Multiple Button Feature Telephones

These instruments are usually proprietary to the manufacturer of the PBX, and are therefore available from only one source. This type of set, examples of which are shown in Figure 14.4, enables the user to activate features by pushing a button rather than flashing the switchhook and dialing a code. For example, to forward calls, the user pushes the "forward all calls" button, which automatically transfers incoming calls to the station designated in software as the destination. The chief advantage of this type of set is that it is unnecessary to remember the special dialing codes. Its disadvantage is its cost (from $100 to $500).

A frequent feature is a message waiting light that shows that the message center or voice mail has received a call for that number. In some systems it is not necessary to have message waiting lights. Stutter dial tone, which is a three-burst interruption of dial tone when the telephone is picked up, can serve as a message-waiting indication.

Multibutton sets should generally be assigned to managers and others who make frequent use of special features such as call forwarding.

FIGURE 14.4
Siemens Saturn Multiple Button Display Telephones

Photograph courtesy of Siemens Corporation
Two display and one non-display sets are shown in this photo.

They can also be assigned to receptionists at central answering positions where budgetary considerations preclude the use of display sets.

Another feature that is frequently required in all types of sets is speaker capability. Some telephones have two-way capability, and others can only listen. Either type of telephone can be used to monitor call progress without lifting the handset. Only a true speaker telephone allows the user to hold a full conversation with the handset in the cradle. Speaker telephones are warranted in most conference rooms, but their use outside private offices should usually be avoided because it is disruptive to other workers.

Multiple Button Display Sets

A multiple button display set, such as the ones shown in Figure 14.4, includes all of the features of the multi-button feature set plus a display that provides calling party identification (within the PBX) and called party identification for calls that forward or transfer from another station. These telephones are advantageous for receptionists and central answering positions, but are often too expensive (up to $800) for all but a restricted few executives. As prices come down on display sets, their use can be justified for a greater number of users.

The display feature should be considered for all answering positions. Most people prefer to have the telephone answered "Mrs. Black's office" instead of the less personal "Accounting Department." To answer a forwarded telephone with a personal response, the answering position must know whose telephone is being answered and ideally should know why the call was forwarded. Many display telephones show not only the number that was actually dialed, but also the reason it was forwarded, such as busy, did not answer, or deliberately forwarded.

Integrated Voice/Data Terminals

Integrated voice/data terminals typified by the one shown in Figure 14.5, integrate a telephone set with a video display terminal. These sets are often used in conjunction with an external messaging system to display actual message text. They usually include the features of a personal computer, and allow users to talk over the telephone while simultaneously using a data terminal. Although these types of instruments will become more common as the price drops, their application should be confined to those who can benefit from the features. At the time of this writing integrated voice/data terminals cost in excess of $1,000.

FIGURE 14.5
Siemens Saturn Integrated Voice/Data Terminal

Photograph courtesy of Siemens Corporation

Trunk Reviews

A trunk review should be performed for all PBX, key, and hybrid key systems before the equipment order is placed. A trunk review determines the correct quantity of trunks of all types based on traffic usage information and growth forecasts. Where an existing system is being replaced, it is often possible to obtain traffic usage information from internal software registers. For example, one of the most popular existing PBXs, the AT&T Dimension, has internal registers that log usage for trunk groups, common equipment such as registers and senders, and selected features. The vendor of your old system should be able to assist in obtaining traffic usage information.

On systems that are not equipped with internal traffic usage registers, external monitors can be attached to the trunks for a study period, or a reasonable approximation can be made from the station message detail register if one exists. Traffic usage recording equipment can be leased, or a study can be performed by a telecommunications management firm. The objective is to obtain a large enough sample of usage to enable you to compute the number of trunks that will give the best balance between cost and service. Chapter 16 explains more fully how to determine trunk quantities.

Many telephone companies offer the option of routing central office trunks on T-1 lines. The advantage to the user is improved transmission and reduced cost of trunk port cards in the PBX. With analog trunks, the loss and noise are proportional to the distance from the central office. Digital lines are essentially noise-free, and have a fixed amount of loss regardless of distance. Moreover, they support 24 trunks at a time, eliminating the need for as many as six analog trunk cards in the PBX. For smaller organizations with a requirement for fewer than about 18 trunks, there may be no cost saving, but larger companies should evaluate the economics of using digital central office trunks.

ESTABLISHING A FALL-BACK PLAN

Most cutovers proceed according to schedule. Vendors are experienced in executing the schedule and are realistic in planning how long it will take, but there is always a chance that something will cause the schedule to slip. Less frequently, an opportunity may occur to advance the schedule.

Perhaps a T-1 carrier becomes available earlier than expected or the vendor can obtain equipment earlier because of cancellation of another customer's order. A well-conceived cutover plan will include a fall-back plan in case the schedule is delayed or if the opportunity to advance it occurs.

In a simple PBX replacement project, the plan can be simple. If nothing else depends on the project, the fall-back plan is as simple as merely delaying the cutover. In many projects, however, something else of importance hinges on the telephone system. The new system may be necessary to support an organizational change, or it may be driven by a move to new quarters, change to a new data processing system, or other such related project.

The following factors should be considered in developing the fall-back plan:

- What other projects are contingent on completion of this plan?
- What penalty will be paid if this plan is delayed?
- What is the probability of delay?
- What factors could cause the cutover to be delayed (strike, equipment failures, shipping delays, delays in a preceding project, etc.)?
- What alternatives can be employed to cause the schedule to hold despite delays?

The fall-back plan should be documented and approved by higher management and those responsible for interrelated projects.

TESTING THE NEW SYSTEM

Another factor the project manager must pay close attention to is acceptance testing of the new system. The vendor will have a manual of performance tests the machine must meet before it is turned up for service. If someone who understands the terminology is available, it is a good idea to observe the vendor's technicians during some of these tests and to review the results of all of them. In addition, some performance tests should also be made. For example, these tests should include:

- Dial through tests to every station and trunk
- Feature tests at each station

- Tests of restriction levels
- Tests of special trunks such as WATS and tie lines
- Tests of automatic route advance
- Tests of console operation and attendant features
- Tests of station message detail recorder and call accounting system

Ideally, the system should be installed, connected to power, and allowed to operate for at least two weeks before cutover. Part of this interval is required for training, but it is also advantageous to allow the system to operate under power to give weak components a chance to fail.

Electronic components are frequently sensitive to heat, which is one of the reasons equipment rooms must be air conditioned or well-ventilated for some types of equipment to function. Before the equipment is turned up for service it may be advisable to heat the equipment room to the maximum operating temperature specified by the manufacturer and to leave it at that temperature for several days. If weak or temperature sensitive components exist, this process will weed them out without causing a service interruption.

POST CUTOVER FOLLOWUP

The first few days following cutover are crucial to the success of any telecommunications system. Trouble reporting or help desk positions should be established and staffed with people who are ready to react rapidly to trouble reports. If the system is capable of locally producing traffic data, it should be monitored carefully for irregularities. On new systems with no history on which to base trunk and common equipment quantities, these are usually based on estimates, and you should be prepared to react quickly to evidences of blockage.

Before the vendors are released, you should verify that all the promised documentation has been produced and turned over. At the least, the documentation prepared at cutover time should include the following:

- System operation and maintenance manuals
- Station wiring diagrams
- Trunk and station crossconnect lists
- Feature assignment lists
- Port assignment lists

- Hunt and pickup group assignments
- User assignment information:
    - Extension number
    - Room number
    - Answering location
    - Features
    - Restrictions

A process must be established internally for maintaining system records. Some systems retain records in memory and are able to produce assignment records when accessed from a terminal. In other systems, microcomputer based data base management systems are an effective way to retain records. A data base management system is also an effective way of keeping trouble reports. In large organizations it is important to be able to analyze trouble reports for patterns of defective cable and wire, equipment blockages, repeated troubles, common equipment trouble, and reports of trouble on long distance trunks and tie lines. Chapter 21 discusses telecommunications record-keeping in greater detail.

SUMMARY

When you select an equipment vendor for a new system, you should discuss with the vendor's references how smooth their cutovers were. Execution of a cutover is something that many managers experience only a few times in their careers, so without expert assistance from the vendor it is easy to overlook critical activities. A cutover requires close cooperation from people representing diverse organizations, from the user to common carriers, to equipment vendors, and even building contractors. Vendors develop specialized management techniques through years of planning and managing cutovers. By taking advantage of their specialized expertise and using the project management techniques presented in the next chapters, you can guide your organization through the rocky shoals of a major transition.

CHAPTER 15

PROJECT MANAGEMENT

Managing a large telecommunications project requires orchestrating the efforts of a diverse set of people, some of whom have a great deal of telecommunications knowledge, and others who have little idea of the technology, but nevertheless have an important role to play. Telecommunications projects range from the simple, which requires a minimum of scheduling and coordination, to the complex, in which the efforts of numerous people in diverse organizations must be directed by an individual who is responsible for completion of the project.

Project management is a discipline with several techniques that have been proven in all types of projects and are equally adaptable to launching a space ship, managing a major building construction project, or coordinating a telecommunications system. Two of the most valuable techniques, the Program Evaluation and Review Technique (PERT) and Critical Path Method (CPM) were developed by the U.S. Navy shortly after World War II to organize and manage large projects. These techniques are explained in this chapter along with a discussion of how to coordinate the critical elements that comprise most complex telecommunications projects.

This chapter presents an overview of project scheduling and management techniques. The chapter will serve as an introduction for those who are unfamiliar with project management, and a review for those who already understand the techniques. The methods are discussed in greater detail in books listed in the bibliography. Although project management and control techniques are common to many types of endeavor, this chapter discusses them specifically with respect to managing a telecommunications project. Figure 15.1 shows in flow chart form the major steps involved in the project management process.

FIGURE 15.1
Flow Chart of The Project Management Process

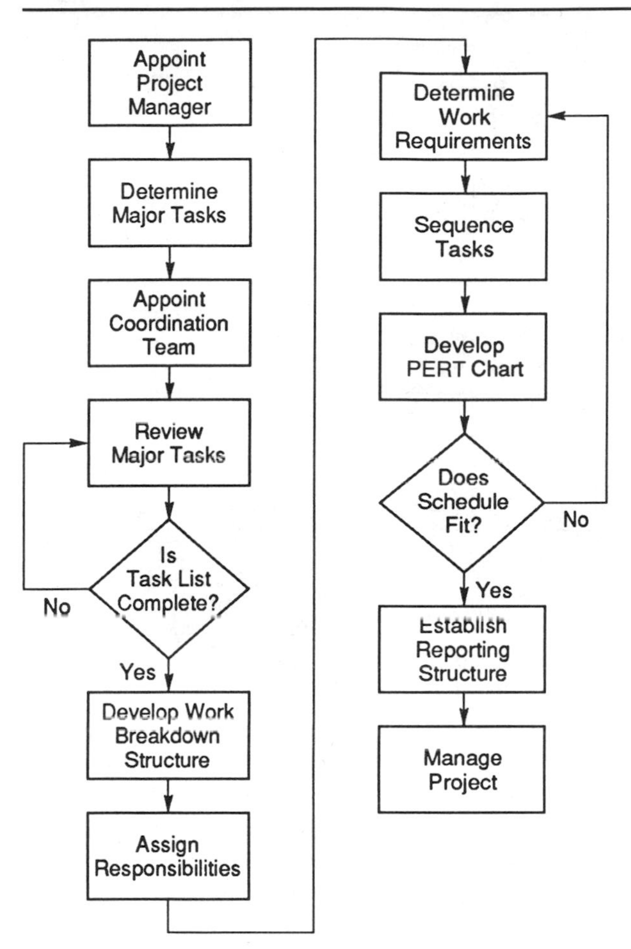

ELEMENTS OF A PROJECT

The relationships between the four main elements of a project are ranked in Figure 15.2. The total project is broken into activities, tasks and subtasks, each of which has a due date and an individual assigned as the responsible manager. Task managers will probably choose to subdivide the tasks into subtasks, but the project manager will usually find it unnecessary to track and schedule this level of detail, provided the task

FIGURE 15.2
Hierarchy of Tasks in Project Management

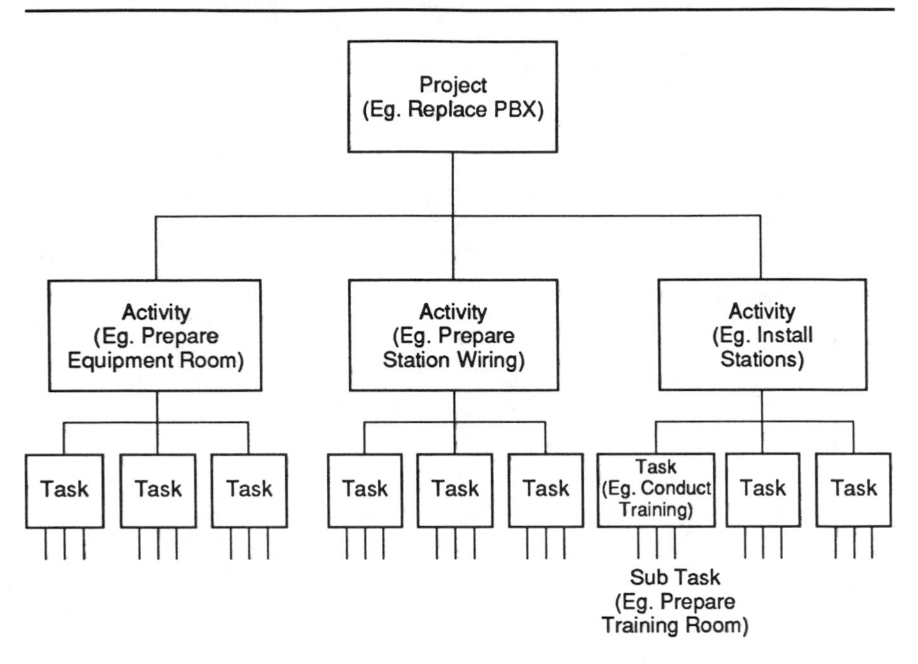

managers fully understand what is expected of them. For example, an activity in most telecommunications projects is preparation of the equipment room. Typical tasks would include completing building, electrical, and heating, ventilating and air conditioning systems. Subtasks under the building task would include framing, sheet rocking, and painting. The project manager would normally not track the subtask detail, but would leave it up to the property manager to complete.

The starting point of any project is a listing of the major activities that must be accomplished to implement the project. Frequently, the project manager is not aware of all the activities required, so these must be determined by knowledgeable members of the project team. Before the project can be selected. however, the project manager must be aware of what types of activities must be performed so the right talent can be selected for the team. The following section furnishes guidelines for selecting a telecommunications project team.

SELECTING A PROJECT TEAM

The need for a project team depends on the magnitude of the project and the complexity of the organization. Small projects in large organizations can often be managed by a handful of people, as can large projects in small organizations. In general, the larger the organization, the more different work groups that are involved in a telecommunications project, and the more it is essential to involve several representatives on the project team.

The primary responsibilities of the team members are:

- Determine a detailed list of tasks and subtasks required for project completion
- Determine the sequence of tasks
- Determine resource requirements for completing tasks
- Accept responsibility for task completion, coordinating within their own work groups as necessary
- Assist in overall schedule planning, including making necessary revisions as the project progresses

The composition of the project team is a function of the organization and the project itself. The following discussion outlines considerations in selecting team members.

Project Leader. The project leader has the key role to play in project management, and should be selected on the basis of expertise in project management rather than because of an organization role. Many companies are organized to manage projects, and in these, selecting a project manager is not difficult. Other companies that are organized on a traditional hierarchical or matrix basis will have to decide whether the project manager should be a member of the user organization, information management, engineering, or some other department with a stake in the project outcome. Smaller companies may engage an outside company to manage the project.

In addition to project management knowledge, the project manager should have enough technical expertise to understand the major tasks, and to select and evaluate key participants based on their ability to handle the tasks. In some cases it may appear that the vendor is in the best position to manage the project. Although vendors often have superior knowledge of what must be done, they are seldom in a position to direct tasks that are not directly related to an equipment transaction. It is usually most effective to appoint someone within the organization to assume overall project responsibility.

Users. Representatives of the using organization should be included on the project team in all instances except when the organization is so small that the team consists of only one or two people. Even though a project is completed satisfactorily, if the system is not properly used, the project cannot be counted as a success. Therefore, key roles of the users are coordinating training, selecting station equipment and features, and developing revised methods and procedures that will be required by the project. Representatives of the using organizations are generally responsible for operational matters. Their representatives should be thoroughly familiar with the users' requirements and should have the authority to speak for all work groups in accepting changes necessitated by the project.

Vendors. Vendors should always be represented on project teams because they understand the application of their equipment better than anyone in the using organization. As stated earlier, however, the vendor should not control the transition except in exceptional circumstances. The vendor's responsibility should be to furnish the schedules and commit to functions related to manufacturing, delivery, installation, testing, and interfacing with other vendors' equipment. Because of their involvement in numerous projects, vendors are usually an excellent source of information for developing lists of essential tasks.

Common Carriers. Representatives from the telephone company and interexchange carriers should be included on any project team. Often, the primary equipment vendor handles the direct interface with these groups, but their functions must be represented. Common carrier responsibilities are to assist in placing orders and establishing due dates for their facilities. They will arrange for forces to complete the central office and customer premises work to change lines and trunks to the new system.

Interfacing organizations. These are the groups in any company that may have a controlling interest in a project because of budgetary, strategic, or key management considerations. This may be the part of the organization that furnishes overall project management. In addition, people should be appointed to carry our the following responsibilities:

- Data processing managers for interface to computer systems
- Managers to arrange for external telecommunications circuits and systems (often the vendor's responsibility)
- Legal representatives (to deal with contracts, licenses and permits)
- Marketing representatives (where systems are being installed to provide service to customers)
- Building space or real property managers
- Office administrative managers for interface to office automation systems
- Others as required by the organization

PROJECT SCHEDULING TECHNIQUES

The most important technique a manager has for controlling a project is a detailed schedule of events. Once the schedule has been compiled, agreed on by the participants, and approved by higher management, it becomes the nucleus around which the various work groups detail their tasks and deploy their resources. Scheduling techniques range from the simple Gantt or bar chart to the PERT chart, which shows relationships between tasks.

The project manager must also determine certain critical dates. Some of these are set as management requirements; others are fixed by vendor schedules. In any case, the completion date is frequently determined by the date of other events, such as the date a major capacity addition is required, a building move is planned, or a major force addition is to occur. The scheduling techniques described in this chapter provide the tools you will need to ensure that the major tasks are completed by the required date.

The amount of detail to track in the project is a matter for the project manager's discretion. Nearly every task on the list will involve several subordinate tasks. Often, these can safely be left to the team members and only the overall result tracked by the project manager. A fundamental principle of managing a project of this kind is that every task must be assigned to an individual who accepts responsibility for completing it on time.

After the tasks have been identified, they must be listed in sequence. For example, in a PBX project the equipment room obviously must be completed before equipment installation can begin, and station wiring and switching equipment must be placed before stations can be tested.

The first step in developing a schedule is to identify the major tasks to be performed. With knowledge of what must be done, the manager knows the disciplines that should be represented on a project coordination team. By putting the right people on the team, a thorough understanding of the details of the work items can be developed. In project management terms, this is called developing the work breakdown structure. The vendors have a key role to play in this operation. Their people have undoubtedly experienced several similar transitions and know many of the tasks required.

Developing a PERT Chart

Interdependencies between projects are best shown by a PERT chart as illustrated in Figure 15.3. Each circle is a "node" on the chart and represents the completion of a major task. Arrows between nodes represent the actual tasks, and the figures in parentheses below the arrows represent the number of days (or weeks) needed to perform the activity. Activities pointing into a node must be completed before those pointing out of the node can begin. Paralleling activities are completed simultaneously and independently of one another.

There are several variations in methods of constructing PERT charts. The one shown lists the task title along the shaft of the arrow; other PERT charts show the task title in the node. Although this chart shows only one figure for the number of days to complete each task, it is frequently advantageous to show three figures: pessimistic, realistic, and optimistic. The task manager often cannot estimate precisely how long it will take, and may feel more comfortable with a range of task times. Providing three separate figures shows the project manager where the schedule may be vulnerable to slippage.

Close attention should be paid to tasks with a wide deviation between pessimistic and realistic figures, because if the pessimistic estimate occurs, the project due date may be missed. It is clear from this chart, for example, that equipment installation cannot begin until the equipment room has been prepared. This constraint is shown by the series relationship of the equipment room and the equipment installation tasks. It is unnecessary, however, to complete equipment installation before

FIGURE 15.3
Simplified PERT Chart of a PBX Cutover

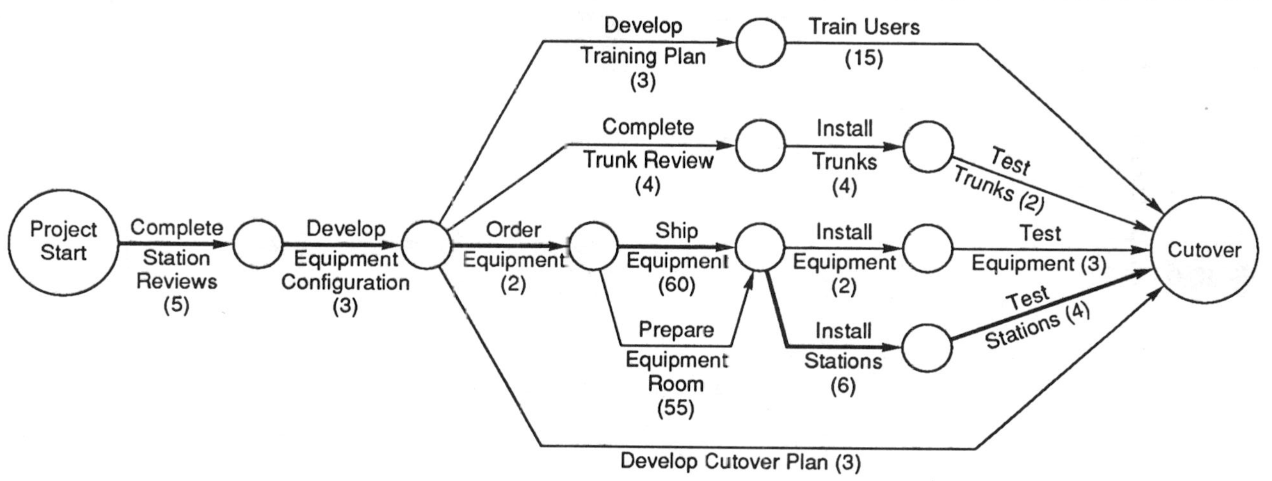

Numbers of work days to complete the tasks are shown in parentheses. The heavy line is the critical path.

station wiring can begin. This is shown by the parallel relationships between the tasks.

Frequently, a project can be shortened by reviewing the sequence of activities and performing them in parallel instead of series. Figure 15.3 is simplified to illustrate the principle of creating a PERT chart. Actual charts will contain many more activities than this illustration.

Calculating the Critical Path

By inspecting a PERT chart you can determine the longest path in elapsed work time from the start of the project until completion. This path, shown by the heavy line in Figure 15.3, is the *critical path.* Any delay in completing a task on the critical path will result in delay of the entire project unless time is made up by shortening a subsequent task. The project team should spend most of its efforts in managing tasks on the critical path because these are the ones with the greatest effect on overall project completion.

The critical path is not a fixed entity. If a task not on the critical path is delayed, the critical path can shift to a new route. Therefore, it is important to pay close attention to tasks that are close to the critical path, such as the task in Figure 15.3 of preparing the equipment room. Its completion interval is only five days longer than the equipment shipping interval, so if completion is delayed, it may shift into the critical path and delay the entire project unless the time is made up somewhere else.

If the number of work days from start to finish exceeds the time allowed for completion, the tasks on the critical path are the place to start deciding where to cut. By inspecting the chart, the project manager can also determine where task leaders may have built in a time cushion.

Drawing PERT charts can be a tedious exercise. Fortunately, many personal computer programs can accept task lists and can construct PERT charts, determine the critical path, and create work assignment lists.

Developing a Gantt Chart

A Gantt chart, as shown in Figure 15.4, is a technique for displaying the starting and ending dates for each activity. This figure displays some of the major tasks from the PERT chart as a bar chart instead of a network. The major weakness of the Gantt chart is that it does not show the dependencies between tasks. In simple projects this lack of detail is of

FIGURE 15.4
Parital Gantt Chart of PBX Cutover

Task	Resp.	Days
Complete Station Reviews	REK	5
Develop Equip. Configuration	OPD	3
Develop Training Plan	KKL	3
Complete Trunk Reviews	REK	4
Order Equipment	JLF	2
Ship Equipment	JLF	90
Prepare Equipment Room	REK	85
Develop Cutover Plan	OPD	3
Install Trunks	REK	4
Train Users	JKL	15
Install Equipment	TRE	2

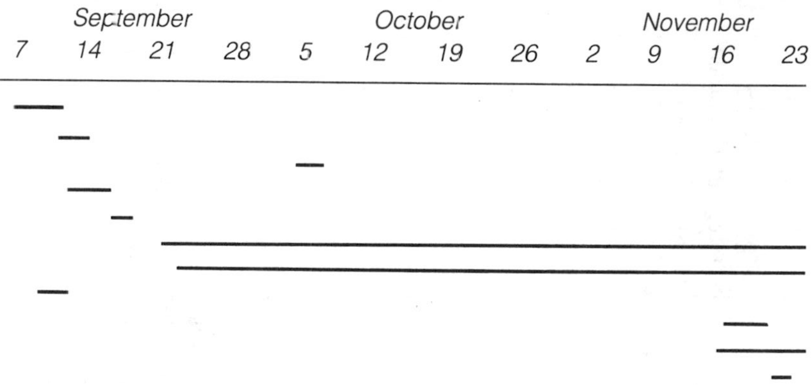

little consequence because it is not difficult for all participants to understand the relationships between tasks, but in complex projects a Gantt chart alone cannot be interpreted as easily as a PERT chart.

The main advantage of the Gantt chart method is its simplicity. Many personal computer scheduling programs support the Gantt chart, and most spreadsheets can be designed to create bar charts with little manual effort.

Developing the Work Breakdown Structure

The work breakdown structure defines in more detail exactly what each task entails. For example, Figure 15.5 is a task assignment sheet for testing a channel bank and T-1 line installation. Enough detail should be shown to eliminate any ambiguity.

The work breakdown structure must be created by someone who understands the technical details of the work to be performed. The amount of detail to show depends on what other kinds of documentation are available. It is unnecessary, for example, to detail the method of installing a PBX because it is included in the manufacturer's practices. It is unnecessary to specify how to pull, secure, and terminate station wiring because these methods are generally accepted practices. It is essential,

FIGURE 15.5
Task Assignment Sheet for Testing Channel Banks and T-1 Line

Total Program: Redesign Telecommunication System
Project: Build New Transmission Network
Activity: Replace Analog Circuits With T-1
Task: Place New T-1 Channel Banks
Work Package: Channel Bank and T-1 Line Acceptance Tests
Responsible: ____________________
W. P. 12-4 Due Date: 3-15 Hours 12
Previous W. P. 12-2 12-3
Following W. P 12-6 12-8

Description of Work: When vendor completes installation of channel banks (W P 12-2), witness all acceptance tests, which are to consist of all tests required in vendors' installation manual. When T-1 lines are installed (W. P. 12-3) perform the following: Feed a quasi-random signal source into one channel of each bank and monitor bit error rate for a period of 72 hours. Test shall continue until a 72 hour period has elapsed with no system failures and each bank provides a bit error rate of 1×10^{-7} or better.

however, that the type of wire and location of the terminations be specified in the work breakdown because they are unique to the particular installation.

The project manager should designate the kind of talents needed from other organizations for the project team based on knowledge of the organization and the nature of the project. Key responsibilities of the team are to review the major tasks identified by the project manager, to augment the list based on their knowledge of special requirements of the organization, to prepare the work breakdown structure, and to carry out assigned tasks.

At the start of the project it is important to document all concerns and assign them to a project team member to coordinate. Task lists can be effectively developed in group sessions without regard to the magnitude of the task or the sequence in which they must be performed. When a complete list of tasks has been developed, the project manager assigns them to team members to develop in more detail.

The project manager should determine which tasks are critical to completion of the project and which subtasks are essential for timely performance of the critical tasks. Subtasks should be delegated to task leaders. The objective is to keep the project simple enough that the project team can avoid getting bogged down in details. For example, in the task list that was discussed in Chapter 14, the task of preparing the equipment room might be diagramed into its own PERT chart. The equipment room task leader is in charge of all the subtasks, but the project manager and the team are interested only that the room is prepared on time, so would not be involved in the level of detail that Figure 15.5 shows.

Sequencing Tasks

The next step is to sequence the tasks, which involves identifying dependencies between tasks. The team should be alert to opportunities to perform tasks in parallel in order to shorten the overall process. It is often advantageous at this stage to break tasks into phases so that dependent tasks can begin before their preceding tasks are complete. For example, it might not be necessary to finish all work on the equipment room before installation begins. It would be possible to begin equipment installation after carpentry work is complete, but before power and air conditioning equipment have been installed.

Developing a schedule is a cut-and-try proposition. There is rarely only one way to make a schedule fit, and the team must continue to work on sequencing tasks until the critical path matches the time required for the objective completion date.

Obtaining Approvals

After the schedule is complete, the task leaders determine the resources required to complete the tasks in the time allotted. The necessary approvals are collected by each task leader, after which the project manager obtains approval for the overall project. Good management practice suggests that the project manager and task leaders have the authority to deploy the resources they have been granted to complete the project on schedule without further approval by higher management.

PROJECT MANAGEMENT AND CONTROL

After the project plan has been approved, the project manager's responsibility is to introduce the controls needed to complete it on schedule. The primary control techniques available to the project manager are:

- Task control sheets
- The schedule
- Project team meetings
- Progress reports
- Budgetary controls

A sample task control sheet is shown in Figure 15.6. The sheet shows the title of the project and a brief description of the work required. The titles of preceding and succeeding tasks are shown together with their required completion dates. The individual responsible for the task and his or her address and telephone number are shown. If appropriate, subtasks are also listed on the form. The bottom of the sheet is reserved for reporting progress, and for showing any roadblocks that jeopardize completion of the task on time. Each team member should receive a copy of the control sheet.

When a roadblock is encountered, it should be documented on the control sheet with an explanation of the action required to prevent

FIGURE 15.6
Sample Task Assignment Form

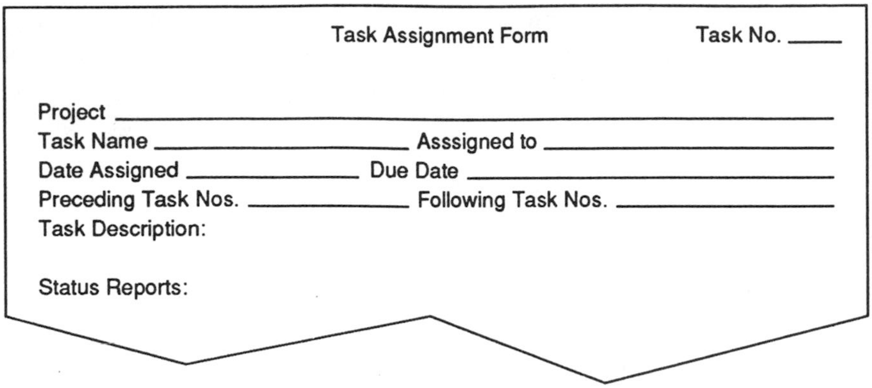

Task Assignment Form Task No. ____

Project ____________________
Task Name ____________ Asssigned to ____________
Date Assigned ____________ Due Date ____________
Preceding Task Nos. ____________ Following Task Nos. ____________
Task Description:

Status Reports:

slippage of the task. The control sheet serves as a chronological record of action taken to ensure that the task is completed on time. It also serves to communicate to other team members the status of tasks on which their tasks depend.

Project Schedule

The project schedule itself is the most important control technique because it represents the commitments of managers to complete their tasks on time. Each team member should have a copy of the schedule, and the schedule should be revised to reflect changes as they occur. Task completions should be reported to the project manager, who can show them on the schedule. Any delayed completions are flagged, and if they are critical to completion of the project, corrective action plans are developed.

Project Team Coordination Meetings

Project team meetings should be scheduled regularly to keep members informed of progress and to resolve problems as they arise. The frequency of meetings and the composition of the team depend on the complexity of the project. The fundamental purpose of the meetings is to update project team members on the status of all tasks and to discuss any concerns or roadblocks that arise. Time should be allotted for task reports. Reports should be documented in meeting minutes, and problems and action taken are documented on the control forms. Project team minutes, distributed to appropriate levels in the organizations represented in the cutover, serve as a useful vehicle to communicate progress and to alert others to potential slippage of the schedule.

Progress Reports

Progress reports should be required of each task leader and presented at team meetings. In many organizations the coordination team meeting minutes are enough to report progress to higher management.

Budgetary Controls

The project budget is often the most powerful control tool the project manager has. If an overall budget for the project has been authorized and parceled out to the project manager and to task leaders, reports of actual expenditures give the project manager insight into how well the project is proceeding.

Another important control technique is obtaining approvals of appropriate managers at designated points in the schedule. Managers should first approve the overall plan, and then as key events occur, they should be asked to approve the result. The equipment order, the station assignment plan, the equipment room design, and other such factors are examples of tasks that should be approved by higher management in most organizations.

SUMMARY

Project management techniques that builders, architects, manufacturers, defense contractors and others use for controlling the work of a diverse group of people within an organization, lend themselves well to managing telecommunications projects. This chapter has presented an overview of the techniques that project managers can use to identify the work to be performed, orchestrate the efforts of many different work groups, and control the completion of tasks for which they have indirect responsibility. Once these techniques are understood, their application becomes second nature to most organizations, and the techniques become an economical and effective way of keeping everyone informed and committed to project completion.

One of the primary objectives of project management is to ease the transition from an existing to a proposed telecommunications system. To avoid disrupting the organization any more than necessary, it is important that the details be worked out in advance and approved by appropriate levels in the organization.

CHAPTER 16

SIZING AND OPTIMIZING VOICE CIRCUITS

In the past few years manufacturers have enhanced PBXs to the point that they are relatively free of trouble, but that does not mean they can be left alone. Although maintenance can largely be ignored until trouble occurs, the network of trunks, tie lines, and long distance circuits must be administered continually. The goal of telecommunications administration is to keep the best balance between service and cost, and this balance is not easy to determine. The most important task is computing the right quantity of trunks. With too many trunks, costs are excessive. With too few, customers and employees will encounter blockage, resulting in lost business or depressed productivity. The cost/service balance point can never be determined exactly because it changes as business conditions change, but with the high cost of trunks, you should monitor load and cost results frequently to be sure the system is in reasonable balance.

Business conditions constantly change with the business cycle or season and as the business grows or contracts. The administrator's job is to assess the tradeoffs between the fixed cost of changing quantities and configurations of trunks, and the variable costs of higher rates, lost productivity, or lost business because of having the incorrect capacity.

FUNDAMENTALS OF TRAFFIC ENGINEERING

To administer a telecommunications system, information about three variables is required: grade of service, traffic load, and quantity of trunks. With data from any two variables, the third can easily be determined.

Grade of Service

Grade of service refers to the percentage of calls that encounter some form of blockage. The amount of blockage that can be tolerated is something that a network manager must determine as a business objective. If the grade of service is too high, (meaning that the probability of blockage approaches zero), many circuits will carry little or no traffic for a large portion of the business day. If the grade of service is too low (meaning that a high percentage of calls are blocked) customers and employees will encounter busy circuits, business opportunities will be lost, and productivity will be depressed.

In most business situations, circuits are designed for a grade of service of between one and five percent blockage; blockage may be higher under certain circumstances. For example, if the switch offers a trunk queuing feature, the system holds users until idle facilities are available and then connects them or rings them back. Queuing is one way to increase the traffic carried on a group of circuits without greatly affecting productivity.

Alternate routing provides another way of increasing the load on a group of circuits. For example, a group of WATS lines may be designed to overflow to MTS. If so, the grade of service level to which the lines are engineered affects only economics, and not the level of service the WATS lines deliver, because the overflow is handled by the MTS group. The decision on the percentage of calls to overflow to the alternate route is based on the difference in cost between the primary and alternate trunk groups.

Traffic Load

Traffic load, or intensity, is expressed as the quantity of traffic presented to a trunk group during the busy hour. Traffic can be measured in minutes, hours, hundreds of call seconds (*CCS*) or *Erlangs*. These terms are used somewhat indiscriminately by traffic designers, and tend to be confusing to some people, but in practice, the conversions are not difficult.

Detail billed long distance services list the time of individual calls in minutes. It is important to remember that telephone bills show *billed* minutes of usage, not *actual* minutes of usage. Calls over MTS are

typically rounded up to the nearest minute so that usage is usually less than the amount billed. This difference may become important in comparing the cost of MTS with bulk-rated services such as WATS that bill in six-second increments. A further distortion is caused by the fact that some usage requires trunk capacity, but is not billed. For example, all calls use trunk capacity during the call setup interval, but billing does not start until the call is completed.

Bulk-rated services such as WATS are typically billed in hours of usage rounded up to the nearest tenth. The Erlang is a unit of traffic intensity that is one hour's worth of usage. Named for the Danish mathematician A. K. Erlang, who pioneered traffic measurements in the early twentieth century, the Erlang is a convenient unit of measurement because it is directly equivalent to hours and can easily be converted to and from minutes by multiplying or dividing by 60.

Traffic measurements are frequently expressed in hundreds of call seconds (CCS), particularly by telephone engineers. CCS is less convenient to use than Erlangs in some cases because of the awkward conversion from minutes. Most switching systems, however, report traffic measurements directly in CCS. To convert among the various measurements, the conversion formulas are:

$$1 \text{ Erlang} = 60 \text{ minutes} = 36 \text{ CCS}$$
$$1 \text{ CCS} = 1/36 \text{ Erlang}$$
$$1 \text{ CCS} = 1.67 \text{ call minutes}$$

Where traffic is measured in small increments, CCS is a more convenient unit to use than Erlangs. For example, a designer can often estimate the total load on a network by knowing something of the calling habits of its users. The typical residence generates about 3 CCS per station during the busy hour, and the typical business about twice that amount. Expressing per-station loads in CCS results in units that are more convenient to handle than Erlangs.

Traffic intensity is expressed as the relationship between the calling rate (called *peg count* in telephone terminology) and the length of an average call (called *holding time*). For example, if a trunk group receives 100 calls in the busy hour and the calls have an average holding time of 180 seconds (three minutes), the trunk group must handle 180 X 100 = 18,000 call seconds. This load is expressed more conveniently as 180 CCS.

Busy Hour

Every network has certain periods of the day when traffic is at its peak. Traffic tables are written to show the amount of traffic that can be handled during the busy hour of the day. Actually, within the busy hour, higher peaks will occur, and blockage will be higher than the design objectives, but it is uneconomical to design trunk groups to handle absolute peaks. Instead, most networks are designed to cause some traffic to be blocked and some to overflow to other services.

Most large processor-controlled PBXs have the capability of measuring traffic usage. If a traffic study is taken over a long enough period, the busy hour can be determined with reasonable accuracy. The study period should be based on management's knowledge of the business. If one season is markedly busier than others, a study should be taken then. For example, a department store would probably get the best view of its busy hour by studying its trunk loads during a week between Thanksgiving and Christmas, and a college would see its peak during registration week. Normally, one week is long enough to study traffic.

One common definition of the busy hour is the average amount of traffic that flows during the ten highest days of the year. To determine the ten highest days, however, presumes that you have a great deal of traffic data and knowledge about seasonal peaks. A more pragmatic approach is simply to schedule traffic studies for a week when you know the load is high.

PBX trunk designers frequently use a *bouncing busy hour,* which is determined by picking the busiest hour of the week for each of several trunk groups on the PBX. For example, the busy hour for local trunks might be at 10:00 AM on Monday, for WATS lines it might be 8:00 AM on Wednesday, and so on.

Offered Versus Carried Load

Network designers must be careful to distinguish between offered and carried load. Toll statements show the amount of traffic actually carried, but these figures are misleading for computing the number of dedicated lines that are required to carry the same amount of load. The difference lies in the time required for call setup and handling incomplete and blocked calls.

Call setup time is the time of network signaling and ringing. The circuit is occupied during setup time, but usage is not billed. Calls are incomplete because of busy, no answer, and long distance network blockage. Where the amount of blockage is significant, traffic formulas and tables are unreliable for determining circuit quantities.

If traffic usage information comes from a long distance bill or a call accounting system statement, it must be adjusted to convert billed minutes to usage minutes. To illustrate, assume that the long distance carrier rounds calls down to the nearest minute if the excess is six seconds more than a full minute, and up if the fractional minute exceeds six seconds. If users have an equal probability of hanging up during any of the 10 six-second intervals in a minute, calls will be rounded down 10 percent and rounded up 90 percent of the time. The average upward rounding will be:

$$\frac{60 \text{ seconds} - 10 \text{ seconds}}{2} = \frac{50}{2} = 25 \text{ seconds}$$

Using the same process, the average downward rounding will be three seconds. The total adjustment will be:

$$\begin{aligned} .1 \times -3 \text{ seconds} &= -0.3 \text{ seconds} \\ .9 \times 25 \text{ seconds} &= \underline{22.5 \text{ seconds}} \\ & \ 22.2 \text{ seconds} \end{aligned}$$

Therefore, 22 seconds per call should be deducted from billed usage to convert to actual usage in determining how many circuits are required.

Next, total usage should be adjusted upward to compensate for setup time and incomplete calls. Figure 16.1 illustrates the process. The upward adjustment of 16 seconds and the downward adjustment of 22 seconds are close to balancing each other out. If you are attempting to determine how many fixed long distance circuits are required to handle traffic that currently flows over MTS, it is often safe to ignore the adjustment.

If you use the adjustment, however, here is an example of how it is applied. Assume that a company uses only MTS for its long distance service, and it wants to determine whether WATS is economical. It must determine how many minutes of traffic would be handled by WATS. If the long distance bill is analyzed, it is found that 5,000 minutes of traffic were billed, and that 1,250 calls were placed for an average holding time of four minutes per call. Using the process in Figure 16.1, we determine

FIGURE 16.1
Calculation of Call Time Adjustment

Answered	*Not Answered*
70 % answered	25 % busy
× 18 seconds setup time	× 5 seconds
12.6 seconds	1.25 seconds
	5 % no answer
	× 36 seconds
	1.8 seconds
Total Usage Adjustment	
Answered call setup time	12.6 sec.
Busy call time	1.25 sec.
No answer time	1.8 sec.
Composite time adjustment	15.7 sec. = 16 seconds

that 16 seconds of holding time per call were used, but not billed for. As discussed earlier, however, 22 seconds per call were billed for but not used. Therefore, we deduct six seconds per call or a total of 125 minutes to compute a total carried load of 4,875 minutes. In a later section we will see how to convert this traffic volume into busy hour volume that can be applied to traffic tables.

Quantity of Trunks

When the load and grade of service are known, the quantity of trunks can be calculated from traffic tables, which are included in Appendix A. Three types of tables are in general use, and are selected on the basis of how users behave when blockage is encountered. In traffic engineering terms the behavior is described as blocked calls held (BCH), blocked calls cleared (BCC), and blocked calls delayed (BCD).

The BCH theory assumes that users encountering blockage immediately redial and continue dialing until the call is completed. (With the automatic last number redial feature on many telephone sets, this assumption is not far from reality). The BCC theory assumes that calls disappear from the system when blockage is encountered. In actual practice they may not disappear, but they may wait long enough before redialing that they can be treated as a new call. The BCD theory assumes that calls are placed in queue and are held until a circuit is available. Figure 16.2 shows the relationship between the three blockage theories and the formulas that are used to compute the quantity of trunks when load and grade of service are known. Appendix A includes traffic tables that have been calculated from all three traffic theories.

In the telecommunications world all three of these theories are applied. To illustrate, assume that a PBX is equipped with WATS lines that are allocated to three classes of users. When WATS blockage is encountered, the first class of callers overflows to MTS and calls are completed. The second class is held in queue until an idle WATS line is available, and the call is then completed. The third class is merely given a fast busy or reorder signal, and must redial the call later.

The first class of users is a BCC situation. Since the blocked calls overflow to another group, a blocked call is completed and does not reappear to the primary trunk group. The trunks serving this class of callers would be designed using Erlang B tables. The second class of callers represents a BCD situation. Calls are held in queue until completed or until the caller hangs up. The trunks serving the second class of users would be designed using Erlang C tables. The third class of

FIGURE 16.2
Blockage Theories and Their Corresponding Formulas

Queue Discipline	*Formula*	
BLOCKED CALLS CLEARED (BCC)	ERLANG B	$P = \dfrac{A^n/n!}{\sum_{x=0}^{n} \dfrac{A^x}{x!}}$
BLOCKED CALLS HELD (BCH)	POISSON	$P = e^{-A} \sum_{x=n}^{\infty} \dfrac{A^x}{x!}$
BLOCKED CALLS DELAYED (BCD)	ERLANG C	$P = \dfrac{\dfrac{A^n}{n!} \dfrac{n}{n-A}}{\sum_{n=0}^{n=1} \dfrac{A^x}{x!} + \dfrac{A^n}{n!} \dfrac{n}{n-A}}$

P = Probability of Blockage
A = Traffic Density in Erlangs
n = Number of Servers
e = Naperian Logarithm Base (2.718+)
X = Number of Busy Channels

users represents a BCH situation, and the trunks serving them would be designed using Poisson tables. The Poisson formula assumes that blocked callers will redial immediately, so more trunks are computed for a given amount of blockage than Erlang B tables.

To illustrate the use of traffic tables, assume that first group generates four Erlangs of traffic during the busy hour. From the Erlang B table in Appendix A.1, a portion of which is reproduced as Table 16.1, it can be seen that ten trunks are required if the grade of service is one percent blockage or eight trunks if five percent blockage is required. To determine the number of trunks required, pick the required grade of service and read down the Erlang or CCS column until you find a number equal to or greater than the traffic load; then read the number of trunks from that row.

By referring to the full Erlang B table in Appendix A-1, it can be seen that eight trunks carries 4.0 Erlangs of traffic at three percent blockage. This means that approximately 3% × 4 Erlangs or .12 Erlangs overflow to MTS.

Assume that the third class of users also generates four Erlangs of traffic. Traffic will be handled over WATS lines while they are available, after which traffic is blocked. From the Poisson tables in Appendix A-2, which are partially reproduced in Table 16.2, it can be seen that ten trunks are required to handle this traffic at one percent blockage and nine trunks at five percent blockage. At the lower blockage level the two tables

TABLE 16.1
Partial Erlang B Table for 1% and 5% Blockage

	1%		*5%*	
Trunks	*Erlangs*	*CCS*	*Erlangs*	*CCS*
1	.01	.4	.05	1.8
2	.15	5.4	.38	13.7
3	.46	16.6	.9	32.4
4	.87	31.3	1.52	54.7
5	1.36	49.0	2.22	79.9
6	1.91	68.8	2.97	107.0
7	2.50	90.0	3.75	135.0
8	3.14	113.0	4.53	163.0
9	3.78	136.0	5.36	193.0
10	4.47	161.0	6.22	224.0

TABLE 16.2
Partial Poisson Table for 1% and 5% Blockage

	1%		*5%*	
Trunks	*Erlangs*	*CCS*	*Erlangs*	*CCS*
1	.011	.4	.05	1.9
2	.15	5.4	.36	12.9
3	.44	15.7	.82	29.4
4	.82	29.6	1.36	49.1
5	1.28	46.1	1.97	70.9
6	1.79	64.4	2.61	94.1
7	2.33	83.9	3.28	118.0
8	2.92	105	3.97	143.0
9	3.50	126	4.69	169.0
10	4.14	149	5.42	195.0

happen to require the same number of trunks, but at higher blockage one more trunk is required for Poisson than for Erlang.

Poisson tables usually require more trunks than Erlang B tables. Many analysts prefer to use only Erlang B for private networks because they believe Poisson tables overstate the number of trunks required. Large private networks and most common carrier networks are constructed with high usage trunk groups that are designed for a high degree of trunk occupancy using Erlang B tables. Final trunk groups that are designed with Poisson tables for a low percent of blockage, carry traffic that overflows from the high usage trunks.

Both Poisson and Erlang tables share a common assumption about the distribution of traffic. They assume that the distribution of arrivals follows a Poissonian distribution, as shown in Figure 16.3, and that service time follows an exponential distribution, as shown in Figure 16.4. These assumptions have been shown to be reasonably accurate for telephone traffic. The same assumptions may hold true for other situations such as data messages arriving at a group of dial-up computer ports. If these assumptions about service and holding time can be accepted, traffic tables can be used to analyze a variety of queuing situations such as computing the quantity of terminals to serve a particular process, the number of modems in a modem pool, or quantities of shared apparatus of all types in telephone and data networks.

A fundamental principle of network design is that of the greater efficiency of large trunk groups. Notice from Table 16.1 that one trunk

FIGURE 16.3
Poissonian Distribution

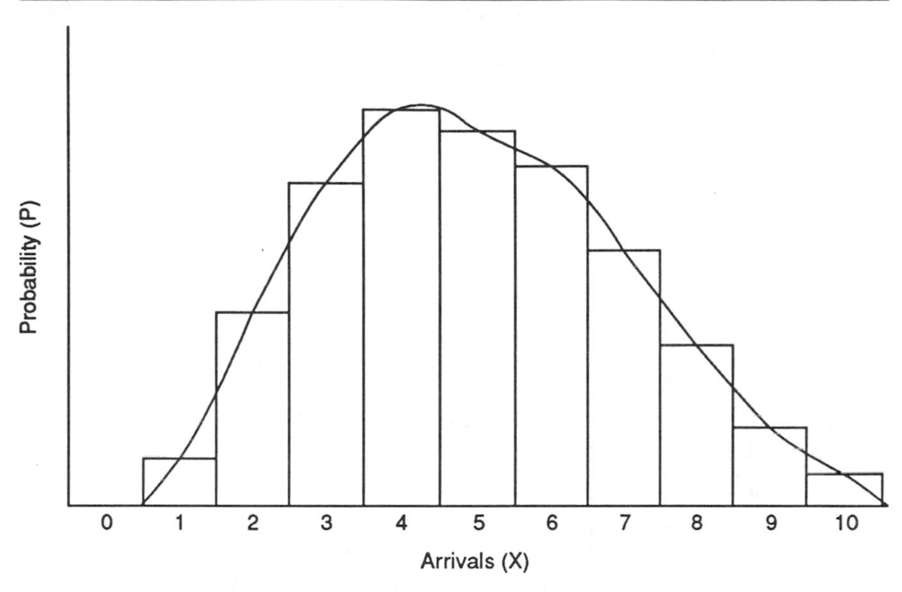

This curve shows the probability (p) that a given number of arrivals (x) will occur.

carries only .05 Erlangs of traffic at five percent blockage, but 10 trunks carry 6.22 Erlangs. If the average capacity per trunk is plotted in a curve such as the one in Figure 16.5, you can see that adding only onc trunk to a group dramatically increases capacity when the size is fewer than about 10 trunks. Beyond that, the traffic-carrying capacity of an added trunk increases, but not so rapidly as in small trunk sizes.

It is important to recognize another assumption implicit in the Poisson and Erlang formulas; the formulas assume that arrivals are independent of one another, and that they are not influenced by external events. The formulas are therefore valid in telecommunications design only when a sufficient number of channels is available to serve users at the designed grade of service. If users encounter excessive blockage, they generally make repeated attempts to access the network. Ineffective attempts add to the offered load, but do not add to the carried load. When the ratio of ineffective to effective attempts becomes high, say more than 10 or 15 percent blockage, the traffic formulas are invalid, and the network consumes an excessive amount of its capacity in handling ineffective attempts.

FIGURE 16.4
Exponential Distribution

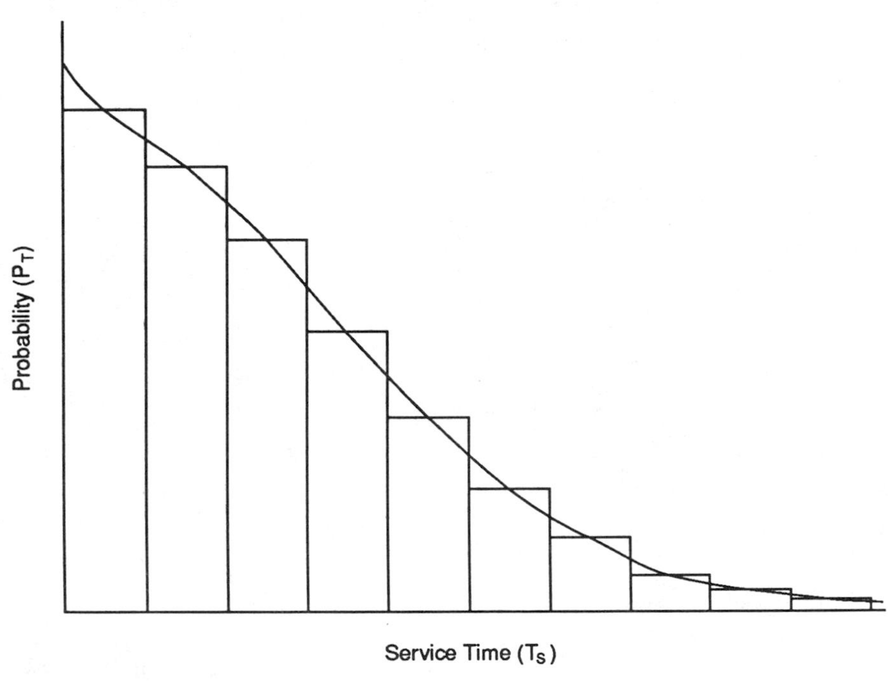

This curve shows the probability (p) that a service time (t_s) will be a given number of seconds.

Many networks exhibit a load/service curve that resembles Figure 16.6. As load approaches the knee of the curve, blockage increases gracefully. A point is reached, however, when service degenerates rapidly, and the design formulas are invalid. Telecommunications managers should be aware that blockage of this order has an adverse effect on the productivity of the work group, and may result in lost business from blocked incoming calls.

Using Erlang C Tables

Erlang C tables, which are included in Appendix A-3, are used to determine the number of lines required in a queueing system for a given load and an objective grade of service. This process is commonly used to

FIGURE 16.5
Carrying Capacity Per Trunk as a Function of Trunk Group Size

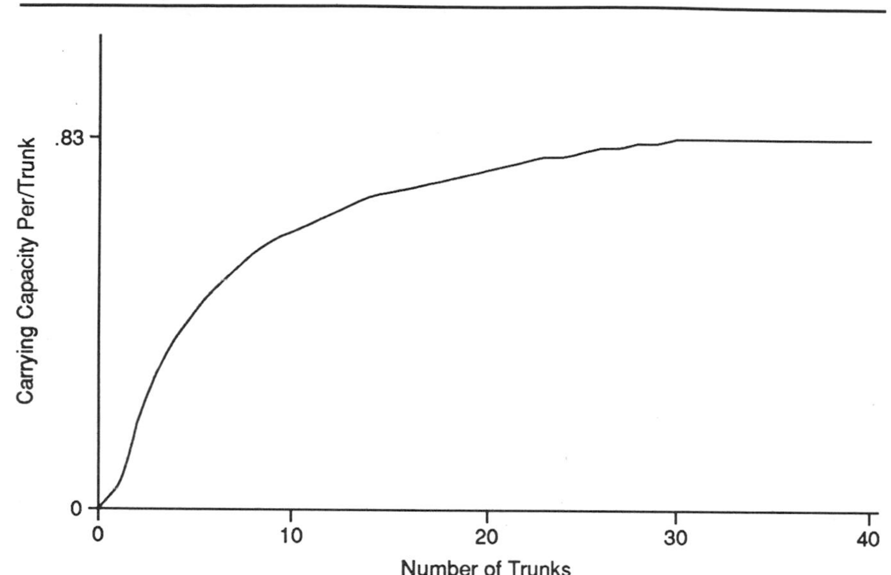

determine the number of trunks required for an automatic call distributor. With the tables you can find the number of trunks (or "servers" as they are called in the Appendix) and from that, determine the average delay of all calls being delayed. Determining the average delay is the primary objective of the analysis.

To illustrate, assume that the expected traffic load to an ACD is 1.5 Erlangs, and that the average holding time per call is 4.0 minutes. Also assume that you have an objective of no more than 15 percent of the calls delayed. From Appendix A-3, a portion of which is shown in Table 16.3, you can see that the table gives a choice in Column N of from two to eight servers. Column P_d shows that if three servers are used, 23.68 percent of the calls will be delayed. Four servers will delay 7.46 percent of the calls, so we will use four to stay within the objective.

To find the average delay for all calls, multiply the average holding time by Column D1. The result is:

$$4.0 \text{ Min.} \times 0.03 = 0.12 \text{ Min.} = 7.2 \text{ seconds}$$

Bear in mind that this figure does not apply to the calls delayed, but to all calls averaged over the system. To determine how long the delayed

FIGURE 16.6
Typical Load/Service Curve of a Network

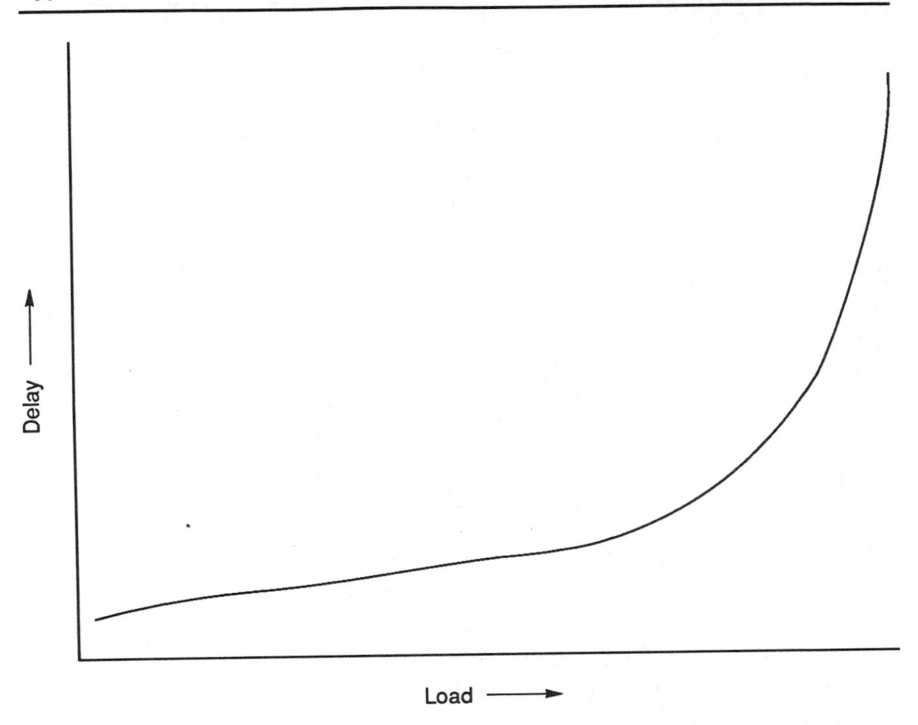

calls are delayed, multiply the average holding time by Column D2. The result is:

4.0 Min. x 0.4 = 1.6 Min.

The probability that calls will be delayed longer than a specific time is given in the remaining columns of the table. For example, the factor 0.045 is shown in the column headed .2. This means that there is a 4.5 percent probability that a call will be delayed longer than 20 percent of the holding time, or 0.8 minutes. With these tables, ACD managers can determine the correct number of positions to have open at any given time to handle an expected workload.

SOURCES OF USAGE INFORMATION

Traffic formulas are sufficiently accurate to enable the designer to do a reasonable job of determining circuit quantities, but a lack of valid data

TABLE 16.3
Partial Erlang C Table

					Probability of Delay (All Calls) of Time "t"						
A	*N*	P_d	D_1	D_2	.2	.4	.6	.8	1	2	3
1.5	2	0.6429	1.29	2	0.582	0.526	0.476	0.431	0.39	0.236	0.143
	3	0.2368	0.158	0.667	0.175	0.13	0.096	0.071	0.053	0.012	0.003
	4	0.0746	0.03	0.4	0.045	0.027	0.017	0.01	0.006	0.001	
	5	0.0201	0.006	0.286	0.01	0.005	0.002	0.001	0.001		
	6	0.0047	0.001	0.222	0.002	0.001	0.001				
	7	0.001		0.182							
	8	0.0002		0.154							

A = Offered Load in Erlangs
N = Number of Servers Required
P_d = Probability that a call will be delayed
D_1 = Average Delay of All Calls
D_2 = Average Delay of Delayed Calls

is the bane of every network designer. The design job is particularly tricky for new businesses where historical data is unavailable. Lacking valid information, the designer must make educated guesses and be prepared to make adjustments when actual usage data can be collected. This section discusses sources of traffic usage data and advises you on methods of interpreting and adjusting it.

Switching System Data

Many modern digital PBXs and tandem switching machines provide traffic usage data routinely. Some systems dump the information on a printer in the form of undigested register readings that require a trained analyst to choose and interpret the readings. A sample of such information is shown in Figure 16.7. Other systems process the readings and display them on a printer or screen in a form that the designer can use to calculate trunk quantities.

FIGURE 16.7
Unformatted Register Readings From a Northern Telecom SL-1 PBX

Trunk Group No.	002 CO		Trunk Group Type
Trunks Equipped	00009	00009	Trunks Working
Incoming Usage	0000052	00028	Incoming Peg Count
Outgoing Usage	0000031	00024	Outgoing Peg Count
Outgoing Overflow	00000	00000	All Trunks Busy
Toll Peg Count	00002		
	003 CO		
	00024	00024	
	0000000	00000	
	0000158	00098	
	00000	00000	
	00102		
	005 CO		
	00004	00004	
	0000000	00000	
	0000075	00048	
	00014	00016	
	00004		
	006 DID		
	00005	00005	
	0000028	00016	
	0000000	00000	
	00000	00000	
	00000		

Older machines such as the AT&T Dimension collect the information, but do not display it locally. The AT&T Remote Maintenance and Testing System (RMATS) Center is able to collect the data remotely and present it in summary form. A portion of a study from an AT&T Dimension ® PBX is shown in Figure 16.8. Other manufacturers also have RMATS centers that can perform remote traffic measurements. Not all PBXs, however, have this capability.

The traffic information collected is useful only if you have knowledge of how the least cost routing is programmed. To gain knowledge of the total outgoing load, for example, you must determine how much traffic is carried on a one-way outgoing trunk group, and how much overflows to the two-way group.

To illustrate, consider the Dimension PBX printout shown in Figure 16.8. The report shows usage data from each of six trunk groups. This particular PBX is designed with no toll restriction. Outgoing toll traffic uses WATS bands 1 or 5, which consist of two trunks each and are assigned to trunk groups 22 and 23. If these trunks are blocked, traffic overflows to one-way outgoing local trunks, and from there to the two-way local trunk group. The "OVL" line shows how much traffic overflows from one group to another, but before the usage study can have any meaning, you must be aware of how the least cost routing is programmed so you can tell where it overflows.

Many older PBXs and key telephone systems have no provision for collecting usage information. In this case, external traffic usage recorders (TURs) can be attached. A TUR is connected to the central office trunks, and measures the number of call attempts and usage in Erlangs or CCS.

Collecting data through the PBX itself or with a TUR is by far the most accurate source of usage information. Since the TUR logs total usage, it is unnecessary to make adjustments for time billed but not used and for time used but not billed. The primary drawback to this method is that only the trunk group is shown, with no clue to the destination. Therefore, without additional information from a station message detail recorder or toll statement, TUR data is not useful for determining which WATS bands are required.

Usage Sensitive Bills

Local measured service, toll, and WATS bills are the second most useful source of usage information. Local measured service bills are generally not detailed, but those produced by some telephone companies show total

FIGURE 16.8
Dimension PBX Partial Traffic Summary Report

Traffic Summary Report

Trunk Group Summary

TRK GRP			*TRK SIZE*	*POLL 1*	*2*	*3*	*4*	*5*	*6*	*7*	*8*	*ABBH*
15	INCOM	CCS	50	89	91	79	78	88	0	0	0	85
15	INCOM	PEG	50	217	195	219	126	162	0	0	0	183
16	DP-DR	CCS	10	37	37	37	37	37	0	0	0	37
16	DP-DR	PEG	10	11	11	11	11	11	0	0	0	11
17	TT-DR	CCS	6	34	41	40	33	38	0	0	0	37
17	TT-DR	PEG	6	455	523	498	481	533	0	0	0	498
18	CO-1WO	CCS	8	88	95	127	101	96	0	0	0	101
18	CO-1WO	PEG	8	80	76	83	79	87	0	0	0	81
18	CO-1WO	OVL	8	0	0	1	0	0	0	0	0	0
19	CO-2W	CCS	13	275	320	283	247	256	0	0	0	276
19	CO-2W	PEG	13	152	173	164	146	70	0	0	0	141
19	CO-2W	OVL	13	0	0	3	0	0	0	0	0	0
19	CO-2W	CCSI	13	216	208	198	203	143	0	0	0	193
20	CO-1WI	CCS	3	18	14	12	14	12	0	0	0	14
20	CO-1WI	PEG	3	5	7	4	5	5	0	0	0	5
21	WT-1WO	CCS	2	63	64	54	58	53	0	0	0	58
21	WT-1WO	PEG	2	24	20	28	1	27	0	0	0	20
21	WT-1WO	OVL	2	36	52	23	2	32	0	0	0	29
22	WT-1WO	CCS	2	41	45	30	32	44	0	0	0	38
22	WT-1WO	PEG	2	35	37	14	7	20	0	0	0	22
22	WT-1WO	OVL	2	13	15	9	2	11	0	0	0	10
23	WT-1WO	CCS	1	12	13	11	4	8	0	0	0	9
23	WT-1WO	PEG	1	4	9	6	4	7	0	0	0	6
23	WT-1WO	OVL	1	1	7	3	5	8	0	0	0	4

Legend:
INCOM = Intercom traffic
DP-DR = Dial Pulse Digit Receiver
TT-DR = Touchtone Digit Receiver
CO-1WO = Central office one-way outgoing trunks
CO-2W = Central office two-way trunks
CO-1WI = Central office one-way incoming trunks
WT-1WO = WATS one-way outgoing trunks
CCS = Hundreds of call seconds
PEG = Peg count (number of attempts)
OVL = Overflow
TRK GRP = Trunk Group
ABBH = Average Bouncing Buy Hour

minutes of usage. WATS bills are also not detailed by some IECs except on special request. When a trunking study is being performed, the detail should be obtained, either from the station message detail recorder (if the PBX is equipped) or by requesting toll detail from the WATS provider. A later section demonstrates how to convert this detail into a traffic design. Within the limitations discussed earlier, detailed toll bills can be used to determine the three most important variables in a networking study: usage, time of day distribution, and cost.

Telephone Company Busy Study

Most telephone companies can perform a busy study on selected trunk groups. This study shows the number of times that callers attempted to access an incoming or two-way trunk group and encountered an all-trunks-busy condition.

A busy study is better than no information at all, but it has several significant drawbacks that limit its usefulness for administering a network. If no busy conditions are logged, it is evident that too many trunks have been provided, but the question of how many are actually required is impossible to answer without disconnecting a trunk or two and rerunning the study. A busy study is useful for incoming and two-way groups, but is of no value on outgoing trunk groups, so these must be measured by another method.

Some telephone companies are capable of doing a full usage study on PBX trunks, usually for a fee. This usage study is as valid as a study made from the PBX itself, but the telephone company may not be able to study non-switched trunk groups such as WATS and tie lines.

Station Message Detail Recorder

The station message detail recorder (SMDR) can also provide reasonably accurate usage information. Some data processing services and call accounting systems are able to read the data stream from the SMDR and record the number of calls and total usage, and, in some cases, also approximate the cost. For this method to be completely effective for sizing all types of trunks, both local and long distance calls must be recorded on the SMDR. Figure 16.9 shows a sample SMDR printout. This can be converted to accurate traffic data by connecting a personal computer to the SMDR port of the switching system, and capturing the

FIGURE 16.9
Sample SMDR Traffic Summary

TRK	*Date*	*Time*	*Duration*	*DGT-Dialed*
01	03/07	10:42:29	00:01:28	5095345211
04	03/07	10:42:32	00:01:29	
06	03/07	10:42:56	00:00:37	2065326766
02	03/07	10:42:59	00:02:55	6022336118
07	03/07	10:43:01	00:01:52	8022763566
08	03/07	11:43:03	00:01:04	
03	03/07	11:43:07	00:04:49	2062425871
11	03/07	10:43:14	00:03:16	9164810870
01	03/07	10:14:25	00:00:53	4157790939
05	03/07	10:14:29	00:00:57	
02	03/07	10:14:44	00:01:31	

An SMDR printout has a minimum amount of formatting. Calls are registered in the sequence in which they are completed. The lines without digits dialed are incoming calls.

call details with a telecommunications program that writes the records to disk. The call details can then be read into a spreadsheet and sorted and summarized to collect peak-hour statistics. Be certain, however, that the switch puts out all calls on the SMDR port, not just those that exceed a particular length.

SAMPLE CASE STUDY

The process of using traffic engineering techniques to optimize a network is best illustrated with a case study. The study starts with a review of bills from the telephone company and the interexchange carrier (IEC). The details are summarized in Figure 16.10. A one-week traffic study results in the busy hour statistics shown in Figure 16.11

Preliminary Analysis of The Traffic Study

From inspecting the telephone bill and the traffic study, it is evident that this PBX has trunking problems. There are too few lines in both WATS groups, which means that excessive traffic is overflowing from Band 1 to 5, and from Band 5 to MTS. There are also too few 800 lines. The

FIGURE 16.10
Networking Case Study
Local and Long Distance Billing Details

Trunk Group	*No. Trunks*	*Cost*	
Local Trunks	12	$ 960	
Tie Trunks	3	$ 360	
Long Distance Services	*Minutes*	*Calls*	*Cost*
IntraLATA long distance	1,200	300	$ 250
InterLATA long distance			
Intrastate	2,200	500	570
Interstate	3,000	850	930
Band 1 WATS	3,690	939	960
Band 5 WATS	9,366	2,383	1,330
Band 5 800	8,694	2,320	$2,080

Note: WATS costs include the fixed line costs

FIGURE 16.11
Networking Case Study
Traffic Study Summary

Trunk Group	*No. Trunks*	*Peak Load (Erlangs)*	*Blockage (Percent)*
Local Trunks			
1-Way Outgoing	3	1.4	11.9
1-Way Incoming	3	1.2	9.0
2-Way	6	3.1	5.8
Band 1 WATS	2	.42	5.8
Band 5 WATS	2	1.2	24.7
Band 5 800	3	1.12	7.9
Tie Lines	2	0.72	13.1

shortage can also be verified by looking at the number of overflows, which show on every 800 service bill. If calls were detailed, which they are not in this case, it would be possible to determine how much traffic is overflowing from each band. Since details are not available, we will make the only possible assumption, that the distribution of traffic on the

WATS lines is identical to that billed by the IEC on the detailed long distance bill.

It is also safe to assume that this PBX has too few local trunks. The traffic study shows how much of the one-way incoming and one-way outgoing traffic overflows to the two-way group. This overflow is by design. The one-way trunks are designed as high usage trunks, and are sized to cause about 10 percent of the traffic to overflow to the two-way group. It is a matter of judgment whether to include one-way groups in the design or to use only two-way groups. The reason for providing one-way trunks it to prevent a large number of calls in one direction from blocking traffic in the other.

The Erlang B table shows that more than five percent of the calls are blocked in the two-way group. To reach a one percent blockage level, we should have eight trunks, but before any changes are made, we must first determine what to do about the long distance traffic. Some long distance is being carried on the local trunks. It may be economical to add WATS lines, in which case the local trunks may be deloaded to the point that six trunks are enough.

Finding Terminating Points

The objective of this step in the process is to determine, for each point of origin, where traffic is terminating. In network design terms, these are called "sources" and "sinks." Sophisticated call accounting systems can produce most of the information you need in machine-readable form, but in some cases, you must develop the information from other sources as discussed in the previous section. To do a complete networking study, you must estimate usage by terminating numbering plan area (NPA).

Since traffic is being carried on WATS now, it is useful to determine its distribution by WATS band. This is accomplished by summarizing traffic by the terminating area code and using overflow to MTS as the pattern of distribution. The area code summary is accomplished by keying a random sample of calls into a spreadsheet. The sample should show area code, number of minutes, and cost. If many calls terminate at a company location or a single city, it is advisable to key the entire telephone number because with telephone number detail it is easy to summarize calls to evaluate the effectiveness of foreign exchange lines and tie lines. In this analysis, we are particularly interested in the amount of overflow from existing tie line groups. For example, Figure 16.11

shows 13.1 percent overflow from the tie line group. By evaluating traffic to that location, you can determine whether the addition of another tie line is justified.

With the long distance calls keyed, the next step is to sort them by area code, count the calls, and summarize the minutes and cost to each area code. If less than a 100 percent sample was taken, the calls should be expanded at this point. The result is shown in Figure 16.12. The percentage distribution of detailed long distance calls is multiplied by the undetailed WATS calls to arrive at a total distribution by WATS band. The reason for keying the calls into a spreadsheet is evident. With the details in machine-readable form, it is easy to sort and expand the details.

From the amount of overflow, you can see that additional WATS trunks are required if you plan to retain WATS. A WATS rating model,

FIGURE 16.12
Distribution of Long Distance calls by WATS band using MTS calls as the key

Interstate MTS Calls Distributed by WATS Band				
WATS Band	*No of Calls*	*No. of Minutes*	*Cost*	*Percentage Minutes*
1	422	1622	$ 458.21	54.1%
2	43	172	47.41	5.7%
3	221	784	253.37	26.1%
4	8	32	12.26	1.1%
5	106	390	158.75	13.0%
Total	800	3000	930.00	100.0%
Total Undetailed WATS				
1	939	3690	$ 960.00	
5	2383	9366	1,330.00	
Total	3322	13056	$2,290.00	
Total WATS Plus MTS Distributed by WATS Band				
1	2218	8681	$1,696.34	
2	233	921	178.70	
3	1089	4196	851.82	
4	43	171	36.69	
5	538	2087	456.45	
Total	4122	16056	$3,220.00	

which you can create on a spreadsheet, is invaluable for determining the quantity and band of WATS lines to choose. The distribution of traffic, heavily concentrated in three bands with little traffic to the other bands, makes it clear that a virtually banded WATS would be advantageous. The method of calculating the efficiency of virtual WATS was illustrated in Chapter 10. One type of virtually banded WATS is "dial-one" WATS, which uses central office trunks to carry long distance traffic. If a long distance analysis indicates that you should change from banded to dial-one WATS, you must determine the impact on central office trunks because all WATS lines will be removed, and the total load handled over local trunks. The question is how much traffic will be added to local trunks. The overflow traffic, of course, already travels over local trunks, so we are concerned only with WATS volumes.

First, we must deduct the traffic volume of MTS calls that were billed for, but not used. For the 850 interLATA MTS calls at 22 seconds per call, an adjustment of 312 minutes is required. Since WATS traffic is billed on the same six-second billing increment as dial one WATS traffic, no adjustment for billing increment is required for WATS. No call setup time adjustment is required for either WATS or MTS calls because we are using machine-collected data that includes call setup time. The total billing adjustment is, therefore, -312 minutes. Assuming there are 22 work days in the month and that 16 percent of the traffic flows during the peak hour, the peak hour adjustment is 2.4 minutes or 0.04 Erlangs.

The total traffic that will be carried over the local trunks will, therefore, be:

Local (including MTS traffic)	5.7 Erlangs
WATS (converted from WATS to MTS)	1.62 Erlangs
Billing Adjustment	-0.04 Erlangs
Total Local Trunk Traffic	7.28 Erlangs

From the Erlang B tables in Appendix A-1, we find that 14 trunks are required to carry this traffic at one percent blockage. Note that this assumes that local traffic is carried on two-way groups. If the existing one-way trunks are retained, and all the long distance traffic is assumed to flow over the two-way group, that group will carry a total of:

Local	3.1 Erlangs
WATS (converted from WATS to MTS)	1.62 Erlangs

Billing Adjustment	-0.04 Erlangs
Total Local Trunk Traffic	4.68 Erlangs

This requires 11 trunks, which makes the total local trunk requirement 17 trunks. This example illustrates the greater traffic-carrying efficiency of larger trunk groups. In this case, three more trunks are required to handle the traffic on one-way groups.

Many PBXs are not capable of summarizing traffic by the busy hour, so some way must be found of inferring the busy hour from a total month's traffic. There are several ways of making this inference. The most accurate is to tally calls from a call accounting system. Details can also be tallied from the long distance statement, but bear in mind that there may be little or no overflow during off-peak hours, and heavy overflow during peaks. The long distance statement may therefore exaggerate the calling peaks.

A third way is to use a rule of thumb as shown in Table 16.4. This table shows what percentage of calls are likely to have occurred during the peak hour of a normal business day. For example, assume that the only information about 800 service was that 144.9 hours were billed during the month. If there were 21 work days in the month, and if the company worked a 12 hour day, we could assume that the peak hour traffic was:

$$144.9 \text{ hours}/21 \text{ days} = 6.9 \text{ hours/day X } 14\%$$
$$= 0.966 \text{ hours peak.}$$

TABLE 16.4
Rules of Thumb for Converting Total Traffic to Busy Hour Traffic

Length of Day in Hours	*Percent of Transactions During Busy Hour*
4	34%
5	27%
6	23%
7	19%
8	17%
9	16%
10	15%
12	14%
24	12%

From the Erlang B table in Appendix A1 it can be seen that four trunks will give a blockage factor of slightly less than two percent.

Tie Line Calculations

The traffic study shows that 13.1 percent is overflowing from the two-way dial repeating tie lines. These trunks carried a peak hour load of 0.72 Erlangs. Assuming they are used for 12 hours per day, we can assume we had a total month's usage of:

$$\frac{0.72 \text{ Erlangs}}{14\%} \times 21 \text{ days} = 108 \text{ hours.}$$

At $120 each, the three lines cost $360. If they carried 108 hours or 6,480 minutes of traffic, their cost was 5.6 cents per minute. The terminating location is in the first mileage band for dial-one carriers, and we will assume that the first mileage band costs 21.1 cents per minute. If 13.1 percent overflows, about 850 minutes of traffic will be carried on dial-one WATS for a cost of $179. Adding a fourth tie line costs $120, but saves $59 so it is justified if you expect the traffic volumes to remain at the present level. From the traffic tables you can determine that four tie lines will carry 1.5 Erlangs of traffic during the busy hour. We can extrapolate that to the total month by the following:

$$\frac{1.5 \text{ Erlangs}}{14\%} \times 21 \text{ days} = 225 \text{ hours.}$$

By adding one trunk we more than double the amount of traffic carried during the month, and should have little overflow to MTS. To verify the actual amount of overflow, you could tally calls from the long distance statement and determine exactly the number of minutes and cost of overflowing calls.

Removing the tie line overflow traffic from the local trunks removes about 0.08 Erlangs of traffic from the local trunks, which is not enough to justify reducing the number of local trunks.

Tie Line Networks

Companies with large amounts of traffic between several locations should consider a network of tie lines. These networks can become complex, but the principles of designing them are the same as those we have discussed.

Traffic volumes are summarized by sources and sinks. Network nodes are selected on the basis of these factors:

- Existing switching machine capacity
- Traffic volumes
- Cost of alternative services

As the network grows more complex, the switching machines at the nodes must have enough intelligence and be programmed to select the most economical route. The designer's objective is to provide enough capacity to minimize long distance cost.

SUMMARY

In designing a network, you are always working with incomplete data and shooting at a moving target. It is possible to be mathematically precise, but still produce a design that works poorly, simply because the assumptions you must make are too broad, and the data is too inaccurate. The moral, of course, is that you must make frequent measurements to monitor the actual results, and adjust trunk quantities accordingly.

All networking studies are made on the basis of history, which may not repeat itself in the future. A number of assumptions must be made, many of which may fail to carry into the future. Therefore, any time major rearrangements are made it is imperative that the results of the change be monitored closely. Frequent traffic studies should be taken, and the results analyzed to keep the right degree of balance between cost and service.

CHAPTER 17

SIZING AND OPTIMIZING DATA CIRCUITS

A telecommunications manager is often faced with the job of designing and rearranging data communications services. The primary data communications alternatives are dial up, polled multidrop, point-to-point, packet switched, or shared by means of time division or statistical multiplexers. Not only are there cost and service penalties to be considered in choosing an alternative, you are also constrained by existing equipment and protocols. The question frequently arises of how an increase in load will affect circuit performance, or what will happen if more terminals are added to a multidrop line or a statistical multiplexer. The answer can be determined in either of two ways: try it and see what happens, or model the circuit mathematically and test it in theory before you go to the expense of connecting it.

Mathematical modeling is an intimidating process to some managers. Fortunately, however, personal computer software can be used to insulate you from the arithmetic. The result is easy to interpret and much less expensive to apply than the cut and try alternative. This chapter explains the use of modeling formulas for the most common data communications networks and shows techniques for creating spreadsheet models that can be used to enable you to predict the consequences of a change before it is actually applied.

The queuing theory techniques discussed in Chapter 16 can also be used to size some data network components. For example, queuing theory is equally applicable to trunks on a PBX or dial-up ports on a computer. Whenever data uses a circuit switched network, it is indistinguishable from voice, and circuits are sized in exactly the same manner. The primary difference in designing circuits is that the number of call attempts and holding time are apt to be significantly different in a data network compared to voice. Therefore, rules of thumb for sizing voice circuits are inaccurate for data.

Packet switched and private line data circuits demand modeling techniques that differ significantly from the queuing formulas used in a voice network. Data terminals usually share a voice grade telephone circuit by one of several methods, as shown in Figure 17.1. The voice circuit can be divided into multiple narrow frequency bands, time shared

FIGURE 17.1
Data Circuit Sharing Methods

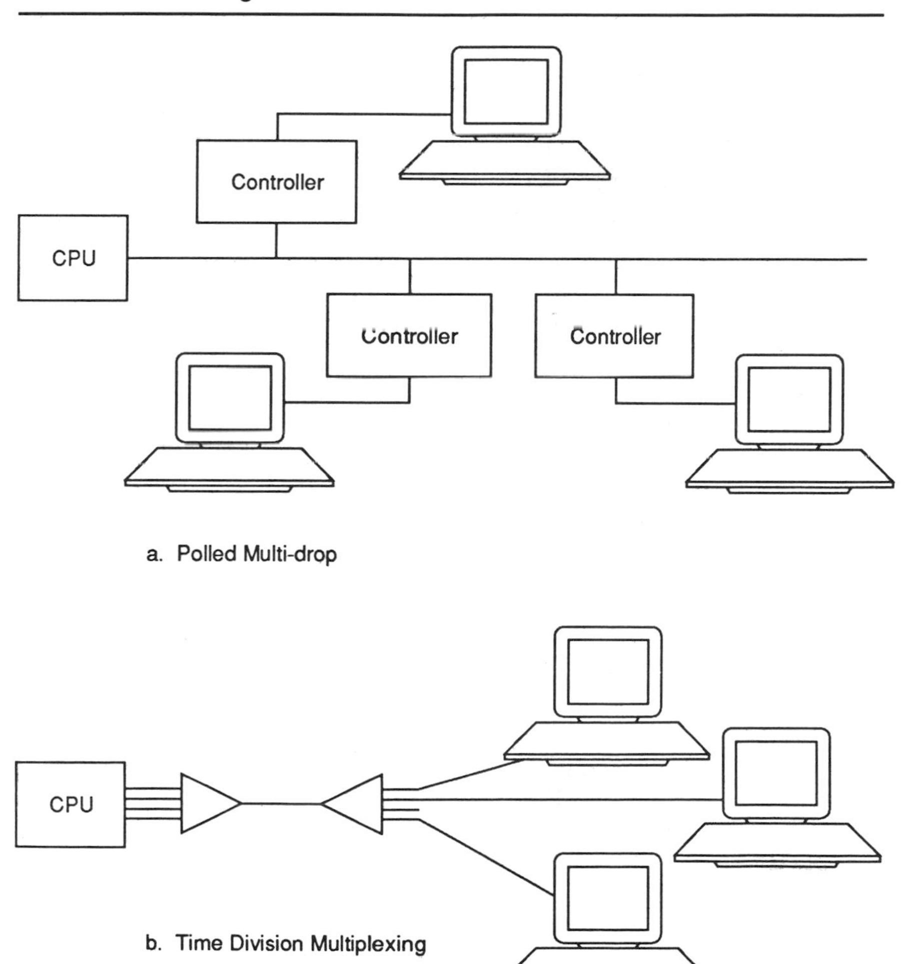

by giving each device access to the full bandwidth of the circuit for a short time, or polled, in which a central controller sends a signal to each device, offering it access to the circuit long enough to send a message.

A fourth method, carrier sense multiple access with collision detection (CSMA/CD), is used in local area networks. Unlike other circuit-sharing methods, CSMA/CD has no central controller. Each station contains the access protocol in its control circuitry. Stations listen to the network to determine if it is idle, and if so, they transmit a packet of data. If packets collide the stations both back off and again attempt to transmit after a random time interval.

Each method of circuit sharing has its advantages and drawbacks, which are more fully discussed in several of the books listed in the bibliography. The purpose of this chapter is not to discuss the pros and cons of different circuit sharing methods, but to give you an overview of modeling techniques that are used by designers to control circuit performance and by managers to prevent response time problems or to help determine their cause when they occur.

INDICATORS OF CIRCUIT PERFORMANCE

Analogous to blockage and transmission impairment in a voice network, data circuit performance is usually measured in terms of response time and error rate. Error rate may be evident to the user in some asynchronous circuits where a parity light flashes on a terminal or incorrect characters appear in text, but synchronous transmission, with its automatic error correction, shields the user from the effects of errors except when they affect response time.

Response Time

Response time is defined as the elapsed interval between sending an input message to a host computer and the arrival of the reply at the terminal device. Figure 17.2 is a diagram of the physical arrangement of a typical circuit showing where the five elements of response time fit. Although only a single terminal, circuit, and host are shown, the circuit may be shared among multiple terminals, with each terminal separated from the others by the protocol.

FIGURE 17.2
Circuit Arrangement For Response Time Model

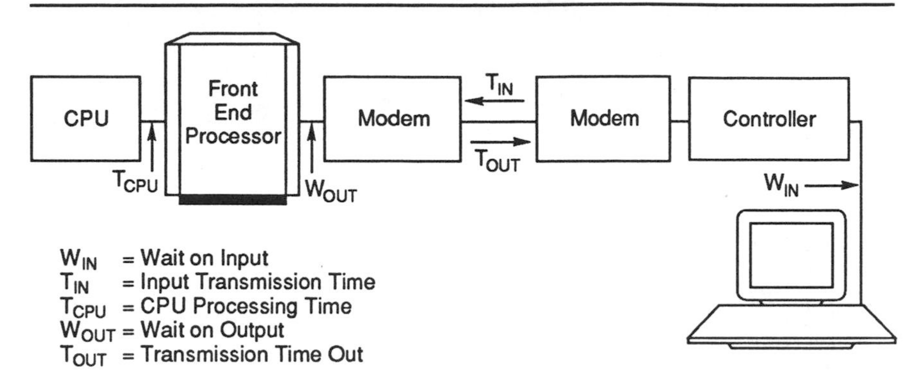

Two different definitions of response time are frequently used; both start when the enter key is pressed to send a screen full of data to the host. One definition measures the elapsed time from pressing the enter key until the first character of the response returns to the terminal. The other is the time from pressing the enter key until the last character of the response returns. From a telecommunications manager's point of view, the first definition is preferable because it removes the variability of the length of the output message from the response time calculation. Figure 17.3 is a block diagram of a data circuit response time model.

Wait on Input (W_{IN}) is the interval from pressing the enter key until the network is ready to accept the message from the terminal. In a point-to-point circuit the terminal does not have to wait, but in a multi-drop circuit a terminal must wait until it is polled before sending a data block.

Input Transmission Time (T_{IN}) consists of three elements: modem delay, circuit propagation delay, and the input message transmission time. Modem delay is the time it takes the modem to condition itself for sending a message. In half-duplex circuits it is called modem turnaround time. Propagation delay is the time it takes a signal traveling over the transmission medium to reach the distant end. Message transmission time is the time it takes the characters in the input message to be transmitted over the circuit at the speed of the modem. For example, if 1,200 bits per second are being transmitted, 120 bits could be transmitted in 120/1200 = 0.1 seconds.

FIGURE 17.3
Data Circuit Response Time Model

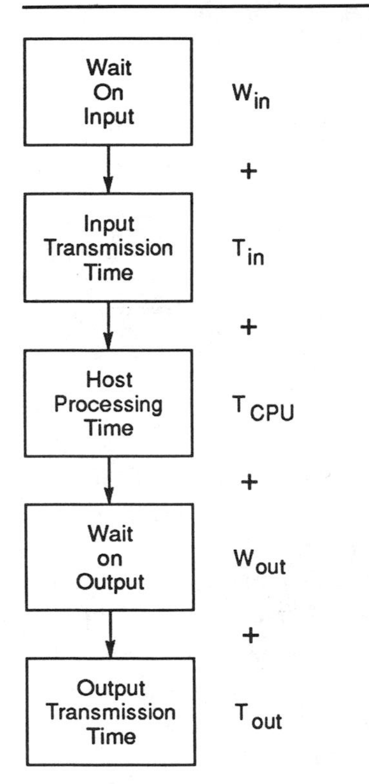

CPU Processing Time (T_{CPU}) is the time it takes the CPU to process the input message and prepare an output message. This time is outside the control of the telecommunications manager. In this chapter we will consider it to be a fixed value of 0.5 seconds. In practice the time can be obtained by measurement or from statistics supplied by the data processing department.

Wait on Output (W_{OUT}) is the counterpart to input waiting time. It is the time the host waits before sending the output message to the terminal. In some protocols the host continues polling while it processes the input message; in others polling is delayed until processing is complete.

Output Transmission Time (T_{OUT}) is the counterpart to input transmission time and consists of the same three components as T_{IN} except for message transmission time, which is computed as the length of time it takes to send the output message.

To illustrate the concept of the response time model, assume that a circuit has the following characteristics:

Wait on Input (W_{IN}): 0.03 seconds
Input Transmission Time (T_{IN}):
Modem Turnaround Time: 0.025 seconds
Circuit Propagation Delay: 0.015 seconds
Input Message Transmission Time: 0.01 seconds
$T_{IN} = 0.025 + 0.015 + 0.01 = 0.05$ seconds
CPU Processing Time (T_{CPU}): 0.5 seconds
Wait on Output (W_{OUT}): 0.03
Output Transmission Time (T_{OUT}):
Modem Turnaround Time: 0.025 seconds
Circuit Propagation Delay: 0.015 seconds
Output Message Transmission time: 0.02 seconds
$T_{OUT} = 0.025 + 0.015 + 0.02 = 0.6$ seconds
Total Response Time $= W_{IN} + T_{IN} + T_{CPU} + W_{OUT} + T_{OUT}$
$= 0.3 + 0.05 + 0.5 + 0.3 + .06$
$= 1.21$ seconds

Response time is easy to calculate once you have accurate values for the variables. In this section we will discuss the variables briefly and cover them in more detail in the design discussions that follow.

Wait on Input and Output

On a point-to-point circuit where the terminal is connected directly to the host, input waiting time is zero, or is so small that it can be ignored. On polled multi-drop circuits, a terminal cannot send data until its turn comes in the polling cycle. Calculation of input waiting time is one of the main design criteria in a multi-drop circuit.

Waiting time is a function of the number of terminals on the circuit and the characteristics of the messages. Data flows at all times in a multi-drop circuit, but a great deal of it is *overhead* or non-information bits. As discussed later, polling messages, which contribute nothing to information transfer, are flowing continually in a multi-drop circuit so that waiting time can never reach zero.

Waiting time is also a function of the line protocol. Three different models are in common use. In a half-duplex line held protocol, the host suspends polling while it processes the input message and prepares the response. In a half-duplex line released protocol, the host continues to poll while it processes the message. In a full-duplex protocol, input and output messages can flow independently of one another. The effect of these protocols on response time is shown later.

Modem Delay

Modems in half-duplex circuits must reverse each time the circuit changes direction. Modem turnaround time is obtained from the manufacturer's specifications. The circuit designer controls this variable by selecting modems with a fast turnaround time. In most applications, this factor can be changed only by changing modems.

Propagation Delay

Propagation delay is the time it takes a signal to travel from one end of the network to the other. Propagation delay is a function of the type of transmission medium, but is generally not controllable by the designer except within narrow limits. If you choose a digital circuit, for example, you have no control over the transmission medium except as a function of the vendor from which you obtain the circuit. On intrastate circuits, the distance is so short that propagation delay can usually be ignored, but in longer circuits, signals propagate at about 80 percent of the speed of light, or about 150,000 miles per second. A 3,000 mile circuit would have a delay of:

3,000 Mi./150,000 Mi. per sec. = .02 seconds or 20 ms.

Satellite circuits have much longer delays because of the distance the signal travels from earth to satellite and back. It is usually safe to assume that propagation delay is 0.25 seconds on a satellite circuit.

Message Transmission Time

Message transmission time is a function of the number of characters in the input and output messages, and the number of characters per second (c/s) the circuit transfers. On synchronous circuits the number of c/s is determined by the modem speed divided by 8, the number of bits per character. On asynchronous circuits, modem speed is divided by 10 because each character has an overhead start and stop bit. For example, if an input message is 20 characters long on an asynchronous circuit operating at 4800 b/s (480 c/s), each character has a transmission time of 1/480 seconds. Twenty characters would have a transmission time of 20/480 or 0.042 seconds.

Throughput

Throughput, or *transfer rate of information bits* (TRIB) is defined as the number of information bits *correctly* transferred per unit of time. The time unit is almost universally the second, so the throughput definition becomes:

$$\frac{\text{Information Bits}}{\text{Seconds}}$$

The throughput formula accepted by American National Standards Institute is:

$$\text{TRIB} = \frac{\text{Number of Information Bits Transferred}}{\text{Time to Get Bits Accepted}}$$

The key words in the throughput definition are "information" and "correctly received." To calculate throughput you must deduct all the non-information bits that flow in the circuit, and the effects of bits or blocks received in error. An asynchronous character, illustrated in Figure 17.4, carries two bits of overhead baggage with every character transmitted. If the circuit is error-free and is fully occupied, an asynchronous circuit can never have a throughput greater than 80 percent of the modem speed.

A synchronous circuit sends information in blocks sandwiched between header and trailer records. As illustrated in Figure 17.5, the header consists of synchronizing or framing, address, and control characters. The trailer consists of *cyclical redundancy check* (CRC) or *block check characters* (BCC) that are used to detect and correct errors. The header and trailer characters are used to route blocks, control errors, delimit frames, and control data flow. Overhead is much lower in a synchronous circuit than in an asynchronous circuit if the block length is

FIGURE 17.4
Asynchronous Character

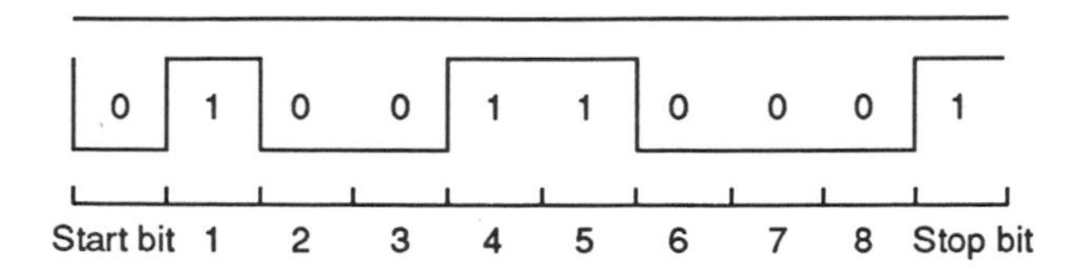

FIGURE 17.5
Synchronous Frame

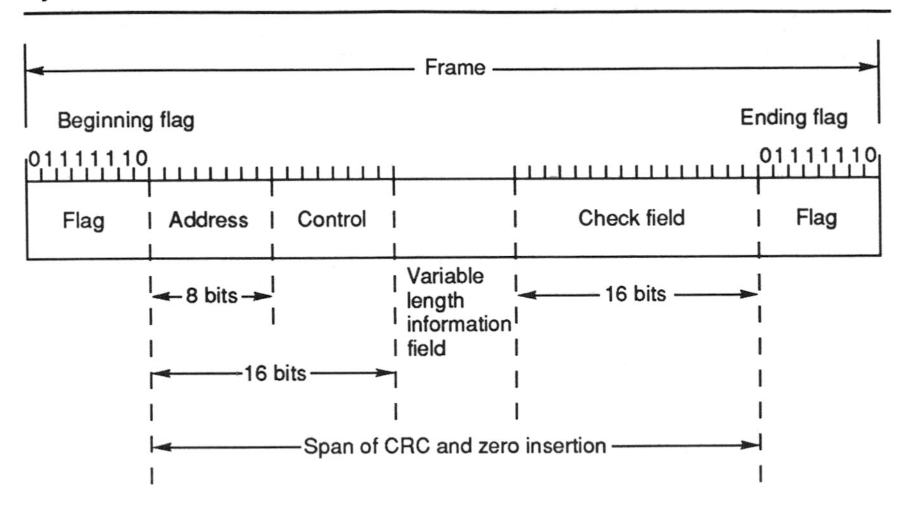

appreciable compared to the header and trailer characters. Part of the design process consists of determining the correct block length.

Throughput formulas are used not only in applications where terminals or other data devices are connected directly to a host, but also are used to compute throughput between multiplexers or concentrators and the host.

OPTIMIZING POINT-TO-POINT CIRCUITS

The generalized throughput formula discussed above is converted to specifics by substituting actual variables that are a function of the circuit characteristics and the protocol. For example, in a synchronous protocol such as Synchronous Data Link Control, the point-to-point circuit throughput would be:

$$\frac{K(M\text{-}C)\ (1\text{-}P)}{M/R + \Delta T}$$

Where:

K = No. of Information Bits per Character
M = Information Block Length in Characters

C = No. of Overhead Characters per Block
P = Probability That Block Will Be Retransmitted Because of Error
R = Transmission Rate in Characters per Second
ΔT = Time Between Blocks in Seconds

Sample Throughput Calculation

The probability of error (P) is calculated from the circuit error rate by the formula:

$$P = 1 - (1\text{-BER})^{(M+C)K}$$

Where:

BER = Error rate

For example, assume that a circuit has a bit error rate (BER) of 1 X 10^{-5} (0.00001), that block length (M) is 900 characters, that overhead characters (C) per block are 6, and that there are 8 information bits per character. The probability of error would be:

$$\begin{aligned} P &= 1 - (1 - 0.00001)^{906 \times 8} \\ &= 1 - (0.99999)^{7248} \\ &= 1 - .93 \\ &= .07 \end{aligned}$$

This formula compensates for blocks that must be retransmitted because of error, and takes into consideration the block length, modem speed, and the number of overhead bits.

The time between blocks, ΔT, is the amount of time consumed because of the characteristics of the circuit. See Figure 17.6 for a graphic representation of ΔT. The modem turnaround time (or RTS/CTS delay as it is sometimes called) is a fixed characteristic of the modem. Two such delays are encountered in a half-duplex transmission, once from terminal to host, and once from host to terminal. Likewise, two propagation delays are encountered in each transaction. Added to this figure is the amount of time required for an acknowledgement to transit the circuit.

Assume that modem turnaround time is 0.15 seconds, and that circuit propagation time is 0.02 seconds. If the circuit is operating at

FIGURE 17.6
Elements of Transmission Time Between Blocks (ΔT)

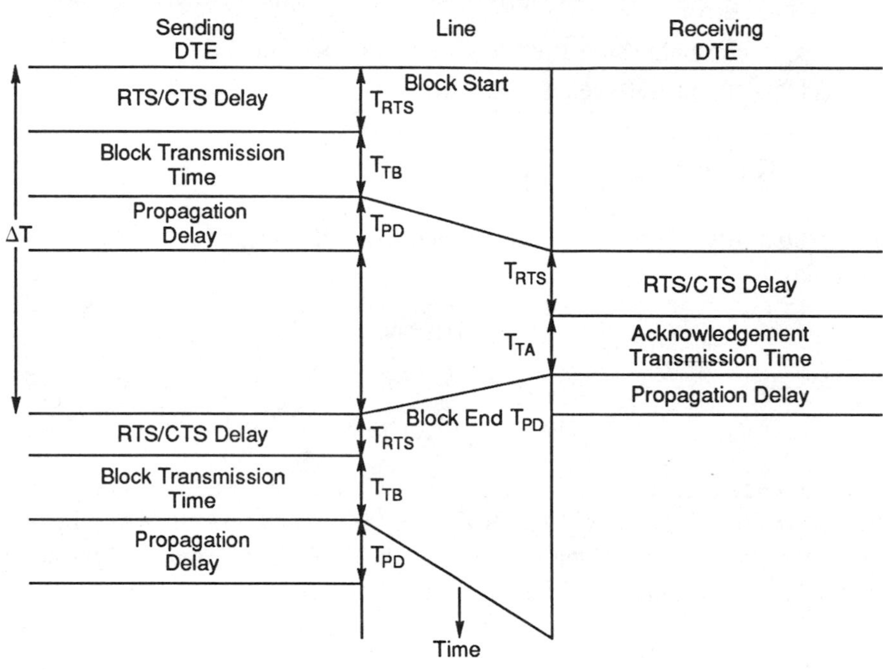

T_{RTS} = RTS/CTS Delay

T_{TB} = Block Transmission Time

T_{PD} = Propagation Delay

T_{TA} = Acknowledgement Transmission Time

4,800 B/S (600 characters per second) and if the acknowledgement message is six characters, the time required for the acknowledgement is 6 characters X 1/600 or 0.01 seconds. The ΔT figure would be:

$$(2 \text{ X } 0.15) + (2 \text{ X } 0.02) + 0.01 = 0.35 \text{ seconds}$$

To illustrate throughput calculation, assume that a circuit configured like the one in Figure 17.2 has the characteristics shown in Figure 17.7. Throughput would be:

$$\text{TRIB} = \frac{8(900\text{-}6)(1\text{-}.07)}{(900/600) + .35} = \frac{6651}{1.85} = 3595 \text{ B/S}$$

FIGURE 17.7
Spreadsheet Model for Calculating Throughput of a Point-to-Point Data Circuit

Throughput Calculation Model

Bit Error Rate (BER) =	0.00001	
Block Length (M) =	900	Characters
Circuit Length =	3000	Miles
Modem Speed =	4800	BPS
Modem Delay =	0.15	Seconds
Throughput =	3595	BPS
Overhead Characters (C) =	6	Characters
Acknowledgement Characters =	6	Characters
Bits per Character (K) =	8	Bits
Probability of Error (P) =	0.070	
Transmission Rate (R) =	600	CPS
Propagation Delay =	0.020	Seconds
Ack Transmission Time =	0.01	Seconds
Time Between Blocks (ΔT) =	0.350	Seconds

Although the throughput formula is complex, it can easily be programmed into a spreadsheet using the above throughput formula. Figure 17.7 is the output table of a spreadsheet model that calculates throughput when any of the variables are changed. The model comprises the throughput formula discussed above. The five top rows are the most frequently changing variables, and can easily be altered to determine the effect on throughput. Propagation delay can automatically be calculated as a function of circuit length, or it can be entered directly into the table. If the spreadsheet has a graphics feature, the response time can be plotted, as shown later, as one of the circuit variables such as information block length (M).

Computing Response Time on a Point-to-Point Circuit

As mentioned earlier, wait time on input and output are zero or so close to zero that they can be ignored in a point-to-point circuit. Response time then becomes the sum of input and output transmission time and central processing unit processing time. Assuming the throughput is 3,600 b/s or 450 c/s, and that CPU processing time is 0.5 seconds, T_{IN} and T_{OUT} would be:

Input Transmission Time (T_{IN}):

Modem Turnaround Time: 0.15 seconds

Circuit Propagation Delay: 0.02 seconds

Message Transmission time: 6 Char./450 c/s = 0.013

$T_{IN} = 0.15 + 0.02 + 0.01 = 0.18$ seconds

Output Transmission Time (T_{OUT}):

Modem Turnaround Time: 0.15 seconds

Circuit Propagation Delay: 0.02 seconds

Message Transmission time: 6 Char./450 c/s = 0.013 Sec

TOUT = 0.15 + 0.02 + 0.01 = 0.18 seconds

Response time is:

$$\text{Total Response Time} = W_{IN} + T_{IN} + T_{CPU} + W_{OUT} + T_{OUT} = 0 + 0.18 + 0.5 + 0 + 0.18 = 0.9 \text{ sec}$$

Controlling Throughput

The circuit designer has control over several variables that affect point-to-point circuit performance. The modem speed is selected by the designer within the limits that the circuit's bandwidth will support. Circuit characteristics can be selected, ranging from low-speed subdivided channels of a voice frequency circuit to a 56 kb/s data circuit. Even wider bandwidth can be chosen under some common carrier service offerings.

The circuit bit error rate is a function of the circuit type. Digital circuits have the lowest error rate, and tend to be the most expensive, although the price of digital circuits is expected to drop rapidly as increased fiber optic circuit capacity becomes available.

The number of information bits per character and number of overhead characters per block are functions of the protocol and are not easily changed. Propagation delay is a function of circuit type, and is generally outside the designer's control once the basic transmission medium is selected. Satellite circuits have a much longer delay than terrestrial circuits, but the amount of propagation delay cannot be controlled by the designer except by choosing the circuit type. Modem turnaround time is a function of the modem and can be controlled, within limits, by the type of modem that is selected. Fast reversal modems can be effective in improving throughput.

The least expensive variable to change is block length. The process of optimizing a circuit consists of changing variables and plotting the results in a throughput curve. A combination of modem speed, block length, and error rate is chosen to yield the maximum throughput. Figure 17.8 shows a curve plotted by the throughput model of Figure 17.7. A table at the bottom of the spreadsheet can enable you to enter a range of block lengths, for which the model calculates the corresponding throughput. By viewing the graph on the spreadsheet you can readily see which block length gives maximum throughput.

OPTIMIZING POLLED MULTI-DROP CIRCUITS

Compared to a point-to-point circuit, a polled multi-drop circuit is complex to design because there are so many more variables to contend

FIGURE 17.8
Throughput Curve Calculated by the Spreadsheet Model of Figure 17.7

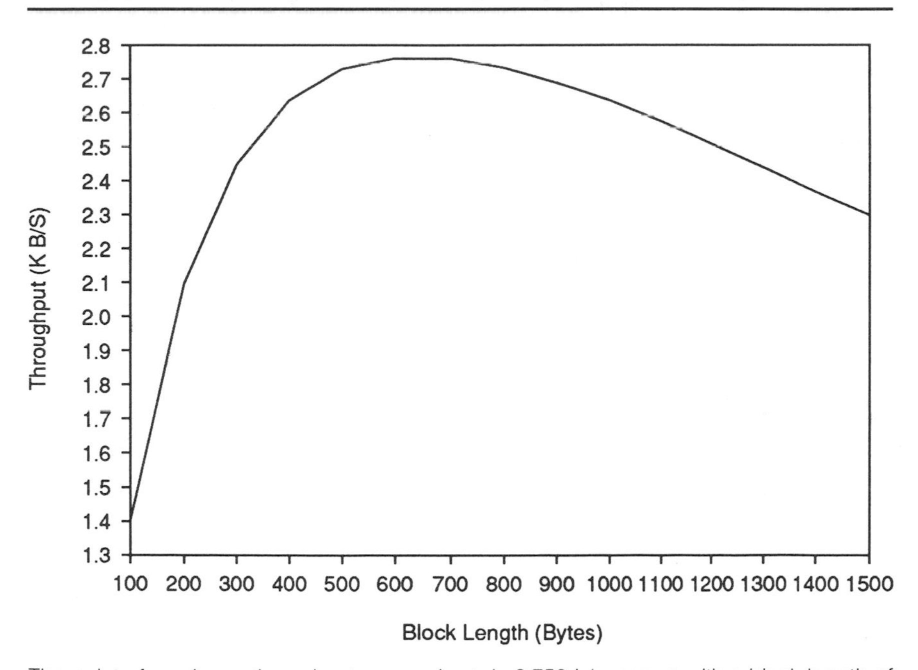

The point of maximum throughput, approximately 2,750 b/s, occurs with a block length of 600 to 700 bytes. The transmission parameters can be varied to maximize throughput.

with. As with other circuit types, some variables are controllable, and some are difficult or border on the impossible to change.

The least controllable variables are a function of the data communications protocol, which is changed only by a software change in the host and controllers. Many computer systems do not support alternate protocols, or upgrades are so expensive as to be prohibitive. In any case, your first task is to determine what protocol is being used, and to identify the characteristics of the polling overheads. Other variables, such as modem speed and circuit type, can be changed within limits.

The most controllable feature is the message volume, which is a function of the quantity and length of the messages and the number of terminals on the circuit. To optimize the circuit, you must determine how many devices the circuit will support without exceeding response time objectives.

Probably the most common method of designing multi-drop circuits is to keep adding terminals until response time gets out of hand, and then cut back. This time honored "seat-of-the- pants" method works if your intuition is reliable and sharpened by several years of experience. It can also be expensive because you may underutilize circuit capacity because you are afraid of response time degeneration, or the circuit may crash because too many terminals are added. A far less costly way of designing circuits is to model circuit performance mathematically. A method of modeling multidrop circuit performance is diagramed in Figure 17.9.

Protocol Analysis

Protocols are classified as half- or full-duplex. With half-duplex protocols, the circuit turns around at the end of each transmission. A common example of a half-duplex protocol is IBM's Binary Synchronous (Bisync).

Transactions between host and terminal are one of three types:

Polling Messages, which are inquiries from the host to determine whether the terminal has traffic to send

Input Messages, which are messages from the terminal to the host

Output Messages, which are messages from the host to the terminal

To analyze the messages, you must determine the exact procedure the host and terminal use to exchange traffic. It helps to diagram the

FIGURE 17.9
Multidrop Circuit Design Process

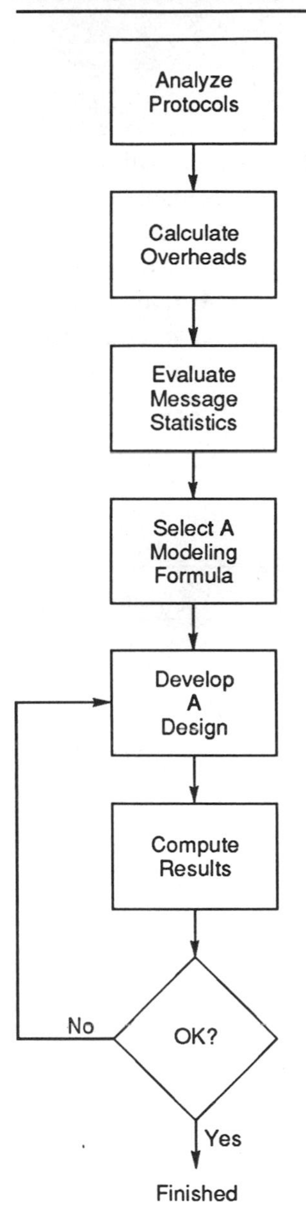

protocol as shown in Figure 17.10 to 17.12. These three figures show three fundamental types of protocols as discussed earlier; half-duplex line held, half-duplex line released, and full duplex.

Calculating Message Overheads

From your knowledge of the protocol and the characteristics of the circuits, it is possible to determine overheads with a reasonable degree of accuracy. A negative poll is, of course, totally overhead because no information bits are transmitted. The overhead bits surrounding the information block on input and output messages are a function of the number of bits transmitted, the modem speed, and the modem turnaround time. Figure 17.13 shows how to analyze a polling sequence to determine how many characters are transmitted in each direction.

FIGURE 17.10
Polling Sequence of Half-Duplex Line Held Protocol

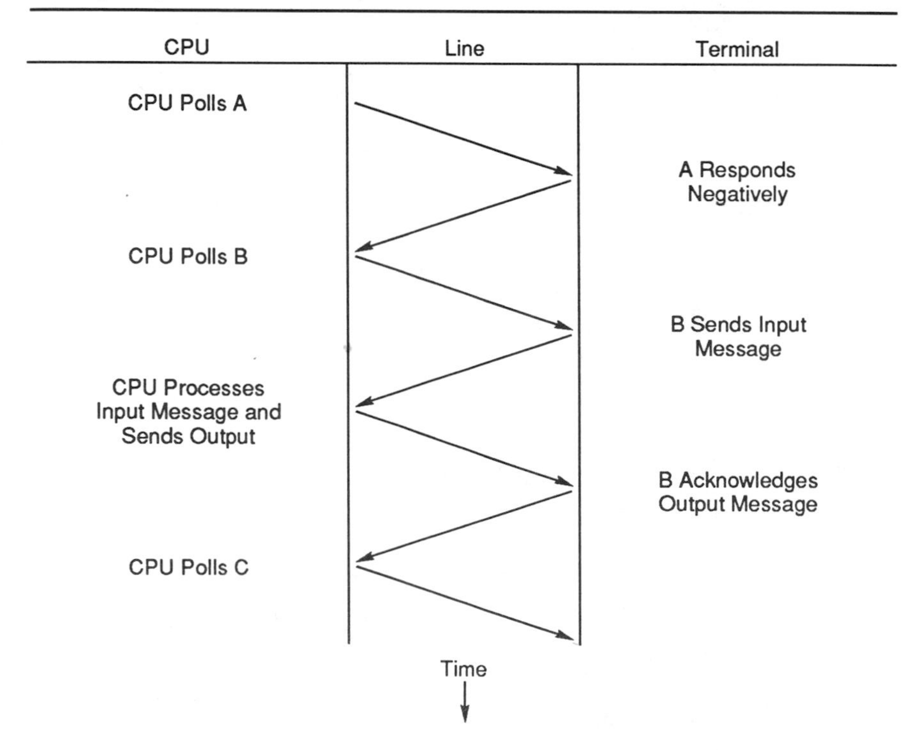

FIGURE 17.11
Polling Sequence of Half-Duplex Line Released Protocol

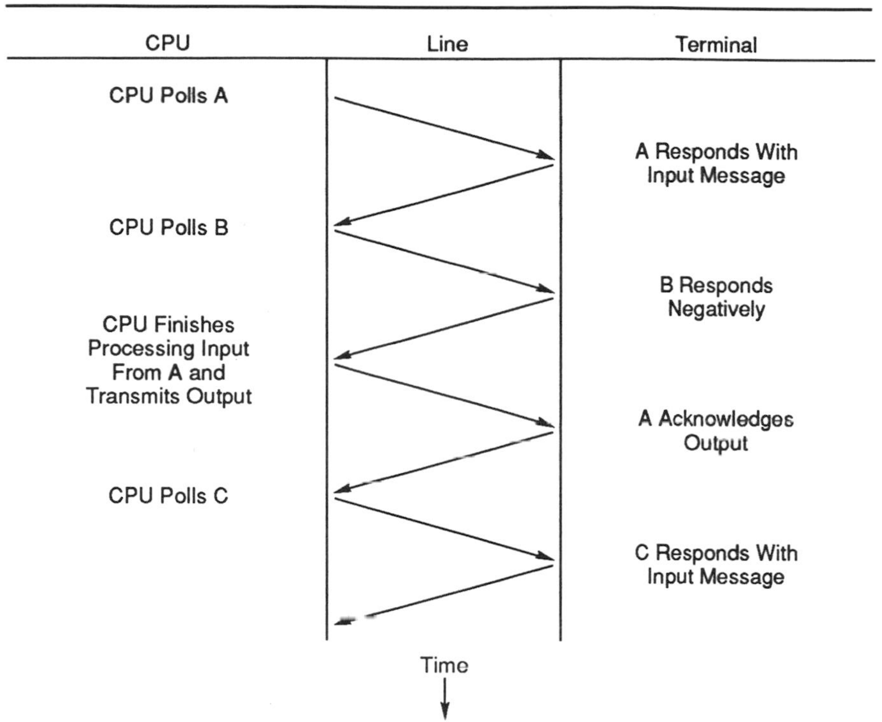

Figure 17.14 shows a spreadsheet model that can be created from the formulas in Figure 17.13 and used to calculate message overheads in a half-duplex circuit. From your analysis of the protocol, the number of overhead characters, modem reversals, and propagation delays are determined. The model calculates the overhead in seconds from the circuit characteristics, which are entered at the top of the spreadsheet. In a full duplex circuit, the overheads are calculated in exactly the same manner except that modem turnaround is reduced or eliminated altogether.

Calculating Message Statistics

Most data networks have a high degree of variability in the length of input and output messages. To calculate response time, you must determine the

FIGURE 17.12
Polling Sequence of Full-Duplex Protocol

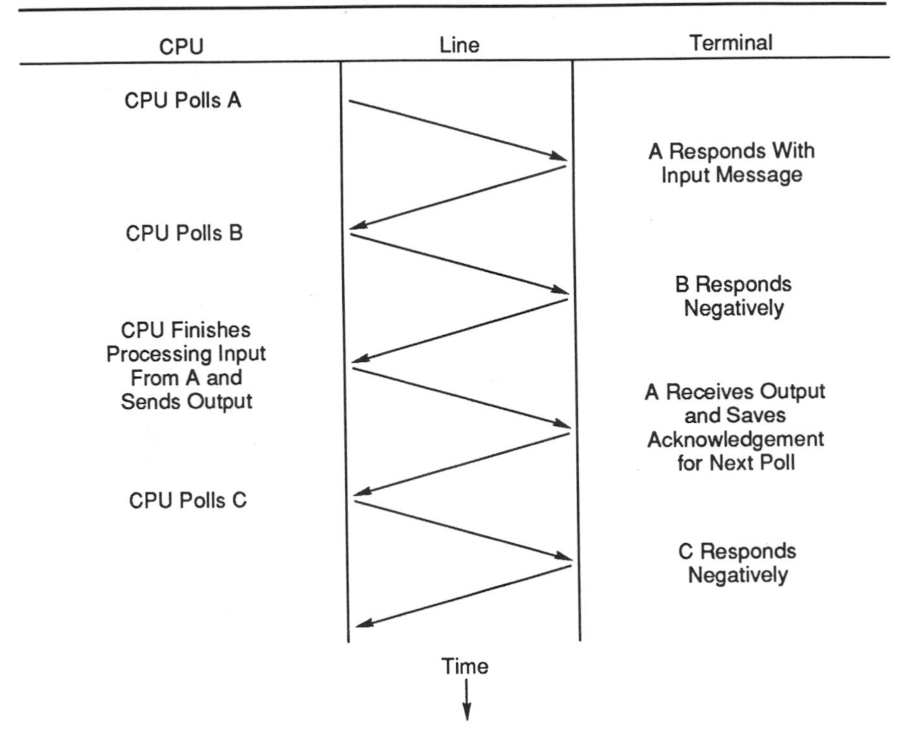

mean, variance, and standard deviation of message length. If these terms are unfamiliar, any introductory statistics book will explain their meaning. This section explains how to calculate the statistics for data traffic.

The most reliable source of message statistics is the host computer, front end processor, or other network element that collects data from actual transactions. If actual data is not available, you will have to obtain information by analyzing transactions manually. The objective of the analysis is to determine the distribution of the number of characters in input and output messages. Figure 17.15 shows a typical message distribution. It is constructed by grouping messages into classes based on size, and calculating a distribution probability. A spreadsheet is ideal for calculating mean, variance, and standard deviation. Figure 17.15 shows the derivation of message statistics in a spreadsheet model that contains both overhead statistics and input and output message statistics.

FIGURE 17.13

Sample Half-Duplex Protocol and Corresponding Overhead Formulas

Sequence									*Formula*
Negative Poll									
CPU:	SYN	SYN	CUA	CUA	DVA	DVA	ENQ		(R_{NP}) (RTS) + (P_{NP}) (PD) + (C_{NP}) (CH)
TERM:	SYN	SYN	EOT						
Input Message									
CPU:	SYN	SYN	CUA	CUA	DVA	DVA	ENQ		(R_{IN}) (RTS) + (P_{IN}) (PD) + (C_{IN}) (CH)
TERM:	SYN	SYN	STX	—TEXT—		ETX	BCC	BCC	
CPU:	SYN	SYN	ACK						
TERM:	SYN	SYN	EOT						
Output Message									
CPU:	SYN	SYN	CUA	CUA	DVA	DVA	ENQ		(R_{OUT}) (RTS) + (P_{OUT}) (PD) + (C_{OUT}) (CH)
TERM:	SYN	SYN	ACK						
CPU:	SYN	SYN	STX	—TEXT—		ETX	BCC	BCC	
TERM:	SYN	SYN	ACK						
CPU:	SYN	SYN	EOT						

Legend

CPU: Central Processing Unit
TERM: Terminal
SYN: Synchronizing characters
ACK: Acknowledgement character
EOT: End of text character
CUA: Control unit address
DVA: Device address

BCC: Block Check Character
ENQ: Inquiry
RTS: Modem reversal time
PD: Propagation delay
CH: Character time
R ( ): No. of modem reversals
P ( ): No of prop. delays
C ( ): No of overhead char.

FIGURE 17.14
Spreadsheet Model for Calculating Overheads

Polling Overheads Model

Modem Reversal Time (RTS) =	0.008	Seconds
Propagation Delay (PD) =	0.01	Seconds
Modem Speed =	2400	BPS
Bits Per Character (K) =	8	Bits
Character Time (T) =	0.0033	Seconds
Negative Poll Overhead Characters (C_{NP}) =	10	
Negative Poll Modem Reversals (R_{NP})	1	
Negative Poll Propagation Delays (P_{NP}) =	2	
Input Message Overhead Characters (C_{IN}) =	19	
Input Message Modem Reversals (R_{IN}) =	2	
Input Message Propagation Delays (P_{IN}) =	4	
Output Message Overhead Characters (C_{OUT}) =	22	
Output Message Modem Reversals (R_{OUT}) =	2	
Output Message Propagation Delays (P_{OUT}) =	5	
Negative Poll Overhead =	0.06	Seconds
Input Message Overhead =	0.06	Seconds
Output Message Overhead =	0.14	Seconds

Choosing an Analytic Model

The overheads and message statistics are entered into an analytic model for computing waiting time, which is the time from pressing the enter key until the terminal is polled and ready to transmit. Waiting time is entered into the response time model of Figure 17.3. Chou[1] proposes the following formula for waiting time:

$$W_{IN} = \frac{1}{2}\left(1 + \frac{V_C}{T_C^2}\right)T_C$$

[1]Chou, Wushow *et al*, *Computer Communications*, Vol. 1 (Englewood Cliffs, NJ: Prentice Hall, Inc., 1983), pp. 423–425. Note: Mathematical notation is changed to be consistent with usage in this book.

FIGURE 17.15
Spreadsheet Model for Calculating Input and Output Message Statistics

Input and Output Message Statistics

Input Message Size	*Probability of Occurrence*	*Output Message Size*	*Probability of Occurrence*
30	0.3	90	0.3
60	0.3	180	0.2
90	0.2	240	0.2
120	0.1	300	0.1
180	0.1	420	0.1
Total	1	540	0.1
		Total	1

Input Line and Message Overhead = 0.06 Seconds
Output Line and Message Overhead = 0.14 Seconds
Modem Speed = 300 CPS

Input Message Statistics

Input Time (T_{IN})	Probability	Weighted T_{IN}	T_{IN}	Weighted T_{IN}
0.158	0.3	0.0474	0.0250	0.0075
0.258	0.3	0.0774	0.0666	0.0200
0.358	0.2	0.0716	0.1282	0.0256
0.458	0.1	0.0458	0.2098	0.0210
0.658	0.1	0.0658	0.4330	0.0433
Average T_{IN}		0.3080		0.1174
Variance of Input Message Size =			0.02	
Standard Deviation =			0.15 Seconds	

Output Message Statistics

Output Time (T_{OUT})	Probability	Weighted T_{OUT}	T_{OUT}	Weighted Tout
0.439	0.3	0.1318	0.1930	0.0579
0.739	0.2	0.1479	0.5466	0.1093
0.939	0.2	0.1879	0.8823	0.1765
1.139	0.1	0.1139	1.2981	0.1298
1.539	0.1	0.1539	2.3695	0.2370
1.939	0.1	0.1939	3.7610	0.3761
Average T_{OUT}			0.9293	1.0866
Variance of Output Message Size =			0.22	
Standard Deviation =			0.47 Seconds	

Where:

W_{IN} = Waiting time on input
V_C = Variance of poll cycle time
T_C = Mean of poll cycle time

Values for the mean and variance of the poll cycle time, T_C and V_C must be calculated to derive the value of waiting time. Three models are proposed by Chou, based on line protocol.

Half-Duplex Line Held

In the half-duplex line held protocol, the host suspends polling while it processes input messages from the terminal. Figure 17.9 illustrates the polling sequence. The formulas for the mean and average poll cycle time in this sequence are:

$$T_C = \frac{M\ T_\rho}{1 - (\rho_{IN} + \rho_{CPU} + \rho_{OUT})}$$

$$V_C = \lambda T_C(V_{IN} + V_{CPU} + V_{OUT}) + \lambda T_C\left(1 - \frac{\lambda T_C}{M}\right)(T_{IN} + T_{CPU} + T_{OUT})^2$$

Where:

T_C, V_C = Mean and Variance of poll cycle time
M = Number of terminals on line
T_ρ = Average time of negative poll
ρ_{IN}, ρ_{CPU} ρ_{OUT} = Average line utilization contributed by input messages, CPU processing time, and output messages
λ = Input Message Arrival Rate
T_{IN}, V_{IN} = Mean and variance of input message transmission time
T_{CPU}, V_{CPU} = Mean and variance of CPU processing time.
T_{OUT}, V_{OUT} = Mean and variance of output message transmission time.

Half-Duplex Line Released

As shown in Figure 17.10, when an input message arrives, the host resumes polling while it processes the message. The analytic model for this protocol is:

$$T_C = \frac{MT\rho}{1 - (\rho_{IN} + \rho_{OUT})}$$

$$V_C = \lambda T_C(V_{IN} + V_{OUT}) + \lambda T_C\left(1 - \frac{\lambda T_C}{M}\right)(T^2_{IN} + T^2_{OUT})$$

Full-Duplex

In the full-duplex model, as shown in Figure 17.11, the host receives an input message from the terminal, and processes it while polling continues in the host-to-terminal direction. The analytic model for this option is:

$$T_C = \frac{MT_\rho}{(1 - \rho_{IN})(1 - \rho_{OUT})}$$

$$V_C = \frac{\lambda T_C(V_{IN} + V_{OUT}) + \lambda T_C\left(1 - \frac{\lambda T_C}{M}\right)(T^2_{IN} + T^2_{OUT})}{(1 + \rho_{IN}\rho_{OUT})^2}$$

Sample Multidrop Optimization

To illustrate application of the analytic model, assume that we have a multidrop data circuit that has polling overheads calculated as described earlier and summarized in Figure 17.14, and message statistics as summarized in Figure 17.15. Also assume that the protocol is half-duplex line released. The variables are entered into the appropriate response time model by using a spreadsheet program employing Chou's formulas as shown in Figure 17.16.

By changing the number of terminals and recalculating, a response time curve can be calculated as shown in Figure 17.17. These formulas are unwieldy to manipulate manually, but lend themselves readily to calculation on a spreadsheet. Although the analytic models could be included in the same spreadsheet that contains the message statistics, they are separated to reduce complexity.

The process just described will aid you in analyzing a multi-drop circuit within limits. Some of the variables used may change over a period of time, so a margin of safety should be built into the calculations. It is not advisable to load a circuit to the maximum on the basis of a modeling formula alone because too many things may change to yield

FIGURE 17.16
Spreadsheet Model for Calculating Response Time in A Multidrop Circuit

Multidrop Response Time Model—Half Duplex Line Released

No of Terminals	30	
Messages Per Hour	1000	
Average Input Time	0.3	
Average Output Time	0.59	
Variance Input Time	0.0056	
Variance Output Time	0.0057	
Average Poll Cycle Time	2.39	Seconds
Average Response Time	2.65	Seconds
	HDX Line Released	
Arrival Rate (A) =	1000	
Input Message Arrival Rate (λ) =	0.278	
Response Time (T_R) =	2.646641	
Variance Input Message Time (V_{IN}) =	0.0056	
Variance Output Message Time (V_{OUT}) =	0.0057	
Variance CPU Time (V_{CPU}) =	0	
Waiting Time In (W_{IN}) =	1.256641	
Output Waiting Time (W_{OUT}) =	0	
Negative Poll Time (TP) =	0.06	
Mean Poll Cycle Time (TC) =	2.391143	
Variable (X) =	0.664206	
Variance of Poll Cycle Time V_C) =	0.292051	
No of Terminals on Line (M) =	30	
Input Transmission Time (T_{IN}) =	0.3	
Output Transmission Time (T_{OUT}) =	0.59	
CPU Processing Time (T_{CPU}) =	0.5	
Utilization (U) =	0.247222	
Ave. Line Utilization Input Messages (ρ_{IN}) =	0.083333	
Ave. Line Utilization Output Messages (ρ_{OUT}) =	0.163888	
Ave. Line Utilization CPU (ρ_{CPU}) =	0.13888	

unsatisfactory results. Moreover, designing to the maximum leaves no room for growth or changes in the message statistics. Unlike voice circuits, which can be added to a PBX or point-to-point data communications network at a low cost, rearranging a multi-drop circuit is expensive and disruptive. Therefore, on an initial design, no more than 75 percent of the calculated capacity should be assigned.

FIGURE 17.17
Multidrop Circuit Response Time Curve

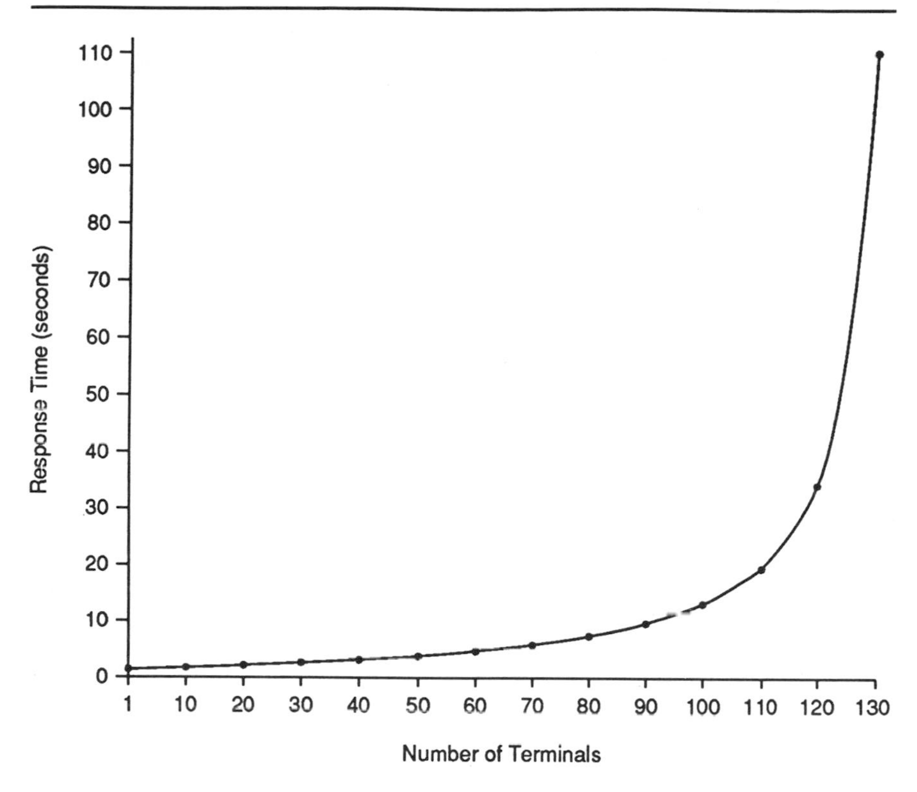

SELECTING DATA COMMUNICATIONS CIRCUITS

When AT&T was the only carrier furnishing long haul circuits, the selection process was straightforward; you could choose circuits either to minimize cost or distance. Generally, the cost minimization was used because of differences between "A" and "B" cities as described in AT&T's tariffs. The tariffs specify three different rates, A to A, A to B, and B to B cities. The A city classification isn't necessarily a function of size, but instead refers to centers AT&T has chosen for its rating hubs. A cities are listed in Appendix I with V and H coordinates for calculating mileage.

Now that competition has entered the market, it is no longer safe to rely on distance as an indicator of cost. Many fiber optic and satellite transmission companies offer services that are insensitive to distance. This distance insensitivity is particularly true between cities that originate and terminate large volumes of traffic. The trend toward distance insensitivity will likely continue in the future, which means that the most reliable way to minimize cost is to get price quotations from vendors.

Calculating Circuit Mileage

Where rates are based on mileage, most carriers use the V & H coordinate system developed by Bell Laboratories. In this system, the United States and Canada are divided into a network of coordinates. If you know both the vertical and horizontal coordinates of two points, you can calculate the airline distance between them with the formula:

$$\text{Distance} = \frac{\sqrt{(V_1 - V_2)^2 + (H_1 - H_2)}}{10}$$

Chapter 10 includes an example of using V & H tables for calculating airline distance.

V & H coordinates for all cities in North America are published in AT&T Tariff No. 264. They are also available from information services such as CCMI McGraw-Hill and on floppy disk from many PBX vendors who use them in call accounting systems to rate long distance calls.

Digital Vs. Analog Circuits

Digital circuits are becoming an increasingly economical alternative for data communications. Available in speeds of 2.4 to 56 kb/s, digital circuits require no digital-to-analog conversion, and have a much lower error rate than analog circuits. Eventually, digital circuits will replace analog entirely, so wherever the cost is comparable, it pays to favor digital circuits.

Data Transmission on Analog Circuits

Analog circuits consist of three principal elements, an originating local loop, an intercity circuit, and a terminating local loop. The intercity portion is inherently a four-wire full-duplex circuit because of the way the carriers multiplex their facilities. The local loop will be two- or four-wire depending on the telephone company's design and whether you order a four-wire circuit. The decision which to use is primarily based on cost.

The makeup of analog channels restricts their bandwidth to approximately 3,000 Hz, which limits the speed of data transmission. An exception to this is a limited distance circuit, which is one that is carried on unloaded twisted pair wire within a single central office area. (*Loading* is a technique of placing inductance in series with a cable pair to limit its bandwidth and improve transmission without using amplifiers).

To transmit data at high speeds over analog circuits, sophisticated modem modulation techniques are used to compress the data into the limited bandwidth. Speeds as high as 19.2 kb/s can be transmitted over a two-wire circuit with current technology. This speed is approaching the theoretical limit, so further increases in speed are not to be expected.

Data Transmission on Digital Circuits

Digital circuits are available in speeds of 2.4, 4.8, 9.6, and 56 kb/s. Although digital circuits do not require a modem, they must be equipped with two devices, which may be combined in a single package, to interface the terminal to the line. The channel service unit (CSU) is a line driver and receiver that terminates a four-wire loop at the user's premises. It equalizes the cable, provides loop-around testing access, and furnishes electrical protection. Other functions that must be provided by either a data service unit (DSU) or the terminal equipment are zero suppression, clock recovery, and signal generation.

Higher bandwidths can be secured by obtaining a T-1 circuit, which is a 1.544 mb/s bit stream that can be subdivided by a channel bank or T-1 multiplexer into sub channels. A channel bank subdivides the bit stream into 24 channels, each of which can carry a voice or data circuit. Unless the voice channel is subdivided by an external multiplexer, each data circuit occupies the entire bandwidth of the channel.

T-1 multiplexers are designed more specifically for data services than are channel banks. They subdivide the 1.544 mb/s channel into multiple data and voice and data channels, using a variety of plug-in cards. T-1 multiplexers reduce or eliminate the need for external multiplexers to subdivide the T-1 bit stream.

SUMMARY

This chapter has presented a brief overview of the data network optimization process. Designers use many other techniques for optimizing the topology of circuit, determining where to locate multiplexers, determining how many packets per second a packet network can support, and other such techniques that are beyond the scope of this book. Additional information is explained in depth in books listed in the bibliography.

Not all telecommunications managers are responsible for both voice and data, but those that carry data communications responsibility must be familiar with the concepts presented in this chapter. Not only are these techniques useful for determining when it is safe to add more load to a circuit, it is also necessary to understand them in locating and clearing trouble. If a data circuit that has been operating satisfactorily suddenly begins exhibiting slow response time, the manager needs to know what factors affect response time as a way of isolating and correcting the problem.

CHAPTER 18

TELEPHONE TRANSMISSION

Most of us are consumers of telephone transmission. We know good transmission when we hear it, and we've grown increasingly demanding of transmission quality. Transmission performance, however, is a highly subjective variable. A telecommunications network must accommodate talker volumes ranging from people with weak voices who don't speak directly into the transmitter to people who scarcely need the telephone for distances of a block or less. People also differ substantially in their hearing ability so that a conversation that is perfectly adequate for one individual might be uncomfortably loud for another and barely perceptible to someone with impaired hearing.

The telephone network can do little to compensate for differences in talkers and listeners, but it must be carefully designed to provide satisfactory transmission under a variety of environmental conditions. As users' sophistication has grown, manufacturers have introduced features and services that can adversely affect user satisfaction unless the system is designed with transmission quality in mind. In pre-divestiture days, the Bell System had end-to-end design control over transmission quality. The Bell Operating Companies and most independent telephone companies followed design principles that were specified by AT&T. The same variables are still being followed, but many changes have occurred to put the ultimate responsibility for transmission performance on the shoulders of the end user.

In some cases regulatory actions have made transmission design an economic matter and placed the responsibility directly on the end user. At least one state utility commission, Oregon, has allowed the telephone company to impose a charge for improved transmission performance. Before this change went into effect in mid-1987, U.S. West Communications charged one rate for a business telephone line and a higher rate for a PBX trunk. PBX trunks were designed with a maximum loss of 5 dB (the concepts of loss and the decibel are terms that will be described later in this chapter). The regulatory decision lowered the PBX trunk cost to the same price as business lines, but permitted the company to impose an extra charge for improved transmission performance for those customers that were equipped with repeaters, which are amplifying devices the telephone company uses on high-loss lines. Users are, of course, free to reject the extra charge, but as we will see later, they risk encountering transmission impairments on some types of calls.

This chapter is intended to introduce you to transmission terminology, some of the fundamental design principles, and to provide enough information that you will be aware when special transmission design precautions should be taken.

TRANSMISSION QUALITY

Quality in a voice communication network is determined by two important criteria, switching service and transmission performance. The measurement of the two is entirely different. Switching is a go/no-go matter; either a call is connected or it isn't. Since switching design is statistically based, service measurements are expressed in absolutes such as the number of ineffective attempts and the number of cut-off calls. Note that there is no subjectivity here. A call either completes or it doesn't; it either remains connected until someone hangs up or it is cut off. With modern switching equipment it is relatively easy to monitor service because quality measurements can be built into the generic program of the system.

Transmission quality, on the other hand, is almost entirely subjective. Obeying the old adage that you can't please everyone all the time, transmission objectives are designed to provide a grade of service that about 80 percent of the users rate as good or excellent. Consider the circuit in Figure 18.1. Some of the switching equipment has been omitted for clarity, but in the AT&T network the intertoll circuit connecting the two users could link as many as nine circuits in tandem. Many transmission impairments are cumulative over distances and multiple links and must be controlled to ensure user satisfaction with the service.

All circuits in both public and private networks have characteristics that impair end-to-end transmission quality. *Loss* is a drop in volume

FIGURE 18.1
Typical Long Haul Circuit

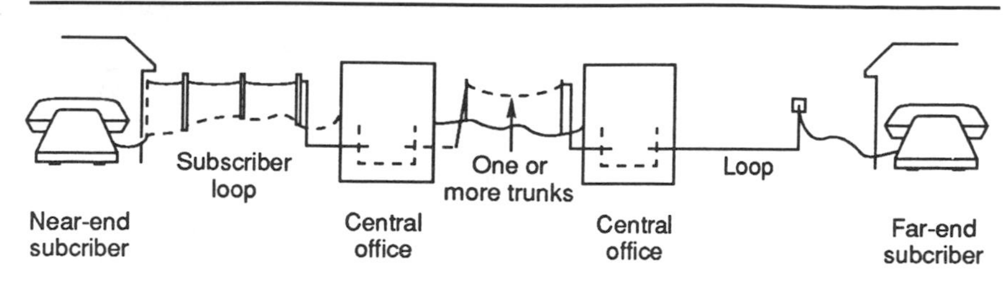

between the two ends of the connection. *Noise* is any unwanted audible signal in the circuit. *Echo* is a reflection of a received signal back to the transmitting end of the circuit. In addition to these three impairments, three other characteristics also affect voice transmission, although to a lesser degree. *Bandwidth* is the range of frequencies a circuit can accommodate. For voice transmission, a range of 200 to 3200 Hz is sufficient; however this speed limits data transmission speed. *Amplitude distortion* describes the varying amount of loss at different frequencies within the circuit's pass band. *Envelope delay distortion* describes the relative propagation delay of different frequencies within the pass band. The last two variables are important for voice grade data, but have little effect on voice transmission.

We will have more to say later about the sources of these impairments. For now it is important to understand that they are always present to some degree, and that the degree to which loss, noise, and echo interfere with communication depends largely on the user's opinion.

Transmission quality standards have been developed from subjective measurements made by Bell Laboratories and CCITT using circuits similar to that shown in Figure 18.2. Varying amounts of loss and noise are introduced into a circuit between two users who are then asked to rate transmission quality as excellent, good, fair, or poor. From these tests several conclusions can be drawn:

- Loss and noise have a opposing effect on quality. A high loss connection is more tolerable if noise is low, and circuit noise is more tolerable if the received signal level is high
- Users are far from unanimous in rating circuit quality. Instead, their evaluations fall into ranges; for example, a hard-of-hearing

FIGURE 18.2
Circuit Connection Model Used for Rating Transmission Quality

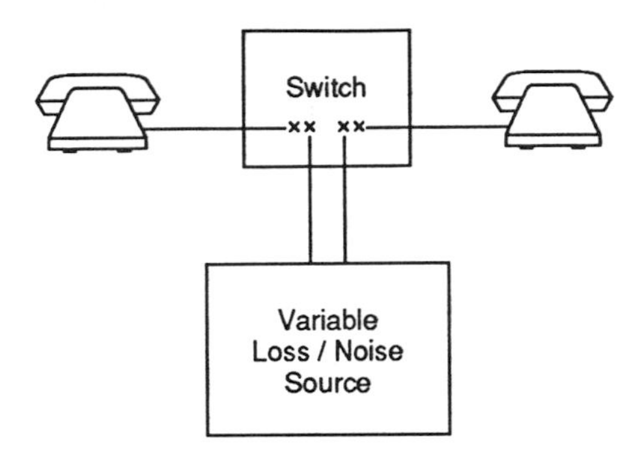

person may rate a low-loss call as "good," while a person with sensitive ears perceives it as too loud

- Quality expectations change over time; as technical advances improve transmission quality, users adjust their expectations to match the service provided

The last point is particularly significant to private network managers. If transmission quality is not approximately equal to that of public networks, users will likely complain and productivity may be affected. Fortunately, good transmission quality isn't necessarily more expensive than mediocre quality. Transmission quality is a function of three factors: the equipment comprising the network, how well it is maintained, and the circuit design quality. Let us review the sources of transmission impairments as an aid to understanding how they are controlled and how the above factors affect quality.

Echo

The vast majority of telecommunications circuits outside the immediate telephone central office serving area travel over multiplex equipment that has separate transmit and receive paths (called four-wire). As shown in Figure 18.3, the local loop is two-wire in all but a few telecommunications services. The two- to four-wire interface requires a *hybrid* circuit, or four-wire terminating set, which is a balancing device that separates and combines the two directions of transmission. The degree of isolation

FIGURE 18.3
Echo Paths in Long Distance Circuits

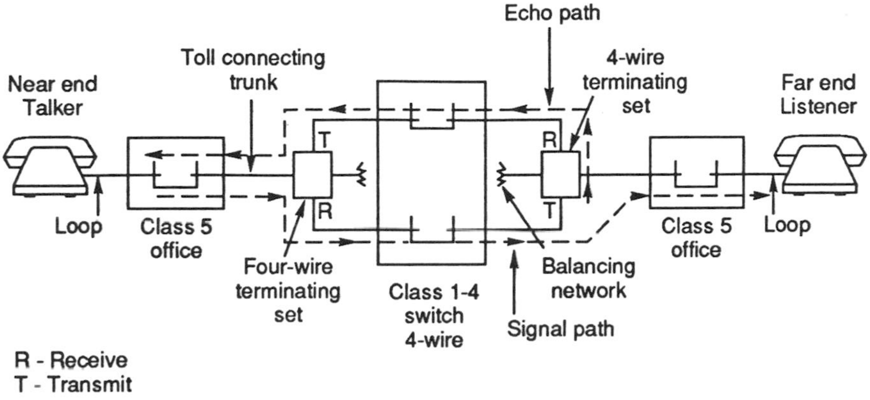

R - Receive
T - Transmit

between the transmit and receive ports of the hybrid is at a maximum when the impedance of the balance network is identical to the impedance of the two-wire loop. On dedicated connections it is possible to obtain a high degree of balance, but on switched connections the balancing network is a compromise circuit that approximates the impedance of the average loop. As a result, some of the transmitted signal always feeds through the hybrid and returns to the transmitting end of the circuit over the echo path shown in Figure 18.3. The degree of isolation between the transmit and receive ports of the hybrid is called the *return loss*. All long haul circuits contain amplification, so in the worst case, if the circuit has no return loss, it oscillates or sings, much like a public address speaker feeding back through a microphone.

When the circuit has enough return loss to prevent singing, echo is of little concern provided it arrives at the talker's receiver at about the same time it leaves the transmitter. In any circuit, a small but finite amount of delay occurs. With little delay, echo is perceived as *sidetone*, the normal feedback that makes a telephone sound "alive." However, if the echo signal is delayed by more than about 45 milliseconds it is considered intolerable by most users. The amount of delay is proportional to circuit length because of the propagation delay of the transmission medium.

The propagation speed of the transmission medium varies with the type of facility, but is in the order of 75 to 80 percent of the speed of light.

In circuits more than 1800 miles long *echo suppressors* or *echo cancelers* are used to control echo. An echo suppressor cuts off the sender's receive path while he or she is talking. An echo canceler is a more sophisticated device that processes the signal to determine whether received energy originated at the near end or the far end of the circuit. It cancels the echo signal but does not affect the received signal. Echo is particularly significant over satellite circuits where the round trip delay is about 0.5 seconds.

Loss

The amount of annoyance from an echoed signal is related not only to delay in the echo path, but is also a function of the level of the reflected signal. Fortunately, before echo reaches the sender the signal has traversed the length of the circuit twice and is thus attenuated by twice the loss of the circuit plus the return loss of the hybrid. In interoffice circuits a controlled amount of loss is deliberately designed into the circuit to attenuate the echoed signal.

The unit of loss is the decibel (dB). A one dB change in level is about the smallest increment that the human ear can detect. A three dB reduction in level is, by definition, a loss of half the receive power, and is readily noticeable to most listeners. The dB is not an absolute measure of volume. Rather, it is a measure of differences between the signal level at various *transmission level points* (TLPs) in a circuit. A TLP is an access point where the signal level can be measured, or a theoretical point such as the output of a switching machine that is used as a reference point for designing circuit loss. Loss is commonly measured with respect to a 1000 Hz tone at a power level of one milliwatt. This level is referred to as 0dBm, or 0dB compared to one milliwatt.

Shorter circuits such as access trunks between a local central office and an interexchange carrier's switch are also designed with loss to guard against singing or the hollow sound that accompanies a near-singing circuit. Telecommunications managers can assure themselves of transmission quality by obtaining circuits designed with a controlled amount of loss. It is important to understand that too little loss can be just as degrading to circuit quality as too much, and that the amount of loss that can be tolerated is directly related to the amount of circuit noise.

Noise

Noise in telecommunications circuits has three primary sources:

- *Thermal* noise which results from electron movement in conductors
- *External interference*, which is the coupling of a signal from one circuit into another
- *Intermodulation* noise, which is noise resulting from nonlinearity in modulating circuits and amplifiers

The circuit designer can do little to control thermal noise because it is inherent in the equipment design and quality of equipment maintenance. Interference from external sources, however, is another matter. Crosstalk from other circuits or hum from power lines are controllable and result from deficiencies in circuit maintenance or poor design. Intermodulation noise is controlled by proper loading of transmission equipment. For example, an analog microwave radio is designed to carry a given number of circuits. Overloading it by applying too many circuits or by applying circuits at too high an input power level will cause noise.

On analog circuits noise accumulates with the length of the circuit; the longer the circuit the greater the noise. On digital circuits the signal is regenerated at each repeater point so that noise is not cumulative. However, long haul digital circuits are more expensive than analog circuits and will likely remain so for the next few years.

Two of the most important noise sources are switching noise in electromechanical central offices and poorly balanced two-wire local loops. Switching machine noise can be heard as short static pops, called *impulse noise*. Although impulse noise is detrimental to data, it is usually tolerable to voice transmission. As stored program control offices replace electromechanical offices, this source of interference is diminishing, at least in metropolitan areas.

A considerable amount of circuit noise results from improperly designed and maintained local loops, including the private cabling on the user's premises. In a two-wire circuit that is electrically balanced so that induced signals are equal on both sides of the pair, induced signals cancel each other and are not audible. However, if the pair has a slight imbalance as shown in Figure 18.4, the interfering signal is detected as audible output. Therefore it is of utmost importance that cable pairs and their

FIGURE 18.4
Circuit Imbalance Results in Audible Noise

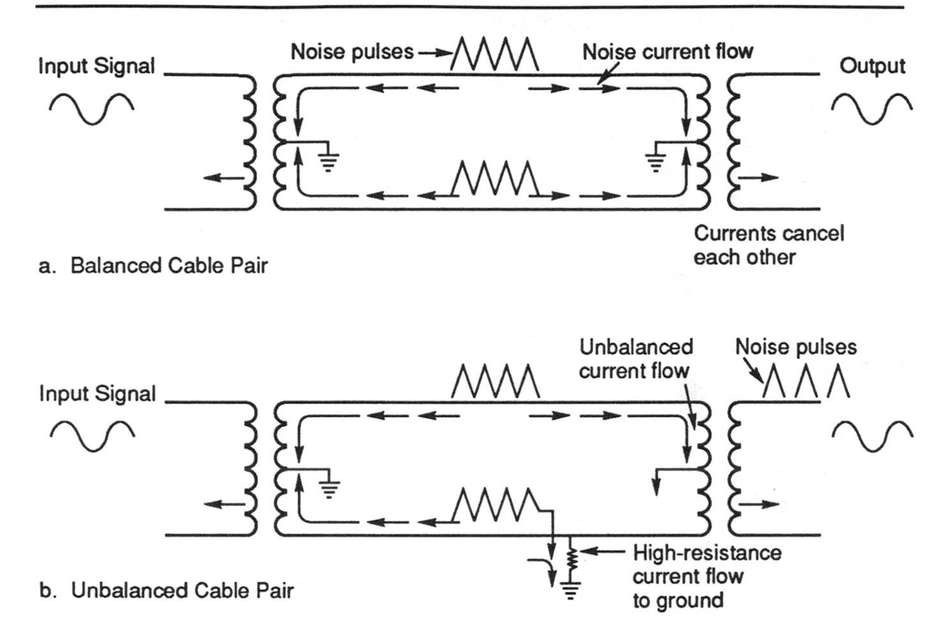

terminating equipment be balanced. Imbalance occurs from current leakage, which often results from wet cable, from high resistance splices, and from unbalanced terminating equipment. Imbalance in cable plant often results from damage or deterioration that allows moisture to penetrate or corrosion to set in.

Crosstalk is the most serious form of noise because if it is intelligible it causes the user to doubt the privacy of the connection. Crosstalk is caused by driving a local loop with an excessively high level signal, or by applying a signal to a poorly balanced cable pair. It is also caused by filter irregularities and nonlinearity in frequency division multiplex equipment.

Noise is measured with reference to the threshold of audibility, -90 dBm, which is equal to 0dBrn or zero decibels of reference noise. A 0dBrn noise power level is inaudible to most listeners. The degree of interference from a noise signal depends not only on the level of the noise, but also on its frequency. Telephone instruments are designed to emphasize frequencies in the middle of the voice frequency range and to attenuate high and low frequencies. The response curve of a telephone is shown in Figure 18-5. Noise measurements are made through a filter known as "C message" weighting to approximate the degree of interference noise will have on a voice channel.

Bandwidth and Distortion

The bandwidth of a telecommunications channel is nominally set by filters in multiplex equipment at 200 to 3200 Hz. As Figure 18.5 shows, this is ample for intelligibility in voice circuits. As all telecommunications equipment is designed to this objective, bandwidth is rarely a problem to voice network users. Special services such as high speed data and high fidelity audio require greater bandwidth that can be obtained with equipment designed for the purpose.

Amplitude and delay distortion are likewise seldom a concern in voice networks. Imperfections in bandwidth limiting filters are the primary cause of these types of distortion. Although amplitude and delay distortion have little effect on voice, they can be detrimental to high speed data and high fidelity audio channels. The effects of distortion are mitigated by applying external equalizing circuits to the channel.

FIGURE 18.5
Response Curve of a Telephone Instrument

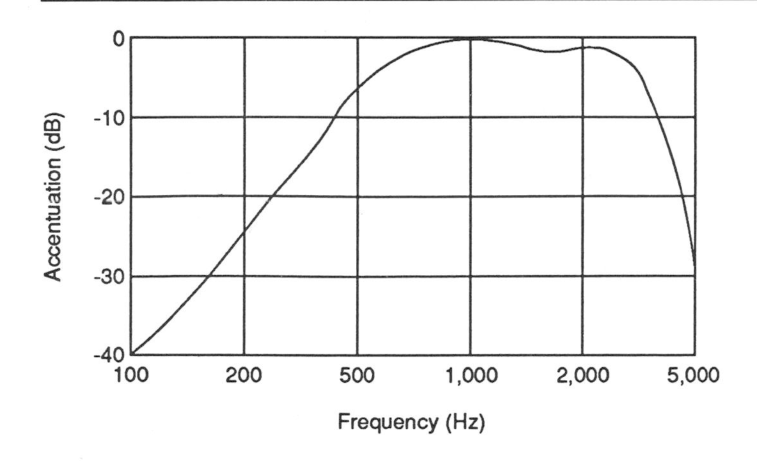

TRANSMISSION DESIGN

The transmission designer's problem is determining how much of the impairment to assign to each circuit element to arrive at an overall connection that is satisfactory to the users. The user's problem is determining which element is at fault when impairments exceed the objectives. Let us first look at the composition of these elements, and discuss how impairments arise.

Local Loop

Transmission design in the U. S. is based on criteria that were established by Bell Laboratories when AT&T controlled all the network elements. With complete design control, the Bell System developed a design that optimized costs within the limits of then-existing technology. The design was based largely on the characteristics of telephone sets and local loops. Because each of the approximately 85 million Bell access lines has one local loop and one or more telephones, that were owned by the Bell companies until recently, the multipliers of any costs associated with loops and telephones were enormous. According to AT&T's publication, *Engineering and Operations in the Bell System* (1983), almost 60 percent of the integrated Bell System's $60 billion investment in transmission facilities was concentrated in loop plant. To achieve the objective of affordable universal service, costs of telephone sets and local loops had to be controlled. The result is a network design that is largely driven by the characteristics of these two elements.

The local loop is designed to one of two criteria, resistance design or long route design. All but a fraction of the users are within 3.4 miles of a serving central office, and are therefore served by loops designed to ensure that telephone sets work without the need for external amplification. Resistance design achieves an economic balance between transmission performance and the use of the finest gauge wire possible in the loop. The smaller the wire gauge the lower the cost, and the more conductors that will fit in a conduit. However, the economies of fine-gauge cable are counterbalanced by its transmission characteristics. Fine-gauge cable has more loss and higher resistance than coarse-gauge cable, and the higher the resistance, the more poorly the telephone set performs.

The characteristics of the telephone set weigh heavily in loop design. The standard telephone handset contains a carbon transmitter that is sensitive to the amount of current flowing in the loop. With too little current, which results from the increased resistance of long and fine-gauge loops, the output of the transmitter drops off, adding to the impairment from the cable loss. A balance is reached in loop design to provide at least 23 milliamps of current over a loop that has a loss objective of eight dB or less.

Long route design is used for the few customers farthest from the central office. External equipment including electronic amplifiers, range extenders, and subscriber loop multiplex equipment, is used to reduce loss and increase current flow in local loops. However, equipment costs preclude their use on the bulk of local loops.

The local loop is also an important noise source. Loops live in a hostile environment where they are subjected to the ravages of weather, backhoes, and target practice. Some users overdrive loops with high-level signals that cause crosstalk, or they connect unbalanced equipment that defeats the inherent noise canceling properties of a balanced cable pair. For these and many other reasons, loops often contribute more than their share of loss and noise to the overall connection.

Loops also share right-of-way and pole lines with electric power distribution systems, which induce noise into telephone circuits. By itself, induction is not necessarily detrimental, however. In rural areas, unshielded open wires often share pole lines with electric power circuits without interference. As explained earlier, the degree of interference depends on the balance of the telephone circuit. When the two wires of a cable pair are identically balanced in resistance and isolation from ground, induced voltages are equal when measured from each side of the pair to ground and therefore the noise is inaudible.

Noise on most subscriber loops should be in the order of 5 or 10 dBrnc0. When the noise exceeds 20 dBrnc0 it is generally an indication that corrective action is needed. The letters "C0" appended to the noise measurement mean that the measurement is made from or adjusted to a 0 transmission level point through a C message weighting filter that approximates the response of the human ear.

One final characteristic must be considered in evaluating the local loop as a transmission medium. The ordinary cable pair has a gain / frequency characteristic shown in Figure 18.6. To reduce loss and improve frequency response within the voice band, telephone companies insert induction or *load coils* in series with the loop. This technique improves voice frequency response, but greatly attenuates frequencies

FIGURE 18.6
Loss of 10,000 feet of a 26 Gauge Telephone Cable Pair

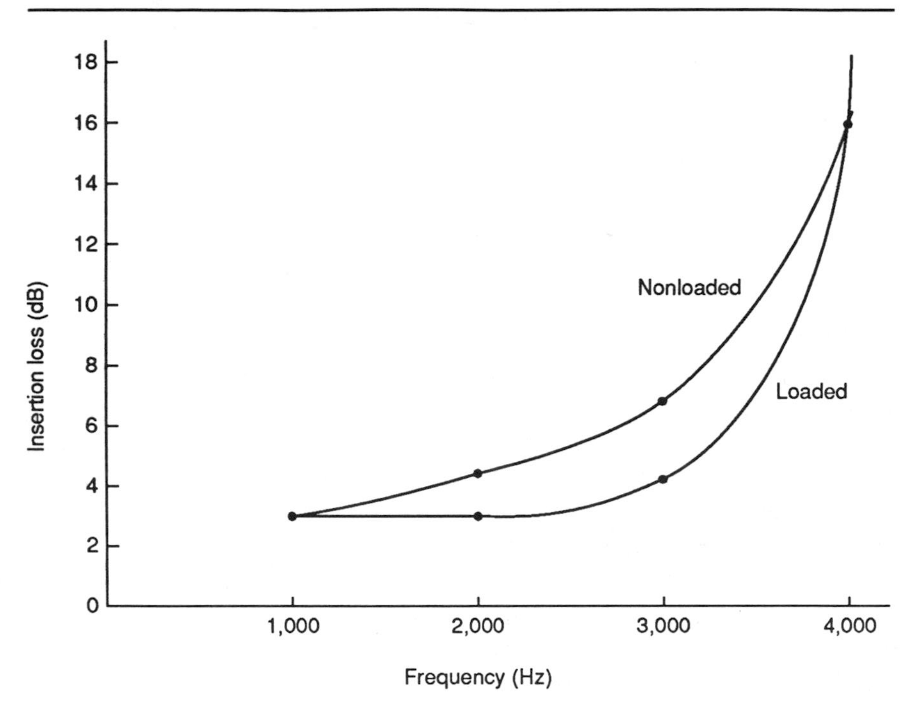

higher than about 3400 Hz. Loading places an upper limit on data transmission speed. As the bandwidth of long haul circuits is limited by multiplex equipment anyway, loading has no effect on long haul data. However, some services that operate at high speed, such as limited distance data service, are impossible to use over loaded circuits.

Access Circuits and Long Haul Transmission

The other two elements of a telecommunications network, access circuits and long haul circuits, are more flexible than the local loop and telephone set when it comes to transmission performance. The vast majority of these circuits are derived over multiplex equipment that contains built-in gain so that loss can be overcome without incremental expenditure for amplifiers. In fact, attenuators are used to introduce loss into circuits that would otherwise have too much gain. However, access lines and trunks contribute noise to the overall connection. Noise is cumulative over the analog facilities used for most long haul voice circuits, so the longer the circuit, the more noise it will introduce into the connection. This problem will be unavoidable until all-digital facilities are in use. The inherent noise advantage of digital circuits will then be available to long haul users.

TRANSMISSION MEASUREMENTS

Proper transmission design is only one element of a transmission quality assurance plan. The other elements are periodic transmission measurements and regular equipment maintenance. Fortunately, most modern equipment is designed to minimize the amount of maintenance required. In privately-owned facilities the manufacturer's specifications should be consulted to determine what, if any, preventive maintenance is required. With facilities acquired through common carriers, equipment maintenance is taken care of by the vendor.

Transmission measurements should be made by either the user, the carrier, or both at the time circuits are initially established, and at regular intervals thereafter. Measurements are also warranted when transmission complaints begin to occur. In the past, such complaints could be referred to the telephone company, but most post-divestiture networks are composed of circuits and facilities from multiple suppliers. The network becomes particularly complex when combinations of interLATA and intraLATA common carrier facilities are combined, with privately owned microwave and PBXs possibly thrown in to further complicate matters. In such networks the user should be equipped to perform some measurements. At the least, loss, noise, and return loss measurements are called for. Data circuits may also require envelope delay, impulse noise, and P/AR (peak-to-average) measurements.

Every network should have a maintenance plan that is designed for transmission quality assurance. If the network is composed entirely of facilities provided by a single vendor, the vendor *may* perform periodic measurements. Rather than assuming this is the case, however, it is worth inquiring what the vendor's policy is. Most telephone companies align circuits when they are initially installed, but do not make end-to-end measurements except when trouble is reported. Moreover, except for large users that bypass the local loop with direct microwave or satellite services, several vendors are usually involved in private networks, and making measurements helps to determine which vendor is at fault.

If the network is composed entirely of privately-owned facilities, your organization will probably have technicians who are equipped and trained for making periodic transmission measurements. For other networks it is a good idea to train or contract with someone to make circuit measurements at least quarterly.

Loss Measurements

Loss is measured with a transmission measuring set (TMS) that consists of a tone generator or oscillator, and a level detector with a meter or digital readout. Many TMSs also include noise measuring apparatus. A more sophisticated device is a Transmission Impairment Measurement Set (TIMS) such as the one shown in Figure 18.7. A TIMS measures not only loss and noise, but also contains a signal source and circuitry for measuring other voice circuit impairments. A TMS or TIMS is used to measure level at specified TLPs and at impedances that match the impedance of the TLP. Loss is nominally specified at 1000 Hz, but in practice, measurements are made at 1004 Hz to prevent interference with digital carrier equipment.

Measurements are made by sending a single tone into the circuit at a transmitting TLP, and measuring the level at a receiving TLP. Levels are adjusted by changing pads or adjusting amplifier gain so the design loss is achieved within specified limits. A TMS is also used to measure frequency distortion of a circuit by sending and receiving a band of frequencies rather than a single tone.

Most common carrier and many private networks are equipped to make automatic transmission measurements on circuits. Both the transmitting and receiving ends are equipped with testing systems that automatically send test tones and measure levels in both directions. Equipment known as a *responder* at the far end records measurements from the near end, and reports the results over a data link. Most automatic test equipment is designed to measure noise and to test signalling.

Noise Measurements

Noise is measured with a noise measuring set that is either a separate instrument or part of a TMS. Noise measurements are made at a TLP with the far end of the circuit terminated in its characteristic impedance. If the

FIGURE 18.7
The Hewlett Packard Transmission Impairment Measuring Set (TIMS)

Photograph courtesy of Hewlett Packard Company
(This set is used for making analog measurements on voice grade circuits.)

circuit is used for voice communication, measurements are made through a C message filter. If the TLP is not a 0 level point, noise measurements are adjusted to 0 and expressed as dBrnc0. For example, if noise measures 27 dbrnc at a +7 TLP, the 7 dB gain would be subtracted and the circuit noise expressed as 20 dBrnc0.

Return Loss Measurements

The amount of return loss in a circuit depends on the degree to which the impedance of a balancing network matches the impedance of the two-wire line. Return loss measurements are made by sending a band of frequencies or a wide band of noise into one port of the hybrid and measuring the level that feeds through to the other port. Return loss measurements are made with a return loss test set. Measurements are made over a band of frequencies from 500 to 2500 Hz. Corresponding measurements are high and low frequency singing return loss tests (SRL-high and SRL-low), which introduce filters to confine the singing point to the high and low ends of the voice frequency band.

Manual Tests and Trouble Reports

The alternative to automatic testing is manual testing, which usually requires people at both ends of a circuit. Signals are exchanged for loss measurements, and noise is measured at the receiving ends. If routine transmission measurements are not feasible because of lack of trained personnel and equipment, about the only way to ensure transmission quality is to rely on trouble reports.

Whether a transmission measuring program is used or not, a plan should be established to record and analyze transmission complaints. Reports must include the originating and terminating number, time of day, and nature of the complaint (eg., noise, crosstalk, singing, can't hear, etc.). From these complaints the network manager can determine which circuit group is faulty. From the type of complaint it may be possible to determine which vendor's facilities are at fault, although even the most experienced technicians often guess wrong.

Data Transmission Tests

Data transmission complaints are usually more difficult to diagnose than voice complaints because no human uses the circuit and expresses the difficulty encountered. Problems manifest themselves as a high error rate with little clue as to the cause. For example, microwave fading may occur long enough to cause several data blocks to be rejected, but may not generate transmission complaints on voice circuits. Other types of data deficiencies such as excessive envelope delay won't be noticed on voice and low speed data circuits at all, but will be disastrous to 9600 b/s data. Data transmission complaints can be evaluated only by special test equipment operated by trained personnel.

WHY TELECOMMUNICATIONS MANAGERS MUST MONITOR TRANSMISSION

If the circuits in a private network are designed to the same loss-noise grade of service objectives that most common carriers use, the result will be satisfactory transmission for most users. However, many services now on the market have built-in traps that can easily result in unsatisfactory transmission. A few common ones are discussed below.

Conferencing

Conferencing is a great convenience to telephone users. However, if the parties are bridged by conference circuits that simply tie the station conductors together, each time a station is added to the conversation, additional loss is added to the connection. The results of "push button" conferencing are impossible to predict because of the variability of the other transmission characteristics in a connection. Sometimes a connection works well, and other times the parties are unable to hear or be heard. The solution to this problem is using a conferencing bridge that adds enough gain to the ports to overcome loss. Conference bridges are accessed manually through an operator or by dialing into the bridge through a PBX, or the local telephone central office.

Remote Access

Many PBXs and tandem switches are arranged for remote access so users at other locations can obtain access to tie lines, WATS lines, and other such facilities. The problem with such arrangements is that an additional local loop and sometimes an interoffice trunk are inserted in the connection. Like push button conferencing, the results of remote access are unpredictable because the user has no control over the transmission characteristics of the connection to the switching machine, and because of the variability of the connection beyond the PBX or tandem switching system. Remote access is a popular adjunct to many PBXs, but users should be aware that transmission may be less than satisfactory.

PBX Switching Network Loss

Another source of transmission complaints is the switching network loss of some PBXs. Electromechanical PBXs were constructed with little or no loss between lines, but newer digital PBXs often insert loss as a way of improving echo performance. Digital PBXs have four-wire switching networks that are interfaced to the two-wire station loop with a hybrid. As with central office switching machines, the hybrid becomes a feedback path that is often controlled by inserting loss in the transmission path. This loss is of little consequence between stations on the same premises, but the extra loss may be detrimental to off-premise stations, stations using remote access, and to stations using tie lines.

Off-Premise Extensions

Off-premise extensions offer an economical way to interconnect an organization that has several locations in the same city. For example, many school districts have a central administration building and high schools with many stations, and elementary schools with fewer stations. Off-premise extensions are less expensive than separate business lines, and are often used to tie the entire organization together.

Anyone constructing such a network must be aware of potential transmission problems. If the off-premise stations connect to each other and to the outside over high-loss lines, transmission may be unsatisfactory on some calls. The contrast in level may be apparent to users who find less loss on calls to users in the central location than to other users in their own location.

SUMMARY

The foregoing is far from a complete list of sources of transmission impairment. The key point to remember is that any service or equipment being tied to either a public or private network is a potential source of impairment. Every change or addition to a network should be evaluated for its effect on transmission performance. The final measure of a telecommunications network is how economically the elements fit together to provide satisfactory service. It is easy to forget that the product you are buying when you acquire a network is transmission, and that transmission performance becomes an important indicator of total network performance.

Transmission performance in both private and public telecommunications networks is a vital indicator of user satisfaction. Proper equipment selection, application, and circuit design are the primary methods of controlling transmission quality. Educating users to be satisfied with a lesser degree of quality is possible in a network with tight control of the users, but where users have a choice they will usually vote against poor transmission quality by choosing an alternate communication service.

The purpose of this chapter has been to show that transmission cannot be taken for granted; it happens only by design. Often equipment or services are offered at cut-rate prices, but with transmission marginally below the objective. It is easy to be lulled into thinking that "only two or three dB" won't make that much difference, but it may end up costing more in unsatisfactory and redialed connections than buying high quality service to begin with.

CHAPTER 19

RESELLING TELECOMMUNICATIONS SERVICES

In the euphoria that surrounded AT&T's divestiture of its operating companies and the opening of telecommunications services to unfettered competition, many companies expected that large profits would result from resale of both long distance and local telephone services. A "smart" building or industrial park offers services that enable tenants to obtain single-point-of-contact telecommunications services at a lower cost than purchasing them on the open market. The concept of a smart building wasn't born with divestiture, but the divestiture decree helped service providers gain approval for some services that were questionable or prohibited in the regulated environment.

The practice of reselling telecommunications services, unfortunately for several large investors, has been less appealing than the theory. Large and well-funded companies such as AT&T and IBM have dipped a toe into the shared tenant service waters, found them icy, and backed off. Companies with substantially inferior financing and experience have likewise backed away from shared tenant services, but other companies find it profitable, if not in direct cash flow, at least in making their properties more attractive to prospective tenants. In this chapter we will examine the elements of shared tenant telecommunications services (STTS), and will describe an approach for determining whether it is an economically feasible venture.

Another business opportunity offered by divestiture is that of privately owned coin telephones. Once an exclusive monopoly of the telephone companies, many states have opened the way for others to connect coin instruments to the network, but the service providers have encountered many unexpected problems. In this chapter we will also discuss the potential benefits and the risks of private coin telephone ownership.

SHARED TENANT TELECOMMUNICATIONS THEORY

STTS is an important element of a so-called "smart building." A smart building is one in which processors control lighting, heating, ventilation, air conditioning, and elevators. The essence of the smart building is the provision of features that enhance energy management and improve the quality of life for its tenants. The brain of the smart building, a central processor, uses the telecommunications system to communicate with sensors and transducers that detect environmental conditions and activate controls. Even though separate channels may be used for voice, data, and video communications, the cables, raceways, and conduits will usually be shared with the smart building's nerve center.

The provision of modern control systems may be enough to attract tenants, but it is reasonable to consider carrying the smart building one step further into the realm of shared tenant telecommunications services. Some building services have little or nothing to do with telecommunications. For example, copy centers, word processing pools, office supply stores, and catering services may relieve tenants of the need to shop for outside services, but it is just a short leap to the concept of sharing computer and telecommunications services.

If the demand for time-shared computer services is great enough, the building owner may be able to profit from the sale of services such as accounting, computer type setting, and security services. These services are particularly attractive to small companies who prefer to invest their money in the business rather than in ancillary services. Computer services are virtually inseparable from telecommunications services because the telephone system becomes the method of linking the tenants to the computer. By providing the backbone facilities for interconnecting these services, the building owner keeps control of the overall design, and prevents the conduit congestion that inevitably results from unbridled terminal-to-computer coaxial cable runs.

The next logical step is providing switched telecommunications services. Several modern digital PBXs are capable of being partitioned so that each tenant effectively has a private switching system. Through the PBX the tenant gains access to shared facilities such as local trunks, long distance lines, modem pools, and the like. The theory of STTS suggests that building tenants will find the service irresistible for a variety of reasons such as these:

- Local and long distance services can be acquired at a lower rate than tenants can obtain them because of economies of scale
- Tenants will be attracted to a single point-of-contact for telecommunications service provided by the building owner
- Tenants are relieved of the administrative costs of keeping records, handling adds, moves, and changes, purchasing services, and other such labor intensive activities
- The telecommunications system can be integrated with other attractive services such as copying and printing, high speed facsimile, computer-aided graphics, and other such services that tenants may not find economically feasible on their own

SHARED TENANT TELECOMMUNICATIONS PRACTICE

In practice, the smart building and STTS concept has been a costly investment for many building owners. Many owners report that the extra costs of telecommunications services have not been repaid, and that tenants seem indifferent to the benefits. STTS equipment represents a large initial investment that often requires years to repay. Many cities began to experience high office vacancies, more or less coinciding with a slump in oil prices, just as shared tenant systems went into service. Many of the financial problems the service providers have experienced are attributable to over-building as much as to STTS, but the capital investment in shared tenant systems is an additional overhead that must be repaid.

Another factor contributing to the STTS problem has been continuing reductions in telecommunications prices. For the past few years, PBX prices have been dropping significantly, so that the building owner may be forced to amortize equipment at rates higher than tenants would have to pay for a new PBX. If the owner requires subscribing to the service as a condition of tenancy, some prospective customers choose a building that offers more freedom to choose their own telephone system. If the owner attempts to attract customers to STTS by offering lower prices, and if the percentage of development fails to materialize, the owner loses money on STTS.

A related factor has been changes in long distance service pricing. A few years ago, long distance resellers profited by purchasing bulk-rated services such as WATS, and aggregating enough volume to benefit from the low per-minute prices of high volume discounts. Shared tenant service providers have reasoned that they can easily resell WATS themselves and still offer a better price than tenants can obtain service for on the open market. There are still volume discounts to be obtained in long distance, but, as discussed in Chapter 10, the long distance market is changing rapidly. A recent trend has been away from WATS and toward other services such as virtually banded WATS and dial-one WATS. AT&T has led the long distance industry with regular price reductions and changes in the volume discount schedule. A host of new services have been introduced that make it easier for users to justify their own discount long distance service, which diminishes the cost advantage previously enjoyed by resellers.

Moreover, to retain a price advantage, resellers increasingly must install dedicated trunks to high-volume locations to reduce their per-minute costs. The effect of this tendency is to make it difficult for small resellers, such as a building owner, to compete because they lack the volume to justify dedicated circuits. Furthermore, resellers that do use WATS services profit by offering their services to residential users who call during off-peak hours when WATS rates are low. This off-peak calling adds to the volume and helps the reseller gain the volume discounts offered by WATS. A building owner finds it difficult to benefit from the lower costs of off-peak pricing because most businesses use little long distance outside normal business hours.

Another restraining factor has been regulatory action by telephone companies, many of which view the resale of switched local service as an encroachment on their exclusive territory. Action by state utility commissions has ranged from outright prohibition of local service resale to requiring that shared lines be metered, or to imposing an additional "joint user" fee for shared lines.

Another factor is the nature of the market itself from the tenant's viewpoint. Many tenants are part of a larger organization that has its own professional telecommunications management. With distributed switching becoming more common, a tenant may install a remote switch unit in place of a PBX to be part of its home telephone system. Even lacking professional telecommunications management, most prospective tenants will carefully compare the cost of the shared system with the price of providing their own. As equipment prices continue to drop, an independently owned system becomes more attractive.

A RATIONAL APPROACH TO STTS

Despite the hazards, STTS may still be an attractive business opportunity if it is approached in a rational manner. A detailed business case study should be undertaken before any decisions to provide the service are made. The owner should approach an STTS project with the assumption that a subscription to the service will be voluntary. Mandatory subscription to STTS may appear to be an attractive way to recover the investment, but the loss of prospective tenants may outweigh the advantages of high occupancy of the telephone system. The service should be made as attractive as possible under a range of assumptions about the percentage of users who will subscribe to STTS. If the service is not potentially profitable under the most pessimistic assumptions, management should decide to limit the extent of shared services.

Figure 19-1 is a summary of a spreadsheet model of STTS feasibility. The model can be expanded to show any quantity of station development. A planner can vary the number of stations per worker, the percentage of subscribing tenants, traffic load for each trunk type, rates for local and toll service, and other such variables that affect the extent of STTS development. A look-up table determines the quantity of trunks required for given assumptions of traffic volume.

Separate sections of the spreadsheet calculate local trunk costs for incoming, outgoing, and DID trunks. Another section determines quantities of long distance trunks based on expected calling volumes. PBX and station costs are calculated in another section of the model. A fourth section enables you to vary the local and toll charges to compute expected revenues. These figures can be computed for a range of scenarios and entered into *pro forma* balance sheets and income statements.

A second model, shown in Figure 19.2, is used for evaluating the service from the tenant's point of view. This model provides space for calculating costs of local and toll service, given different assumptions of

FIGURE 19.1
Spreadsheet Model of Shared Tenant Service Feasibility

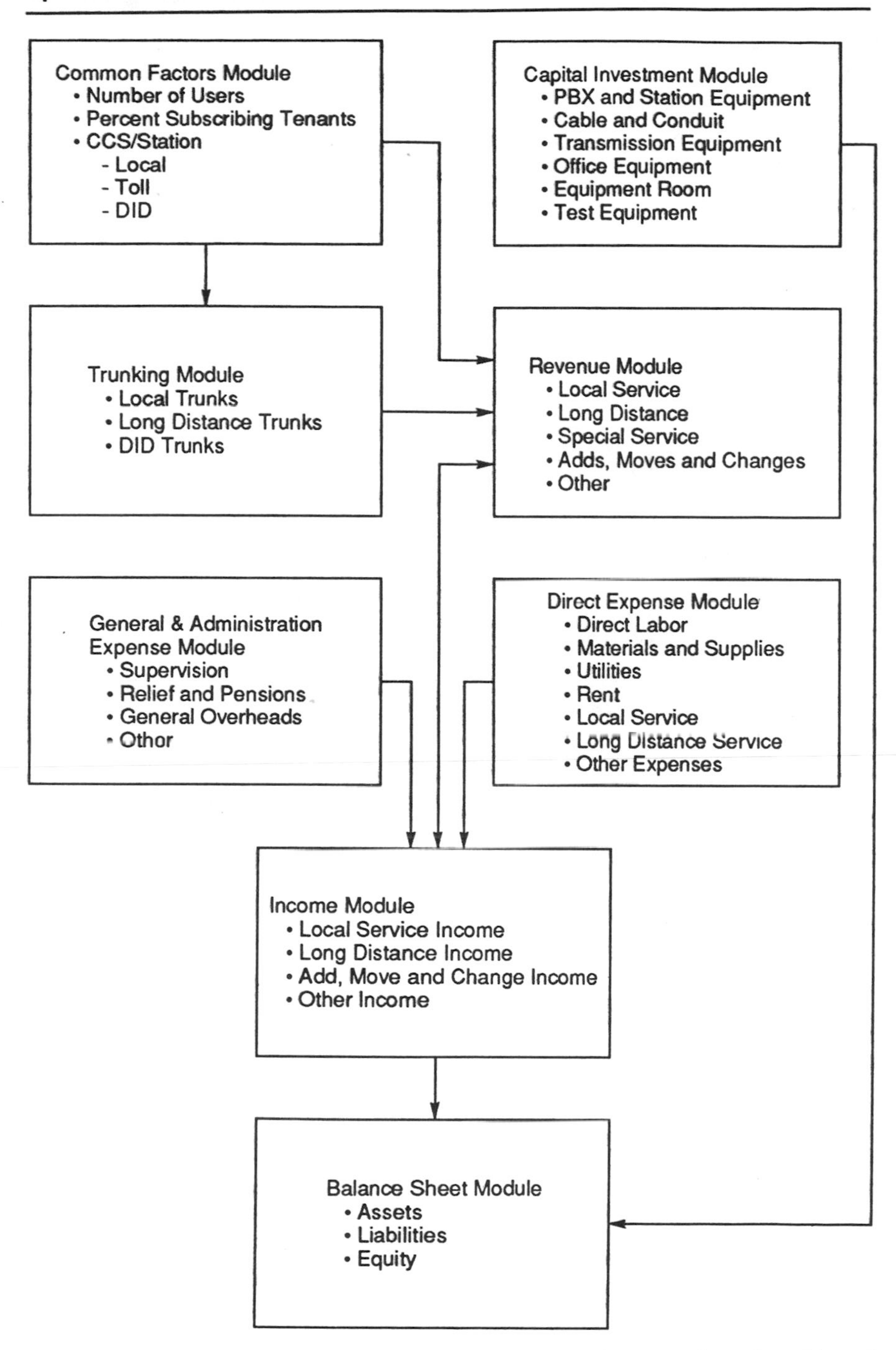

A spreadsheet model can be made to approximate costs and revenues from shared tenant telecommunications service. If the modules are linked, assumptions can be varied to compute the profitability of the service.

FIGURE 19.2
Shared Tenant Service From Tenant's Point of View

Step 1: Enter Usage Level:	A	
Enter H for heavy,		
A for average		
Or L for Light		
Step 2: Enter Station Profile		
Total Quantity of Stations	145	
Percent DID Lines	69%	
No. DID Stations	100	
Step 3: Enter Telephone Set Profile		
% Display	10%	
% Multi-button Feature	30%	
% Single Line	60%	
Total	100%	
Step 4: Enter Common Costs		
Inflation	5.0%	
Power Cost/KWH	$0.05	
Maintenance Cost/Port/Month	$3.00	
Cost of Money	9.0%	
Service Life in Years	7	
Salvage Value	7.0%	
Floor Space $ per sq. ft.	$15.90	
Incremental Tax Rate	40.0%	
Step 5: Evaluate Financial Effects		
	PBX	STS
Capital Investment	($92,861)	$0
Investment per Sation	($640)	$0
Annual Expense	($45,465)	($42,850)
Expense per Station	($314)	($296)
After Tax Expense	($27,279)	($25,710)
Net Present Value	($262,051)	($137,211)
NPV STS - PBX		$124,840

This model is a five-step process that a STTS provider can use to evaluate service from the tenant's standpoint. Tables that are not shown in this figure calculate the cost of a PBX and expenses the tenant will incur for trunking, maintenance, and other costs of PBX ownership. The net present value is calculated for each alternative.

station development. By comparing costs of providing STTS with costs tenants will incur themselves, the first order of evaluation is complete.

Value Added Services

Even though STTS may not appear initially attractive, the owner should approach the project from the standpoint of providing services the tenant will find valuable enough to justify the extra cost. If tenants perceive enough value from central services, they may elect shared tenant services as opposed to owning their own system. If the owner provides STTS, some of the services described below can be offered to tenants, including those who own their own telecommunications systems. The following lists only telecommunications-related services. Many other services such as a copy center can be provided.

Telephone Answering Service

A natural adjunct to STTS is the provision of centralized attendant service and telephone answering. Tenants who are too small to justify a full-time receptionist will pay for the service from the service provider. Answering services can also be provided outside normal business hours or when tenants are away from the office. The shared tenant service provider has a built-in advantage over other telephone answering services because of its proximity and the ability to render a single bill for its full range of services.

Facsimile

The cost of facsimile machines has dropped to the point that most larger businesses now have them. There are still many smaller businesses that have such infrequent need for facsimiles, that they are willing to pay a premium for service from the STTS service provider. This service will become particularly important when Group 4 facsimile, which transmits pages in sub-minute times at 56 kb/s, becomes more widely available.

Word Processing and Publishing Services

The advent of communicating word processors makes the provision of secretarial services natural for many buildings and industrial parks. Tenants can deliver source documents manually or by facsimile, have them typed, and returned by building mail or electronic mail. A centralized service can offer the benefits of high-cost printers and electronic type-setting and publishing services. Many smaller companies produce instruction manuals and promotional material that can be published inexpensively by a central service. The proximity of the service provider gives it a competitive advantage over others.

Teleconferencing and Video Conferencing Services

Video conferencing is another service that has not been as profitable as many service providers expected it to be. Nevertheless, under the right conditions it is a service worth considering. The service provider furnishes equipment and meeting room facilities, which the tenant purchases by the hour. The advent of lower cost slow-scan or freeze-frame video equipment may make this service worth considering.

Telephone conferencing facilities are another potentially profitable service in conjunction with shared conference rooms. The service provider furnishes the speakers and microphones so a group too large to use an ordinary speaker telephone can hold a telephone conference. It may also be profitable to provide attendant-controlled conference service or to allow tenants to dial conference calls through an amplified conference bridge.

Telecommunications Consulting Services

Since the service provider has an administrative staff to handle telecommunications matters, it is natural to sell this service to tenants or outsiders who need advice about telecommunications services that are not offered by the building owner. These services could include data communications circuit design, office automation system design, and other such telecommunications-related design services.

Sale of Telephone Equipment

Whether tenants subscribe to STTS or not, they most likely will purchase telecommunications equipment such as telephones, modems, and miscellaneous supplies. These can be stocked and resold to tenants.

Security Services

Alarm systems, entry control systems, and security services such as closed circuit television make use of the telecommunications system, and are logical services to offer to tenants. The scope of this service can range from owning and leasing the service to offering design services or purchasing such equipment to the tenant's specification.

Marketing Plan

Any owner of a building or industrial park who plans to offer shared tenant services should prepare a marketing plan to demonstrate the value of the services to the prospective tenant and to demonstrate to tenants the alternative cost of providing services themselves. The tenant costing model in Figure 19.2 can be used as the nucleus of a presentation. Tenants who consider owning their own telecommunications systems are apt to compare direct costs of equipment and services, but overlook the less obvious administrative costs.

The marketing plan should express the advantages of shared tenant service, which include the following:

- *A single point of contact*: The service provider is an expert on telecommunications, and handles all the tenant's needs. This relieves the tenant of the need to determine what vendors to call or which vendor is responsible when a failure occurs on a multi-vendor service.
- *Responsiveness*: Since the service provider will have personnel on site or easily reached, tenants will be able to receive rapid response to request for service connection and maintenance.
- *Elimination of capital investment*: The tenant is relieved of the need to invest in a telecommunications system. This is particularly advantageous for a tenant that is growing rapidly and is uncertain of future requirements.
- *Cost savings*: STTS can provide local and long distance service for less than other alternatives, particularly for companies with a smaller number of stations.
- *Consolidated billing*: Without STTS, tenants may, at a minimum, receive separate bills for equipment lease or purchase, long distance service, and local telephone service. STTS consolidates bills so tenants can easily see what they are paying for total service.
- *Up-to-date equipment*: The service provider is responsible for keeping equipment updated and for adding newer features as they become available. The tenant is protected from obsolescence.
- *Reliability*: The security offered by backup power systems and duplicated processors and switching equipment is expensive for larger users and prohibitive for smaller ones. The shared tenant

system offers the reliability of the telephone company with the flexibility of a privately-owned system.

- *Freedom from administrative concerns*: Most tenants don't want to know *how* telecommunications works, only that it does work. Instead of spending time monitoring trunk usage, service levels, and costs, shared tenant service lets them spend their efforts on their own business and customers.

Analysis is Needed

Shortly after divestiture several books, articles, and seminars appeared, advocating resale of telephone service as the financial opportunity of the century. When the initial promise was not realized, many experts portrayed resale as a swamp that was destined to lost money. The truth is somewhere between these two extremes. There are opportunities in both shared tenant service and privately owned coin telephones, but they should be approached as any other venture would be: by carefully analyzing the costs and risks against the potential profits.

A detailed business plan is required before any investment is undertaken, bearing in mind that heavy initial capital investments are required, and that payback may not occur for several years. The companies most apt to profit from STTS are those that require the PBX and support services for their own businesses, and do not have to attribute the entire cost of the system to the STTS business. If the telecommunications is required anyway, the incremental investment in STTS may be low enough that the business is profitable.

PRIVATELY OWNED PAY TELEPHONES

The last few years have seen the demise of the monopoly in many services that were once the exclusive province of AT&T and its operating telephone companies. Among these are coin operated telephones.

When state regulatory commissions first authorized private operation of coin telephones, many companies entered the business, only to learn what the telephone companies had known for years; the pay telephone business is not without its problems. Fraud and vandalism rank high on the list of difficulties with coin telephones. Customers tend to be irate when they lose money in telephones, and vent their wrath on the instrument. Many private coin telephone owners failed to recognize the need for rugged construction, and were disillusioned to find that even expensive instruments didn't survive.

Several issues should be considered by anyone contemplating ownership of pay telephones. This section discusses the principal issues and the ways of coping with the problems.

Lack of Answer Supervision

The principal issue in administering private coin telephones is the lack of answer supervision, which precludes accurate call timing. As we discussed in Chapter 10, in discussing long distance services, the telephone network does not relay an indication of call answer through the local loop. This means that some variable must be used to determine when to collect and when to refund coins. The telephone companies get around this problem by putting coin control circuitry in the central office, where answer supervision is present. Coin collect and refund signals are sent from the central office to circuitry in the telephone company's telephones, but these signals are not extended to privately owned telephones.

Not only does lack of answer supervision result in user complaints of coins being collected in error, it also may result in loss of revenues. With so-called "software" answer supervision, it may be possible for people to exchange very short messages without the telephone's registering a completed call. Users became disenchanted with privately owned pay telephones when they discovered they were being charged for calls that were previously free, such as those to directory assistance and disconnected numbers. They also became dissatisfied with long distance charges that are often several times higher than the telephone company charges.

There are three principal ways that coin telephones register call completion. The first, (and least accurate), is based on call timing; after a certain number of seconds elapses, the coin is collected. The problem with this method is that it often collects for unanswered calls and fails to collect for short calls. The second alternative is to register completion when voice energy is detected on the line. This method collects for certain calls that telephone companies do not collect for, such as calls to intercepted numbers, directory assistance, and certain recorded announcements. The third method of registering call completion is by a combination of time and call progress signals. More sophisticated coin telephones can detect progress tones such as ringing, special identification tones, re-order, and busy. The more sophisticated (and expensive) the instrument, the more accurate it is possible to be, but in the final analysis no privately owned coin telephone can compete with the telephone companies' instruments for accuracy.

Maintenance and Vandalism

Keeping coin telephones in good repair is a demanding task and is one that requires rugged instruments to begin with. Many instruments are equipped with self-diagnostics that can be accessed over a telephone line with a terminal. These diagnostics can indicate the presence of damage and full coin boxes as well as providing call accounting features, as discussed later. It is important to evaluate the temperature sensitivity of coin telephones that are exposed to the elements. Outdoor booths are subject to extreme temperature fluctuations, and some electronic apparatus may not survive outside a controlled environment.

The physical structure of the telephone is important. Armored telephone cords are required to keep the handsets from being jerked from the instrument. Transmitter and receiver caps must be cemented in place, and the instrument itself should be made of a high-impact housing. The instrument must be securely mounted on studs that can be removed only from inside a locked housing.

Call Rating

Pay telephones that are connected to the long distance network require some method of rating telephone calls. The more expensive instruments have rating tables built in, and when a number is dialed, the charge is displayed on a read-out panel or announced by a voice synthesizer. If rating tables are not contained in the telephone, the phone must either be toll-restricted or connected to a service such as an alternate operator service (AOS) that rates the calls.

Some pay phones are equipped with credit card readers, and in others the credit card number must be dialed into the instrument. To gain the most versatility, the phone should be capable of accessing a variety of long distance services, although some phones are deliberately restricted to one long distance carrier. Since the carrier will collect the long distance charges from the pay phone owner, it is important that enough money is collected.

Rate changes are ideally handled from remote locations by dialing the phone and loading new rates into an erasable, programmable read-only memory (EPROM). In some instruments it is necessary to visit the telephone and change the chip on which rate tables are loaded.

Another important rating feature with private pay telephones is how to collect for special types of calls and avoid charges for others. Information providers charge varying rates for services such as time, weather, and other types of services that are accessed over the 976 prefix. Charges may range from $.25 for time to $2.00 or more for other types of services. Calls to these numbers should probably be blocked because of the difficulty of rating them. Equally important is the need to block collect calls. This is usually done by having the telephone answer incoming calls with a recording stating that it is a coin telephone and collect calls will not be accepted.

Coin Collection

Collecting coins at proper intervals has always been a problem for telephone companies. If the coins are collected too frequently, collection costs are excessive, and if they are not collected frequently enough, full boxes put the telephone out of service and result in lost revenues. Telephones that contain call accounting circuitry can alleviate this problem by reporting the number of coins deposited and alerting the owner so coins can be collected before the box is full. A call accounting system also makes it feasible for a building owner to be given a key to collect coins. The telephone itself reports how much money is contained in the box, so the building owner can be held accountable for a specified amount of revenue.

Inside the telephone is a device known as a *totalizer* that registers coin denominations. Mechanical totalizers are often unreliable and should be avoided if possible. Be certain, however, that an electronic totalizer is not temperature sensitive if the telephone is mounted outdoors.

State Regulation

The lack of uniformity in state regulation can be a problem to private pay phone owners. Before investing in private pay phones be certain that you understand the regulations of your jurisdiction. Some states have approved private coin telephones, others have turned it down, and in still other states the issue is pending before the utility commission.

Features

Pay telephones have a variety of features, some of which have been mentioned previously, that make them easier to use and administer. Generally, these features add to cost, but may increase profitability.

- *Call accounting*: A call accounting system reports the details of coins collected and long distance calls placed. It can be helpful in combating fraud, determining collection intervals, gaining current operating statistics on each instrument, and paying commissions to property owners.
- *Credit card reader*: This feature enables the telephone to read magnetic stripes on the back of bank cards, and makes it possible for users to charge calls to their credit cards.
- *Display*: Liquid crystal or light emitting diode (LED) read-outs display information such as the amount of money to deposit, time remaining for money deposited, and instructions to the caller. The display may also be used for advertising numbers such as taxi cabs and motels.
- *Hearing aid compatibility*: Some types of telephone receivers are incompatible with hearing aids, which limits their usability by hearing-impaired persons. An adjustable volume control also facilitates use of the telephone, and may be necessary in noisy areas even for persons with normal hearing;
- *Management packages*: Many systems offer a software package that assists in management functions such as collections, billing, long distance optimization, and other such factors involved in managing a group of pay telephones.
- *Real time clock*: Instruments with built-in call rating tables require a clock to adjust rates to time of day, day of the week, and holidays. The ability to remotely reset the clock is a desirable feature to avoid the necessity of visiting each telephone to implement daylight savings time changes.
- *Security*: Features include hardened case, alarms, and automatic call to a predefined number in case of vandalism.

- *Self diagnostics*: Some pay telephones can be accessed by a computer for diagnosis of trouble conditions from a remote location.
- *Toll restriction*: Where it is not feasible to permit long distance calls because of the difficulty of rating and collecting for them, this feature denies access to long distance carriers.
- *Voice synthesizer*: This unit may be used in lieu of digital read-outs to provide dialing and coin deposit instructions.

SUMMARY

There is a critical mass in the resale business. The capital investment in switching equipment and cable, and the expense of setting up administrative systems for handling orders, maintaining records, billing, and collecting means that a certain minimum number of customers must be attracted. PBXs generally begin to be cost effective for a single user at about 100 stations. For multiple tenants, the break even point is much higher; possibly as many as 1,000 stations are needed to make the venture profitable.

The privately owned pay telephone business likewise must reach a certain size before it is profitable. A single owner who has a group of pay telephones in a single location where they can be observed and controlled, and where coins can regularly be collected, may find it worthwhile to maintain a few telephones. If the telephones are located all over a metropolitan area, you must plan to invest in high-quality instruments and the administrative systems to make them profitable. Many motels, shopping malls, and other such establishments have installed private pay telephones briefly, then taken them out because of the administrative problems, but that does not mean they can't be made profitable under the right conditions.

This chapter has shown some of the benefits and hazards of the business of reselling telephone service. There are still opportunities to make a profit either directly through resale or indirectly through providing a service that tenants find an attractive alternative to obtaining service on their own. The decision should be approached with a thoroughly researched business plan.

CHAPTER 20

CONTROLLING TELECOMMUNICATIONS SYSTEMS COSTS

Telecommunications costs are a significant part of every organization's budget, and as such they should be regularly evaluated for accuracy and appropriateness of charges. A major part of the job of any manager with telecommunications responsibilities should be to establish certain routines for controlling costs and action level points for adjusting them when they exceed reasonable levels. These controls range from a routine method of adding and rearranging services to regular verification of bills and control of unauthorized use of telecommunications services.

Common carrier tariffs and rates are complex, change frequently, and are difficult for the carriers' employees to interpret, let alone their customers. It is probably futile for most telecommunications managers to understand tariffs, but a systematic method of checking bills and questioning fluctuations will help make some order from the complexity.

ANALYZING BILLS FOR TELECOMMUNICATIONS SERVICES

Billed items are often listed in telephone companies' service records by universal service order code (USOC). This is a special mnemonic language that allows machines, and occasionally people, to read the service details. Unfortunately for the multi-company service user, the word "universal" is something of a misnomer. Some designators are fairly uniform across ex-AT&T companies and have even penetrated some independent telephone companies. For example, the code "1FB" refers to a single party flat rate business line in most companies. Other designations are unique to the company that employs them, and the names offers little clue to their meaning.

When AT&T and the telephone companies divided customer services among themselves, the billing records were, in most cases, transferred without a field verification. For each service under your jurisdiction, you should request, at least annually, a copy of the service record and compare it against your own records. The billing company should be asked to explain any charges that are not obvious, and where the service or equipment is not actually provided, a refund should be requested back to the start of the misbilling period.

Telephone company bills are divided into five main components: local service, ancillary services, intra-LATA long distance, services billed for others, and taxes. You should become acquainted with the principal elements of each section for the bills rendered by your suppliers.

Local Service Bills

Local service bills contain billing for several components. First is billing for local lines and trunks. The total cost of a local line or trunk is generally the sum of the basic line charge, access charges, charges for auxiliary services such as DTMF dialing, and other fees that have been authorized by the utilities commission. Among these are charges for extended area service, directory listings, trunk mileage, transmission improvement, and other such miscellaneous items.

Every such cost on the bill should be questioned. For example, some telephone companies have different rates for business lines and PBX lines. The criteria for applying the higher PBX rate is usually a function of the type of switching system. Key telephone systems, in which users manually select incoming and outgoing calls with a button, are presumed by some telephone companies to have lower usage than PBXs, in which people dial "9" for the same operations. Since the rate difference between these alternate services is frequently significant, it pays to question exactly what criteria is used for making the distinction, and to determine whether each of your lines meets the criteria. The difference between manual- and machine-selected trunks becomes indistinct with hybrid key telephone systems, many of which can be used in either the manual or machine trunk selection mode.

Some jurisdictions authorize the telephone company to bill for metered usage within the local calling area. Calls may be detailed or billed in bulk. If your system has a station message detail register (SMDR), it is not difficult to validate the telephone company's billing accuracy. The SMDR printout and the telephone bill will not balance exactly, but significant differences between the two should be investigated.

Bills for leased equipment should be checked for reasonableness. The number of telephones should compare to the number of telephone-using employees, the number of line cards should balance against the number of stations, and the number of trunk cards should balance against the trunks on the telephone company and IEC bills.

Taxes and Access Charges

Taxes and access charges should also be questioned. Many cities levy taxes on telephone service, but it is easy for the telephone company to charge tax in error to companies outside the jurisdiction's boundaries. They may also inadvertently levy taxes on organizations that are exempt from the particular tax. For example, any company that resells services must charge tax to the end user, and is probably tax-exempt itself. Also, common carriers such as taxicab, trucking, and shipping companies and non-profit organizations are exempt from the federal excise tax on telephone service.

Access charges are levied on lines and trunks that can access the public switched telephone network. You should verify that each trunk on which access charges are levied is actually subject to the charge.

In addition, a special access surcharge, sometimes called a "leaky PBX" charge, is levied on private lines that can be switched to the public telephone network. Where these charges appear, they should be investigated so you can be certain they are correctly applied. Generally, the special access surcharge applies if the local network can be bypassed over private lines, or if WATS or private lines can be accessed by the direct inward station access (DISA) feature on a PBX. The rules for taxes and access charges are complex, and are frequently interpreted by people whose understanding is less than perfect, so take nothing for granted.

Ancillary Services Bills

Local bills contain charges with or without detail for many miscellaneous services such as line transfer circuits, tie lines, directory advertising, and special credit cards. If the telephone company does not automatically provide periodic details of local billing, you should request it. The detail should be examined at least annually and compared to your own inventory of services.

Pay particular attention to inside wiring charges on both local and long distance bills. Many companies have begun billing separately for maintaining inside wiring that is not owned by the customer. First, be certain that you do not already own the wire, and second, that it is a good business decision to pay the carrier to maintain it.

Long Distance Service Bills

Long distance charges on the local bill will be of two types: intraLATA long distance furnished by the telephone company, and interLATA long distance billed by the telephone company as billing agent for an interexchange carrier. This long distance is further subdivided into paid, collect, third number, and credit card calls. Long distance bills should be analyzed every month to detect unfavorable trends. The best way is to keep the call details on a running list such as the one in Figure 20.1. If the number of minutes is summarized on the bill, it is advisable to keep track of the cost per minute.

The common carrier may also issue separate bills for some types of long distance service. For example, WATS services may be billed by the telephone company, the IEC, or both. Most IECs render their own bills for WATS and WATS-like services, but the telephone company often issues bills for directly dialed long distance calls carried by the IEC's MTS network.

You should pay particular attention to long distance calls that are one minute or less in duration. As explained in Chapter 10, many IECs do not use answer supervision to determine the start and completion of telephone calls. After a specified time elapses they begin billing for the call even if it was not completed. In a large organization it is difficult to keep track of uncompleted calls, but if a pattern of short calls begins to emerge, they should be questioned.

FIGURE 20.1
Long Distance Cost Summary

Service		*January*	*February*	*March*	*April*	*May*	*June*
DDD	Min.	2365	3456	2321	2346	3112	3432
	Cost	$683.49	$953.86	$652.20	$642.80	$774.89	$912.91
	Cost/Min.	$0.289	$0.276	$0.281	$0.274	$0.249	$0.266
Band 1WATS	Min.	4002	4523	5423	4322	5423	4327
	Cost	$956.48	$1,085.52	$1,290.67	$972.45	$1,279.83	$1,064.44
	Cost/Min.	0.239	0.24	0.238	0.225	0.236	0.246
Band 5 WATS	Min.	4922	5563	6670	5316	6670	5322
	Cost	$1,426.23	$1,618.65	$1,924.56	$1,450.05	$1,908.39	$1,587.22
	Cost/Min.	$0.290	$0.291	$0.289	$0.273	$0.286	$0.298
800 Service	Min.	3888	5682	3816	3857	5116	5642
	Cost.	$999.23	$1,443.14	$950.12	$1,022.06	$1,248.34	$1,472.62
	Cost/Min.	0.257	$0.254	$0.249	$0.265	$0.244	$0.261

The monthly service charges for discounted long distance should be monitored. Most IECs charge for dial-one WATS based on a billing number. If your service has been separated into two or more billing numbers, the charge will be levied more than once. Check to see if services can be combined to take advantage of a single charge. For example, many systems have separate local telephone lines for facsimile or dial-up data. It may pay to combine these numbers with a master group of PBX trunks so they can access long distance at a lower cost.

Credit card calls should be checked on each bill. Send all users copies of their bill and ask them to verify that they made the call. Check to sure they still work for the company. Review the quantity and cost of calls back to the home office. When the volume of calls to the office reaches the cost of an 800 number, you can save money by using 800 service for intracompany calls.

When the volume of direct dialed long distance approaches the $150-$200 range, it is a signal that you should begin considering bulk-billed services. Therefore, the trend in long distance costs should be monitored, and an investigation triggered when the cost exceeds a threshold. Action level indicators can be set for both high and low cost levels as shown in Figure 20.1. If you already have bulk-billed services, calls exceeding the threshold indicate problems with least cost routing or too few trunks. If overflow is below the threshold, it probably indicates

July	August	September	October	November	December	Average	Action level
2331	3219	3234	1321	2128	4443	2809	4000
$652.68	$904.54	$899.05	$355.35	$597.97	$1,235.15	$772.07	$1200
$0.280	$0.281	$0.278	$0.269	$0.281	$0.278	$0.275	$0.29
5674	2345	5433	2367	4675	3432	4329	6000
$1,299.35	$579.22	$1,265.89	$591.75	$1,140.70	$854.57	$1,031.74	$1500
0.229	0.247	0.233	0.25	0.244	0.249	$0.238	$0.25
6979	2884	6683	2911	5750	4221	5324	7000
$1,937.49	$863.68	$1,887.60	$882.38	$1,700.93	$1,274.27	$1,538.45	$2100
$0.278	$0.299	$0.282	$0.303	$0.296	$0.302	$0.289	$0.30
3832	5292	5317	2172	3498	7304	4618	6000
$1,046.18	$1,264.80	$1,350.44	$536.42	$836.13	$1,862.59	$1,169.34	$1600
$0.273	$0.239	$0.254	$0.247	$0.239	$0.255	$0.253	$0.27

too many trunks. As a rule of thumb, long distance that is billed on the telephone bill, which indicates overflow, should be between five and 10 percent of bulk-billed toll.

At least annually, you should obtain a printout of the call routing table from the PBX, translate it into English, and check it for reasonableness. Unless the PBX experiences a major transition, the tables will not have changed, but the services themselves may have changed so that the routing decisions made earlier are no longer valid.

CALL ACCOUNTING SYSTEMS

A call accounting system is a computer program that is attached to the station message detail recorder (SMDR) channel of a telephone system, and analyzes call details for the following purposes:

- To control unauthorized use of telephone services
- To allocate costs to departments or cost centers
- To detect evidence of misuse of telephone services
- To sell or resell telephone service
- To bill clients for telephone costs
- To evaluate how employees spend their time
- To verify the accuracy of long distance bills

- To select long distance services
- To obtain trunk usage statistics

Although the terms are sometimes used interchangeably, a call accounting system and an SMDR are not the same thing. SMDR is the circuitry in a PBX or key telephone system that registers the call details and sends them forward over an RS-232-C port. As each call completes, the PBX reports to the SMDR the station number, trunk number, dialed digits, time of day, and elapsed minutes. The use of the SMDR in statistical analysis for trunk usage is covered in Chapter 16. The other uses, which are concerned with telecommunications cost control, are covered in this section.

Controlling Unauthorized Use of Telephone Services

Studies, including one conducted by the author, have shown that 25 percent or more of long distance telephone usage is for unauthorized calls. The cost of the telephone service is only part of the problem; the wasted time often costs more than the service. It is, therefore, advantageous for management to use a call accounting system as one way of controlling unauthorized use.

Before any corrective action is taken, you should be certain that your organization has a published policy about personal use of telecommunications facilities. Many employees have the mistaken belief that WATS lines and tie lines are free. If no policy exists to the contrary, it is easy for many people to convince themselves that the facilities are there anyway, and that using them for personal calls costs the organization nothing.

The second line of defense against unauthorized calls is a detailed long distance statement with the name of the employee placing the call clearly identified. A sample report is shown in Figure 20.2. At the very least, each station user should receive a regular statement of calls placed from that telephone. It may also be a good idea to print local calls for those for whom regular local calling is not a job requirement. Even if local usage is not billed, the call detail report can identify users who are spending excessive time on the telephone. The call detail report should be distributed through appropriate lines of organization.

Allocating Long Distance Telephone Costs

Most organizations find that telephone costs are best controlled by distributing them to organizational units that are responsible for budget-

FIGURE 20.2
Typical Station User Report From a Call Accounting System

Station Detail Report
Period: 6/13/88 to 7/12/88
Extension: 101 Johnson, Sally Accounting
Number of Calls: 13 Duration: 1:25:13 Cost: 22.41
Average Duration: 0:6:55 Average Cost: 1.72

Date	*Time*	*Duration*	*Dialed Number*	*Place Called*	*Tk Gp*	*Cost*
6/14/88	08:57	0:05:08	202–671–1534	Washington DC	5	1.60
6/15/88	09:12	0:06:33	404–458–6543	Atlanta GA	5	1.85
6/15/88	09:44	0:03:40	206–244–5442	Seattle WA	4	2.05
6/17/88	15:43	0:22:12	305–595–8909	Miami FL	7	5.55
6/18/88	12:16	0:12:05	617–965–3456	Newton MA	4	3.45
6/22/88	10:17	0:03:45	615–544–4729	Knoxville TN	3	.60
6/24/88	09:27	0:03:22	301–268–7438	Annapolis MD	5	.75
6/28/88	11:33	0:02:34	704–598–3478	Charlotte NC	3	.44
6/29/88	14:09	0:01:23	206–244–5442	Seattle WA	4	.60
7/2/88	16:22	0:03:12	404–355–4235	Atlanta GA	3	.78
7/6/88	10:10	0:04:55	617–357–3452	Boston MA	6	1.22
7/12/88	09:03	0:04:23	213–673–3453	Inglewood CA	3	1.08
7/12/88	15:00	0:12:01	615–544–4729	Knoxville TN	3	2.44

ing. The call accounting system provides an effective way of distributing costs. Charges are identified either by telephone line or by an account code dialed at the time the long distance call is placed. Calls can then be sorted and billed by account code, with a journal entry furnished to each department.

A primary advantage of charging to departmental accounts is the facility it offers for budgeting telephone services by cost center. If the service is budgeted only by the telecommunications department, other users tend to be unaware of the cost and are less concerned with controlling it.

Detecting Evidence of Misuse

Some telephone costs are incurred simply because the user is not using the telephone service properly. Older systems that lack full least cost routing capabilities are particularly susceptible to misuse. In these systems, users are often instructed to dial access codes for tie lines, WATS lines and other lower cost facilities. In systems with up-to-date least cost routing capability, this feature can detect failures of users to heed a call warning tone that indicates they are about to advance to a high-cost facility. It is also useful for detecting when users are calling at the wrong time of day or are using paid long distance services to a location that has an 800 number, and other such cases of improper use.

Reports such as those shown in Figure 20.3 are useful for detecting misuse. Most call accounting systems can show excessively long calls, calls placed outside normal business hours (which might indicate a janitor was using the telephone without authorization), or calls that exceed a set cost threshold.

Reselling Telephone Services

Another similar use of the SMDR is found in residential units such as apartment houses, hotels and motels, and shared tenant services. In these applications, the call rating feature of the SMDR is used to mark up each call enough to pay the overhead costs of the telephone system as well as the direct costs of the toll service. Specialized SMDRs for these kinds of

FIGURE 20.3
Typical Reports From Call Accounting System Used for Detecting Potential Cases of Abuse

Calls to Specific Numbers

Date: 6/13/88 to 6/20/88

Date	Time	Duration	Dialed Number	Name	EXT	Cost
6/13/88	17:45	0:05:08	976–655–3547	Smithers	125	5.15
6/15/88	09:12	0:06:33	381–047–4322	Brothers	233	11.85

Calls Over One Hour

Date: 6/13/88 to 6/20/88

Date	Time	Duration	Dialed Number	Name	EXT	Cost
6/15/88	09:44	1:03:20	206–244–5442	Woods	212	18.05
6/17/88	12:43	2:22:12	305–595–8909	Schell	127	31.55

Calls Outside Normal Business Hours

Date: 6/13/88 to 6/20/88

Date	Time	Duration	Dialed Number	Name	EXT	Cost
6/18/88	22:26	0:12:05	617–965–3265	Houser	124	2.65
6/22/88	20:27	0:03:45	615–544–4378	Rhodes	143	.66

applications can provide toll statements on demand for clients who are checking out or moving. They also may include hotel/motel management software that shows when rooms have been made up, and may operate in conjunction with a guest wakeup service.

Shared tenant services would be impractical without some form of call accounting system. Specialized shared tenant software is available from some companies to bill local service as well as long distance costs.

Billing Clients for Telephone Costs

Client code dialing is another SMDR feature that is useful to professional organizations that bill telephone charges to clients. If a client code is dialed, details are registered at the time a call is placed, which is a more effective way of allocating costs than the alternative of identifying calls when the long distance statement is received. Where non-detailed services such as WATS and tie lines are connected to the PBX, the SMDR is essential for identifying long distance costs because it is the only call record that exists. A feature available in some PBXs forces users to dial account codes so calls cannot be placed without being allocated to cost centers or clients.

Evaluating How Employees Spend Their Time

A call detail recorder is useful for determining how the time of some employees is spent. This is particularly appropriate for customer contact groups. For example, by sorting calls from a telemarketing group by area code, it is possible to determine areas of the country that are not given enough attention. By looking at the distribution of call length, it may be possible to evaluate the effectiveness of sales people. For example, a salesperson with lower than average sales but with longer than average call holding time may need additional training, or a person with low sales and short call holding time may not be doing a thorough enough job on each contact. You can also determine when sales people are making excessive calls outside their assigned territories.

The data produced by a call accounting system, in conjunction with other indicators, can be an effective method of collecting data about telephone usage when this information is useful for determining the cause of less than satisfactory performance. You can also use call detail records to measure the effectiveness of other types of employees. People whose jobs require them to make numerous telephone contacts can be evaluated to some degree by comparing statistics of numbers of calls placed and average length of call with other effectiveness indicators.

Selecting Long Distance Services

A call accounting system produces information about the long distance service over which each call was placed. It shows by hour of the day when too many or too few long distance trunks are in place. By evaluating the cost of overflow, you can make adjustments in trunk quantities by a simple cost comparison.

SELECTING A CALL ACCOUNTING SYSTEM

There is a great deal of difference in the capabilities of call accounting systems on the market today. Most systems connect to personal computers, although some use stand alone processors, minicomputers, and mainframes. There are generally five methods for obtaining call accounting reports for a telephone system:

- A stand alone call accounting system
- Call accounting software for a mainframe or minicomputer
- Call accounting software for a personal computer
- A service bureau
- Integrated equipment in the PBX

There is little difference in the report capabilities of the different types of system. The decision of which to select is based on other factors such as capacity, availability of existing systems, company preference, and other such criteria. Cost is inevitably a concern in selecting a call accounting system, but it should be one of the least important factors considered. This section discusses the primary criteria for selecting a system.

Report-Generating Capability

The most important factor in selecting a call accounting system is the type and structure of the reports that it generates. The most basic systems are not call accounting systems at all. They are merely a printer attached to the telephone system's SMDR channel. The printer lists the details of each call as it is completed. This type of SMDR has no ability to generate reports, so any analysis is manual. The presence of the SMDR deters some abuse, but it requires considerably more manual effort than call accounting systems.

More effective and more expensive call accounting systems can process calls to look for patterns of abuse or to charge telephone costs to users' departments. You should base the selection decision on how well the reports meet your requirements.

The report should show all the call detail that is included in a regular telephone company long distance bill; time of day, equipment or facility used, telephone number dialed, duration of the call, and cost of the call. If the recording equipment is capable of maintaining a list of frequently called numbers, it is also advisable to identify the called party. If this capability is not available, the called city and state should be identified from telephone number if possible. The equipment or facility that was used on the call is useful for evaluating whether users are observing call warning tones or merely allowing the call to advance to a more expensive route. Supplementary reports that may point to trouble spots should also be provided from the recorder.

Typical report requirements are:

- Call details by area code or state
- Call details by cost center or department
- Call details by extension number
- Call details by long distance carrier. The system should show, for example, how much traffic is overflowing from WATS to MTS
- Repetitive calls to the same telephone number, indicating a potential advantage of using an alternate long distance facility such as a tie line
- Unusually short calls indicating the possibility of defective equipment or trunks
- "Watchdog" reports of calls to designated numbers such as to information services
- Calls outside normal business hours, possibly indicating that an unauthorized person is using the telephone
- Calls to locations not normally called in the course of business
- Usage statistics by trunk number, the system should show enough information that you can determine when trunks are over-or under-used, see Figure 20.4 for a sample report

FIGURE 20.4
Trunk Utilization Report from Call Accounting system

Trunk Utilization Report

From 6/13/88 to 6/20/88

Trunk	Incoming		Outgoing		Total		
	No.	Length	No.	Length	No.	Length	Ave
101	0	0	221	876	221	876	3.96
102	0	0	134	520	134	520	3.88
103	122	500	101	332	223	832	3.73
104	98	412	133	515	231	927	4.01
105	131	12	201	9	332	21	0.06
106	141	531	0	0	141	531	3.76
107	139	512	0	0	139	512	3.68

- Analysis reports by trunk for incoming and outgoing calls and percent usage by hour of the day
- Exception reports for high cost, long duration, or short duration calls

The short call analysis is one of the most valuable methods of detecting trouble in telecommunications systems, and is the principal reason for entering the equipment number on the call detail report. If the same PBX trunk, for example, consistently shows up on the short call list, it is a good indication of trunk trouble because users hang up when they encounter trouble. This check is not possible, however, with call detail recorders that round all calls to the nearest minute because short calls will be either not printed at all, or shown as one minute duration. Trunk 105 in Figure 20.4 is an example of a trunk that is probably defective, as indicated by the short holding time.

The call accounting system should be capable of identifying the cost center to which each extension belongs. It should also be capable of registering costs to an account code dialed by the user. Pay particular attention to how easily these are changed. The most effective systems have a menu-driven routine that leads the operator through the process of changing assignments.

The most effective call accounting systems allow you to design your own reports, store the report-generating details, and retrieve them from a menu. A particularly effective feature is the ability to output call details in a format that can be read and massaged by a spreadsheet. Most spreadsheets support a Document Interchange Format (DIF) file, so if the call accounting system also supports DIF, call details can be downloaded to a spreadsheet. With this feature, report-generating capability is virtually unlimited.

Call Timing and Rating

Because the local central office does not signal answer supervision from the distant end over the subscriber loop, the call accounting system has no way of determining when a call is actually answered. It must rely, therefore, on some method of software answer supervision. The least complex method is to start timing after the call has been ringing for a specified number of seconds. This method is satisfactory for most systems unless you are reselling long distance, in which case a more accurate method may be required.

Call accounting systems employ several different methods of rating calls. The most accurate is a full table of vertical and horizontal (V&H) coordinates. A full V&H table lists every area code and prefix in the United States and Canada, and enables the system to calculate the airline mileage of the call (see Chapter 10 for an explanation of V&H tables and their use). For companies that resell long distance, this method is the most accurate. The tables must be updated regularly to list new prefixes as they are added. If your company places many international calls, it may also be necessary to subscribe to international rates.

A less accurate system is to use V&H coordinates only for calls within the home state and neighboring states. Because the mileage bands used in rating long distance calls are so broad, it is reasonably accurate to consider all of an area code to be within a single band. This method is inaccurate only when the mileage band splits an area code; even then, the difference is only a few cents. The rates are sufficiently accurate to distribute costs within a company where a few cents error makes little difference.

The rating table itself should be user-adjustable. Some firms charge back their long distance costs to clients, and mark up the cost of the call to account for charges that can't be easily identified to a client. For example, monthly service charges or the charge for a group of 800 numbers might be recovered by marking up the cost of long distance that is registered by the SMDR. Hotels and motels almost invariably mark up charges to recover the cost of the telephone system itself. The call accounting system should provide a facility for easily changing the rating tables. Some systems enable the vendor to dial into the call accounting system and change the rates remotely.

Centralized vs. Decentralized Systems

If you have a multi-PBX network, or if you plan to have your call detail reports prepared by a service bureau, the SMDR should be capable of decentralized operation. Calls can be stored in a buffer similar to the one shown in Figure 20.5, and retrieved by a remote processor.

System Management Capabilities

Many SMDR software packages include system management capability. With this feature, a computer assigned to the SMDR can also support system management functions such as these:

- Wire and station assignments
- Equipment inventory
- Work order process
- Telephone directory

FIGURE 20.5
Western Telematic's Pollcat Call Storage Unit

Photograph courtesy of Western Telematic

These system management features are discussed in more detail in Chapter 21.

Multi-Tasking Capability

A useful feature on personal computer-based systems is the ability to operate the personal computer while calls are being recorded in the background. This eliminates the need for dedicating a personal computer to call detail recording. Be wary, however, of the possibility that the call accounting software may conflict with another program that you regularly use. Also, be sure call details are not lost if the user reboots the personal computer. You should require the right to return the software if a conflict develops. Some systems use a plug-in board with a separate processor for buffering and processing call records, which should eliminate software conflicts. This type of system buffers calls in its internal memory and accesses a hard disk on the personal computer to store call details when the buffer is full.

Size of the System

It is important to know how many call records per month the telephone system will generate because that affects the size of the call accounting system. Systems are limited to the number of stations and number of call records they can handle, and how many records per hour they can process. The storage capacity of the system is affected by how long it is required to store call records. Many systems can store more than one month's records, although records can normally be off-loaded to a floppy disk to prevent exceeding capacity.

Records Protection Features

If calls are stored in a buffer, it is important that the buffer be equipped with a backup battery supply to prevent loss of records if the power fails. If the system is not equipped with backup battery, records should be downloaded to permanent storage at regular intervals.

PBX FEATURES FOR CONTROLLING COSTS

Most PBXs have several features that are designed to assist the manager in controlling costs. The features range from warning tones to timed reminders and trunk queuing systems.

Call Warning Tones

Many PBXs and tandem switches provide a feature to alert users when their calls are about to advance to a more costly route. For example, a PBX may have tie lines and WATS lines attached to it. The least cost routing would first attempt to connect a long distance call over a tie line. If these were busy it would next try WATS lines. If the call was about to advance to MTS, a warning tone would alert the caller that the call was about to be completed over a high-cost facility. The caller would have the option of allowing the call to complete, or hanging up and trying later. Where employees are cost-conscious, this is an inexpensive and effective way of limiting high-cost long distance.

Trunk Queuing Systems

A higher degree of usage per trunk can be achieved if the trunks are equipped with a feature that holds the user in queue if all trunks are busy, and completes the calls in priority order. The priority may be the order in which calls are received, or the system may be equipped to give precedence to certain numbers or classes of service. Queuing can significantly increase the load carried on trunk facilities, but unless it is properly administered, the gain can be offset by lost productivity of people held in queue. A switching system that is equipped for queuing should provide statistics about the number of calls queued during the peak hour and the average duration of time in queue.

Queuing systems generally have two methods of handling a blocked user. *Call-back queuing* allows users to hang up. When the facility is available, the system calls them back and places the call. *Hang-on queuing* requires the caller to hold the connection until a trunk is idle and the call can be attempted.

Least Cost Routing (LCR)

The LCR feature, applied to the entire switching system, chooses the lowest cost available route to complete a call. To be most effective, the system should be capable of routing on at least six digits; ie. area code and prefix, because many calls to a nearby area code are cheaper to complete on MTS, while others to the same area code are cheaper on

WATS. Some LCR systems are capable of changing their routing based on time of day. This allows you to take advantage of off-peak prices offered by some common carriers. The objective of LCR is to shield the user from deciding which route to choose for completing a long distance call.

Station Restrictions

The station restriction feature prevents callers from calling any number that is not permitted by their restriction class. Unauthorized call attempts are blocked or routed to the attendant. Generally, the attendant can be accessed for special billing such as a personal credit card, and can connect the line to a long distance operator. A number of restriction classes are provided with most systems:

- *Unrestricted*: Permits the station to complete any call over any long distance or local facility
- *MTS Restriction*: Permits calls to be placed over low cost long distance facilities such as tie lines and WATS, but prevents calling over high-cost facilities such as MTS
- *Toll Restriction*: Permits placing local calls, but prevents any long distance calls from being placed from the station
- *Outgoing Restriction*: Prevents station from placing outgoing calls, but permits unrestricted incoming access
- *Code Restriction*: Blocks calls to specified area code or telephone number; telephone number restriction is useful for blocking calls to services for which there is a charge
- *Incoming Restriction*: Blocks the station from receiving calls from outside the system; this feature is useful for ensuring that telephones needed for emergency access are not tied up with unnecessary outside calls

Administering Station Restrictions

Station restriction levels should be tested and reviewed periodically. Begin your evaluation of restriction levels by reviewing classes of users in your organization. Access to telephones in public areas, waiting rooms, cafeterias, hallways, and other such areas is usually not controlled and no one is assigned custody of the service. In these areas full toll restriction is usually required. If local service is measured, you may also want to restrict local access.

Another class of users may be assigned telephones, but because they have no reason to call long distance, they are restricted from long distance. For example, accounting and information clerks, order desk, production shop, and other such functions may have no requirement for long distance, and restricting the stations can prevent abuse. Of course, calls can always be placed through the attendant, who can connect the party to an outside trunk for legitimate business, or can place a call to the operator or dial credit card calls.

Emergency telephones may be restricted from incoming access to ensure that the telephone is available for outgoing calls. In some organizations, it may be desirable to prevent incoming calls from reaching production workers' telephones. They may be given access for legitimate outgoing calls, but denied incoming access to prevent loss of productivity.

Partial toll restriction may be warranted for users who have limited need to place long distance calls, and who can afford to wait if all low cost facilities are full. Such users might be permitted to call on tie lines, but not WATS lines, or they might be permitted to call on all facilities except MTS. Bear in mind, however, that such restrictions may not gain as much as it seems because calls placed by unrestricted users may overflow to MTS, and therefore defeat any saving that results from the restriction.

Restriction levels should be reviewed and tested at least annually. Tests are made outside normal working hours by going to each telephone, attempting to dial unauthorized numbers, and verifying that they are blocked. Trunk-level restrictions can be checked by busying out all the trunks of a certain class and being certain that users are blocked or permitted to advance, whichever is programmed.

Variable Restriction

Some PBXs are capable of varying the class of service based on time of day. This can be a valuable feature if, for example, managers' offices are occupied during the day, but unoccupied outside of normal working hours. Workers who have access only to restricted telephones can be prevented from using the managers' telephones if the restriction class is changed automatically.

A feature in most PBXs permits certain people who have authorization to override the restriction level by dialing an access code. A related feature enables the attendant to enable or disable restriction class from the console.

Automatic Call Distribution Systems

Automatic call distribution (ACD) systems, which are covered in more detail in Chapter 9, are an effective feature for controlling personnel and telephone costs. Personnel costs are controlled by automatically routing calls to an agent without the intervention of an attendant. In this way, agents are kept more fully occupied than they are with manual distribution systems. Telecommunications costs are controlled by reductions in circuit holding time.

In telemarketing systems, where calls are being received over 800 lines, an ACD reduces circuit costs. A full range of management reports enables the system administrator to schedule force and evaluate employee performance based on statistics produced by the ACD.

Centralized Attendant Service

Centralized attendant service (CAS) is a PBX feature that enables one attendant or a group of attendants to handle calls for more than one PBX. CAS enables attendants to route calls to the proper office or store because all outside calls come to a central point. It enables attendants to cover for one another during absence or relief periods, and gains the efficiency that results from a larger work group.

Automatic Circuit Assurance

Automatic circuit assurance provides some of the features available from a call accounting system for detecting long and short holding times. This PBX feature detects when specific trunks have unusual average holding times. It enables the administrator to detect trunk trouble and get circuits restored quickly.

CIRCUIT SHARING

Costs can be reduced by enabling voice and data communications services to share the same group of circuits. Circuits can sometimes be used for voice during normal working hours and transferred to the data processing department for file transfer or batch printing operations during the rest of the time. Since dedicated circuits are available 24 hours per day, anytime they can be shared, use is practically free.

It may also be possible to use excess capacity in a data communications network to carry facsimile. Facsimile normally takes about one minute at 9,600 b/s per printed page. If the volume is not large, this load can possibly be applied to a data network without overloading it.

Private lines can also be shared on occasion for video conferences. For example, if you have a T-1 circuit between two points, it may be cheaper to divert it temporarily for a video conference and let the traffic be carried over the MTS network than it would be to set up a special circuit for the conference. Voice private lines can similarly be diverted for freeze-frame conferencing. The effect of diverting circuits is least harmful when the conference is scheduled for hours in which the normal traffic load is lowest.

SUMMARY

Without management attention, telecommunications costs are invariably higher than they need to be. Billing errors by service providers, the provision of too many or too few circuits, selection of the wrong kinds of services, and improper or abusive actions by users are some of the ways in which costs get out of hand. Every company should have a systematic method of reviewing telecommunications costs and bringing them back in line. If the company is large enough, it will pay to invest in cost-controlling systems, or researching which call accounting system is the most effective. It is not uncommon for companies to reduce their telecommunications cost by 25 percent or even more by judicious application of the techniques discussed in this chapter.

CHAPTER 21

DEVELOPING A TELECOMMUNICATIONS OPERATIONS PLAN

Up to this point we have discussed tools and techniques that managers can use to administer an organization's telecommunications system. How extensively these tools are applied is a function of the size of the organization and the scope of the telecommunications manager's responsibility. In large organizations, specialists may be assigned to some of the functions, but in small organizations, some functions may take only a small amount of time, and others will not be required at all.

This chapter pulls together some of the major operational functions a telecommunications manager should perform, and discusses how they can be organized and controlled. Such diverse topics as evaluating service, maintaining records, and receiving trouble reports are elements of a telecommunications operations plan. This chapter presents forms, procedures, and techniques that can be applied if your system is complex enough to require them.

This chapter also discusses a topic that is too frequently neglected, planning for disaster recovery. A disaster can be anything from a power failure to total destruction of the telecommunications system. Too often, managers think about recovery only in the midst of disaster, when it is too late to plan. This chapter offers suggestions for preparing fall-back plans.

SERVICE ORDERING PROCESS

Unless all requests for telecommunications service funnel through a single location, you will find it difficult to maintain control of costs and assignments. Regardless of whether you use a contractor or perform the work yourself, a work order form such as the one in Figure 21.1 should be used.

One of the most important blocks on the work order is the one containing assignment fields. It should be clear to all technicians that any discrepancies in the records or changes in the assignment must be recorded. The technician should be required to return the work order after completion so that records, which are discussed in the next section, can be posted.

The work order has four primary purposes:

- To relay to a technician a summary of the work to be done and the assignments to use
- To communicate the work done from the technician back to the records center
- To serve as a document for verifying the accuracy of vendors' bills
- To charge the cost of the work to the party incurring it (if your company uses a charge-back process)

FIGURE 21.1
Telecommunications Work Order Form

Telecommunications Work Order

Requested By ____________ Department ____________
Requested Date ________ Due Date ________ Expense Code ____
Telephone or Equipment No. ____________
Work Requested ____________

Reason for Request ____________

Referred to:

☐ Telphone Co.	Date ________	Promised Date ________
☐ Equipment Vendor	Date ________	Promised Date ________
☐ IEC	Date ________	Promised Date ________
☐ Other	Date ________	Promised Date ________

Assignment
Line Equip No. ________

Cable ________	Pair ________	Color Code
Cable ________	Pair ________	Color Code

Complete Date ________ User Notified ☐ Records Posted ☐

MAINTAINING RECORDS

An often-neglected aspect of telecommunications management is that of keeping records of equipment, wiring, and software. The problems caused by a lack of a record-keeping system are most obvious when changes and rearrangements are required. It is usually possible to trace wires and crossconnects back to the source, but doing so is more costly than creating the records at the time of initial installation and keeping them current with changes.

Telecommunications records can be classified as the following:

- Location records
- Wiring and assignment records
- Station type and feature assignment records
- Equipment wiring and assignment records
- Major components inventory

Although these records are interrelated, they should be maintained separately for all but the smallest systems. They should be created from vendor records of the initial installation, or, if that is no longer possible, from a special inventory taken for the purpose of record creation.

Location Records

The purpose of location records is to document where equipment is physically installed. The degree to which location records are required is a function of the size of the organization and the complexity of the telecommunications system. In small-to-medium-sized companies, it is hardly likely that a PBX or key system controller will be misplaced, but in large companies with several complex installations, the location of every item of equipment should be documented.

Equipment documentation is particularly important when its location is concealed. Conduit runs, buried cable, and cabling concealed in walls and ceilings should be documented in map or list form, or both. Wiring closet or terminal locations should be documented and cross referenced to cable numbers and terminal types and size.

It is a matter of policy whether to maintain location records on equipment that is easily movable, such as telephones, but they are usually unnecessary. The assignment records, described later, may include the type of station and the room location. Because people move frequently and may take their telephone instruments with them, it is usually preferable to maintain station records on the assignment sheet and eliminate the location record. This advice doesn't necessarily apply to expensive terminal equipment such as video display terminals, integrated voice/data terminals, printers, modems, and other semi-fixed equipment. Every company will have a policy on maintaining property records on capital equipment. This policy will probably suffice for station equipment. Data transmission equipment such as concentrators and multiplexers, and voice transmission equipment such as channel banks, repeaters, and signaling equipment should be documented on location records for maintenance purposes.

Location records are of two types, maps and lists. Cabling records are essential for any organization. The route of every cable should be recorded on a map in case it must later be located. Splices and connectors are potential trouble sources, the location of which should be documented. Figure 21.2 shows a typical cable map. Like architectural drawings for wiring and plumbing, a cable map should be diagramed on a floor plan or on a specially prepared plan of areas outside the building. The cable map should show the location of all wiring closets, terminals, and major items of telecommunications equipment such as the PBX and distributing frames. It should also show the type of cable, the number of pairs, and the gauge.

Each cable should be assigned an alphabetic or numeric designator. For example cables could be designated as A, B, C... or 1, 2, 3... An individual cable number should be assigned to each major cable that routes from the main system terminal to wiring closets. Riser cables, which are cables that route between floors in a building, may be crossconnected to the branch cables at each floor in a master wiring terminal. Other items of equipment, particularly those that are moved with some frequency, can be documented on lists.

Location records are especially important in large scale local area networks. Broadband networks, in particular, are composed of numerous components, any of which may need to be found on short notice. Amplifiers, couplers, splitters, and other active and inactive components are the most probable source of trouble in a LAN, and are likely to be physically concealed. Their location should be documented on a building floor plan.

FIGURE 21.2
Outside Cable Map

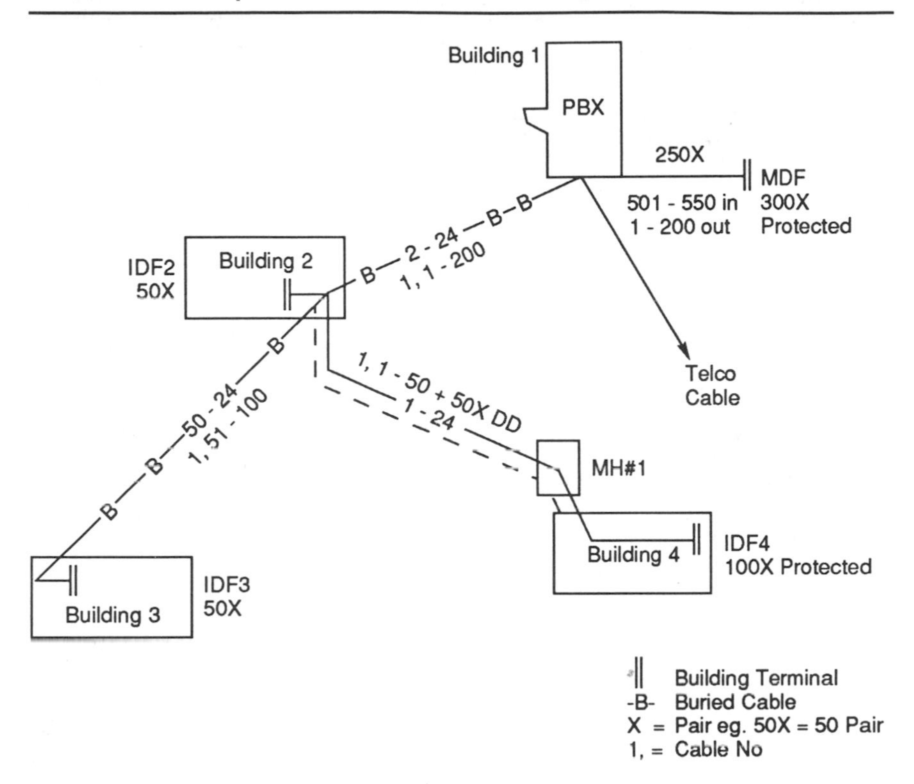

Wiring and Assignment Records

The assignment record is a listing of all the equipment, cabling, station equipment, restrictions, and features associated with each user's service. In larger systems, assignment records should be maintained on a data base management system for ease of administration. Figure 21.3 shows the major elements of an assignment data base and how they are interrelated. Not all fields will be required for every system, and in some specialized situations additional fields may be required.

Records in this data base consist of both inventory and assignment records. The inventory records contain a complete listing of all cables, equipment ports, and extension numbers. These records are created at the

FIGURE 21.3
Elements of an Assignment Data Base

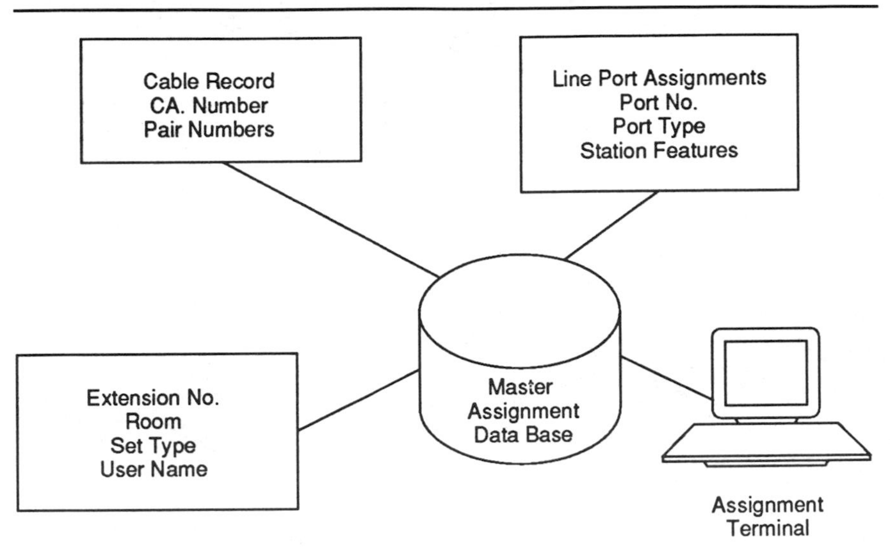

time cable or ports are added, and are used to build a configuration of equipment that can be crossconnected to provide service for each user. A relational data base management system is ideal for keeping the record on larger systems. For smaller systems the records can be kept on paper.

The first record is a listing of all the line ports in the system. It lists the port number and port type (analog, digital, trunk, etc.). As ports are assigned, it is easy to determine from this record how many vacant ports remain. The port table is created at the time the system is installed, and is updated when additional ports are installed in the system.

A second table lists all the cables in the system. Separate numbers are assigned to each cable that terminates in a distributing frame or terminal in a wiring closet. To make an assignment in the cable record, it is necessary to know the wiring closet or room in which the cable terminates. This information is contained in the location record. It is not necessary to maintain cable numbers on horizontal station wiring. Station wiring is merely designated with the number of the wall or floor jack on which it terminates. If riser cables are not used and all jacks are wired directly to the distributing frame in the equipment room, the only wiring record that needs to be maintained is a record of what port each jack is crossconnected to.

The third record in the data base is an inventory of the extension numbers contained in the system. If the port number and extension number are permanently associated in the system, this table can be included with the port records.

An assignment is made by selecting a vacant port from the line port record, a vacant extension number from the extension number table, a vacant riser cable pair, and determining the user's jack number. These assignments are listed in the master assignment record. Assignments can be posted to the data base either before or after the work is completed. If precompletion posting is used, however, every work order should flow past the assignment desk after completion to ensure that any changes in assignments are posted.

For each user, the following minimum information is listed:

- Jack number
- Riser cable number and pair
- Extension number
- Port number

If additional information is kept in the master assignment record, it will be possible to obtain more reports from the system. For example, if name, room number, and department are listed in a data base management system, telephone directories and personnel locater guides for the attendant's use can easily be printed. If the master assignment record lists the type of telephone instrument, features, button assignments, pickup and hunt group assignments, restriction class, and other such information, it serves as a single service record for each extension.

By linking the assignment records to inventory records, lists of vacant ports, extension numbers, and cable pairs can be obtained for making assignments, verifying the accuracy of records, or determining when you are about to run out of capacity. Reports such as the following can be produced by the system if it is kept on a computer:

- Cable pair assignments
- Port assignments
- Feature assignments
- Restriction lists

- Telephone directory
- User lists

Wiring and assignment records are used for maintaining and for installing and rearranging equipment. Technicians who install or move services should be given a work order that shows which equipment and cable pairs to use. Otherwise, they merely choose the facilities to use and may deviate from the plan the designer intended. Figure 21.4 shows a wiring closet that was administered without records or control.

Station Type and Feature Assignment Records

A record should be maintained for each station of the type of telephone set or terminal, the pattern of assigning features to buttons, restrictions, and station assignment information such as the hunt group, pickup group, and the destination to which calls forward when busy or unanswered. Record keeping is simplified if stations are assigned in patterns. Pattern assignment makes it easier for people to use other than their own telephones, and to retain records by a single designation instead of individual station-by-station assignments.

Equipment Wiring and Option Records.

Major items of equipment are wired to the distributing frame where they can be crossconnected to other equipment and common carrier interfaces. Such equipment normally has numerous wiring options, a record of which should be retained in a master wiring list. Figure 21.5 is a sample of a wiring list for a digital channel bank. This wiring list is supported by manufacturers' drawings. The wiring list itself shows that figures A and E of the manufacturer's drawing, T106432, were wired. The manufacturer's drawing shows that these options wire the channel bank for mode 3 and that voice channels are cabled to the distributing frame with four transmission leads and four signaling leads.

A similar wiring record should be retained for every category of equipment, including data communications equipment. On the RS-232-C side of modems, multiplexers, and other such equipment, numerous options can be wired. The record should show which leads are cabled through and whether the equipment is wired as data circuit-terminating equipment (DCE) or data terminal equipment (DTE).

FIGURE 21.4
Wiring Closet That Lacks Administrative Control

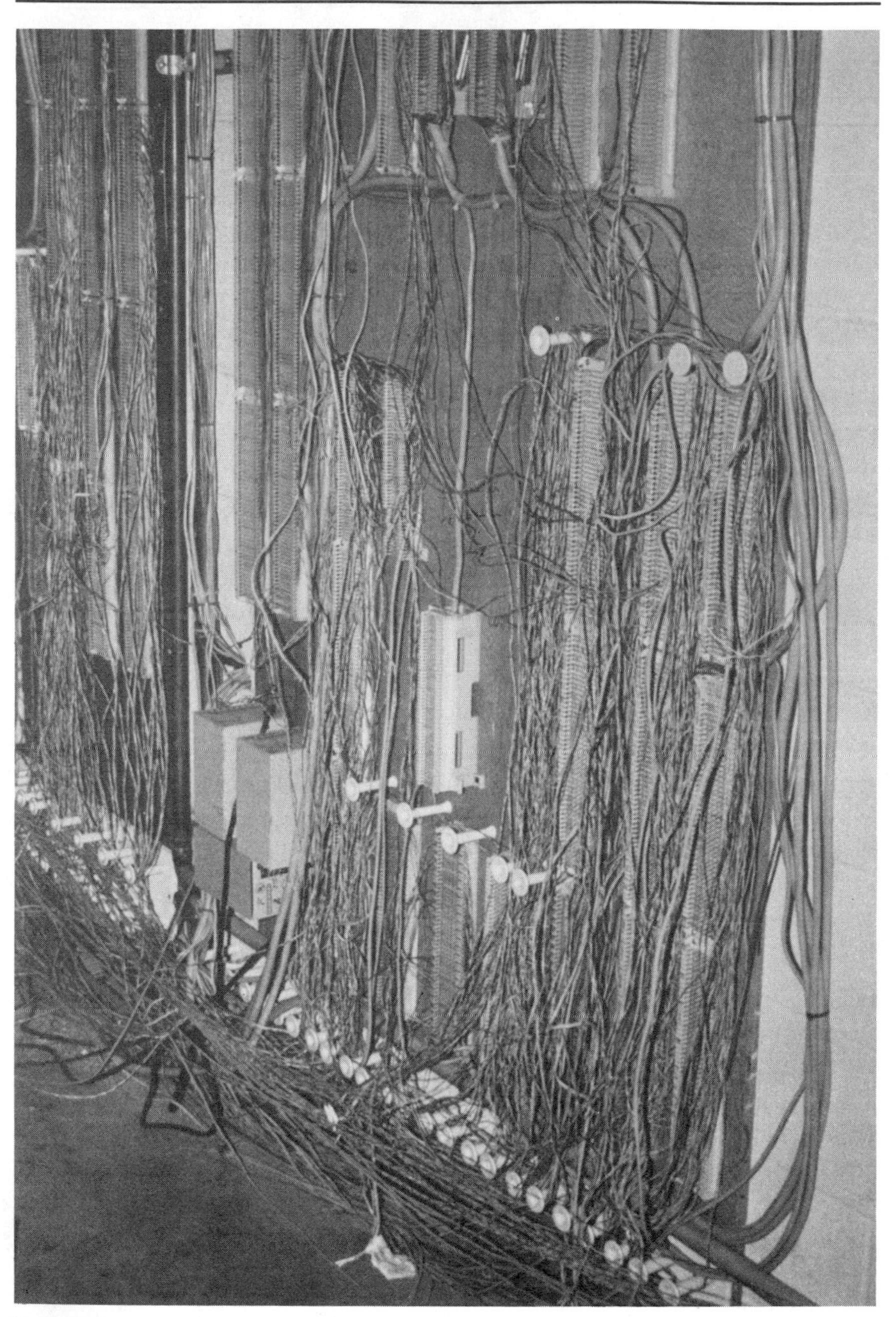

Photograph courtesy of author

FIGURE 21.5
Sample Wiring List for a Digital Channel Bank

		Wiring List Digital Channel Bank			
Drawing No.	*Description*	*Figure*	*Option*	*Quantity*	*Bay*
T106432	Channel Bank	A, E	2	3	12
T104434	Power Unit	B, J, T	1	3	12
T342984	Common Equip	G, B	0	3	12
T432893	Patch Panel	F	1	1	12
T436278	Fuse Panel	J	0	1	12

Distributing Frame Drawings

Records should be kept of the distributing frame layout. The installer should designate every block on the distributing frame, and in addition, a drawing should be kept of block assignments. Some companies install 66-type blocks on different colored plastic backboards, with each color representing a different wiring area. Figure 21.6 shows standard backboard color coding. If your equipment room is wired with colored backboards, you should label them because not all installers follow the same color scheme, and if the wiring plan is to have any value, the coding must be consistent.

To further document that wiring scheme, cable and pair numbers should be shown on the blocks, and on the block drawings. The objective of the distributing frame wiring plan is to ensure consistency and record integrity, and this can be accomplished only be correctly labeling the blocks.

FIGURE 21.6
Standard Backboard Color Plan:

Green = Central Office Trunks
Red = Key Service
Blue = Station Wiring
Purple = PBX Ports
Yellow = Accessories
White = Wire Distribution

Inventory of Major Components

An inventory record should be kept of major pieces of apparatus in the system. This inventory has two principal purposes, to serve as a continuing property record for tax and management audits, and to serve as a record for administering manufacturers' upgrades. The record should consist of all depreciable assets, and may include records of some items, such as software, that are charged to operating expense. The record should, at a minimum, include the following information:

- Equipment description
- Manufacturer
- Model number
- Serial number
- Quantity in service
- Date of purchase
- Purchase price
- Asset account number

The equipment inventory can be kept in a card file, spreadsheet, word processing file, or microcomputer data base management system. If it is kept in a mechanized record, it is easier to sort and summarize for producing special reports.

Record Integrity Checks

Despite your best efforts to keep records intact, discrepancies invariably creep in. The primary cause is workers who change cable pairs or equipment assignments without notifying the person in charge of records. Discrepancies are also caused by people who change assignments to clear trouble, but don't follow through with records changes. A third cause of discrepancies is wiring errors.

Records discrepancies tend to snowball. When a technician first encounters a discrepancy, there is a strong likelihood that, rather than calling the record center, another pair will be chosen, which generates another discrepancy when that pair is assigned, and so on. The remedy is to enforce the discipline of calling for assignment changes, requiring the changes to be shown on the work order, and making periodic comparisons between the records and the way equipment is wired in the field.

It is usually impractical to verify every assignment during a records check. Instead, the most feasible procedure is to check discrepancies. A discrepancy is defined as a pair or port that is shown in the records as idle, but which has wiring attached to it, and equipment or pairs that are shown as working, but are unwired.

The practice of leaving vacant pairs wired through to the main distributing frame can confuse the verification process, but if station terminations are left crossconnected through wiring closets and riser cables, they can be reconnected by crossconnecting the wiring to an idle port on the switch. It is also feasible, if the switch has enough ports, to leave the port connected and merely mark it as nonworking in the PBX's data base. If this practice is followed, the wiring records should be marked accordingly.

MONITORING TELECOMMUNICATIONS SERVICE LEVELS

As we have mentioned several times in this book, the telecommunications manager's primary job is balancing service and cost. Costs are more easily quantifiable than service because we have a universal scorecard for measuring costs (dollars), but the criteria for measuring services is often subjective. Users may lack objective guidelines for measuring service, but they do know when they are not satisfied with the result.

Everyone who has worked around telecommunications knows that people tolerate some service deficiencies up to a point. When the limits of their patience are reached, a service crisis rapidly flares, and when a crisis point is reached, people become hypersensitive to the slightest defect. The moral for the telecommunications manager is to solve service problems before they reach crisis proportions. When do routine problems grow into a crisis? No one can predict, but your reputation as a manager depends on your ability to prevent crises from erupting. This means establishing several processes:

- A regular program of collecting service information
- A method of logging and monitoring trends
- Objectives or indicators that trigger action when limits are exceeded
- A system for keeping people informed of key service indicators
- A process for escalating to higher authority when troubles are not cleared in a reasonable time or recur too frequently

Not all service indicators relate to trouble (except in the eyes of those who are experiencing them). Users expect promptness in administering adds, moves, and changes, and are sensitive to differences in the quality of long distance circuits. For example, dissatisfaction may be expressed over delay in satellite circuits when the circuit is working normally.

While the telephone companies have been measuring service for many years, there are few guidelines to follow when determining what service indicators to track or what service levels to expect. Telephone company measurement plans were designed to deal with a mass public. A few key criteria were set, and indicators established at levels the public tended to accept. One of the phenomena of service measuring plans is that as service improves, expectations rise accordingly. Service levels that were superb by yesterday's standards are mediocre by today's, so you must be prepared to readjust action levels. There are no absolutes in measuring service. You must adjust the service level to the organization's culture and take steps to keep complaints within reasonable bounds.

Speed of Telephone Answering

One indicator of how well your organization is perceived by customers is the number of rings before calls are answered. Most callers are willing to tolerate four or five rings, but more than that tends to indicate a problem. This indicator is easy to measure. You can make an informal tally from your own telephone calls within the organization, and you can also tell by listening to the number of rings during visits to various parts of the organization.

An excessive number of rings usually indicates failure to use message-handling features in the system. For example, users should be trained to use the forward-all-calls button to cause calls to transfer immediately instead of waiting for the third ring.

Measuring the Quality of Switching Service

Switching service is one of the easiest criteria to measure. Either a connection is made or it is not. If a connection is made, there remains the problem of assessing its transmission quality, but that is a subject that is discussed in the next section.

A fundamental principle of telecommunications system design is that every call should reach some destination. Either the call is completed or a tone informs the caller why it was not. Call progress tones take the form of busy, reorder (fast busy), and vacant number and level tones and recordings. No call should simply go into limbo (in telephone vernacular, "high and dry").

Switching service should be measured not only from the standpoint of internal users, but also from the eyes of an outsider looking in. Customers often evaluate the effectiveness of an organization by how accessible its employees are through the telephone system. Therefore, it is important to measure incoming service as well as outgoing.

Monitoring Trunk Statistics

As discussed in Chapter 16, a regular program of monitoring trunk usage is essential for both service and cost control. Figure 21.7 shows traffic statistics from several trunk groups in a PBX and the corresponding blockage level. Trunk statistics from most machines show the quantity of trunks and usage. You can read the approximate amount of blockage from traffic tables. Beware, however, that the traffic tables are invalid when blockage generates an excessive number of repeated call attempts. Also, be aware that blockage for short time periods is apt to be higher than indicated by the tables. A third factor you must consider is the routing pattern in the switching system. If long distance calls overflow to MTS, a high level of blockage doesn't necessarily mean user dissatisfaction; the result may be excessive costs.

A second indicator of switching service is a blocked call or all trunks busy count provided by many systems. Most PBXs tally occasions when all trunks in a group are busy. Consult the manufacturer's documentation to determine whether the register measures only the number of times all trunks were busy, or the number of times all trunks were busy *and* an attempted call could not be completed. The most valid service indicator is a count of when users actually encountered blockage when attempting to place a call.

Incoming call statistics are more difficult to interpret. A traffic study tells the number of calls and total usage that were carried, but does not show the actual offered load. The telephone company can make a blocked call count of incoming calls. Usually, there is a charge for this service.

Bills for 800 service show the quantity of blocked calls at the bottom of the bill. This statistic should be monitored monthly, and ideally it should be plotted so you can observe trends.

If your business depends on your customers' reaching the company, as most do, a regular process of observing incoming service is essential.

FIGURE 21.7

Trunk Group Measurements Report from an AT&T System 75 PBX

Peak Hour For All Trunk Groups: 1200 - 1300

Grp No.	*Grp Size*	*Grp Type*	*Grp Dir*	*Meas Hour*	*Total Usage*	*Total Calls*	*Inc. Call*	*Grp Ovfl*	*Que Size*	*Calls Quad*	*Que Ovf*	*Que Abd*	*Out Serv*	*% ATB*	*%Out BLX*
1	23	CO	two	1300	656	323	170	17	0	0	0	0	0	2	0
2	2	fx	inc	1300	24	70	70	0	0	0	0	0	0	0	0
3	7	fx	out	1300	143	48	0	7	2	6	1	1	1	0	0
4	4	Wats	out	1300	73	81	0	5	1	0	0	0	0	0	0

The table on the next page indicates the meaning of the columns in the above report.

Figure courtesy of AT&T Corp.

FIGURE 21.7—*Continued*
Trunk Group Measurements

Measurement	*Explanation*
Peak Hour for All Trunk Groups	The measurement hour during the day with the largest total usage for all trunk groups.
Group Number	A group number between 1 and 50 (V1) and 1 and 60 (V2) that identifies each trunk group associated with the accumulated date.
Group Size	The number of trunks in a trunk group. A maximum of 60 trunks can occupy a trunk group.
Group Type	The type of trunk associated with the accumulated data. The system monitors the following trunks: ▪ Central Office (CO) ▪ Client-Provided Equipment (Auxiliary Trunk) ▪ Foreign Exchange (FX) ▪ Wide Area Telecommunications Service (WATS) ▪ Direct Inward Dialing (DID) ▪ Tie Trunk ▪ Advanced Private Line Termination (APLT) ▪ Tandem ▪ Digital Multiplexed Interface (DMI)
Trunk Group Direction	Identifies whether the assigned trunk groups are incoming (INC), outgoing (OUT), or 2-way incoming and outgoing (TWO).
Measurement Hour	The clock hour in which measurement was taken. A measurement hour starts and ends on the hour. For example, data recorded at 10 a.m. represents data accumulated from 10 a.m. to 11 a.m.
Total Usage in CCS	Total usage in CCS for all trunks in a trunk group. (Represents the total time the trunks are busy serving calls during the 1-hour period.)
Total Calls	The number of calls carried by a trunk group. The Average Holding Time (AHT) can be determined as follows: $AHT = \frac{Total\ Usage\ CCS}{Total\ Calls} \times 100.$

FIGURE 21.7—*Concluded*
Trunk Group Measurement

Measurement	*Explanation*
Incoming Calls	The total incoming calls via the trunk group to the system. These calls may be incoming from a 1-way trunk group or the incoming portion of a 2-way trunk group. This measurement is zero (0) for outgoing trunk groups.
Trunk Group Overflow	The number of calls that reach a trunk group and find all the trunks in the group busy. Calls directed to queue are considered as trunk group overflow calls.
Trunk Queue Size	A number between 0 and 100 that identifies the slots assigned to respective trunk group queue. The maximum queue slots that can be assigned to the system is 100.
Calls Queued	The number of calls that enter the trunk group queue after finding all trunks busy.
Queue Overflow	The number of trunks that were not queued because all slots in the queue were occupied.
Queue Abandoned (Abd)	The number of calls that are removed from the queue either by the system for staying in the queue for the maximum allowed time (currently 30 minutes) or when the user cancels the auto call-back.
Out-of-Service	The number of trunks in the trunk group (listed as maintenance busy) that are out of service at the time data is collected. This measurement is obtained by sampling each trunk in the system during the hourly interval.
Percent (%) All Trunks Busy	The percentage of time within the polling interval that all trunks in the trunk group were unavailable for use.
Percentage (%) Outgoing Blocking	The percentage of outgoing calls not carried on a trunk group to the outgoing calls offered. For trunk groups with no queue, the calls not carried are those calls that find all trunks busy. For trunk groups with queues, the calls not carried are those calls that find all trunks busy and cannot be queued because the queue is full. An "*" is displayed for incoming trunks because this column pertains to outgoing or two-way trunks.

Many callers will not wait. If they are placed on hold for excessive lengths of time or if they frequently encounter busy signals, they move to the competition. If you have knowledge of conditions during the busy hour, there are numerous corrective measures that can be taken including adding more trunks and taking steps to shift part of the load to lower usage hours.

Automatic Call Distributors (ACDs) produce numerous indicators of service. The statistics produced by an ACD should be regularly monitored to determine the service levels of both the equipment and the people who serve the public.

Console Statistics

Most PBXs and some hybrid systems tally statistical information from the attendant console. If you know how to interpret the information, you can make valid inferences about several key service indicators. Most consoles register the number of calls and average call handling time during peak hours. Using the queuing formula below, you can calculate how long the average caller waits. This information may be tallied by the console, but if it is not, it is easy to calculate with queuing theory. The formulas for calculating waiting time and average queue length with single-server queuing is:

$$\text{Average Waiting Time} = \frac{\text{Average Arrival Rate}}{\text{Svc Rate x (Svc Rate - Arrival Rate)}}$$

$$\text{Average Queue Length} = \frac{(\text{Average Arrival Rate})^2}{\text{Svc Rate x (Svc Rate - Arrival Rate)}}$$

For example, assume that 60 calls arrive during the busy hour and the average handling time is 20 seconds. The arrival rate would be 60 calls per hour. The service rate is three times the arrival rate, or 180 calls per hour. The average waiting time would be:

$$\text{AWT} = \frac{60}{180 \times (180 - 60)} = \frac{60}{21600} = 0.0028 \text{ Hr.} = 10 \text{ sec}$$

The average queue length would be:

$$\text{QL} = \frac{60^2}{180 \times (180 - 60)} = \frac{3600}{21600} = 0.17 \text{ callers}$$

Attendant console service can be improved by adding another console, redirecting calls around the console with direct inward dialing, or reducing the attendant's handling time. Handling time can be reduced in some systems by operational changes such as changing paging methods or relieving the operator of message-handling duties.

Message handling is one of the more frequent causes of poor console service. A substantial part of the load on the attendant may be users calling in to retrieve messages, or unanswered calls returning to the console. By establishing intermediate answering positions and using call forwarding, it is often possible to relieve the attendant of a substantial portion of the call-handling load.

Console statistics also usually provide several other service indicators. Some systems show the volume of abandoned calls. A few abandoned calls are to be expected; callers frequently hang up after one or two rings because they are interrupted or realize they have called the wrong number, but an unfavorable trend in abandoned calls can be a clear indicator of unsatisfactory service. Many consoles also register the number of times the attendant is accessed for recalls, user assistance, or to retrieve messages. Figure 21.8 shows a PBX console statistical report.

Monitoring Trunk Queuing Statistics

Systems equipped with trunk queuing often produce statistics about the length of time callers remain in queue. Excessive queuing times can be alleviated by augmenting trunk quantities or with call-back queuing in which the system calls the user when a trunk is available. Time in queue is not only an indicator of service, it is also an important cost indicator because callers tend to be non-productive while they are waiting for a vacant trunk.

Data Network Service Indicators

Front end processors and multiplexers usually tally important service indicators. If they do not, a protocol analyzer can be attached to the circuit to measure service levels.

Response Time

Response time measurements are defined as the time between pressing the enter key on a terminal and when either the first or last bit of the response returns from the host. Response time objectives should be set for the

FIGURE 21.8
Attendant Console Results from an AT&T System 75 PBX

Attendant Group Measurements

Date:- -3:00 pm FRI MAR 22,

Grp Size	*Meas Hour*	*Total Usage*	*Calls Ans.*	*Calls Aban.*	*Calls Queued*	*Time Avail*	*Speed Serv.*
1	1100	16	105	8	60	18	10 Yesterday's Peak
1	1100	36	24	6	7	36	6 Today's Peak
2	1400	30	21	5	6	42	8 Last Hour

Legend:
Grp. Size = Number of attendant consoles
Meas Hour = Hour of the day with greatest total usage (except last hour)
Total Usage = Active time in CCS all attendants spent on calls
Calls Ans. = Total calls answered by all attendants
Calls Aban = Total calls that hang up before the attendant answers
Call Queued = Total calls that are held in queue because attendants are busy
Time Avail = Time in CCS that attendants are available to handle calls
Speed Serv. = Average interval between call reaching attendant and being answered

(Courtesy of AT&T Co.)

network and regularly measured. A typical objective is a response time of three seconds or less on ninety-five percent of the transactions. Chapter 17 gives more information on calculating and improving response time.

Block Reject Rate

The rate at which blocks are rejected because of errors is tallied by many data communications devices, and offers a useful clue to response time problems. By observing a gradually increasing block reject rate, you may be able to detect trouble before it drives response time outside objective limits.

Installation Service

User satisfaction can also be measured by how long it takes to process requests for adds, moves, and changes. Users frequently forget about telephone service until it is needed, and therefore may be impatient if it isn't delivered immediately. Objective installation intervals should be published, and met wherever possible. As discussed in Chapter 5, a valid forecasting system is needed to insure that vacant ports and pairs are available when required.

Trouble Report Analysis

In the next section we will discuss a process for receiving, processing, and analyzing trouble reports. Trouble report details provide valuable indicators of how your users perceive telecommunications service.

The overall level of trouble reports is one indicator of service, but most users are more concerned about their personal experience than they are about how good or bad service was in the company as a whole. If the report level is too high or is trending upward, it may be a better indicator of cost than of service as perceived by the users.

Repeated troubles, i.e., troubles that recur on the same station within a 30 to 60 day period are probably the best indicator of customer dissatisfaction. Users expect an occasional problem, but when the trouble is either not fixed or recurs, they tend to lose their tolerance.

The percentage of reports on which no trouble was found (NTF) offers a useful indicator of service. NTF reports can be tracked by the company or technician who did the work to see if some individuals have particularly good or poor records of finding trouble. NTF reports are frequently indicators of intermittent troubles or they may indicate the need for additional operator training or a better user interface.

Trouble report clearing time is another effective indicator of user satisfaction. Users are particularly intolerant of excessive clearing time when the service is completely inoperative. The clearing time objective has to be judged by the demands of the organization, but two hours for out-of-service conditions and 24 hours for impairments is a reasonable starting place. The clearing time should be measured from the time the report is received until the user has been informed that it is cleared. It is very important that each report be closed by informing the person who reported trouble that it has been cleared.

Vendor Service Evaluation

You should maintain performance records on every vendor that provides service on telecommunications equipment. The same kinds of variables that were discussed under trouble report analysis above should be tracked by the vendor providing the service.

Vendors tend to quote response times in their maintenance agreements, but these are objectives and may not be met in practice. Moreover, response time is only one measure of user satisfaction. More important is clearing time, a statistic that is rarely quoted by vendors. As discussed later, the trouble report form should contain a field for logging the vendor responsible so objective data can be summarized.

Transmission Service

Transmission service is more difficult to measure than switching service because it is more subjective. As discussed in Chapter 18, there is no single indicator of transmission quality. The three primary indicators are loss, noise, and echo. These three variables are closely related, and have a significant effect on one another. For example, a noisy circuit is more tolerable if the loss is low (ie. the volume is high), and conversely, high loss is more tolerable if the noise is low.

Loss can be measured on local trunks with a transmission measuring set by dialing the office's milliwatt supply. The milliwatt supply number is usually the three-digit office prefix plus 1000; for example, 234-1000. The milliwatt number furnishes a 1004 Hz tone at 0 dBm (one milliwatt) for a few seconds, and then terminates the line so a noise measurement can be made. If you have a transmission measuring set, it is a good idea to measure loss and noise on local trunks when they are first installed and to check them once or twice a year or when complaints of excessive noise or low volume occur. The loss should not exceed five or six dB.

If you have tie trunks between PBXs, you should also measure loss and noise on these at regular intervals. It is possible to make one-way transmission measurements by dialing over the tie trunk to the milliwatt supply at the central office serving the remote switching machine. The loss of the tie trunk should be approximately three to five dB plus the loss of the local trunk. Two-way measurements can be made either by stationing people at both ends of the connection, or by making independent one-way measurements as described above.

Echo can be measured accurately, but it is usually beyond the scope of most private telecommunications systems. If echo is loud enough to be objectionable, it indicates trouble on the part of the interexchange carrier, and should be reported.

A regular program of observing key service indicators will do a lot to forestall reports of user dissatisfaction. The type and extent of tests to be made depend on the nature of your network. The more complex the switching and trunking arrangements, the more essential it is to have a regular process for measuring user satisfaction. In less extensive networks, including most PBXs and key systems, statistical information should be collected and regularly monitored as a way as measuring user satisfaction. The greater the degree to which you can detect troubles before they are reported by users, the higher the degree of satisfaction and the more effective your job of managing the telecommunications system will be.

RECEIVING AND ANALYZING TROUBLE REPORTS

A central location, sometimes called a "help desk," should be set up in each company for receiving, recording, and forwarding trouble reports. The trouble ticket shown in Figure 21.9 contains information that most companies will need for recording and analyzing troubles. The same form can be used for both data and voice reports. If your company is large enough, it may pay to log trouble reports in a spreadsheet or data base management system to facilitate analysis, as discussed later.

Report Receipt

When the help desk receives a trouble report, the following information should be obtained:

- Date and time of receipt
- Name and telephone number of the reporting party
- Identification of the equipment or service on which trouble was experienced
- Nature of the trouble

Diagnosis

The trouble report should be given a preliminary diagnosis by the help desk. Trouble diagnosis is covered in more detail in Chapter 23. The person who is assigned this responsibility may have little idea what is actually causing the trouble, but the diagnosis is the basis for determining which vendor should be given the report.

FIGURE 21.9
Sample Trouble Ticket

Trouble Ticket

Date ______ Time ______ Circuit or Equip Id. No. ______
Reported by ______ Received By ______ Diagnosed By ______
Trouble Reported ______
Diagnosis ______
Referred to: ______ Date ______ Time ______
Trouble Found: ______
Work Done: ______
Cause: ______
Cleared Date ______ Time ______ User Notified ______
Clearing Time: ______ Work Time ______ Materials Used: ______
Records Posted By: ______ Date: ______

Trouble Reported

NDT No Dial Tone
CRO Can't reach other station
CLO Can't log on
CBC Can't be called
SIG Signaling
DD Dead Data
MMX Modem or multiplexer
NSY Noisy
ERR High Error Rate
FEA Feature
SR Slow response Modem or multiplexer

Trouble Found

PBX
CPU
WIR Wiring
TEQ Terminal Equipment
NTF No trouble found
CKT Trunk or data circuity
SFW Software
CC Came clear

Causes

UA User action
EF Equipment failure
CF Circuit failure
UNK unknown
BUG Software fault
WEA Weather
OUT Outside action

Referred To

The third section of the trouble ticket indicates the date and time and to whom the report was referred. If the report is referred to the vendor, an estimate should be obtained of when a technician will be dispatched. The help desk should be instructed to follow up if a technician does not arrive within a reasonable time, and if trouble is not cleared within an objective interval.

Clearance

When the report is cleared, the following information should be obtained from the technician:

- Date and time service was restored
- Nature of the trouble
- Work done to clear the trouble
- Number of hours spent
- Material used
- Cause of the trouble

The help desk should call the person who reported the trouble and pass the information that it has been cleared. The details of this final report are shown in the last block of the clearance section.

Analyzing Trouble Reports

The help desk should close each report by entering information that relates to service quality. The following information should be registered on each trouble ticket:

- Out of service or impaired
- Clearing time--defined as the number of working hours between receipt of the report and restoral
- Cause--the help desk should assign one of the cause codes to the trouble

The telecommunications manager should evaluate trouble reports periodically, giving particular attention to analysis when there is evidence of poor service or excessive cost. If the reports are entered in a data base, it is a simple matter to sort them by such indicators as cause, clearing time, responsible party, etc. As discussed in the service evaluation section, a thorough analysis of trouble reports is one of the most effective ways of evaluating service quality.

DISASTER PLANNING

As telecommunications systems become more sophisticated, they become increasingly susceptible to catastrophic failure, and as your company relies more heavily on its telecommunications systems, the impact of a failure is increasingly serious. It is difficult to predict where disaster will strike, but every company with more than a few hundred telephone stations, or a data communications system that lies at the heart of its business strategy, should prepare a documented disaster plan. The elements of the plan are unique to each business, but may include the following:

- Loss of a PBX or tandem switch
- Loss of a major trunk group
- Loss of a key device such as a front end processor or multiplexer
- Loss of power
- Loss of a major cable to another company facility or to the telephone company

Some general guidelines for protecting against catastrophic failure are discussed below to guide you in preparing a disaster plan.

Loss of a PBX or Tandem Switch

Electromechanical switching machines may have been limited in features, but they had one significant advantage over their electronic successors: they were rugged and robust. About the only thing that could cause the total system to fail was loss of power or common apparatus such as a ringing machine. Electronic switches are far more susceptible to failure because they contain so much common equipment, much of which is too expensive to duplicate. Here are some precautions you should take to minimize the likelihood of failure:

- Keep the equipment room tightly secured
- Install environmental alarms--smoke, heat, high water, or alarms for other potential hazards
- Separate automatic sprinklers from sources that may inadvertently activate them
- Keep equipment rooms clean and free of flammable materials
- Provide electrical circuits of sufficient capacity and dedicated to specific families of equipment

Despite all the environmental precautions you take, total system failure may occasionally occur, and good planning can minimize the impact. The first line of defense is to establish a relationship with a reliable source of repair, either on the vendor's staff or your own. Catastrophic failure can vary from a simple case of trouble that defies normal trouble shooting techniques to complete destruction of the system by fire, flood, or vandalism. In case of extraordinary trouble conditions, the strategy is to escalate trouble rapidly to someone who can repair it. In case of destruction, the objective is to take temporary action to alleviate the situation while the system is replaced.

Install Telephones on Central Office Lines

If it is possible to manually activate the power fail transfer option on a switching system, trunks can be accessed from standard telephones. Of course, the capacity is reduced to about 10 percent of normal, but incoming calls will not go unanswered, and emergency outgoing calls can be placed. If it is impossible to use the power fail transfer, telephones can be temporarily wired to the cable pair of loop start trunks. If the outage will be prolonged, the telephone company can be requested to rewire ground start trunks to loop start so ordinary telephones can be used, or a temporary start switch can be wired to the telephone. The telephone company can also add more lines and change one-way trunks to two-way to help alleviate the loss of a PBX.

Power Failure

A protracted power outage can render a telecommunications system inoperative even if battery backup is provided. The first line of defense is power fail transfer telephones. As discussed in the previous section, these are automatically connected to central office trunks when a failure occurs, providing a minimum level of incoming and outgoing service.

The second line of defense is an emergency power source. If the building is situated so a generator can be mounted outside, it is a simple matter to equip the building with an emergency receptacle that can be connected to a generator to supply power to the telecommunications system and other vital equipment. A backup power plant is expensive, and is justified only if the penalty of failure is severe and the probability of prolonged outage is high. Most systems can tolerate a short outage, and can protect against outages of one or two hours with battery backup at a reasonable cost.

Temporary Services

It is often possible to bypass failed switching systems by using a surplus PBX or key system that a vendor has available. Another possible way to alleviate a switching system failure is to use cellular or mobile radio service.

Cable Failures

A cable failure can be disastrous to a telecommunications system, particularly in buildings with large riser cables and a company with distributed buildings and privately owned cable. Outside cables are vulnerable to damage by excavations, weather, vandalism, and other such sources. To minimize the effect of cable failures, the following precautions should be taken:

- Keep a good set of records to aid in restoration
- Place protective shields and conduit around exposed runs in all hazardous areas
- Maintain spare cable or establish a source of immediate supply
- Establish a plan for rerouting around failed sections (particularly in the case of coaxial LANs)

Restoration of most major trunk failures is the responsibility of the telephone company or common carrier, and normally can be accomplished quickly by patching to spare equipment. Increasingly, large companies are bypassing the telephone company and going directly to the interexchange carrier to avoid the expensive "last mile." This link, which is often unprotected, is a potential source of disastrous failure. Furthermore, as companies increasingly use T-1 carrier, many circuits are contained in one package, increasing the vulnerability to failure. In both cases, the best solution for preventing total failure is diversity, ie. routing a portion of the facilities over an alternate route. This is frequently beyond the control of the user, however, which leaves your network vulnerable.

THE TELECOMMUNICATIONS OPERATIONS REVIEW

An operations review is a technique that guides you through an evaluation of the existing telecommunications management system. Where procedures exist, the operations reviewer obtains copies of relevant documents for further analysis. Where procedures do not exist, a notation is made of a need for further action. Appendix L contains an operations review questionnaire that you can use to perform the initial scan. The operations review is a straight forward technique. Ask to have a process or procedure explained or demonstrated, and collect sample copies. The bills are of particular interest. You should collect a copy of a full month's bills. If detailed toll statements are more than 10 pages long, a sample of every fifth sheet is enough. Sample copies are also enough for repetitive forms such as wiring records, restriction lists, and station assignment records. Service records should be obtained for local telephone service and any leased equipment or lines.

If the system includes a station message detail recorder or call accounting system, a full week's printout should be obtained. If the PBX produces traffic usage information, obtain a copy of a recent study listing usage for every trunk group terminating on the machine. Data communications systems often produce usage information from front end processors, multiplexers, or the host computer. If usage information is not available, obtain copies of transaction records and estimates of volumes of each type of record.

The section on anticipated growth is essential. If the organization produces any regular business forecasts, obtain copies of these for possible correlation with volumes of transactions. If forecasts are not prepared, growth estimates must be obtained from officials who are in a position to judge growth.

The material collected is analyzed to determine the answers to the following questions:

- Are the proper procedures in place to enable management to reach a reasonable balance between cost and service?
- Are the procedures documented?
- Are the procedures being used?
- Is there a single point of contact for all service requests?
- Are purchased services regularly evaluated to be certain the company is getting its money's worth?
- Are the various elements of telecommunications organized for maximum effectiveness?

The operations review is an audit of the effectiveness of telecommunications management. Its purpose is not to cast blame, but to gain a quick look at whether corrective action will likely result in improved performance. It is not necessary, or even desirable to have an outsider conduct the review. The questionnaire is designed to be self-administered by management that wants to maintain peak performance.

SUMMARY

This chapter has presented a variety of topics, all of which relate to the administrative as opposed to the technical side of a telecommunications manager's job. If a company has lacked professional telecommunications management in the past, these functions are probably not being performed. Technicians probably choose their own assignments and leave behind an undocumented tangle of spaghetti-like wiring. The time it takes to add or rearrange a station is two or three times as great as it needs to be, simply because a technician must trace wires or run new ones even though vacant pairs are already in place. In such a company, the work not only takes longer, it must also be assigned to an outside technician because of the complexity, when it could easily be done with the company's own work forces and a moderate amount of training.

Without a process for systematically monitoring service, crises tend to erupt without warning, and no objective data exists from which to analyze the problem. Vendors may be paid more than once for doing the same work, or they may be paid for working on equipment that is still under warranty simply because the necessary service records do not exist. Forecasting is a haphazard exercise because records of vacant equipment do not exist and shortages are discovered only after the last vacant port or pair is used.

Not all of the administrative techniques a telecommunications manager should apply are discussed in this chapter. Security is important enough to warrant a chapter of its own. As the complexity of private networks grows, more and more companies are relying on mechanized systems to support their operation, a topic that is discussed in the final chapter.

CHAPTER 22

TELECOMMUNICATIONS SECURITY

We've all read the horror stories of damage done by mischievous or malicious hackers, who are those people that make a hobby or avocation of gaining unauthorized access to private networks. Security issues go much deeper, however, than denying access to the teenager with a personal computer, a modem, and an automatic dialer. Organizational security is at risk from wire tapping, electronic eavesdropping, unauthorized reception of microwave or satellite communications, malicious damage to facilities and equipment, and exploration of corporate files and data bases by people who have legitimate access to the network but who are not authorized unlimited access to corporate files.

It is very difficult to guarantee the absolute privacy of communications. Even though circuits are closely guarded, it isn't always necessary to get direct access to a circuit to reach the information it carries. Most electronic devices radiate energy to some degree, and with the appropriate apparatus information can be retrieved from free space. For example, video display terminals use sweep circuits to "paint" characters on the screen. Unless terminals are properly shielded, they radiate energy that can be detected over a radius of 100 feet or more. The Department of Defense has developed requirements known as "Tempest" specifications, to which devices used by its agencies and contractors must conform to prevent the leakage of sensitive information from terminals. Many commercial devices also conform to Tempest specifications and offer at least one level of protection against electronic eavesdropping.

Eavesdropping from radiating terminals is only a small part of the overall security problem. Telecommunications managers must constantly guard against security breaches. It is easy to be complacent, but wherever the motivation exists, whether from competitors seeking proprietary information, foreign governments seeking defense secrets, or employees merely seeking to satisfy their own curiosity, potential for damage exists in most organizations.

Most organizations have a need to protect product plans, competitive strategies, and sensitive internal information such as payroll and employee evaluations. One vendor of satellite down-link equipment reported monitoring a video conference of a major auto manufacturer in which plans for introducing new models were being discussed with no apparent security precautions. Not even voice scrambling is enough to ensure the privacy of information this sensitive.

OBJECTIVES OF THE TELECOMMUNICATIONS SECURITY

Security systems must balance operational expediency against effectiveness. The value of information lies in its accessibility, but to safeguard information it must be made inaccessible to all except those who are duly authorized. Unfortunately, as we limit accessibility we make it more time consuming and difficult for authorized persons to use the information.

In the interest of expediency, users often take shortcuts. Passwords are disabled, displayed prominently, or buried in batch files so that it is easy to log on to the system. Passwords may be closely guarded, but embedded in a telecommunications program that is designed for automatic log-on. Given access to the terminal, anyone who knows how to use the program can log on the system so password security is compromised.

The first objective of a security system is to reach an optimum balance between security effectiveness and how difficult the system is to use.

The second objective of a security system is to make it difficult for unauthorized persons to intrude. For example, dedicated circuits make a computer more secure from intrusion than dial-up circuits.

The third security objective is to detect any instances of intrusion as quickly as possible. If we can determine quickly that intrusion has occurred, we can take defensive measures before the intruder damages valuable files, or we can trace the intrusion to its source.

The final objective of security systems is to prevent an intruder from altering information. Far less damage is done if an intruder merely browses through files than if files are altered. A security system should make it as difficult as possible for any but authorized persons to write to a vital file.

SOURCES OF VULNERABILITY

The telecommunications operational review, which is discussed in Chapter 21, is an excellent method for determining where your telecommunications systems are vulnerable to intrusion. You should periodically look at your network through the eyes of an outsider to see where leakage might occur. It may even be worthwhile to hire a specialist to review the precautions you have taken to see how easily a determined person might breach security. Questions such as these should be asked:

- Are equipment rooms and wiring closets kept locked?
- Are cable routes and manholes protected from unauthorized access?
- Is access to records limited to persons with a genuine need to know?
- Are sensitive communications carried over radio?
- Are passwords guarded and changed frequently?

A more complete list of questions is included in the sample operational review questionnaire in Appendix L. The more sensitive the information traveling on your network, the more imperative it is that you undertake such a review periodically.

Records Security

Telecommunications systems gain some measure of security simply because sensitive information is lost in a jumble of meaningless information. For example, it might be fairly easy for an intruder to tap a single pair in the middle of a 1200 pair cable, except that hours of listening would be required to know which pair carried sensitive information. An intruder's job would obviously be simplified by access to assignment records, so such records must be closely guarded. If kept on paper, they should be kept in a locked cabinet; if kept in a data base, it should be protected by password access.

Equipment Rooms and Wiring Closets

As concentration points for large numbers of circuits, equipment rooms and wiring closets are particularly vulnerable. They should be kept locked, and the keys given only to authorized personnel. Be careful about leaving master records in such rooms. As cabling comes closer to the terminal, it becomes increasingly easy to determine which one to tap, so sensitive lines should not be tagged in such locations. Most organizations are also vulnerable in these locations to malicious damage, which makes it imperative that access be limited to authorized persons.

Password Security

Most data communications systems restrict access by assigning user numbers and passwords. If passwords are to be effective they must be guarded closely and changed at regular and unannounced intervals. People have a tendency to choose passwords that are easy to remember. If an intruder knows enough about an individual, the probability of guessing the password is greatly increased.

Many telecommunications programs are designed for automatic log-on, which means that the password is stored on a disk. Anyone who is familiar with the program would find it easy to read the password directly from the disk. Even if access to the sensitive portion of the disk is password-protected, it still may be possible to use a utility program to read the information on the disk. Automatic log-on programs should be avoided except where they are used to access information that is not sensitive.

Dial-Up Circuits

Apparatus connected to the public switched telephone network is vulnerable to unauthorized access. It is difficult to keep telephone numbers from becoming known. For example, if an intruder wants a list of telephone numbers of dial-up terminals, all that is needed is a copy of the long distance bill, and repetitive calls to the same number will immediately be apparent.

Some measure of security can be gained by using dial-back modems. A caller dials into a host, sends the proper identification, and the host dials the caller back over a separate circuit. This method is relatively secure, but it suffers from the disadvantage of extra time, incurring the cost of two telephone calls for each transaction, and possibly being limited in the number of stations that can be called from the host.

Dial-up service has the advantage of being difficult to intercept over the public network. An intruder can easily eavesdrop on a microwave channel or a satellite down link, but the route taken by calls over the public network is so diverse that it is virtually impossible to detect calls in the center of the network. Private line services, however, which use the same channel consistently, can be eavesdropped on easily after the sensitive channel is located.

Direct Inward System Access

Direct Inward System Access (DISA) is a feature used in many PBXs to enable callers outside the PBX to dial in and use system features such as WATS lines. DISA is a convenient feature, but it is vulnerable to intrusion. Although intruders are not apt to do any damage, they may incur unauthorized expenditures.

In many PBXs, a separate trunk group is dedicated to DISA, and the privacy of the trunk number is the only guard against unauthorized access. Anyone dialing the number is admitted to the PBX and can dial its features. In some systems this is used as a "poor man's DID," and if that is the purpose, the PBX should be programmed to prevent DISA callers from using long distance services. If the purpose of the feature is to give company employees access to long distance services, it is important to have a system that requires password access, and to monitor usage of the feature through the call accounting system. Most call accounting systems can be programmed to show ineffective incoming calls to a particular trunk group. If DISA is assigned its own group, you should monitor ineffective calls every day to determine if someone is trying to break the system password. Also, DISA passwords should be changed at regular intervals or when a major personnel change occurs.

Data Base Security Levels

Security levels in data bases that are accessible over a telecommunications system should be established for different classes of users. Typical classifications are:

- Personal files that are restricted from access by all but the owner of the file
- Private files that are restricted to a list of authorized passwords, who are permitted to read or write
- Shared files that can be read by all, but written only by a limited group
- Public files that can be read or written by all who have access to the data base

Access to files can be restricted to certain terminals; some manufacturers electronically embed terminal identification numbers at the time of manufacture. The data base can be programmed to accept file access only from authorized terminals. It is difficult, however, to keep unauthorized persons from using terminals, so password protection is needed in addition to terminal identification codes.

Shared Transmission Facilities

Local area networks and multidrop networks are designed to enable stations to share a common transmission medium. Devices on the network are capable of copying all messages, and are programmed to discard messages addressed to other devices. The address exclusion method offers some degree of protection, but a knowledgeable user can copy all messages without a great deal of difficulty. Protocol analyzers (see Chapter 23) can monitor all traffic on the network, and can easily display the contents of messages.

Since LANs and multidrop networks operate over dedicated facilities, the facilities must be tapped in order for anyone to gain access to the information. Tapping may be difficult for outsiders, but it is easily accomplished by employees. One of the dangers is tapping by employees who are authorized access to the network, but are not permitted unrestricted access to the information it conveys. As discussed later, encryption is the most effective method of preventing unauthorized access to the information.

Radio Transmission

It is impossible to keep unauthorized persons from receiving radio transmissions. This applies both to terrestrial radio links and satellite down links. As explained earlier, it is almost impossible for an intruder to receive sensitive information from the middle of a common carrier switched network because it is interspersed with so many other messages. The closer to the originating or terminating points the transmission is received, however, the greater the likelihood that sensitive information can be intercepted. Private microwave links are vulnerable, and cellular radio is so public you should assume that all transmissions are being monitored. Although there are federal laws against revealing radio messages intended for another, enforcement is almost non-existent, and scanners are readily available to enable eavesdroppers to monitor conversations.

The only way to guard sensitive information on radio is to encrypt data or scramble voice. Either method assumes that both parties share a common key, which means that these methods are effective only between parties who have identical equipment and who regularly communicate encryption or scrambling keys. This leaves a large amount of information vulnerable to interception because communications with the public at large cannot be effectively encrypted or scrambled.

The most effective way of guarding against interception is to use a medium that is less vulnerable to eavesdropping. Fiber optics is presently the most secure communication medium, but it obviously is not feasible for every application. Where radio must be used, assume the information is being intercepted or take steps to scramble or encrypt it.

SECURITY METHODS

In developing a telecommunications security plan, it is helpful to subdivide security into seven different classifications:

- Operational safeguards
- Encryption
- Key management
- Authentication
- Message sequence numbering
- Log-on procedures and passwords
- Transaction audit trails

An effective telecommunications security plan will make use of elements of most of these strategies.

Operational Safeguards

Operational safeguards are the methods of providing physical security for the telecommunications system. We have already discussed the need to keep equipment rooms and wiring closets locked and records secure from unauthorized access. Other physical security methods include:

- Keeping terminals locked when they are not in use
- Using personal identifiers such as magnetic cards for operator authentication
- Separating work functions so the independent work of two persons is necessary to complete a transaction
- Reconciling data so that the sum of individual transactions is verified against an independently developed and transmitted batch total
- Limiting the value of information that can be obtained over the network; for example, automatic teller machines limit the dollar volume of transactions permitted in a 24-hour period.

Physical security of telecommunications transactions is analogous to documenting security methods; sensitive documents are kept in locked enclosures and only authorized persons have the key.

Encryption

Not even the best physical security methods prevent information from falling into the hands of unauthorized people. As we discussed earlier, some transmission media by their very nature are easily intercepted. Also, some data networks broadcast information to all terminals, and the primary problem is keeping information from those who are authorized to access the network but not to read all the information it conveys. For these applications, encryption is the most effective way to secure sensitive information.

Most encryption systems work with a simple principle; the sender and receiver privately agree on a key, which is a combination of characters used to scramble the transmission. The encoder and decoder are embedded in software or a "black box" at both ends of the circuit. As long as both are equipped with the same key, the transmission is perfectly intelligible at the input and output, but indecipherable in between.

People have been developing codes for centuries. One of the most effective is the "book code," in which the encoder assigns a numerical value to each word. The value is the page, line, and word number of the word's appearance in a particular book. The decoder, using a copy of the same volume, converts a string of numbers into the original text. Such codes are not foolproof; experts can make inferences based on their knowledge of word frequency in the language.

Care must be taken in encryption that the characters that are created do not interfere with the network itself. For example, special characters mark the ends of frames in synchronous protocols, and packet switching networks may be incapable of handling non-ASCII characters.

A National Bureau of Standards encryption method, the Data Encryption Standard (DES) is becoming widely accepted as an encryption medium. To facilitate exchange of information, it is important that users adopt compatible methods. Just as cyclical redundancy checking is almost universally used as an error checking and correction method, DES is designed into devices so users can communicate by exchanging the key. For example, the American Bankers Association for Electronic Funds Transfer has adopted the DES standard. The DES algorithm can be designed into a co-processor chip that is plugged into devices for a minimal investment. The DES standard is almost unbreakable. It offers 72×10^{12} code combinations, which would be next to impossible to break empirically.

The voice equivalent of an encryption message is the voice scrambler. Both parties must have identical scrambling apparatus, which encodes and decodes speech, turning it into meaningless gibberish beyond the terminals. Scrambling is generally feasible only among parties who communicate regularly because scrambling equipment tends to be sold and administered in pairs.

Key Management

If encryption is adopted as a way of life, it is obviously essential that the encryption key be restricted to authorized users. Key management involves the creation, assignment, distribution, cancellation, and change of the encryption keys. It also involves controlling the key so only authorized persons have access to it. The most effective encryption system provides for changing keys frequently. If an unauthorized person should get the key, the amount of damage that can be done is limited if the key is changed the next day.

The key obviously cannot be transmitted over the same circuit it is designed to protect. Some companies transmit the key by telephone, but of course wire tapping can compromise security. In some applications the key is delivered daily by bonded courier. One of the most effective methods is for each party to an encrypted session to independently generate a key and send it in the clear to an encryptor so the total key is stored only in the encryption device. Neither party to the communication has the entire key, and the key cannot be independently read. Whatever method is used, the communication is only as secure as the method of managing the key.

Authentication

Authentication messages are similar to the mysterious exchanges of meaningless phrases that are the stuff of spy stories, except that they are exchanged by machines instead of people. If a host receives a message from a terminal on a dial- up circuit, it may be important to know that the terminal is really entitled to conduct a transaction. The host sends a test word or message to the remote, and on the basis of that message and the previously agreed upon key, the terminal generates an authentication message. The host likewise generates an authentication message and compares the two. If the two messages are identical, the terminal is deemed to be properly authenticated.

Message Sequence Numbering

In transaction-oriented systems it is important to know that transactions have not been lost or that unauthorized transactions have not been inserted. For example, in a fund transfer system it would be costly to lose a message charging a user's bank account after the funds had been paid. It would likewise be costly to post a deposit that an unauthorized person had independently generated. One method of preventing these unauthorized transactions is to sequentially number all messages. Terminals can acknowledge messages by number so the host can check for missing numbers in a sequence.

Log-On Procedures and Passwords

In the course of logging onto a network, a user encounters a primitive level of network security. The primary purpose of the log-on message is to identify the user at the first level, to determine which of several services on the network are authorized to be accessed, and to establish charging and billing accounts. The log-on is accepted only in conjunction with a valid password.

Passwords must not appear on the screen when they are typed. When people change jobs the password should always be changed and never left with the new user. Since the password does not appear on the screen in a properly designed system, the operator may easily mistype the password and be rejected by the system. To make it easy for operators to use, the system should accept two or three attempts to enter the password, but because hackers usually operate by repeatedly trying passwords until they are finally successful, the system should limit the number of ineffective attempts before it drops the connection. The system should also record ineffective attempts so you can tell when someone is trying to intrude. The most effective way of deterring a persistent hacker may be to allow him or her to log on, accept the password, and hold the connection long enough to trace it to the source through the telephone company.

The log-on can be authenticated by routines built into hardware or firmware in the terminal that can be read by the host. The host can also send a random message to the terminal, the terminal can encrypt it, and return it to the host. When the host decrypts the message it must be identical to the message it originally sent.

Security is enhanced by a few simple precautions such as changing passwords frequently, restricting knowledge of passwords to authorized persons, guarding automatic log-on software vigorously, and ensuring that passwords are not chosen merely because they are easy to remember.

Transaction Audit Trails

Security of the telecommunications system can be improved from the audit trails that are built into many effective data processing applications. Messages are time- and date-stamped with the identification of both the terminal and the user attached to each message. Back-up tapes are kept so transactions can be regenerated if the network fails. Audit checks are performed on fields to ensure not only that the transaction is entered properly but also to help authenticate the originator.

Records should be kept of all network maintenance activities that could affect message integrity. Failure and restoral times, patches to back up circuits and equipment, and other such activities, should be logged for several purposes, one of which is to aid in maintaining an audit trail.

SUMMARY

Security should never be an afterthought in designing a telecommunications system; it should be part of the integral network design. It is a rare organization that has no secrets to protect. The threats of security breach come not only from outsiders, but from employees and contractors who may have some motivation for acquiring confidential information, or who may have grievances to vent. Security problems are inherent to both voice and data communications, but except for physical safeguards, the methods of ensuring security are much different for voice and data.

Humans have entirely different methods of authentication than machines, and have enough versatility and intelligence to recognize attempts to break security. Machines, on the other hand, are programmed to be absolutely predictable. When another machine passes the test, the communication is allowed. Therefore, methods must be inherent in the design to make machines predictably unpredictable. That is, they must authenticate one another, and when that is not enough they must encrypt to ensure that messages do not fall into unauthorized hands. In deference to their human operators, who lack the infinite patience of machine, the security systems must not get in the way of operations. If your security system costs people productivity, they will take shortcuts that compromise the ultimate objective.

CHAPTER 23

NETWORK MAINTENANCE AND TESTING METHODS

In developing a network operations plan you must select a strategy for how much trouble testing and repair you will do before referring the problem to a vendor. Most companies should be prepared to do a certain minimum level of testing, if for no other reason than to expedite trouble clearance by referring it to the proper vendor. Basic tests can be very simple. For example, you might plug a working telephone into a jack at the demarcation point with the telephone company to find out whether dial tone is present, or you can check for loose power and RS-232-C cords in an inoperative terminal.

At the other extreme is a company that maintains equipment itself. Even though it may have a highly trained staff and be well equipped with test equipment, some types of repair will always be referred to a vendor, simply because it is too expensive to staff and train for troubles that may occur only once in a decade.

Your operations plan should include a testing plan. Every business that uses telecommunications extensively should evaluate testing strategies, develop a plan, and either write the contracts or acquire the staff and equipment necessary to execute the plan. This chapter discusses testing strategies and examines some of the implications from choosing different maintenance alternatives. In this chapter we are primarily concerned with manual testing methods and equipment. Chapter 24 discusses a closely related subject, network management systems, and the various methods of gaining access to circuits to test them.

MAINTENANCE STRATEGIES

A company can adopt one or a combination of strategies for maintaining its telecommunications equipment and services. The strategy does not have to be uniform across all classes of equipment. It might, for example, to be feasible to perform most data communications maintenance with your own staff, but to contract PBX maintenance to a vendor. Strategies most frequently employed are:

- No self-maintenance; all trouble is referred to a vendor
- Limited self-maintenance; obvious troubles are cleared in- house with just enough testing to be certain the right vendor is called
- Self-maintenance for obvious troubles, but call a vendor when your own efforts fail
- Full self-maintenance on one class of equipment, but limited or no maintenance on other classes
- Self-maintenance on all but the most obscure troubles

TROUBLESHOOTING TECHNIQUES

If you have ever watched an expert troubleshooter in operation, you were aware that he or she has a highly developed technique. On closer analysis, you will find that nearly all troubleshooters use the same basic approach, although they are probably not even aware that they have an established process. Troubleshooters almost always use inductive reasoning techniques; that is, reasoning from effect to cause. This is the Sherlock Holmes technique; asking "what kind of system could have created the kind of effect that I am observing?" A typical process for detecting, isolating, and correcting troubles is shown in Figure 23.1.

Define the Problem

Even though the old saying, "a problem correctly defined is half solved," is a cliche, it is nonetheless true. Ineffective troubleshooters tend to use the "fly at a screen door" technique; that is, they repeatedly bang against the screen attempting to find the hole that will let finally them through. In the urgency of a failure condition, the impulse to do something is almost irresistible. The most professional troubleshooters discipline themselves

FIGURE 23.1
A Typical Process for Detecting, Isolating, and Correcting Troubles

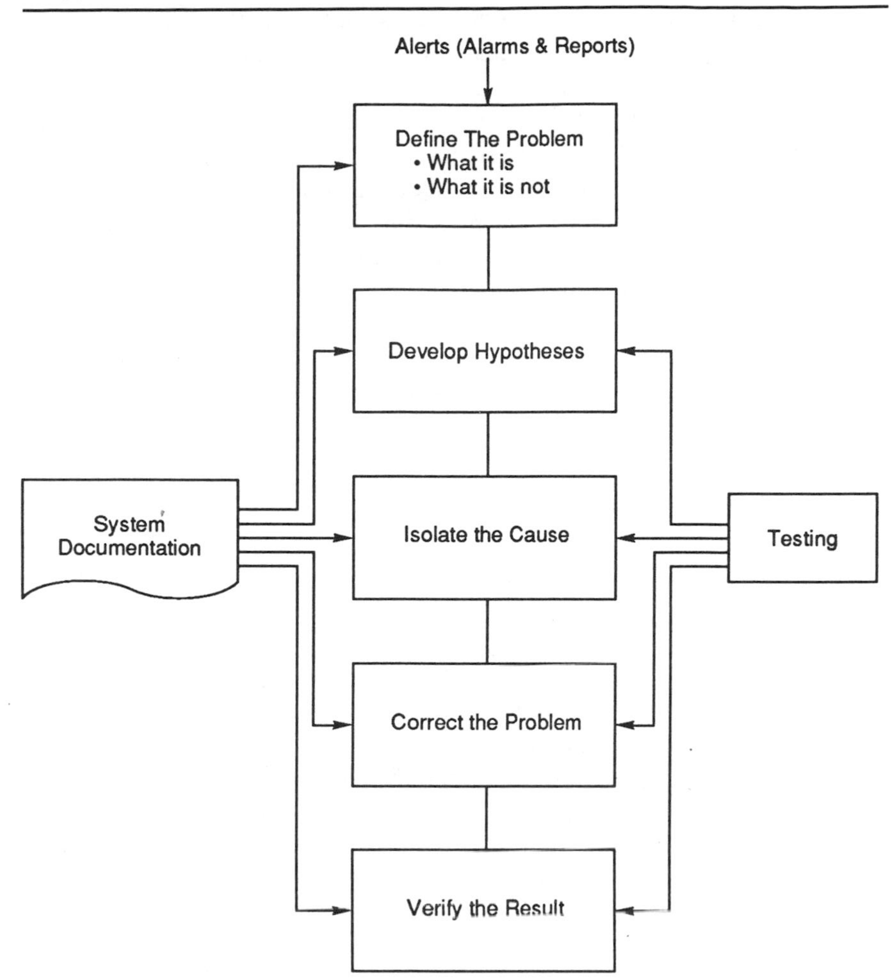

No troubleshooting system is complete without a testing system to determine the extent and location of troubles, and a documentation system to assist in troubleshooting and to show how the system should appear when it is operating properly.

to spend the time to collect the information they need to understand what the problem really is and set about to clear it systematically.

Often, the moment the problem is defined, it is immediately apparent what is causing it. Almost invariably, however, the problem is expressed differently by people who see it from different perspectives. For example, the user may say the problem is "slow response time," the customer service manager may perceive the problem as "irate customers caused by slow processing time," the technician may see the problem as an unreasonable customer services manager, and higher management may see the problem as excessive cost. One thing is certain; if you clear the trouble without understanding the problem, it isn't troubleshooting, it's just good luck.

A doctor always diagnoses a patient's problems by attempting to discover the symptoms. So it is with a network technician. Network equipment and facilities have several ways of reporting symptoms to the outside world:

- Alarms
- Status reports
- User reports
- On-line status monitoring results

One of the axioms of troubleshooting is to look for what has changed. This means you should document how things look when all is operating properly. For example, what lights on modems are illuminated, what color are certain status lamps when they are operating normally, and what is the normal condition of switches and patches? The status of lights on multiplexers, PBX line and trunk cards, and other devices that display lead conditions with lamp signals should be recorded. The operators of protocol analyzers need to know how the circuit looks when it operates properly. Unless the meaning of status lamps is intuitively obvious, you should keep a record of how they look when all is operating properly. It is also essential to keep a record of changes and rearrangements because frequently a problem is related to activity by a technician who just finished working on something else.

Finally, it is just as important to know what a problem is not, as to know what it is. For example, if the reported condition is slow response time, there is obviously no point in looking for an open circuit condition, so you're saved from the effort of making continuity measurements or looking for failed modems.

To do an effective troubleshooting job, you must also know exactly how the network is configured. System configurations range from a PBX or key system with central office station lines terminated on it to a complex combination of switches, multiplexers, circuits, modems, front-end processors, and wiring. Frequently, the wiring diagram exists only in someone's head, and when that person is unavailable, someone else must trace connections to find out how the network is configured. This is not only costly, but it also delays restoration. A network must, therefore, be diagramed and all equipment labeled clearly.

Develop Hypotheses

With an understanding of what the problem is and is not, the trouble-shooter is in a good position to develop theories on what is causing the problem. Like a doctor, a network troubleshooter knows that more than one thing can cause the same problem. For example, in the case of slow response time, the problem could be an overloaded circuit, excessive central processing unit processing time, incorrect block length, a high circuit bit error rate, or operator error. Obviously, it requires a certain level of understanding on the part of the technician to develop these hypotheses or else some kind of troubleshooting aid must be provided.

Isolate the Cause

By using inductive reasoning, the troubleshooter has developed a hypothesis consisting of possible causes that could produce the observed effect. Now it is necessary to eliminate all the incorrect causes, and the one that remains will be the one that is causing the problem. Trouble-shooters have numerous techniques for isolating the cause. Among these are:

Substitution. By substituting a known good device for one that is suspect, the cause is either confirmed or disproved. Substitution is one of the most effective ways of clearing trouble. Although the ultimate cause may not be not discovered in the substitution process, the service is restored, and the trouble can be cleared without the pressure of operating with degraded service. For example, in the case of a low response rate problem, a technician could use a protocol analyzer at different points in the circuit to emulate the central processing unit, the circuit, the load sensitive devices such as multiplexers, and other such circuit elements to determine which one is at fault. Substitution is not effective for some types of troubles. For example, with the possible exception of clearing a high bit-error condition, the substitution method is a poor technique for solving a response-time problem. It is very effective, however, when a service has completely failed.

Testing. If you know how conditions should be on a properly operating service, you can use test equipment to find the deviation. Just as a doctor looks for temperature, pulse rate, dilated pupils, and other such symptoms to isolate the cause of a disease, a technician tests such factors as level, noise, continuity, bit-error rate, and other such variables to isolate the cause of the problem. A technician may use test equipment to detect irregularities such as a high bit-error rate, excessive envelope delay, impulse noise, or other deviations that impair data circuit performance.

Sectionalizing. In sectionalizing trouble the technician narrows the problem down to progressively smaller sections until finally the section remaining is the one with the trouble. A modem loop around test, as shown in Figure 23.2, is a good example of sectionalizing trouble. By looping the circuit back on itself at different points, as shown in the figure, it is possible to determine which portion of the circuit is at fault.

A no-dial-tone report in a PBX trunk can also be cleared by using sectionalization. A technician first goes to the demarcation strip and tests

FIGURE 23.2
Modem Loopback Testing To Determine Defective Circuit Element

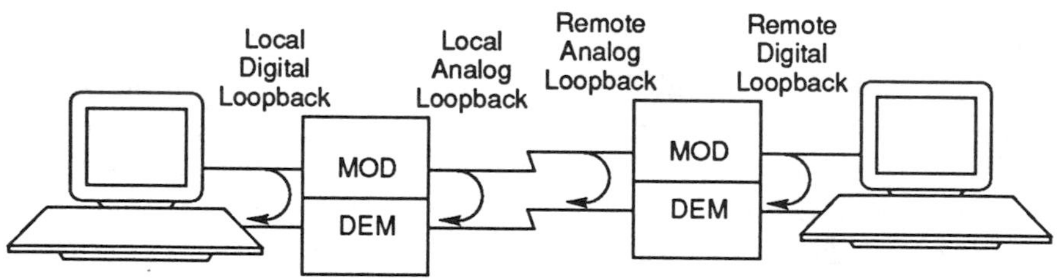

By progressively looping a circuit, it is possible to detect defective modems at near and far ends of the circuit, defective circuits, and defective terminal or modem-to-terminal connection.

to determine whether dial tone can be drawn from there. If dial tone cannot be detected, the problem is referred to the telephone company (which also goes through a process of sectionalization to determine whether the problem is in cable plant, distributing frame, or central office). If dial tone is obtained at the demarcation point, the technician checks to see whether it is arriving at the PBX trunk port as a means of isolating wiring trouble. Using a combination of substitution and sectionalization, the technician narrows the problem to defective wiring, a bad PBX trunk port, or software problems.

Pure Reasoning. It is sometimes possible to clear trouble by pure reasoning. By following a circuit diagram and observing lights and indicators, a technician may be able to isolate the cause without testing. Some equipment has internal diagnostic capability that can direct a technician in a high percentage of cases to the precise equipment (usually a plug-in card) that is causing the trouble. Troubleshooting by reasoning requires intimate knowledge of how the circuit functions, and excellent documentation.

Take Corrective Action

After the cause of the trouble has been isolated, the next step is to correct it. Corrective action may be any one or a combination of the following:

- Refer it to another responsible group (for example the telephone company)
- Change the defective equipment or card
- Correct wiring defects
- Substitute known good equipment
- Patch to a known good circuit
- Redesign to remove the cause of overload

The bulk of troubleshooting time is spent on the first three steps of defining the problem, developing hypotheses, and isolating the cause. After the cause is known, corrective action and restoral usually take only a short time. There are several aids you should have at your disposal to assist in corrective action. The first of these is a list of all suppliers, carriers, and contracting organizations to whom you can refer trouble. Don't wait until the first case of trouble occurs before finding out who you can call to correct it.

Next, you should retain a log on the trouble ticket or an external document of exactly who the trouble was reported to and when. If trouble is not corrected in a reasonable period of time, escalate. Every network control center should have a formal escalation procedure indicating who is to be notified and after what length of time on each type of outage. Besides escalating, in every organization certain people should be informed of progress. These include both higher management and the users. If both groups are kept informed of progress, it will save the time that you otherwise spend answering questions.

Verify That The Trouble is Cleared

Before the services are restored to the users, you should make tests to verify that the corrective action actually cleared the trouble. The type of tests to be made depend on the nature of the trouble, but will usually be a repetition of the tests made to isolate the cause.

Restore the Service to the Users

A final step in the trouble handling process is to restore the service to the users. As discussed in the next section, this normally should be done by the help desk that received the report initially. In every case, the users should be consulted to ensure that the problems that they reported have been cleared.

MATCHING TEST EQUIPMENT TO THE STRATEGY

Test equipment on the market today is highly sophisticated, and may easily outstrip the ability of its users to apply it. There is little reason to invest in complex equipment if your staff is incapable of using all its functions. Unfortunately, a considerable amount of equipment is left idle because of the problem of test equipment overkill. Either someone purchased equipment that wasn't appropriate to the situation (such as a protocol analyzer in an all voice network), which no one knew how to operate, or which could not be properly applied because of the circuit configuration. The guiding principle should be to spend only as much as you can afford on test equipment you truly need.

Test Equipment Standards

Many (if not most) networks have not emerged from a unified and deliberate design—they have evolved over a long period of time from circuits and devices of uncertain parentage. The networks themselves lack standards, and not surprisingly the testing systems suffer from the same deficiencies. Before you invest in a control center, consider some of the following:

Fixed or portable. Whether test equipment is rack mounted in a centralized facility or is portable equipment that can be carried out into the field depends on your objectives and your network design philosophy.

Programmability. Data test equipment is often programmable to allow the application of certain standard tests. Programmability is needed, not only to repeat tests, but to build in traps for error conditions, storage of test results, and conformance to changes in the protocol.

Software or Protocol Interface. Data test equipment is usually compatible only with particular protocols and must be chosen to support the ones your network uses.

Functionality. An inevitable question is whether you should buy several single-purpose test sets or a few multi-purpose sets. The answer lies in the skill of your work force, and your own testing strategy and requirements. Multi-purpose sets are usually less expensive per function, but may offer functions you have no need for. Single purpose sets may result in your missing an occasionally-needed function, and furthermore, you will be faced with a conglomeration of patch cords if you have to connect several single purpose sets to handle a function that a multi-purpose set supports with ease. For example, the transmission impairment measuring set shown in Figure 23.3 replaces as many as six single-purpose test sets.

Ease of Use. One feature everyone wants is ease of use, but that is a subjective characteristic that depends on the skill of the operator and how often the function is performed. Test equipment should be thoroughly evaluated by the people who will use it before it is purchased.

FIGURE 23.3
The Hewlett Packard HP 4948A in service Transmission Impairment Measuring Set

Photograph courtesy of Hewlett Packard Company

Speed. Transmission speed is important in data communications test sets. Many sets have an upper limit of 19.2 kb/s, which makes them unacceptable for high speed digital circuits. Higher speeds, however, usually mean higher cost, so there is little reason to spend extra money for a speed that is not used.

Vendor Independence. One of the major hazards in constructing a telecommunications testing system today is becoming excessively dependent on a single vendor. There are several vendors with products sophisticated enough to justify your relying on them, but there is no assurance that any single vendor will offer the best solution when it comes time to make a substantial change. Moreover, no single vendor consistently offers the best cost performance, the best delivery dates, the best documentation and training, or other such features that are so vital to success in operating the network. There are no solutions to the problem of excessive vendor dependence. There are only precautions to be observed.

Host-Based Testing

Data communications networks are tested either by distributing testing functions at each individual circuit termination point, or by a host-based system such as IBM's Netview, which is discussed in Chapter 24. Both methods have their advantages. Host-based systems are often highly effective if the network equipment is all supplied by the same vendor, usually the manufacturer of the host computer. Very few networks, however, are composed of equipment from only one vendor, which usually means that alternative testing strategies are required.

The major drawback to host-based systems is the fact that they usually do not work with other vendors' equipment. A second drawback is the lack of unification in a system as diverse as IBM's Systems Network Architecture. Third party vendors also are preparing products to improve the user interface of host-based testing systems. Several products currently on the market receive reports from the testing system and display them in graphic form. The network map, in such systems, is drawn in color on a computer screen. Several levels of display may be provided. An overall map shows how the network is configured. If all is well, a node is shown in green. If an alarm condition exists, the operator can zoom in on the node for greater detail. This detail may show all the major components and how they are interconnected. The component in trouble is displayed in red. By zooming in on that component, the operator receives details about the alarm. Third party systems such as this can potentially receive reports from many vendors' equipment and assist in overcoming the problems of interpreting the meaning of alarm and status signals.

Outboard Testing Systems

The alternative to a host-based system is using either stand-alone equipment on single trouble-at-a-time tests or using an alternative network testing system. Some non-host-based systems are provided by data communications equipment vendors. As discussed in Chapter 24, these systems use some kind of reporting mechanism to relay information about network health back to the control center where network maps, status monitors, and support systems aid in testing and diagnosing trouble.

Another testing alternative is the use of a completely outboard testing system. In systems such as these, test access and switching points are connected to switching equipment that brings them out to spare equipment, test equipment, and responders. External circuits, often dial-up, are used to connect to the circuit under test.

Selecting the Network Testing Strategy

With some of the preceding alternatives and pitfalls in mind, you should approach the network control center design project as a task at least as complex as the design of the overall network, and preferably, if it is not too late, as an element of the total network design.

TEST EQUIPMENT

The dividing line between test equipment and the network management equipment discussed in the next chapter is not distinct. As test equipment becomes progressively more complex, it becomes capable of performing analyses that may substitute for human training and experience. Your organization must carefully consider its maintenance strategy before selecting such equipment, but if the capability of the work force and the demands of the service support it, sophisticated test equipment can be an excellent investment.

Data Analyzers

Data or protocol analyzers are a family of equipment that can look at the contents of a data signal and allow an experienced operator to diagnose problems. Low end analyzers are constructed on a board that plugs into a personal computer and displays the data signal on a personal computer screen. At the other end of the scale are expensive devices that can be used to certify compatibility of a device with a public data network. Figure 23.4 shows a specialized protocol analyzer. The following is a brief explanation of the major functions that data analyzers typically perform.

Device emulation. The analyzer can be substituted for any device on the network. With proper programming it can behave like a host, a controller, or a terminal.

Trapping. The analyzer can be programmed to look for a particular data pattern and trap one or more blocks preceding and following. For example, the analyzer could look for cyclical relundancy check

FIGURE 23.4
The Hewlett Packard HP 4951C Protocol Analyzer

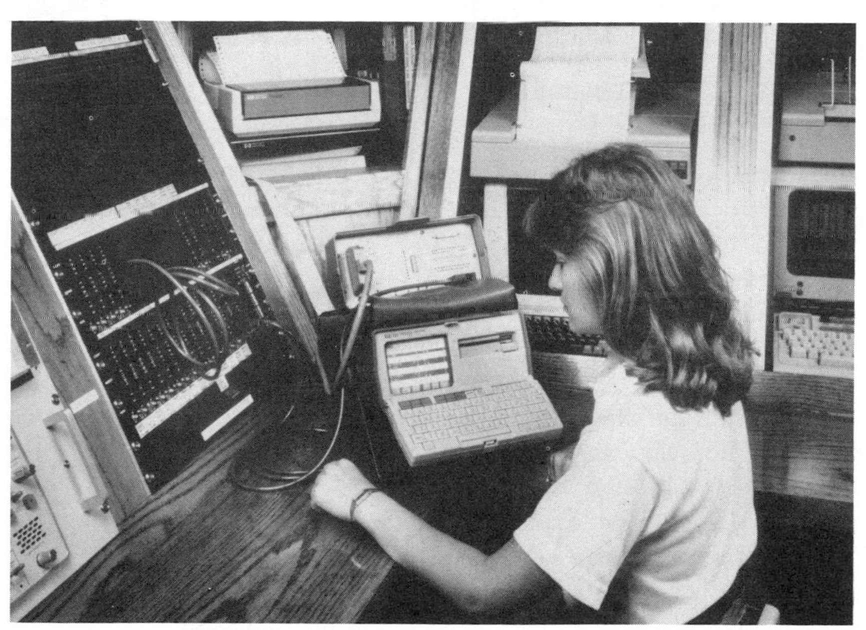

Photograph courtesy of Hewlett Packard Company

This portable unit is used for installation, maintenance and design of data communications networks up to 19.2 kb/s.

errors, poll and response to a specific station, or other such bit patterns on the line.

Performance Measurements. The analyzer can be programmed to perform selected performance checks such as measuring response time, block error rate, or line utilization.

Lead Status. Analyzers can show the status of control leads on an interface, and the transition between states. They also can measure timing between events to detect timing problems and measure modem turn-around time.

Analog Circuit Tests

Analog tests are made on both voice and data circuits. Multiple function sets can measure all the variables, including some, such as envelope delay, that have no effect on voice circuits. The following is a brief description of the principal analog circuit tests.

Continuity. By sending a tone over a circuit and listening for it or detecting it with a meter at the distant end, circuit continuity can be determined.

Level. If a tone of a precise level and frequency (usually one milliwatt at 1004 Hz) is sent over a circuit, a calibrated meter at the distant end can measure the transmission loss of the circuit.

Gain-Frequency Response. A frequency response curve is made by sending tones at 100 Hz intervals over the pass band (nominally 300-3300 Hz) of the circuit.

Steady State Noise. A noise measuring set inserted into a test point measures the steady state noise of the circuit.

Impulse Noise. An impulse noise set measures noise bursts of short duration that may cause data errors. Noise bursts are measured against a particular level threshold to determine what portion exceed the threshold objective.

Peak to Average Ratio (P/AR). A P/Ar measurement is a qualitative test of circuit condition. A complex analog signal is sent down a channel and detected with a special measuring set at the far end. The P/AR measurement is a qualitative indication of the capability of the circuit to handle data.

Harmonic Distortion. Harmonic distortion is a measure of the ability of the circuit to handle high speed data signals. It is measured with a specialized test set that detects the amount of second and third harmonic distortion in a circuit.

Envelope Delay. Envelope delay is a test that detects the relative delay of different frequencies propagated over an analog circuit. Data modulation methods that employ phase shifts to encode signals are sensitive to variations in the envelope delay. The tests are made with a pair of specialized sets that measure relative delay over the pass band of the circuit.

Phase Jitter. Phase jitter is any variation in the phase of a signal. Phase jitter test sets inject a steady-state tone into one end of a circuit and detect the amount of phase variation arriving at the other. Jitter is important in any data circuit that uses phase modulation. It is also important in T-1 and digital data circuits.

SUMMARY

A certain minimum level of testing capability is necessary for most companies for two reasons, speed of service restoral and economy. In most companies, maintenance strategies tend to grow and evolve, and are related to the company's problem history and the training and experience of its personnel instead of resulting from a deliberate plan. Testing and maintenance procedures should, however, be part of every company's network operations plan. This chapter has outlined some of the principles of establishing a testing plan. Chapter 24 enlarges on these concepts with a discussion of network management systems.

CHAPTER 24

NETWORK MANAGEMENT SYSTEMS

Network management is one of the most ambiguous concepts in the telecommunications world. Despite the impression that some vendors convey, a network management system is not something that you buy and install like a piece of equipment. It is a set of tools, techniques, policies, and procedures that you develop for your own network. Since every network is unique, it is impractical, given the present state of technology, to develop a universal network management system. What can be done and what must be done in every company is to develop a management system that achieves the company's objectives for the network. Such a system will have most of the following elements, several of which have been discussed in earlier chapters:

- An inventory of all circuits and equipment
- A trouble report receiving process
- A trouble history file
- A trouble testing and isolation facility and procedures
- A hierarchy of trouble clearance and escalation procedures
- An activity log for retaining records of all major changes
- An alarm reporting and processing facility

Not every network management system will have all of the elements listed above, and in some networks some of the elements may be rather primitive, but most systems will contain the above functions to some degree. The more complex the network, the more likely it will be that you can justify automating the function or supporting it with some kind of mechanized system.

FUNCTIONS OF A NETWORK MANAGEMENT SYSTEM

Network management is a collection of activities that are required to plan, design, organize, control, maintain, and expand the network. A network management and control system consists of a collection of techniques, policies and procedures, and systems that are integrated to ensure that the network delivers its intended functions. At the heart of the system is a database of information, either mechanized or on paper. The data base consists of several related files that allow the network managers to have the information they need to exercise control over its functions.

A network control system has five major functions:

- Monitoring circuits and equipment on the network
- Isolating trouble when it occurs
- Restoring service to end users
- Managing network performance
- Managing network information

In addition to the above, other functions may be assigned to the control center. Functions such as designing the network, and monitoring and maintaining network security must be performed by someone. In small networks, these functions are likely to be assigned to the network control center or contracted to outsiders, but in larger networks they will be assigned to specialized groups.

Inventory

The major file in the network management data base is an inventory of the circuits and equipment that make up the network. The equipment inventory was discussed in Chapter 21. A circuit inventory, a sample of which is shown in Figure 24.1, lists the circuit identification number, end

FIGURE 24.1
Sample Circuit Inventory (Partial)

Circuit Inventory

Circuit No.	*A Terminal*	*Z Terminal*	*Type*	*Use*	*Length*	*Signaling Type*	*Test Level*	*Loss*		*Noise*	
								A-Z	*Z-A*	*A-Z*	*Z-A*
1T1	New York	Phila.	T1	Tie Trk	550	In Band	-2.3	0	0	7	8
2T1	New York	Phila.	T1	Tie Trk	550	In Band	-2.3	0	0	7	9
3T1	New York	Phila.	T1	Tie Trk	550	In Band	-2.3	0	0	5	6
4T1	New York	Phila.	T1	Tie Trk	550	In Band	-2.3	0	0	8	11
5T1	New York	Phila.	T1	Tie Trk	550	In Band	-2.3	0	0	12	6
6T1	New York	Phila.	T1	Data	550	None	-2.3	0	0	7	13
7T1	New York	Phila.	T1	Data	550	None	-2.3	0	0	5	7
8T1	New York	Phila.	T1	Data	550	None	-2.3	0	0	8	9

points, and several other items that may or may not be required, depending on the purpose of the circuit. These include:

Facility type. This record should indicate whether the circuit is analog or digital. If it rides on a company-controlled backbone facility such as T-1, it should include the identification of the underlying broadband facility. If the circuit is derived from a sub-multiplexed voice frequency facility, the identification of that facility should be included.

Circuit Use. Indicate whether the circuit is used for voice, data, facsimile, video, or other type of service.

Length. The length of the circuit in miles may be useful for calculating delay in administering data communications networks. It also serves as a reasonableness test in billing verification. Note, however, that length for billing purposes and length for delay purposes may not be the same thing. Delay is a function of how the IEC routes the circuit, but billing is a function of the number of airline miles between terminating points.

Test Levels. The measured transmission levels at each test level point should be shown. Identification of jacks, switched access connectors, or other test points that are associated with the circuit (not the equipment) should be shown. The identification of test level points is of particular importance on analog circuits. The results of transmission measurements should be recorded. Separate fields are required for loss and noise on voice circuits. On data circuits, P/AR, envelope delay, frequency response, and harmonic distortion may also be recorded. On digital circuits it is advisable to make and record bit-error rate and jitter measurements.

Signaling Method. The signaling method employed on the circuit (E & M, loop, ringdown, in-band digital, etc.) should be recorded.

Interconnecting circuits. If the circuit connects to another circuit to form a longer facility, the identification of interconnecting facilities should be included.

The objective of the circuit inventory is to enable a technician to determine how the circuit is configured and what measurements were obtained when it was initially installed. As discussed in Chapter 23, in clearing trouble it is important to know how these variables measured when the circuit was working properly.

Equipment Inventory

An inventory of every item of data or voice communication equipment should be retained in the data base. As discussed in Chapter 21, the purpose of the inventory is:

- To serve as a property record for all telecommunications equipment owned by the company
- To retain records of the equipment vintage for administering equipment changes and upgrades
- To retain a record of the location of equipment
- To assist in forecasting and administering growth requirements

If the equipment record is kept in a relational data base, it can also be used to assemble a complete configuration of any circuit from the port on a PBX or front-end processor to the far-end terminal equipment.

Cost Information

A file should be retained of billing information on each individual circuit. Like the equipment file, this file can be linked to the circuit file by the circuit identification number. If a single circuit is billed as a part of a backbone facility such as a T-1, the individual circuits should be shown as sub-sets of the primary file. Figure 24.2 shows a typical cost record. In addition to the circuit identification and backbone facility identification, the cost record should also include:

- Vendor identification including name and address
- Billing date
- Payment due date
- History of several months of past billing levels

If the bill is for an equipment lease it should be linked to the equipment file by equipment ID number. The purpose of the cost file is threefold: to track cost history on the service, to determine overall costs for a particular circuit, and to serve as a master inventory of costs for preparing budgets.

FIGURE 24.2
Sample Circuit Cost Record

Circuit No.	Vendor	Bill Date	Payment Due	January	February	March	April	May	June	Cumulative	Average
1T1	AT&T	1st	22nd	$4,650.37	$4,650.37	$4,650.37	$4,650.37	$4,650.37	$4,650.37	$27,902.22	$4,650.37
3KP122	Telco	10th	31st	$127.55	$127.55	$127.55	$171.55	$127.55	$127.55	$809.30	$134.88
4KD453	Telco	10th	31st	$543.43	$543.43	$543.43	$543.43	$597.77	$657.55	$3,429.04	$571.51
103234	MCI	1st	5th	$689.54	$689.54	$689.54	675.75	$662.23	$648.99	$4,055.59	$675.93
.	.	.	.	.	.	.	.	.	.	.	.
.	.	.	.	.	.	.	.	.	.	.	.
			Total	$6,010.89	$6,010.89	$6,010.89	$6,041.10	$6,037.93	$6,084.46	$36,196.16	$6,032.69

Trouble History File

As described in Chapter 21, a complete trouble history record is required on each circuit and item of equipment. Separate trouble history records should be kept on circuits and on equipment items to aid in determining trends and clearing chronic troubles. Refer to Chapter 21 and to Figure 21.9 for typical trouble tickets, and for more details in administering a trouble report system. This file may also be linked to the circuit and equipment files for the following purposes:

- Isolating trouble patterns to a particular circuit, backbone facility, or family of equipment
- Detecting chronic trouble conditions
- Determining past causes as an aid to current troubleshooting

NETWORK CONTROL

Another function of a network management system is the network control center that enables you to perform diagnostics, monitor the network, isolate trouble, and restore service from a centralized location. The center is also responsible for verifying circuit performance after installation but before it is turned over to end users. The control center is at the focal point of periodic performance measurements, and contains the people, the access points, and the test equipment to verify circuit operation.

Purposes of a Network Control Center

A network control center has several purposes, but in the end they can be summarized as this: the network was developed and installed to aid the effectiveness of the organization either in controlling costs or improving the productivity of its people. Like other organizational resources, the network must be managed, and the control center is the focal point for this management. Increasingly, as discussed in Chapter 2, organizations are recognizing the strategic importance of the network in meeting business objectives. If the network has strategic importance, it goes without saying that the organization's relationships with its employees, suppliers, distributors, or customers are adversely affected whenever any part of the network is down.

Even though the control center is not in the direct link between the organization and its users and clients, it is the facility that orchestrates the operation of the network. The control center is also important because of one very practical consideration; it is difficult to obtain skilled people to manage the network. A control center and its related equipment brings together the aids and information that enable you to manage the network with lower skill levels than would otherwise be required.

AT&T and the former Bell Operating Companies used a similar strategy in administering their networks. The nationwide Bell System network was by far the most complex network in the world. It consisted of sophisticated equipment using numerous signaling methods and complex protocols. Yet, it was, and in segments still is, managed by technicians who were skilled in the workings of parts of the network, but not the whole. The successful operation of the network was possible because of these factors:

- Extensive documentation of all aspects of network operation
- Numerous data bases to store information about the network
- A complex set of mechanized support systems to assist in alarming, diagnosis, and control
- Robustness enhanced by alternate routes, spare equipment, and testing and patching facilities

Network Control Strategies

The primary issue in developing a network control strategy is whether control is centralized or distributed. Each method has its advantages in certain situations. You must analyze your own company's culture, its objectives, the degree to which it is geographically dispersed, its budget, the skill level and location of the people available, and other such factors in deciding whether to centralize or distribute network control. The three primary strategies that you can adopt are centralized management and control, centralized management and distributed control, and distributed management and control.

Centralized Management and Control

A centralized management and control facility places all of the equipment, data bases, and personnel resources in one location and gives them control over all aspects of the network. If the network is not widely dispersed geographically, this is obviously the best solution because it makes the most efficient use of limited personnel resources. The centralized management and control system avoids the duplication that inevitably occurs when test equipment and personnel are based at other locations. In a centralized system, everyone operates from a common data base. Records are centrally stored and therefore easier to keep updated. With the proper equipment and training, a small group of people can become skilled in operation of the entire network. Large networks supporting companies operating twenty-four hours a day can be supervised by a continually-manned control center, where staffing distributed centers might be prohibitively expensive.

Although the centralized network control center is often desirable, it has drawbacks. Probably the most important drawback is the need for expensive terminal equipment to test and restore the network. Responders, access switches, matrix switches, and other such devices are needed to deploy a fully centralized system. Furthermore, the network control center people are more removed from the users, which may result in their being less sensitive to user problems, or at least displaying the impression of remoteness and lack of sensitivity. A remote control center is heavily dependent on records. Unless records are flawless, restoration will be impeded.

Decentralized Control and Management

In a completely decentralized system, each center has its own management, testing equipment, and data bases. An obvious problem is keeping the data bases in synchronism. Any time change orders are worked, all the data bases must be updated simultaneously. Preferably, all data bases are updated from a single source; if not, the danger of error is always present. Furthermore, functions are duplicated between control centers. Besides the obvious expense of duplication, the previously mentioned problem of lack of skilled technicians is more acute under this method.

Despite its disadvantages, decentralized control and management has a number of offsetting advantages. Because the technicians are closer to the end users, they are available to solve minor problems that may take longer to fix from a central site. They can perform physical tasks such as setting options on terminals, reconnecting disconnected plugs, replacing circuit cards, and generally soothing users who may be upset about the trouble. They can test and monitor the network from inexpensive jack panels and can carry test equipment such as volt meters, transmission sets, and protocol analyzers out to the end users' locations.

Centralized Management and Distributed Control

In some ways, the last strategy offers some of the best aspects of both centralization and decentralization. Technicians are close to the users and can use simple and inexpensive testing techniques, but everyone relies on a central data base and centralized management of the testing and repair force. The primary factor in choosing this strategy is cost. If it is less expensive to distribute the work force, it should be distributed, but if test equipment is less expensive than personnel, the work force should probably be centralized. This method works particularly well in companies that have departmental networks and need someone on-site for assistance.

For example, if a company has departmental LANs, it is probably most effective to leave control of them in the hands of the users. However, a centralized group may have the knowledge and equipment required to collect and analyze performance information about the LAN. Another variation of the centralized/decentralized strategy is to train some of the users in elementary troubleshooting techniques and restoration functions such as patching in spare equipment so they can perform limited maintenance tasks under central direction.

Organization of the Network Control Center

Organization structure is a highly individual thing in most companies, and the network management and control center is no exception. The functions discussed in this section are found in most control centers, but the number of people required to perform them depends on many factors that are unique to the organization. The functions discussed here may be vested in one or several people, or they may be contracted to a vendor.

Help Desk

The help desk is the principal point of contact between the user and the control center. This position is responsible for the following:

- Receiving trouble reports from users
- Suggesting possible causes to aid users in clearing reports themselves

- Preparing initial records on trouble tickets or logs
- Relaying intermediate progress reports to users
- Closing trouble with users
- Preparing final trouble ticket or log entries

Many companies find that as much as 75 to 85 percent of all trouble reports are cleared by the help desk. These are troubles that are caused by simple actions of the users themselves: switches not turned on, improper log-on procedures, improper use of telephone instruments, and cables or cords broken or disconnected. The person at the help desk should, therefore, have a good working knowledge of how the telecommunications system functions. As troubles are cleared, this person becomes familiar with the most common causes and can tactfully suggest ways the users can handle their own problems.

It is usually a good idea to have all progress reports and completion reports funnel through the help desk. This establishes the help position in the minds of the end users as one that is integral to the solution of their problems. It is not necessary that this person be technically oriented. More important than technical knowledge is skill in handling people.

Technical Assistance

The person responsible for providing technical assistance to end users will handle the majority of the difficult trouble cases. Most of the test equipment, documentation, and support systems are used by this person to clear more complex troubles, and so this function normally handles 10-15% of the trouble reports. Typical functions are:

- Testing and isolating trouble conditions
- Monitoring network performance
- Substituting spare equipment or circuits for defective ones
- Replacing defective circuit cards
- Repairing wiring defects
- Installing equipment and performing installation acceptance tests
- Adding and moving telephone and data terminals

Many smaller companies will refer these tasks to vendors' repair forces. If there is not enough of this kind of work to warrant a full-time position, another strategy is to train personnel who have other responsibilities so they are available for occasional or emergency assistance.

High-Level Technical Support

Exceptional troubles are handled by some higher level of technical support that, by virtue of training, experience, or specialized equipment and information, is in a position to clear troubles that are beyond the ability of the technical assistance staff. These people often are employed on other functions such as programming, or they may be employed by vendors to lend expert technical assistance. Only a small percentage of troubles, usually five percent or fewer, are handled by a high-level technician.

Data Base Update Desk

It is essential that the network control center's data base be kept current with changes and trouble-clearing activity. Someone, therefore, must be designated to keep this information posted on a daily basis. If posting is delayed, the trouble clearing effort can be seriously impaired because of inaccurate records. All trouble tickets and installation and change orders must flow through this desk on the way to file.

This position may also be responsible for preparing summary reports and conducting searches of the data base for special information requests. This function is logically assigned to the help desk.

Analysis Desk

When information we have described so far in this chapter has been entered into a data base, a great deal of analysis is possible. Unless you have a specific purpose in mind, however, analysis is likely to be a waste of time. Whoever performs this function must therefore have an understanding of the overall network, the problems that are being encountered, the structure of the database, and how the information can be extracted to yield useful summaries. As discussed in Chapter 21, analysis can be carried on for several different purposes and has the objective of detecting the underlying trends that indicate how well (or how poorly) the network, its components, and its managers are performing. The analysis function is often the direct responsibility of the telecommunications manager.

Network Management and Control Equipment

We have discussed two of the three legs of a network control center; its personnel and its data base. In this section we will discuss the hardware elements that tie them together. The choice of hardware is closely related to your overall design strategy, centralized or decentralized, the skill level of your people, the nature and geographic spread of the network, and other such factors. This section discusses some of those factors and describes some of the principal testing methods available for both voice and data networks.

Network Control Equipment

Network control equipment is used to collect information about the status of the network, reconfigure the network to restore or rearrange it, and test it to isolate faults. A great deal of the equipment on the market is vendor dependent, and if so, to use it the network must be composed of components from a single vendor or equipment specifically designed to be compatible. From a telecommunication manager's standpoint, an ideal network control system would have the following characteristics:

- It would return network status information from every component
- The information would be readable and would point unerringly at faulty components
- Abnormalities would be displayed in such a manner that maintenance forces could not easily overlook them
- Troubles would be self-correcting by the network with a minimum need for human intervention
- Service would be restored following failures with a minimum of lost time
- The cost of the system would be justified in improved productivity of users and maintenance staff
- The system would prevent failures by alerting maintenance forces to potential trouble conditions before users detect the problems
- Equipment would support the protocols and network equipment of all vendors

These ideals are not, of course, all met in practice, but sophisticated equipment is available that accomplishes many of the objectives listed above. In this section we will review the purposes and characteristics of the main generic types of network control equipment.

Dial Back-up Equipment

The least expensive way of protecting private line data circuits is over the public switched network. Dial back-up of voice lines is usually automatic. If a tie-line fails, the PBX either automatically overflows to long distance or can quickly be programmed to do so, but data circuits must be equipped with auxiliary equipment to attach them to the switched network.

Dial back-up equipment is technically simple and inexpensive. Figure 24.3 shows schematically how dial back-up works. Either one or two circuits are required, depending on whether the circuit to be replaced is 2-wire or 4-wire. It may be necessary to operate at a slower speed on a switched basis, particularly if modems are not equipped with adaptive equalizers.

Patching

If circuits and equipment are wired through patch jacks, an operator on the premises can reconfigure circuits and rearrange equipment with relative ease. Figure 24.4 shows schematically how patching can be used to restore a failed circuit. Patching is a manual process; personnel are needed on-site to patch circuits, but with a patch panel circuits can be

FIGURE 24.3
Dial Backup Equipment For Data Circuit Restoration

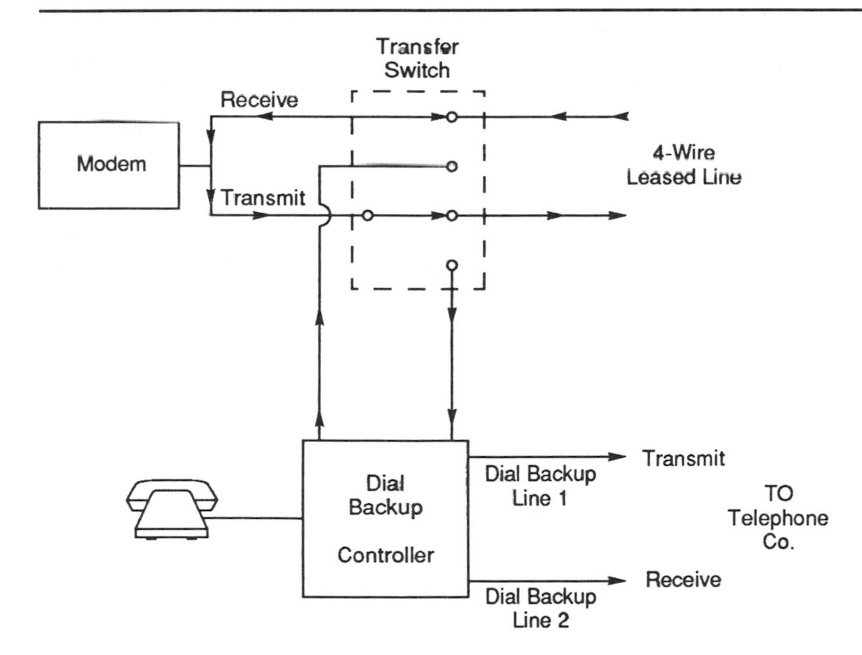

FIGURE 24.4
Restoring Failed Equipment by Patching

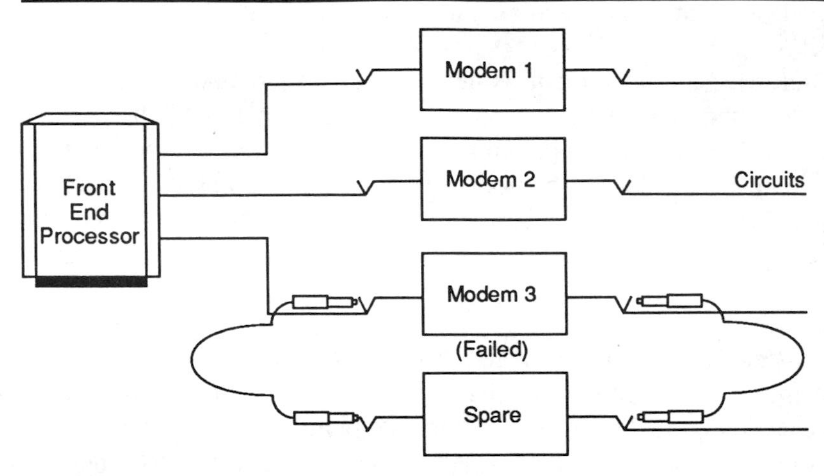

Circuits are normally connected through the jacks. The patch cord disconnects the failed equipment and replaces it with the spare.

reconfigured with far less time and effort than the alternative, which is rewiring. Patching is a poor substitute for rewiring on a long term basis, however. A panel draped with patch cords is difficult to administer and hazardous to service. When equipment or circuits have been by-passed with patches, the records are temporarily inaccurate and must be supplemented with entries from the patch log. Compared to other forms of control, patching is labor intensive and may be less economical on a long term basis than other alternatives.

Patch panels offer a great deal of flexibility for administering circuits within a building. In Chapter 13 we discussed the principle of connecting PBX ports and stations to a patch panel for ease of administration. Data circuits can also be routed through patch panels to enable technicians to replace modems, multiplexers, or other apparatus with a minimum of time and without the need to rewire.

Switching

Network components can be protected with switching equipment that provides access for testing and monitoring circuits or patching in spares. Switches can be configured for both local and remote operation, and can be accessed from either a dedicated line or over a dial-up circuit.

Switching equipment can also be used to bridge circuits for monitoring purposes or to open them to perform end-to-end tests. A simple continuity check, for example, can be made by bridging a circuit and monitoring to find whether a voice or data signal or a test tone can be received from the other end. By inserting a responder into the circuit it is possible to make a complete set of transmission measurements from a central position. Although switching is not an exotic technology, it is an effective and reliable method of accessing circuits and equipment from a remote location with a minimum requirement for personnel. It is, however, not inexpensive.

Matrix Switches

The matrix switch is a special type of switch that can be used for circuit restoration, reconfiguration, or monitoring. Under software control, a matrix switch can interconnect any combination of ports or the circuits or equipment that are wired to them. Besides the obvious function of restoring failed equipment, a matrix switch is useful for initiating temporary circuit reconfigurations, even on a daily basis. For example, a circuit might be used as a voice tie-line during prime working hours and switched to a data center for a file transfer at night. The switch could be programmed to perform this change daily at a given hour. Figure 24.5 shows a typical matrix switch.

Digital Cost Connect Systems

A digital crossconnect system (DCS) is a special type of matrix switch that is designed for T-1 circuits. As shown in Figure 24.6, a bit stream can be divided into several components without the use of a channel bank. Individual DS-0 channels can be routed over other T-1 circuits. The network configuration can be changed by time of day, or instructions can be sent from a control center to the DCS to permanently or temporarily reroute circuits or even entire T-1s. Although the DCS is designed for T-1 service, it is possible to connect analog circuits to it through a channel bank.

Intelligent Modems

Not to be confused with modems that respond to keyboard commands for dialing and setting configuration, intelligent modems are designed to relay status information to a central site. The site is usually equipped with a processor that interprets the information, analyzes performance, and

FIGURE 24.5
Telenex Autonex Matrix Switch

Photograph courtesy of Telenex Company

FIGURE 24.6
A Digital Crossconnect System Divides the Bit Stream Without the Use of Channel Banks

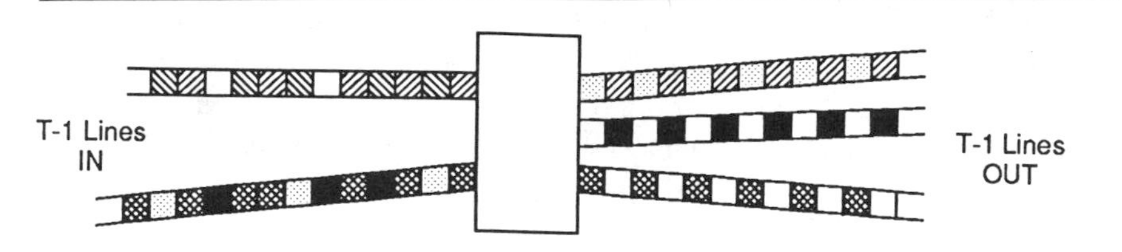

diagnoses trouble. The modems use a slow speed subchannel to transmit information from the modem to the central site. Such information as the transmit-receive signal levels can be used to perform a qualitative evaluation of the circuit.

HOST-BASED NETWORK MANAGEMENT

Network management exhibits the free-for-all characteristics that have afflicted data communications since its inception. There are no universal network management systems; most systems on the market are designed to manage particular vendors' products. The International Standards Organization may eventually recommend network management standards, but no standards are imminent at the time of this writing.

Several companies are marketing host-based network management systems. In these systems, the network management software resides in the host computer or front end processor that operates the network. The theory is excellent, but host-based systems may require that all equipment be furnished by the vendor that supplies the host.

Netview and Netview/PC

As an example of a complete network management system, we will review IBM's Netview and Netview/PC, products that some observers believe may become the network management standard of the future. Since IBM Systems Network Architecture (SNA) is, in effect, a *de facto* standard, IBM is a logical company to introduce such a network management product.

Netview and its companion product Netview/PC were introduced in 1986 as products designed to manage both SNA and non-SNA networks. Netview is a centralized network management system that resides in an IBM System/370 mainframe host. It combines the functions of three IBM network management products; Network Problem Determination Application (NPDA), Network Communication Control Facility (NCCF), and Network Logical Data Manager (NLDM). The goals of Netview are to simplify the process of installing a network, unify the tools used in administering the network, guide users in operating and maintaining it, and to funnel information through a single centralized point that facilitates a higher degree of automation. The purpose of Netview is to manage and diagnose the network from a central location.

Netview consists of a number of functions that were not unified under IBM's previous network management systems:

Hardware status monitoring. The system communicates with SNA hardware to determine status, detect failures, and detect major events. Status information is displayed on a CRT in a form that users can customize.

Session status monitoring. The session status monitors Logical Units (LUs), Physical Units (PUs), and System Service Control Points (SSCPs) in the network. The session status monitor measures response time and detects session failures in addition to monitoring the logical components of the network.

The control function. This function operates the network using Virtual Terminal Access Method (VTAM) and Netview commands. Devices such as protocol converters and modems can be controlled and reconfigured from the central site.

Status monitoring. This function displays network status on a CRT. This feature is also used to reactivate the network following a failure.

The customization function. This function enables users to configure Netview to fit their own operations. Users can devise their own messages on a status monitor. Statistical information, which is collected by Netview, can be customized to fit the operation. Colors can be changed on the display panel to indicate the potential severity of an alert. It is possible, through the customization option, to automate certain functions if human intervention is not required.

A help display. This feature guides operators in deciding what action to take. Help messages inform the operator of the meaning of various alerts, and what commands and codes are needed to take corrective action.

Netview contains the functions that are needed to monitor an SNA network, and as it matures, it will undoubtedly be widely accepted by companies that use SNA. A large number of non-SNA data networks and an even larger number of voice networks are in existence, and if Netview is to approach universality, these networks must somehow be addressed. IBM's approach to the non-SNA world is Netview/PC.

Netview/PC

Netview/PC is a monitoring and control system for non-SNA products, including some data products made by IBM, and voice networks supported by Rolm PBXs. Netview/PC is also designed to introduce network management functions to the IBM Token Ring to allow the host computer to see into an attached local area network. Through its Application Processor Interface (API) Netview/PC is designed to support products made by other manufacturers. The interface specifications and codes for Netview/PC are published so that any vendor can design its equipment with status and control circuitry and can write an interface program to implement Netview/PC.

Netview/PC either acts on a stand alone basis or communicates through Netview and Customer Interface Control System (CICS). Control and problem isolation can either be localized or centralized. From a centralized location it is possible to bypass failed components and reconfigure the network if the proper interfaces are deployed. Figure 24.7 shows the major modules of Netview and Netview/PC and how they interface the various systems and components to form a total management and control system.

The network management system is divided into three components; the Focal Point, the Entry Point, and the Service Point.

FIGURE 24.7
The Major Components of Netview and Netview/PC

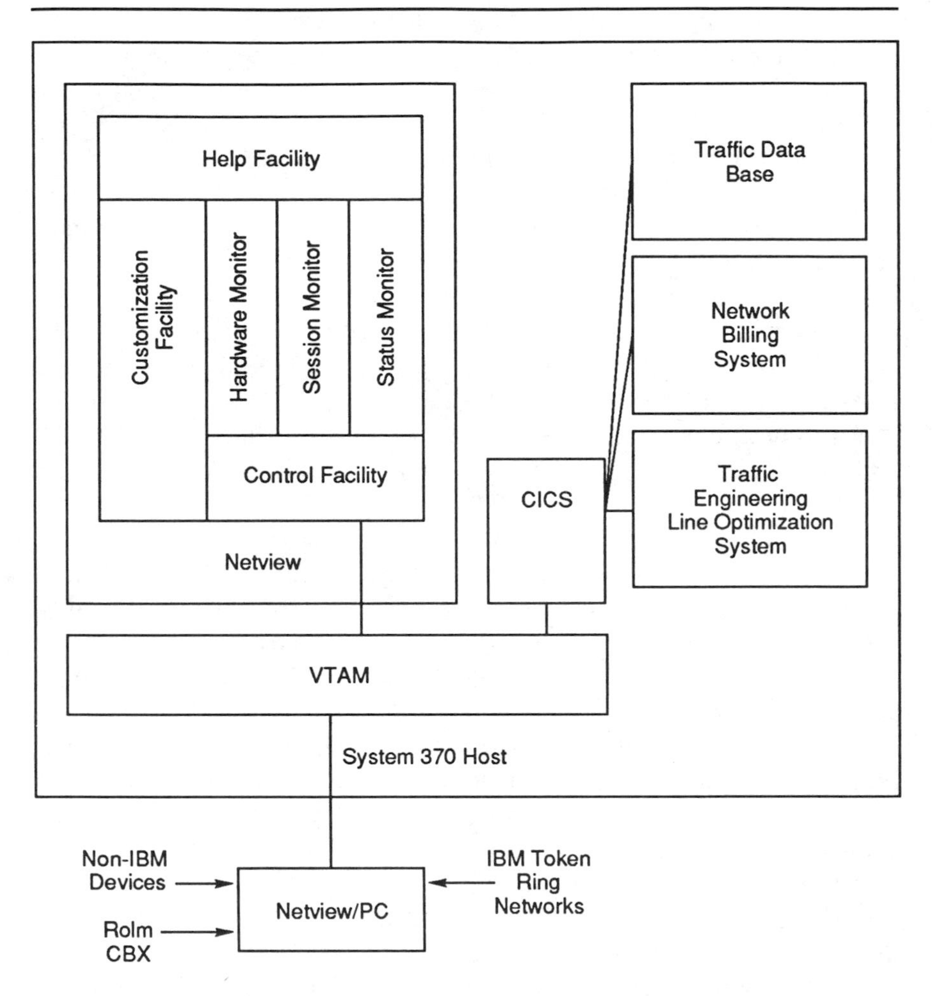

The Focal Point is the center for network management and control. Its functions are problem analysis, tracking, operational management and control, and change management and control. The Netview program itself resides at the focal point. Other software packages include these:

- The Netview Tariff Data Base, which contains the common carrier tariffs used for rating and billing long distance calls

- The Network Traffic Engineering Line Optimization System, which receives traffic usage information from voice switches and determines the optimum mix of circuits
- The Netview Network Billing System, which produces bills to distribute costs by department, equipment, and long distance carrier
- The Network Performance Monitor, which receives usage and error and failure information, and uses it to convey status and performance information to the operators

The Entry Point is the interface for relaying network management information from SNA devices to the focal point. Typical software packages connect IBM System/36, System/38, and System/88 information into the network.

The Service Point performs Entry Point functions for non-SNA devices. The Entry Point and Service Point receive commands from the Focal Point and return equipment identity, and status and performance information to the Focal Point in response to its commands. The Service Point translates Focal Point commands to whatever form is required by the attached devices. Typical software residing in the service point includes Netview/PC, the Rolm Alert Monitor and the Rolm Call Detail Recorder. For all such devices, the Service Point performs the following functions:

Alert management. The system records when alerts were received, logs them, and determines the origin, severity, and probable cause of the malfunction. With proper definition from the user, the alert management system recommends corrective action for the operator to take.

Problem recording. The system tracks the status of problem devices, collects statistical information, and analyzes the statistics.

The remote control facility. Enables operators at Netview locations to send orders through Netview/PC to the attached devices.

Netview/PC is IBM's first foray into management of voice network devices. In its initial implementation, Netview/PC is capable of performing limited network functions for Rolm CBXs. It is capable of centralized troubleshooting of CBXs, including tracking problems and monitoring alerts. The Rolm Alert Monitor enables a central site to manage and receive alarms from up to ten CBXs. As discussed earlier, the central site is able to perform call detail recording functions for CBXs by polling them to obtain call detail and usage information.

Netview is also capable of managing the network through IBM modems that are configured for network management. Netview can perform tests to check line quality and to analyze line condition for signal-to-noise ratio, phase distortion, and frequency shift. Netview can instruct modems to send and evaluate test patterns and can reconfigure the modem, control its speed, and activate a switched back-up condition.

Netview/PC also performs network management functions for IBM Token Ring networks. The system can aid in problem determination, recovery from error conditions, and can record errors and events in its trouble log.

It is too soon to tell whether Netview will become accepted as a network management standard. The essential functions are contained in Netview, are planned, or can be implemented by outside vendors. The open architecture of Netview/PC provides the framework on which other manufacturers can build their network management systems. The early signs are that many other vendors will follow IBM's lead and will support Netview. If this occurs, Netview may become, in effect, a network management standard.

SUMMARY

Several trends are evident that have a significant impact on the future of network management. First, more and more companies are assembling their own private networks, both as a cost control measure and as a strategy to make themselves more accessible to employees, customers, and agents. As these networks grow, the company's ability to tolerate outages diminishes in direct proportion. A second trend is toward the multiple vendor solution. No longer is it prudent, or even possible, to vest all of your network resources in a single vendor. Even though some interexchange carriers offer to manage your facilities through the tangle of local telephone companies and equipment vendors, it is usually more cost effective for a large company to manage them itself.

Another unmistakable trend is a continuing reduction in the cost of processing power. With desktop machines approaching the power of yesterday's mainframes, it is becoming more feasible to use computers in conjunction with low-cost long distance facilities to manage networks. This trend is occurring at a time that the complexity of networks is outstripping the ability of many people to grasp them, so the marriage of technician and computer is a particularly advantageous one.

As processing power drops in price, it is predictable that expert systems, that branch of artificial intelligence that enables experts to guide non-experts through a question and answer process, will flourish. Taken together, these trends make it safe to predict that network management will continue to grow in importance.

APPENDIX A

TRAFFIC TABLES

APPENDIX A-1
Erlang B Table

	P.01		*P.03*		*P.05*		*P.10*	
Trunks	*Erlangs*	*CCS*	*Erlangs*	*CCS*	*Erlangs*	*CCS*	*Erlangs*	*CCS*
1	0.01	0.4	0.03	1.1	0.05	1.8	0.11	4
2	0.15	5.4	0.28	10.1	0.38	13.7	0.60	21.6
3	0.46	16.6	0.75	26.9	0.90	32.4	1.27	45.7
4	0.87	31.3	1.26	45.5	1.52	54.7	2.05	73.8
5	1.36	49	1.88	67.7	2.22	79.9	2.89	104
6	1.91	68.8	2.54	91.4	2.97	107	3.75	135
7	2.50	90	3.25	117	3.75	135	4.67	168
8	3.14	113	4.00	144	4.53	163	5.61	202
9	3.78	136	4.75	171	5.36	193	6.56	236
10	4.47	161	5.53	199	6.22	224	7.50	270
11	5.17	186	6.33	228	7.08	255	8.50	306
12	5.89	212	7.14	257	7.94	286	9.47	341
13	6.61	238	7.97	287	8.83	318	10.47	377
14	7.36	265	8.81	317	9.72	350	11.47	413
15	8.11	292	9.64	347	10.64	383	12.48	449
16	8.86	319	10.50	378	11.53	415	13.50	486
17	9.64	347	11.36	409	12.47	449	14.53	523
18	10.44	376	12.25	441	13.39	482	15.56	560
19	11.22	404	13.11	472	14.31	515	16.58	597
20	12.03	433	14.00	504	15.25	549	17.61	634
21	12.83	462	14.89	536	16.19	583	18.64	671
22	13.64	491	15.78	568	17.14	617	19.69	709
23	14.47	521	16.67	600	18.08	651	20.78	748
24	15.28	550	17.58	633	19.03	685	21.78	784
25	16.11	580	18.47	665	20.00	720	22.83	822
26	16.97	611	19.39	698	20.94	754	23.89	860
27	17.81	641	20.31	731	21.89	788	24.94	898
28	18.64	671	21.22	764	22.86	823	26.00	936
29	19.50	702	22.14	797	23.83	858	27.06	974
30	20.33	732	23.06	830	24.81	893	28.11	1012
31	21.19	763	24.00	864	25.78	928	29.17	1050
32	22.06	794	24.92	897	26.75	963	30.22	1088

APPENDIX A-1—Continued

	P.01		P.03		P.05		P.10	
Trunks	Erlangs	CCS	Erlangs	CCS	Erlangs	CCS	Erlangs	CCS
33	22.92	825	25.83	930	27.72	998	31.31	1127
34	23.78	856	26.78	964	28.69	1033	32.36	1165
35	24.64	887	27.72	998	29.67	1068	33.42	1203
36	25.50	918	28.64	1031	30.67	1104	34.22	1232
37	26.39	950	29.58	1065	31.64	1139	35.58	1281
38	27.25	981	30.50	1098	32.64	1175	36.64	1319
39	28.14	1013	31.44	1132	33.61	1210	37.72	1358
40	29.00	1044	32.39	1166	34.61	1246	38.78	1396
41	29.89	1076	33.33	1200	35.58	1281	39.86	1435
42	30.78	1108	34.28	1234	36.58	1317	40.94	1474
43	31.67	1140	35.25	1269	37.58	1353	42.00	1512
44	32.53	1171	36.19	1303	38.56	1388	43.08	1551
45	33.42	1203	37.11	1336	39.56	1424	44.17	1590
46	34.33	1236	38.11	1372	40.53	1459	45.25	1629
47	35.22	1268	39.06	1406	41.53	1495	46.33	1668
48	36.11	1300	40.00	1440	42.53	1531	47.39	1706
49	37.00	1332	40.97	1475	43.53	1567	48.47	1745
50	37.89	1364	41.92	1509	44.53	1603	49.56	1784
51	38.81	1397	42.89	1544	45.50	1638	50.61	1822
52	39.69	1429	43.89	1580	46.50	1674	51.69	1861
53	40.61	1462	44.81	1613	47.50	1710	52.81	1901
54	41.50	1494	45.81	1649	48.50	1746	53.89	1940
55	42.39	1526	46.69	1681	49.50	1782	55.00	1980
56	43.31	1559	47.69	1717	50.50	1818	56.11	2020
57	44.19	1591	48.69	1753	51.50	1854	57.11	2056
58	45.11	1624	49.61	1786	52.61	1894	58.19	2095
59	46.00	1656	50.61	1822	53.61	1930	59.31	2135
60	46.89	1688	51.61	1858	54.61	1966	60.39	2174
61	47.89	1724	52.50	1890	55.61	2002	61.50	2214
62	48.81	1757	53.50	1926	56.61	2038	62.61	2254
63	49.69	1789	54.50	1962	57.61	2074	63.69	2293
64	50.61	1822	55.39	1994	58.61	2110	64.81	2333
65	51.50	1854	56.39	2030	59.61	2146	65.81	2369
66	52.39	1886	57.39	2066	60.61	2182	66.89	2408
67	55.33	1992	58.39	2102	61.61	2218	68.00	2448
68	54.31	1955	59.31	2135	62.61	2254	68.00	2448
69	55.19	1987	60.31	2171	63.69	2293	70.19	2527
70	56.11	2020	61.31	2207	64.69	2329	71.31	2567
71	57.00	2052	62.31	2243	65.69	2365	72.39	2606
72	58.00	2088	63.19	2275	66.69	2401	73.50	2646
73	58.89	2120	64.19	2311	67.69	2437	74.61	2686
74	59.81	2153	65.19	2347	68.69	2473	75.61	2722
75	60.69	2185	66.19	2383	69.69	2509	76.69	2761
76	61.69	2221	67.19	2419	70.81	2549	77.81	2801

APPENDIX A-1—Concluded

	P.01		*P.03*		*P.05*		*P.10*	
Trunks	*Erlangs*	*CCS*	*Erlangs*	*CCS*	*Erlangs*	*CCS*	*Erlangs*	*CCS*
77	62.61	2254	68.11	2452	71.81	2585	78.89	2840
78	63.50	2286	69.11	2488	72.81	2621	80.00	2880
79	64.39	2318	70.11	2524	73.81	2657	81.11	2920
80	65.39	2354	71.11	2560	74.81	2693	82.19	2959
81	66.31	2387	72.11	2596	75.81	2729	83.31	2999
82	67.19	2419	73.00	2628	76.89	2768	84.39	3038
83	68.19	2455	74.00	2664	77.89	2804	85.50	3078
84	69.11	2488	75.00	2700	78.89	2840	86.61	3118
85	70.00	2520	76.00	2736	79.89	2876	87.69	3157
86	70.89	2552	77.00	2772	80.89	2912	88.81	3197
87	71.89	2588	78.00	2808	82.00	2952	89.89	3236
88	72.81	2621	78.89	2840	83.00	2988	91.00	3276
89	73.69	2653	79.89	2876	84.00	3024	92.11	3316
90	74.69	2689	80.89	2912	85.00	3060	93.11	3352
91	75.61	2722	81.89	2948	86.00	3096	94.19	3391
92	76.61	2758	82.89	2984	87.11	3136	95.31	3431
93	77.50	2790	83.89	3020	88.11	3172	96.39	3470
94	78.39	2822	84.89	3056	89.11	3208	97.50	3510
95	79.39	2858	85.89	3092	90.11	3244	98.61	3550
96	80.31	2891	86.81	3125	91.11	3280	99.69	3589
97	81.19	2923	87.81	3161	92.19	3319	100.81	3629
98	82.19	2959	88.81	3197	93.19	3355	101.89	3668
99	83.11	2992	89.81	3233	94.19	3391	103.00	3708
100	84.11	3028	90.81	3269	95.19	3427	104.11	3748

APPENDIX A-2
Poisson Table

	P.01		*P.03*		*P.05*		*P.10*	
Trunks	*Erlangs*	*CCS*	*Erlangs*	*CCS*	*Erlangs*	*CCS*	*Erlangs*	*CCS*
1	0.01	0.4	0.03	1.1	0.05	1.9	0.11	3.8
2	0.15	5.4	0.27	9.7	0.36	12.9	0.53	19.1
3	0.44	15.7	0.67	24	0.82	29.4	1.10	39.6
4	0.82	29.6	1.16	41.6	1.36	49.1	1.75	63
5	1.28	46.1	1.71	61.6	1.97	70.9	2.44	88
6	1.79	64.4	2.30	82.8	2.61	94.1	3.14	113
7	2.33	83.9	2.92	105	3.28	118	3.89	140
8	2.92	105	3.58	129	3.97	143	4.67	168
9	3.50	126	4.25	153	4.69	169	5.42	195
10	4.14	149	4.94	178	5.42	195	6.22	224

APPENDIX A-2—Continued

	P.01		*P.03*		*P.05*		*P.10*	
Trunks	*Erlangs*	*CCS*	*Erlangs*	*CCS*	*Erlangs*	*CCS*	*Erlangs*	*CCS*
11	4.78	172	5.67	204	6.17	222	7.03	253
12	5.42	195	6.39	230	6.92	249	7.83	282
13	6.11	220	7.11	256	7.69	277	8.64	311
14	6.78	244	7.86	283	8.47	305	9.47	341
15	7.47	269	8.61	310	9.25	333	10.28	370
16	8.17	294	9.36	337	10.06	362	11.14	401
17	8.89	320	10.14	365	10.83	390	11.97	431
18	9.61	346	10.89	392	11.64	419	12.83	462
19	10.36	373	11.67	420	12.44	448	13.67	492
20	11.08	399	12.47	449	13.25	477	14.53	523
21	11.83	426	13.28	478	14.08	507	15.39	554
22	12.58	453	14.08	507	14.89	536	16.25	585
23	13.33	480	14.89	536	15.72	566	17.11	616
24	14.08	507	15.67	564	16.56	596	17.97	647
25	14.86	535	16.47	593	17.39	626	18.83	678
26	15.61	562	17.31	623	18.22	656	19.72	710
27	16.39	590	18.11	652	19.06	686	20.58	741
28	17.17	618	18.94	682	19.92	717	21.47	773
29	17.97	647	19.75	711	20.75	747	22.36	805
30	18.75	675	20.58	741	21.61	778	23.22	836
31	19.53	703	21.42	771	22.47	809	24.11	868
32	20.33	732	22.25	801	23.33	840	25.00	900
33	21.11	760	23.08	831	24.19	871	25.89	932
34	21.92	789	23.92	861	25.06	902	26.78	964
35	22.72	818	24.75	891	25.92	933	27.67	996
36	23.53	847	25.61	922	26.78	964	28.56	1028
37	24.33	876	26.44	952	27.64	995	29.44	1060
38	25.14	905	27.28	982	28.50	1026	30.33	1092
39	25.97	935	28.14	1013	29.36	1057	31.25	1125
40	26.78	964	28.97	1043	30.22	1088	32.14	1157
41	27.58	993	29.83	1074	31.11	1120	33.06	1190
42	28.42	1023	30.67	1104	31.97	1151	33.94	1222
43	29.22	1052	31.53	1135	32.83	1182	34.86	1255
44	30.06	1082	32.39	1166	33.72	1214	35.75	1287
45	30.89	1112	33.25	1197	34.61	1246	36.67	1320
46	31.72	1142	34.11	1228	35.47	1277	37.56	1352
47	32.53	1171	34.97	1259	36.36	1309	38.47	1385
48	33.36	1201	35.86	1291	37.22	1340	39.36	1417
49	34.19	1231	36.72	1322	38.11	1372	40.28	1450
50	35.03	1261	37.58	1353	38.97	1403	41.17	1482
51	35.86	1291	38.44	1384	39.86	1435	42.08	1515
52	36.72	1322	39.33	1416	40.75	1467	43.00	1548
53	37.56	1352	40.19	1447	41.64	1499	43.92	1581
54	38.39	1382	41.06	1478	42.53	1531	44.83	1614
55	39.22	1412	41.92	1509	43.42	1563	45.72	1646

APPENDIX A-2—Concluded

	P.01		*P.03*		*P.05*		*P.10*	
Trunks	*Erlangs*	*CCS*	*Erlangs*	*CCS*	*Erlangs*	*CCS*	*Erlangs*	*CCS*
56	40.08	1443	42.81	1541	44.31	1595	46.64	1679
57	40.92	1473	43.67	1572	45.19	1627	47.56	1712
58	41.78	1504	44.56	1604	46.08	1659	48.47	1745
59	42.61	1534	45.42	1635	46.97	1691	49.39	1778
60	43.47	1565	46.31	1667	47.86	1723	50.31	1811
61	44.31	1595	47.17	1698	48.75	1755	51.22	1844
62	45.17	1626	48.06	1730	49.64	1787	52.14	1877
63	46.03	1657	48.94	1762	50.53	1819	53.06	1910
64	46.86	1687	49.83	1794	51.42	1851	53.97	1943
65	47.72	1718	50.69	1825	52.33	1884	54.89	1976
66	48.58	1749	51.58	1857	53.22	1916	55.81	2009
67	55.00	1980	52.47	1889	54.11	1948	56.72	2042
68	50.31	1811	53.36	1921	55.03	1981	57.67	2076
69	51.17	1842	54.25	1953	55.92	2013	58.58	2109
70	52.03	1873	55.14	1985	56.83	2046	59.50	2142
71	52.89	1904	56.03	2017	57.72	2078	60.42	2175
72	53.75	1935	56.89	2048	58.64	2111	61.36	2209
73	54.61	1966	57.78	2080	59.53	2143	62.28	2242
74	55.47	1997	58.67	2112	60.44	2176	63.22	2276
75	56.33	2028	59.58	2145	61.33	2208	64.14	2309
76	57.17	2058	60.44	2176	62.25	2241	65.06	2342
77	50.00	2001	61.36	2209	63.17	2274	66.00	2376
78	58.94	2122	62.25	2241	64.06	2306	66.94	2410
79	59.81	2153	63.14	2273	64.97	2339	67.86	2443
80	60.67	2184	64.03	2305	65.89	2372	68.81	2477
81	61.53	2215	64.92	2337	66.81	2405	69.72	2510
82	62.42	2247	65.83	2370	67.69	2437	70.64	2543
83	63.28	2278	66.72	2402	68.61	2470	71.58	2577
84	64.17	2310	67.64	2435	69.53	2503	72.50	2610
85	65.03	2341	68.53	2467	70.44	2536	73.44	2644
86	65.92	2373	69.42	2499	71.36	2569	74.39	2678
87	66.78	2404	70.33	2532	72.25	2601	75.31	2711
88	67.67	2436	71.22	2564	73.17	2634	76.25	2745
89	68.53	2467	72.11	2596	74.08	2667	77.17	2778
90	69.42	2499	73.03	2629	75.00	2700	78.11	2812
91	70.28	2530	73.92	2733	79.06	2733	79.06	2846
92	71.17	2562	74.83	2694	76.83	2766	80.00	2880
93	72.06	2594	75.72	2726	77.72	2798	80.92	2913
94	72.92	2625	76.64	2759	78.64	2831	81.86	2947
95	73.81	2657	77.53	2791	79.56	2864	82.81	2981
96	74.69	2689	78.44	2824	80.47	2897	83.72	3014
97	75.58	2721	79.36	2857	81.39	2930	84.67	3048
98	76.44	2752	80.25	2889	82.31	2963	85.61	3082
99	77.33	2784	81.14	2921	83.22	2996	86.56	3116
100	78.22	2816	82.06	2954	84.14	3029	87.47	3149

APPENDIX A-3
Erlang C Table

					Probability of Delay (All Calls) of Time "t"						
A	*N*	P_d	D_1	D_2	*.2*	*.4*	*.6*	*.8*	*1*	*2*	*3*
.05	1	.0500	.053	1.05	.041	.034	.028	.023	.019	.007	.003
	2	.0012	.001	.513	.001	.001					
.10	1	.1000	.111	1.11	.084	.070	.058	.049	.041	.017	.007
	2	.0048	.003	.526	.003	.002	.002	.001	.001		
	3	.0002		.345							
.15	1	.1500	.176	1.18	.127	.107	.090	.076	.064	.027	.012
	2	.0105	.006	.541	.007	.005	.003	.002	.002		
	3	.0005		.351							
.20	1	.2000	.250	1.25	.170	.145	.124	.105	.090	.040	.018
	2	.0182	.010	.556	.013	.009	.006	.004	.003		
	3	.0012		.357	.001						
	4	.0001		.263							
.25	1	.2500	.333	1.33	.215	.185	.159	.137	.118	.056	.026
	2	.0278	.016	.571	.020	.014	.010	.007	.005	.001	
	3	.0022	.001	.364	.001	.001					
	4	.0001		.267							
.30	1	.3000	.429	1.43	.261	.227	.197	.171	.149	.074	.037
	2	.0391	.023	.588	.028	.020	.014	.010	.007	.001	
	3	.0037	.001	.370	.002	.001	.001				
	4	.0003		.270							
.35	1	.3500	.538	1.54	.307	.270	.237	.208	.183	.095	.050
	2	.0521	.032	.606	.037	.027	.019	.014	.010	.002	
	3	.0057	.002	.377	.003	.002	.001	.001			
	4	.0005		.274							
.40	1	.4000	.667	1.67	.355	.315	.279	.248	.220	.120	.066
	2	.0667	.042	.625	.048	.035	.026	.019	.013	.003	.001
	3	.0082	.003	.385	.005	.002	.001	.001			
	4	.0008		.278							
	5	.0001		.217							
.45	1	.4500	.818	1.82	.403	.361	.324	.290	.260	.150	.086
	2	.0827	.053	.645	.061	.044	.033	.024	.018	.004	.001
	3	.0114	.004	.392	.007	.004	.002	.001	.001		
	4	.0012		.282	.0001						
	5	.0001		.220							
.50	1	.5000	1.00	2.00	.452	.409	.370	.335	.303	.184	.112
	2	.1000	.067	.667	.074	.055	.041	.030	.022	.005	.001
	3	.0152	.006	.400	.009	.006	.003	.002	.001		
	4	.0018	.001	.286	.001						
	5	.0002		.222							

A = Offered load in Erlangs
N = Number of trunks (servers) required
P_d = Probability that a call will be delayed

D1 = Average delay of all calls
D2 = Average delay of delayed calls

APPENDIX A-3—Continued

					Probability of Delay (All Calls) of Time "t"						
A	*N*	P_d	D_1	D_2	*.2*	*.4*	*.6*	*.8*	*1*	*2*	*3*
.55	1	.5500	1.22	2.22	.503	.459	.420	.384	.351	.224	.143
	2	.1186	.082	.690	.089	.066	.050	.037	.028	.007	.002
	3	.0196	.008	.408	.012	.007	.004	.003	.002		
	4	.0026	.001	.290	.001	.001					
	5	.0003		.225							
.60	1	.6000	1.50	2.50	.554	.511	.472	.436	.402	.270	.181
	2	.1385	.099	.714	.105	.079	.060	.045	.034	.008	.002
	3	.0247	.010	.417	.015	.009	.006	.004	.002		
	4	.0035	.001	.294	.002	.001					
	5	.0004		.227							
.65	1	.6500	1.86	2.86	.606	.565	.527	.491	.458	.323	.227
	2	.1594	.118	.741	.122	.093	.071	.054	.041	.011	.003
	3	.0304	.013	.426	.019	.032	.007	.005	.003		
	4	.0046	.001	.299	.002	.001	.001				
	5	.0006		.230							
	6	.0001		.187							
.70	1	.7000	2.33	3.33	.659	.621	.585	.551	.519	.384	.285
	2	.1815	.140	.769	.140	.108	.083	.064	.049	.013	.004
	3	.0369	.016	.435	.023	.015	.009	.006	.004		
	4	.0060	.002	.303	.003	.002	.001				
	5	.0008		.233							
	6	.0001		.189							
.75	1	.7500	3.00	4.00	.713	.679	.646	.614	.584	.455	.354
	2	.2045	.164	.800	.159	.124	.097	.075	.059	.017	.005
	3	.0441	.020	.444	.028	.018	.011	.007	.005		
	4	.0077	.002	.308	.004	.002	.001	.001			
	5	.0011		.235							
	6	.0001		.190							
.80	1	.8000	4.00	5.00	.769	.738	.710	.682	.655	.536	.439
	2	.2286	.190	.833	.180	.141	.111	.088	.069	.021	.006
	3	.0520	.024	.455	.034	.022	.014	.009	.006	.001	
	4	.0096	.003	.312	.005	.003	.001	.001			
	5	.0015		.238	.001						
	6	.0002		.192							
.85	1	.3500	5.67	6.67	.825	.800	.777	.754	.732	.630	.542
	2	.2535	.220	.870	.201	.160	.127	.101	.080	.025	.008
	3	.0607	.028	.465	.039	.026	.017	.011	.007	.001	
	4	.0118	.004	.317	.006	.003	.002	.001	.001		
	5	.0019		.241	.001						
	6	.0003		.194							

A = Offered load in Erlangs
N = Number of trunks (servers) required
P_d = Probability that a call will be delayed

D1 = Average delay of all calls
D2 = Average delay of delayed calls

APPENDIX A-3—Continued

					Probability of Delay (All Calls) of Time "t"						
A	*N*	P_d	D_1	D_2	.2	.4	.6	.8	1	2	3
.90	1	.9000	9.00	10.0	.882	.865	.848	.831	.814	.737	.667
	2	.2793	.254	.909	.224	.180	.144	.116	.093	.031	.010
	3	.0700	.033	.476	.046	.030	.020	.013	.009	.001	
	4	.0143	.005	.323	.008	.004	.002	.001	.001		
	5	.0024	.001	.244	.001						
	6	.0004		.196							
.95	1	.9500	19.0	20.0	.941	.931	.922	.913	.904	.860	.818
	2	.3059	.291	.952	.248	.201	.163	.132	.107	.037	.013
	3	.0801	.039	.488	.053	.035	.023	.016	.010	.001	
	4	.0172	.006	.328	.009	.005	.003	.001	.001		
	5	.0031	.001	.247	.001						
	6	.0005		.198							
	7	.0001		.165							
1.0	2	.3333	.333	1.00	.273	.223	.183	.150	.123	.045	.017
	3	.0909	.045	.500	.061	.041	.027	.018	.012	.002	
	4	.0204	.007	.333	.011	.006	.003	.002	.001		
	5	.0038	.001	.250	.002	.001					
	6	.0006		.200							
	7	.0001		.167							
1.1	2	.3903	.434	1.11	.326	.272	.227	.190	.159	.065	.026
	3	.1146	.060	.526	.078	.054	.037	.025	.017	.003	
	4	.0279	.010	.345	.016	.009	.005	.003	.002		
	5	.0057	.001	.256	.003	.001	.001				
	6	.0010		.204							
	7	.0002		.169							
1.2	2	.4500	.562	1.25	.383	.327	.278	.237	.202	.091	.041
	3	.1412	.078	.556	.098	.069	.048	.033	.023	.004	.001
	4	.0370	.013	.357	.021	.012	.007	.004	.002		
	5	.0082	.002	.263	.004	.002	.001				
	6	.0016		.208	.001						
	7	.0003		.172							
1.3	2	.5121	.732	1.43	.445	.387	.336	.293	.254	.126	.063
	3	.1704	.100	.588	.121	.085	.061	.044	.031	.006	.001
	4	.0478	.018	.370	.028	.016	.009	.006	.003		
	5	.0114	.003	.270	.005	.003	.001	.001			
	6	.0023		.213	.001						
	7	.0004		.175							
	8	.0001		.149							

A = Offered load in Erlangs
N = Number of trunks (servers) required
P_d = Probability that a call will be delayed

D1 = Average delay of all calls
D2 = Average delay of delayed calls

APPENDIX A-3—Continued

					Probability of Delay (All Calls) of Time "t"						
A	*N*	P_d	D_1	D_2	*.2*	*.4*	*.6*	*.8*	*1*	*2*	*3*
1.4	2	.5765	.961	1.67	.511	.453	.402	.357	.316	.174	.095
	3	.2024	.126	.625	.147	.107	.077	.056	.041	.008	.002
	4	.0603	.023	.385	.036	.021	.013	.008	.004		
	5	.0153	.004	.278	.007	.004	.002	.001			
	6	.0034	.001	.217	.001	.001	.001				
	7	.0006		.179							
	8	.0001									
1.5	2	.6429	1.29	2.00	.582	.526	.476	.431	.390	.236	.143
	3	.2368	.158	.667	.175	.130	.096	.071	.053	.012	.003
	4	.0746	.030	.400	.045	.027	.017	.010	.006	.001	
	5	.0201	.006	.286	.010	.005	.002	.001	.001		
	6	.0047	.001	.222	.002	.001					
	7	.0010		.182							
	8	.0002		.154							
1.6	2	.7111	1.78	2.50	.656	.606	.559	.516	.477	.320	.214
	3	.2738	.196	.714	.207	.156	.118	.089	.068	.017	.004
	4	.0907	.038	.417	.056	.035	.021	.013	.008	.001	
	5	.0259	.008	.294	.013	.007	.003	.002	.001		
	6	.0064	.001	.227	.003	.001					
	7	.0014		.185							
	8	.0003		.156							
1.7	2	.7811	2.60	3.33	.736	.693	.652	.614	.579	.429	.318
	3	.3131	.241	.769	.241	.186	.144	.111	.085	.023	.006
	4	.1087	.047	.435	.069	.043	.027	.017	.011	.001	
	5	.0326	.010	.303	.017	.009	.005	.002	.001		
	6	.0085	.002	.233	.004	.002	.001				
	7	.0020		.189	.001						
	8	.0004		.159							
	9	.0001		.137							
1.8	2	.8526	4.26	5.00	.819	.787	.756	.727	.698	.572	.468
	3	.3547	.296	.833	.279	.220	.173	.136	.107	.032	.010
	4	.1285	.058	.455	.083	.053	.034	.022	.014	.002	
	5	.0405	.013	.312	.021	.011	.006	.003	.002		
	6	.0111	.003	.238	.005	.002	.001				
	7	.0027	.001	.192	.001						
	8	.0006		.161							
	9	.0001		.139							

A = Offered load in Erlangs
N = Number of trunks (servers) required
P_d = Probability that a call will be delayed

D1 = Average delay of all calls
D2 = Average delay of delayed calls

APPENDIX A-3—Continued

A	N	P_d	D_1	D_2	Probability of Delay (All Calls) of Time "t"						
					.2	.4	.6	.8	1	2	3
1.9	2	.9256	9.26	10.0	.907	.889	.872	.854	.838	.758	.686
	3	.3985	.362	.909	.320	.257	.206	.165	.133	.044	.015
	4	.1503	.072	.476	.099	.065	.043	.028	.018	.002	
	5	.0495	.016	.323	.027	.014	.008	.004	.002		
	6	.0143	.003	.244	.006	.003	.001	.001			
	7	.0036	.001	.196	.001						
2.0	3	.4444	.444	1.00	.364	.298	.244	.200	.164	.060	.022
	4	.1739	.087	.500	.117	.078	.052	.035	.024	.003	
	5	.0597	.020	.333	.033	.018	.010	.005	.003		
	6	.0180	.005	.250	.008	.004	.002	.001			
	7	.0048	.001	.200	.002	.001					
	8	.0011		.167							
	9	.0002		.143							
2.1	3	.4923	.547	1.11	.411	.343	.287	.240	.200	.081	.033
	4	.1994	.105	.526	.136	.093	.064	.044	.030	.004	.001
	5	.0712	.025	.345	.040	.022	.012	.007	.004		
	6	.0224	.006	.256	.010	.005	.002	.001			
	7	.0062	.001	.204	.002	.001					
	8	.0016		.169							
	9	.0003		.145							
	10	.0001		.127							
2.2	3	.5422	.678	1.25	.462	.394	.335	.286	.244	.109	.049
	4	.2268	.126	.556	.158	.110	.077	.054	.037	.006	.001
	5	.0839	.030	.357	.048	.027	.016	.009	.005		
	6	.0275	.007	.263	.013	.006	.003	.001	.001		
	7	.0080	.002	.208	.003	.001					
	8	.0021		.172	.001						
	9	.0005		.147							
	10	.0001		.128							
2.3	3	.5938	.848	1.43	.516	.449	.390	.339	.295	.146	.073
	4	.2560	.151	.588	.182	.130	.092	.066	.047	.009	.002
	5	.0980	.036	.370	.057	.033	.019	.011	.007		
	6	.0333	.009	.270	.016	.008	.004	.002	.001		
	7	.0101	.002	.213	.004	.002	.001				
	8	.0027		.175	.001						
	9	.0007		.149							
	10	.0001		.130							

A = Offered load in Erlangs
N = Number of trunks (servers) required
P_d = Probability that a call will be delayed

D1 = Average delay of all calls
D2 = Average delay of delayed calls

APPENDIX A-3—Continued

					Probability of Delay (All Calls) of Time "t"						
A	*N*	P_d	D_1	D_2	*.2*	*.4*	*.6*	*.8*	*1*	*2*	*3*
2.4	3	.6472	1.08	1.67	.574	.509	.452	.400	.355	.195	.107
	4	.2870	.179	.625	.208	.151	.110	.080	.058	.012	.002
	5	.1135	.044	.385	.040	.024	.014	.008	.001		
	6	.0400	.011	.278	.019	.009	.005	.002	.001		
	7	.0126	.003	.217	.005	.002	.001				
	8	.0035	.001	.179	.001						
	9	.0009		.152							
	10	.0002		.132							
2.5	3	.7022	1.40	2.00	.635	.575	.520	.471	.426	.258	.157
	4	.3199	.213	.667	.237	.176	.130	.096	.071	.016	.004
	5	.1304	.052	.400	.079	.048	.029	.018	.011	.001	
	6	.0474	.014	.286	.024	.012	.006	.003	.001		
	7	.0154	.003	.222	.006	.003	.001				
	8	.0043	.001	.182	.002	.001					
	9	.0012		.154							
	10	.0003		.133							
	11	.0001		.118							
2.6	3	.7589	1.90	2.50	.701	.647	.597	.551	.509	.341	.229
	4	.3544	.253	.714	.268	.202	.153	.116	.087	.022	.005
	5	.1487	.062	.417	.092	.057	.035	.022	.013	.001	
	6	.0550	.016	.294	.028	.014	.007	.004	.002		
	7	.0188	.004	.227	.008	.003	.001	.001			
	8	.0057	.001	.185	.002	.001					
	9	.0016		.156							
	10	.0004		.135							
	11	.0001		.119							
2.7	3	.8171	2.72	3.33	.769	.725	.682	.643	.605	.448	.332
	4	.3907	.301	.769	.301	.232	.179	.138	.106	.029	.008
	5	.1684	.073	.435	.106	.067	.042	.017	.002		
	6	.0652	.020	.303	.034	.017	.009	.005	.002		
	7	.0227	.005	.233	.010	.004	.002	.001			
	8	.0071	.001	.189	.002	.001					
	9	.0020		.159	.001						
	10	.0005		.137							
	11	.0001		.120							

A = Offered load in Erlangs
N = Number of trunks (servers) required
P_d = Probability that a call will be delayed

D1 = Average delay of all calls
D2 = Average delay of delayed calls

APPENDIX A-3—Continued

A	N	P_d	D_1	D_2	Probability of Delay (All Calls) of Time "t"						
					.2	.4	.6	.8	1	2	3
2.8	3	.8767	4.38	5.00	.842	.309	.778	.747	.718	.588	.481
	4	.4287	.357	.833	.337	.263	.209	.164	.129	.039	.012
	5	.1895	.086	.455	.122	.079	.051	.033	.021	.002	
	6	.0755	.024	.312	.040	.021	.011	.006	.003		
	7	.0271	.006	.238	.012	.005	.002	.001			
	8	.0088	.002	.192	.003	.001					
	9	.0026		.161	.001						
2.9	3	.9377	9.38	10.0	.919	.901	.883	.866	.848	.768	.695
	4	.4682	.426	.909	.376	.302	.242	.194	.156	.052	.017
	5	.2121	.101	.476	.139	.092	.060	.040	.026	.003	
	6	.0868	.028	.323	.047	.025	.014	.007	.004		
	7	.0320	.008	.244	.014	.006	.003	.001	.001		
	8	.0107	.002	.196	.004	.001	.001				
	9	.0032	.001	.164	.001						
3.0	4	.5094	.509	1.00	.461	.417	.377	.341	.309	.241	.187
	5	.2362	.118	.500	.193	.158	.130	.106	.087	.053	.032
	6	.0991	.033	.333	.073	.054	.040	.030	.022	.010	.005
	7	.0376	.009	.250	.025	.017	.011	.008	.005	.002	.001
	8	.0129	.003	.200	.008	.005	.003	.002	.001		
	9	.0040	.001	.167	.001	.001					
3.1	4	.5522	.614	1.11	.505	.461	.422	.385	.352	.281	.225
	5	.2616	.138	.526	.216	.179	.148	.122	.101	.063	.039
	6	.1126	.039	.345	.084	.063	.047	.035	.026	.013	.006
	7	.0439	.011	.256	.030	.020	.014	.009	.006	.002	.001
	8	.0155	.003	.204	.009	.006	.004	.002	.001		
	9	.0050	.001	.169	.003	.002	.001				
3.2	4	.5964	.746	1.25	.551	.508	.469	.433	.400	.327	.268
	5	.2886	.160	.556	.241	.201	.168	.140	.117	.075	.048
	6	.1271	.045	.357	.096	.073	.055	.041	.031	.016	.008
	7	.0509	.013	.263	.035	.024	.016	.011	.008	.003	.001
	8	.0185	.004	.208	.011	.007	.004	.003	.002	.001	
	9	.0061	.001	.172	.003	.002	.001	.001			
3.3	4	.6422	.917	1.43	.599	.558	.521	.485	.453	.380	.319
	5	.3169	.186	.588	.267	.226	.190	.161	.135	.089	.058
	6	.1427	.053	.370	.109	.083	.063	.048	.037	.019	.010
	7	.0585	.016	.270	.040	.028	.019	.013	.009	.004	.001
	8	.0219	.005	.213	.014	.009	.005	.003	.002	.001	
	9	.0074	.001	.175	.004	.002	.001	.001			
	10	.0023		.149	.001	.001					

A = Offered load in Erlangs
N = Number of trunks (servers) required
P_d = Probability that a call will be delayed

D1 = Average delay of all calls
D2 = Average delay of delayed calls

APPENDIX A-3—Continued

A	N	P_d	D_1	D_2	Probability of Delay (All Calls) of Time "t" .2	.4	.6	.8	1	2	3
3.4	4	.6893	1.15	1.67	.649	.611	.576	.542	.511	.440	.378
	5	.3467	.217	.625	.295	.262	.215	.183	.156	.104	.070
	6	.1595	.061	.385	.123	.095	.073	.056	.043	.023	.012
	7	.0670	.019	.278	.047	.033	.023	.016	.011	.005	.002
	8	.0256	.006	.217	.016	.010	.006	.004	.003	.001	
	9	.0090	.002	.179	.005	.003	.002	.001	.001		
	10	.0029		.152	.001	.001					
3.5	4	.7379	1.48	2.00	.702	.668	.635	.604	.575	.507	.448
	5	.3778	.252	.667	.325	.280	.241	.207	.178	.123	.084
	6	.1775	.071	.400	.138	.108	.084	.065	.051	.027	.015
	7	.0762	.022	.286	.054	.038	.027	.019	.013	.006	.002
	8	.0299	.007	.222	.022	.014	.009	.006	.004	.001	
	9	.0107	.002	.182	.007	.004	.003	.001	.001		
	10	.0035	.001	.154	.002	.001	.001				
	11	.0011		.133	.001						
3.6	4	.7878	1.97	2.50	.757	.727	.699	.671	.645	.584	.528
	5	.4104	.293	.714	.357	.310	.270	.234	.204	.144	.101
	6	.1966	.082	.417	.153	.122	.096	.075	.059	.032	.018
	7	.0862	.025	.294	.061	.044	.031	.022	.016	.007	.003
	8	.0346	.008	.227	.022	.014	.009	.006	.004	.001	
	9	.0127	.002	.185	.007	.004	.003	.002	.001		
	10	.0043	.001	.156	.002	.001	.001				
3.7	4	.8390	2.80	3.33	.814	.790	.767	.744	.722	.670	.622
	5	.4443	.342	.769	.390	.343	.301	.264	.232	.168	.121
	6	.2168	.094	.435	.172	.137	.109	.086	.069	.039	.022
	7	.0971	.029	.303	.070	.050	.036	.026	.019	.008	.004
	8	.0399	.009	.233	.026	.017	.011	.007	.005	.002	.001
	9	.0150	.003	.189	.009	.005	.003	.002	.001		
	10	.0052	.001	.159	.003	.001	.001				
	11	.0017		.137	.001						
3.8	4	.8914	4.46	5.00	.874	.856	.840	.823	.807	.767	.730
	5	.4796	.400	.833	.425	.377	.335	.297	.263	.195	.144
	6	.2383	.108	.455	.191	.153	.123	.099	.079	.046	.026
	7	.1089	.034	.312	.079	.057	.042	.030	.022	.010	.004
	8	.0457	.011	.238	.030	.020	.013	.009	.006	.002	.001
	9	.0176	.003	.192	.010	.006	.004	.002	.001		
	10	.0062	.001	.161	.003	.002	.001	.001			
	11	.0020		.139	.001						

A = Offered load in Erlangs
N = Number of trunks (servers) required
P_d = Probability that a call will be delayed

D1 = Average delay of all calls
D2 = Average delay of delayed calls

APPENDIX A-3—Continued

					Probability of Delay (All Calls) of Time "t"						
A	N	P_d	D_1	D_2	.2	.4	.6	.8	1	2	3
3.9	4	.9451	9.45	10.0	.936	.926	.917	.908	.899	.877	.855
	5	.5162	.469	.909	.462	.414	.371	.332	.298	.226	.172
	6	.2609	.124	.476	.212	.271	.139	.113	.091	.058	.032
	7	.1215	.039	.323	.089	.065	.048	.035	.026	.012	.005
	8	.0521	.013	.244	.035	.023	.015	.010	.007	.002	.001
	9	.0205	.004	.196	.012	.007	.004	.003	.002		
	10	.0074	.001	.164	.004	.002	.001	.001			
	11	.0025		.141	.001	.001					
4.0	5	.5541	.334	1.00	.501	.454	.410	.371	.336	.262	.204
	6	.2848	.142	.500	.233	.191	.156	.128	.105	.064	.039
	7	.1351	.045	.333	.100	.074	.053	.041	.030	.014	.007
	8	.0590	.015	.250	.040	.027	.018	.012	.008	.003	.001
	9	.0238	.205	.200	.014	.009	.005	.003	.002	.002	
	10	.0088	.001	.167	.005	.003	.001	.001			
	11	.0030		.143	.002	.001					
4.1	5	.5933	.659	1.11	.542	.496	.453	.414	.378	.302	.241
	6	.3098	.263	.526	.256	.212	.176	.145	.120	.075	.046
	7	.1496	.052	.343	.112	.084	.063	.047	.035	.017	.008
	8	.0667	.017	.256	.045	.031	.021	.014	.009	.004	.001
4.1	9	.0274	.006	.204	.017	.010	.006	.004	.002	.001	
	10	.0104	.002	.169	.006	.003	.002	.001	.001		
	11	.0036	.001	.145	.002	.001					
	12	.0012		.127	.001						
4.2	5	.6338	.792	1.25	.585	.540	.499	.460	.425	.348	.285
	6	.3360	.187	.556	.281	.234	.196	.164	.137	.087	.056
	7	.1651	.059	.357	.125	.094	.071	.054	.041	.020	.010
	8	.0749	.020	.263	.051	.035	.024	.016	.011	.004	.002
	9	.0314	.007	.208	.019	.012	.007	.005	.003	.001	
	10	.0122	.002	.172	.007	.004	.002	.001	.001		
	11	.0044	.001	.147	.002	.001	.001				
	12	.0015		.128	.001						
4.3	5	.6755	.965	1.43	.630	.587	.548	.511	.476	.400	.335
	6	.3634	.214	.588	.307	.259	.218	.184	.155	.102	.066
	7	.1815	.067	.370	.139	.106	.081	.062	.047	.024	.012
	8	.0839	.023	.270	.058	.040	.028	.019	.013	.005	.002
	9	.0358	.008	.213	.022	.014	.009	.005	.003	.001	
	10	.0142	.002	.175	.008	.005	.003	.001	.001		
	11	.0052	.001	.149	.003	.001	.001				
	12	.0018		.130	.001						

A = Offered load in Erlangs
N = Number of trunks (servers) required
P_d = Probability that a call will be delayed

D1 = Average delay of all calls
D2 = Average delay of delayed calls

APPENDIX A-3—Continued

					Probability of Delay (All Calls) of Time "t"						
A	*N*	P_d	D_1	D_2	*.2*	*.4*	*.6*	*.8*	*1*	*2*	*3*
4.4	5	.7184	1.20	1.67	.677	.637	.600	.565	.532	.458	.394
	6	.3919	.245	.625	.334	.285	.243	.207	.176	.118	.079
	7	.1988	.076	.385	.153	.118	.091	.070	.054	.028	.015
	8	.0935	.026	.278	.065	.046	.032	.022	.015	.006	.003
	9	.0407	.009	.217	.026	.016	.010	.006	.004	.001	
	10	.0164	.003	.179	.009	.005	.003	.002	.001		
	11	.0061	.001	.152	.003	.002	.001				
	12	.0021		.132	.001						
4.5	5	.7625	1.52	2.00	.725	.690	.656	.625	.594	.524	.462
	6	.4217	.281	.667	.363	.312	.269	.231	.199	.137	.094
	7	.2172	.087	.400	.169	.132	.103	.080	.062	.033	.018
	8	.1039	.030	.286	.073	.052	.036	.026	.018	.008	.003
	9	.0460	.010	.222	.029	.019	.012	.008	.005	.002	.001
	10	.0189	.003	.182	.011	.006	.004	.002	.001		
	11	.0072	.001	.154	.004	.002	.001	.001			
	12	.0026		.133	.001	.001					
4.6	5	.8078	2.02	2.50	.776	.746	.716	.688	.661	.598	.541
	6	.4525	.323	.714	.393	.342	.297	.258	.225	.158	.112
	7	.2366	.099	.417	.186	.146	.115	.091	.071	.039	.021
	8	.1150	.034	.294	.082	.058	.041	.030	.021	.009	.004
	9	.0519	.012	.227	.033	.022	.014	.009	.006	.002	.001
	10	.0217	.004	.185	.013	.007	.004	.003	.001		
	11	.0084	.001	.156	.004	.002	.001	.001			
	12	.0031		.135	.001	.001					
4.7	5	.8542	2.85	3.33	.829	.804	.781	.758	.735	.682	.633
	6	.4846	.373	.769	.426	.374	.328	.288	.253	.183	.132
	7	.2570	.112	.435	.204	.162	.129	.102	.081	.066	.026
	8	.1269	.038	.303	.091	.066	.047	.034	.024	.011	.005
	9	.0582	.014	.233	.038	.025	.016	.010	.006	.002	.001
	10	.0248	.005	.189	.015	.009	.005	.003	.002		
	11	.0098	.002	.159	.005	.003	.001	.001			
	12	.0036		.137	.002	.001					
	13	.0012		.120	.001						

A = Offered load in Erlangs
N = Number of trunks (servers) required
P_d = Probability that a call will be delayed

D1 = Average delay of all calls
D2 = Average delay of delayed calls

APPENDIX A-Concluded

					Probability of Delay (All Calls) of Time "t"					
A	*N*	P_d	D_1	D_2	*.2*	*.4*	*.6*	*.8*	*1*	*2*	*3*
4.8	5	.9017	4.51	5.00	.884	.866	.849	.832	.816	.776	.738
	6	.5178	.431	.833	.459	.407	.361	.320	.284	.211	.156
	7	.2783	.127	.455	.223	.179	.144	.115	.093	.053	.031
	8	.1395	.044	.312	.101	.074	.053	.039	.028	.013	.006
	9	.0651	.015	.238	.043	.028	.018	.012	.008	.003	.001
	10	.0282	.005	.192	.017	.010	.006	.004	.002	.001	
	11	.0114	.002	.161	.006	.003	.002	.001	.001		
	12	.0043	.001	.139	.002	.001					
	13	.0015		.122	.001						
4.9	5	.9503	9.50	10.0	.941	.931	.922	.913	.904	.882	.860
	6	.5521	.502	.909	.495	.443	.397	.356	.319	.242	.184
	7	.3007	.143	.476	.244	.198	.160	.130	.105	.062	.037
	8	.1530	.049	.323	.112	.082	.060	.044	.032	.015	.007
	9	.0725	.018	.244	.048	.032	.021	.014	.009	.003	.001
	10	.0320	.006	.196	.019	.012	.007	.004	.002	.001	
	11	.0131	.002	.164	.007	.004	.002	.001	.001		
	12	.0050	.001	.141	.002	.001	.001				
	13	.0018		.123	.001						
5.0	6	.5875	.588	1.00	.532	.481	.435	.394	.356	.278	.216
	7	.3241	.162	.500	.265	.217	.178	.146	.119	.072	.044
	8	.1673	.056	.333	.124	.092	.068	.050	.037	.018	.008
	9	.0805	.020	.250	.054	.036	.024	.016	.011	.004	.001
	10	.0361	.007	.200	.022	.013	.008	.005	.003	.001	
	11	.0151	.003	.167	.008	.005	.002	.001	.001		
	12	.0059	.001	.143	.003	.001	.001				
	13	.0021		.125	.001						
5.1	6	.6241	.693	1.11	.570	.521	.476	.435	.398	.318	.254
	7	.3486	.183	.526	.288	.238	.197	.163	.135	.084	.052
	8	.1824	.063	.345	.136	.102	.076	.057	.043	.021	.010
	9	.0891	.023	.256	.060	.041	.028	.019	.013	.005	.002
	10	.0406	.008	.204	.025	.015	.009	.006	.004	.001	

A = Offered load in Erlangs
N = Number of trunks (servers) required
P_d = Probability that a call will be delayed

D1 = Average delay of all calls
D2 = Average delay of delayed calls

APPENDIX B

FINANCIAL TABLES

Compound Sum of $1

Period	*1%*	*2%*	*3%*	*4%*	*5%*	*6%*	*7%*	*8%*
1	1.0100	1.0200	1.0300	1.0400	1.0500	1.0600	1.0700	1.0800
2	1.0201	1.0404	1.0609	1.0816	1.1025	1.1236	1.1449	1.1664
3	1.0303	1.0612	1.0927	1.1249	1.1576	1.1910	1.2250	1.2597
4	1.0406	1.0824	1.1255	1.1699	1.2155	1.2625	1.3108	1.3605
5	1.0510	1.1041	1.1593	1.2167	1.2763	1.3382	1.4026	1.4693
6	1.0615	1.1262	1.1941	1.2653	1.3401	1.4185	1.5007	1.5869
7	1.0721	1.1487	1.2299	1.3159	1.4071	1.5036	1.6058	1.7138
8	1.0829	1.1717	1.2668	1.3686	1.4775	1.5938	1.7182	1.8509
9	1.0937	1.1951	1.3048	1.4233	1.5513	1.6895	1.8385	1.9990
10	1.1046	1.2190	1.3439	1.4802	1.6289	1.7908	1.9672	2.1589
11	1.1157	1.2434	1.3842	1.5395	1.7103	1.8983	2.1049	2.3316
12	1.1268	1.2682	1.4258	1.6010	1.7959	2.0122	2.2522	2.5182
13	1.1381	1.2936	1.4685	1.6651	1.8856	2.1329	2.4098	2.7196
14	1.1495	1.3195	1.5126	1.7317	1.9799	2.2609	2.5785	2.9372
15	1.1610	1.3459	1.5580	1.8009	2.0789	2.3966	2.7590	3.1722
16	1.1726	1.3728	1.6047	1.8730	2.1829	2.5404	2.9522	3.4259
17	1.1843	1.4002	1.6528	1.9479	2.2920	2.6928	3.1588	3.7000
18	1.1961	1.4282	1.7024	2.0258	2.4066	2.8543	3.3799	3.9960
19	1.2081	1.4568	1.7535	2.1068	2.5270	3.0256	3.6165	4.3157
20	1.2202	1.4859	1.8061	2.1911	2.6533	3.2071	3.8697	4.6610
21	1.2324	1.5157	1.8603	2.2788	2.7860	3.3996	4.1406	5.0338
22	1.2447	1.5460	1.9161	2.3699	2.9253	3.6035	4.4304	5.4365
23	1.2527	1.5769	1.9736	2.4647	3.0715	3.8197	4.7405	5.8715
24	1.2697	1.6084	2.0328	2.5633	3.2251	4.0489	5.0724	6.3412
25	1.2824	1.6406	2.0938	2.6658	3.3864	4.2919	5.4274	6.8485
26	1.2953	1.6734	2.1566	2.7725	3.5557	4.5494	5.8074	7.3964
27	1.3082	1.7069	2.2213	2.8834	3.7335	4.8223	6.2139	7.9881
28	1.3213	1.7410	2.2879	2.9987	3.9201	5.1117	6.6488	8.6271
29	1.3345	1.7758	2.3566	3.1187	4.1161	5.4184	7.1143	9.3173
30	1.3478	1.8114	2.4273	3.2434	4.3219	5.7435	7.6123	10.0627
31	1.3613	1.8476	2.5001	3.3731	4.5380	6.0881	8.1451	10.8677
32	1.3749	1.8845	2.5751	3.5081	4.7649	6.4534	8.7153	11.7371
33	1.3887	1.9222	2.6523	3.6484	5.0032	6.8406	9.3253	12.6760
34	1.4026	1.9607	2.7319	3.7943	5.2533	7.2510	9.9781	13.6901
35	1.4166	1.9999	2.8139	3.9461	5.5160	7.6861	10.6766	14.7853
36	1.4308	2.0399	2.8983	4.1039	5.7918	8.1473	11.4239	15.9682
37	1.4451	2.0807	2.9852	4.2681	6.0814	8.6361	12.2236	17.2456
38	1.4595	2.1223	3.0748	4.4388	6.3855	9.1543	13.0793	18.6253
39	1.4741	2.1647	3.1670	4.6164	6.7048	9.7035	13.9948	20.1153

Compound Sum of $1 (CVIF)
This table shows the sum to which an initial investment of one dollar will grow over a period of years. To use this table, select the interest rate from the top row and read down the column to the appropriate number of years. For example, one dollar at 10 percent grows to $2.59 in ten years.

COMPOUND SUM OF $1—Continued

Period	*9%*	*10%*	*11%*	*12%*	*13%*	*14%*	*15%*
1	1.0900	1.1000	1.1100	1.1200	1.1300	1.1400	1.1500
2	1.1881	1.2100	1.2321	1.2544	1.2769	1.2996	1.3225
3	1.2950	1.3310	1.3676	1.4049	1.4429	1.4815	1.5209
4	1.4116	1.4641	1.5181	1.5735	1.6305	1.6890	1.7490
5	1.5386	1.6105	1.6851	1.7623	1.8424	1.9254	2.0114
6	1.6771	1.7716	1.8704	1.9738	2.0820	2.1950	2.3131
7	1.8280	1.9487	2.0762	2.2107	2.3526	2.5023	2.6600
8	1.9926	2.1436	2.3045	2.4760	2.6584	2.8526	3.0590
9	2.1719	2.3579	2.5580	2.7731	3.0040	3.2519	3.5179
10	2.3674	2.5937	2.8394	3.1058	3.3946	3.7072	4.0456
11	2.5804	2.8531	3.1518	3.4785	3.8359	4.2262	4.6524
12	2.8127	3.1384	3.4985	3.8960	4.3345	4.8179	5.3503
13	3.0658	3.4523	3.8833	4.3635	4.8980	5.4924	6.1528
14	3.3417	3.7975	4.3104	4.8871	5.5348	6.2613	7.0757
15	3.6425	4.1772	4.7846	5.4736	6.2543	7.1379	8.1371
16	3.9703	4.5950	5.3109	6.1304	7.0673	8.1372	9.3576
17	4.3276	5.0545	5.8951	6.8660	7.9861	9.2765	10.7613
18	4.7171	5.5599	6.5436	7.6900	9.0243	10.5752	12.3755
19	5.1417	6.1159	7.2633	8.6128	10.1974	12.0557	14.2318
20	5.6044	6.7275	8.0623	9.6463	11.5231	13.7435	16.3665
21	6.1088	7.4002	8.9492	10.8038	13.0211	15.6676	18.8215
22	6.6586	8.1403	9.9336	12.1003	14.7138	17.8610	21.6447
23	7.2579	8.9543	11.0263	13.5523	16.6266	20.3616	24.8915
24	7.9111	9.8497	12.2392	15.1786	18.7881	23.2122	28.6252
25	8.6231	10.8347	13.5855	17.0001	21.2305	26.4619	32.9190
26	9.3992	11.9182	15.0799	19.0401	23.9905	30.1666	37.8568
27	10.2451	13.1100	16.7386	21.3249	27.1093	34.3899	43.5353
28	11.1671	14.4210	18.5799	23.8839	30.6335	39.2045	50.0656
29	12.1722	15.8631	20.6237	26.7499	34.6158	44.6931	57.5755
30	13.2677	17.4494	22.8923	29.9599	39.1159	50.9502	66.2118
31	14.4618	19.1943	25.4104	33.5551	44.2010	58.0832	76.1435
32	15.7633	21.1138	28.2056	37.5817	49.9471	66.2148	87.5651
33	17.1820	23.2252	31.3082	42.0915	56.4402	75.4849	100.6998
34	18.7284	25.5477	34.7521	47.1425	63.7774	86.0528	115.8048
35	20.4140	28.1024	38.5749	52.7996	72.0685	98.1002	133.1755
36	22.2512	30.9127	42.8181	59.1356	81.4374	111.8342	153.1519
37	24.2538	34.0039	47.5281	66.2318	92.0243	127.4910	176.1246
38	26.4367	37.4043	52.7562	74.1797	103.9874	145.3397	202.5433
39	28.8160	41.1448	58.5593	83.0812	117.5058	165.6873	232.9248

Present Value of $1

Period	*1%*	*2%*	*3%*	*4%*	*5%*	*6%*	*7%*	*8%*
1	0.9901	0.9804	0.9709	0.9615	0.9524	0.9434	0.9346	0.9259
2	0.9803	0.9612	0.9426	0.9246	0.9070	0.8900	0.8734	0.8573
3	0.9706	0.9423	0.9151	0.8890	0.8638	0.8396	0.8163	0.7938
4	0.9610	0.9238	0.8885	0.8548	0.8227	0.7921	0.7629	0.7350
5	0.9515	0.9057	0.8626	0.8219	0.7835	0.7473	0.7130	0.6806
6	0.9420	0.8880	0.8375	0.7903	0.7462	0.7050	0.6663	0.6302
7	0.9327	0.8706	0.8131	0.7599	0.7107	0.6651	0.6227	0.5835
8	0.9235	0.8535	0.7894	0.7307	0.6768	0.6274	0.5820	0.5403
9	0.9143	0.8368	0.7664	0.7026	0.6446	0.5919	0.5439	0.5002
10	0.9053	0.8203	0.7441	0.6756	0.6139	0.5584	0.5083	0.4632
11	0.8963	0.8043	0.7224	0.6496	0.5847	0.5268	0.4751	0.4289
12	0.8874	0.7885	0.7014	0.6246	0.5568	0.4970	0.4440	0.3971
13	0.8787	0.7730	0.6810	0.6006	0.5303	0.4688	0.4150	0.3677
14	0.8700	0.7579	0.6611	0.5775	0.5051	0.4423	0.3878	0.3405
15	0.8613	0.7430	0.6419	0.5553	0.4810	0.4173	0.3624	0.3152
16	0.8528	0.7284	0.6232	0.5339	0.4581	0.3936	0.3387	0.2919
17	0.8444	0.7142	0.6050	0.5134	0.4363	0.3714	0.3166	0.2703
18	0.8360	0.7002	0.5874	0.4936	0.4155	0.3503	0.2959	0.2502
19	0.8277	0.6864	0.5703	0.4746	0.3957	0.3305	0.2765	0.2317
20	0.8195	0.6730	0.5537	0.4564	0.3769	0.3118	0.2584	0.2145
21	0.8114	0.6598	0.5375	0.4388	0.3589	0.2942	0.2415	0.1987
22	0.8034	0.6468	0.5219	0.4220	0.3418	0.2775	0.2257	0.1839
23	0.7954	0.6342	0.5067	0.4057	0.3256	0.2618	0.2109	0.1703
24	0.7876	0.6217	0.4919	0.3901	0.3101	0.2470	0.1971	0.1577
25	0.7798	0.6095	0.4776	0.3751	0.2953	0.2330	0.1842	0.1460
26	0.7720	0.5976	0.4637	0.3607	0.2812	0.2198	0.1722	0.1352
27	0.7644	0.5859	0.4502	0.3468	0.2678	0.2074	0.1609	0.1252
28	0.7568	0.5744	0.4371	0.3335	0.2551	0.1956	0.1504	0.1159
29	0.7493	0.5631	0.4243	0.3207	0.2429	0.1846	0.1406	0.1073
30	0.7419	0.5521	0.4120	0.3083	0.2314	0.1741	0.1314	0.0994
31	0.7346	0.5412	0.4000	0.2965	0.2204	0.1643	0.1228	0.0920
32	0.7273	0.5306	0.3883	0.2851	0.2099	0.1550	0.1147	0.0852
33	0.7201	0.5202	0.3770	0.2741	0.1999	0.1462	0.1072	0.0789
34	0.7130	0.5100	0.3660	0.2636	0.1904	0.1379	0.1002	0.0730
35	0.7059	0.5000	0.3554	0.2534	0.1813	0.1301	0.0937	0.0676
36	0.6989	0.4902	0.3450	0.2437	0.1727	0.1227	0.0875	0.0626
37	0.6920	0.4806	0.3350	0.2343	0.1644	0.1158	0.0818	0.0580
38	0.6852	0.4712	0.3252	0.2253	0.1566	0.1092	0.0765	0.0537
39	0.6784	0.4619	0.3158	0.2166	0.1491	0.1031	0.0715	0.0497

Present Value of $1 (PVIF)
This table shows the present value of one dollar received at a future time period. To use this table, select the interest rate from the top row and read down the column to the appropriate number of years. For example, the present value of one dollar discounted at 10 percent for ten years is $.3855.

Present Value of $1—Concluded

Period	*9%*	*10%*	*11%*	*12%*	*13%*	*14%*	*15%*
1	0.9174	0.9091	0.9009	0.8929	0.8850	0.8772	0.8696
2	0.8417	0.8264	0.8116	0.7972	0.7831	0.7695	0.7561
3	0.7722	0.7513	0.7312	0.7118	0.6931	0.6750	0.6575
4	0.7084	0.6830	0.6587	0.6355	0.6133	0.5921	0.5718
5	0.6499	0.6209	0.5935	0.5674	0.5428	0.5194	0.4972
6	0.5963	0.5645	0.5346	0.5066	0.4803	0.4556	0.4323
7	0.5470	0.5132	0.4817	0.4523	0.4251	0.3996	0.3759
8	0.5019	0.4665	0.4339	0.4039	0.3762	0.3506	0.3269
9	0.4604	0.4241	0.3909	0.3606	0.3329	0.3075	0.2843
10	0.4224	0.3855	0.3522	0.3220	0.2946	0.2697	0.2472
11	0.3875	0.3505	0.3173	0.2875	0.2607	0.2366	0.2149
12	0.3555	0.3186	0.2858	0.2567	0.2307	0.2076	0.1869
13	0.3262	0.2897	0.2575	0.2292	0.2042	0.1821	0.1625
14	0.2992	0.2633	0.2320	0.2046	0.1807	0.1597	0.1413
15	0.2745	0.2394	0.2090	0.1827	0.1599	0.1401	0.1229
16	0.2519	0.2176	0.1883	0.1631	0.1415	0.1229	0.1069
17	0.2311	0.1978	0.1696	0.1456	0.1252	0.1078	0.0929
18	0.2120	0.1799	0.1528	0.1300	0.1108	0.0946	0.0808
19	0.1945	0.1635	0.1377	0.1161	0.0981	0.0829	0.0703
20	0.1784	0.1486	0.1240	0.1037	0.0868	0.0728	0.0611
21	0.1637	0.1351	0.1117	0.0926	0.0768	0.0638	0.0531
22	0.1502	0.1228	0.1007	0.0826	0.0680	0.0560	0.0462
23	0.1378	0.1117	0.0907	0.0738	0.0601	0.0491	0.0402
24	0.1264	0.1015	0.0817	0.0659	0.0532	0.0431	0.0349
25	0.1160	0.0923	0.0736	0.0588	0.0471	0.0378	0.0304
26	0.1064	0.0839	0.0663	0.0525	0.0417	0.0331	0.0264
27	0.0976	0.0763	0.0597	0.0469	0.0369	9.0291	0.0230
28	0.0895	0.0693	0.0538	0.0419	0.0326	0.0255	0.0200
29	0.0822	0.0630	0.0485	0.0374	0.0289	0.0224	0.0174
30	0.0754	0.0573	0.0437	0.0334	0.0256	0.0196	0.0151
31	0.0691	0.0521	0.0394	0.0298	0.0226	0.0172	0.0131
32	0.0634	0.0474	0.0355	0.0266	0.0200	0.0151	0.0114
33	0.0582	0.0431	0.0319	0.0238	0.0177	0.0132	0.0099
34	0.0534	0.0391	0.0288	0.0212	0.0157	0.0116	0.0086
35	0.0490	0.0356	0.0259	0.0189	0.0139	0.0102	0.0075
36	0.0449	0.0323	0.0234	0.0169	0.0123	0.0089	0.0065
37	0.0412	0.0294	0.0210	0.0151	0.0109	0.0078	0.0057
38	0.0378	0.0267	0.0190	0.0135	0.0096	0.0069	0.0049
39	0.0347	0.0243	0.0171	0.0120	0.0085	0.0060	0.0043

APPENDIX C

SAMPLE REQUEST FOR PROPOSALS FOR PBX SYSTEM

This sample is presented as one approach to issuing a PBX RFP to vendors. Readers are permitted to copy and modify this RFP for their own use; however, neither the author, Pacific Netcom, Inc., nor the publishers makes any warranty that it will fit all cases either legally or technically, nor do the author, Pacific Netcom, Inc., or publishers intend to give legal or technical advice. Any persons who use this document do so at their own risk. The author and publishers specifically disclaim any responsibility for the results. Any company or persons using this RFP or any portion therof are required to include on the first page of the document, the following statement:

"All or portions of this document are copyrighted by Pacific Netcom, Inc, and are used with its permission."

REQUEST FOR PROPOSAL FOR PBX SYSTEM
INTRODUCTION

Insert actual information in place of italicized paragraphs or phrases.

Insert here a description of the using organization and the project. Substitute name of the issuing organization for the word "customer" throughout.

I. GENERAL INFORMATION

A. Bidding Information

Customer reserves the right to reject any or all proposals received in response to this RFP. We further reserve the right to select any part of the proposals to develop the most effective configuration of equipment.

Reprinted with permission. Copyright © 1988, Pacific Netcom, Inc.

B. Amendments to RFP

If it becomes necessary to revise any part of this RFP, amendments will be provided to all vendors who receive the basic RFP, and who attend the vendor conference.

C. Vendor Conference

A preproposal conference will be held *Enter time and place of vendor conference if held. If not, delete this paragraph.* We will hold a question and answer session, and then will visit sites. The purposes of this conference are to provide interested vendors with an opportunity to ask questions arising from their review of the system specifications and a site inspection of Customer's buildings. Questions asked, but not answered at the conference will be documented, answered in summary, and sent to all vendors who attend. Attendance at this conference is (*optional or mandatory*); no other site visits will be arranged for vendors who do not attend.

D. Response Date

In order to be considered, sealed proposals must arrive at *Enter time and place of proposal delivery.*

E. Schedule of Events

The schedule of events for this RFP is: *Enter dates in the table below. Add or delete events as necessary*

	Date
RFP issued	
Vendor Conference	
Proposals due	
Evaluation of proposals complete	
Contract negotiations complete	
Equipment Orders Placed	
Installation Complete	

F. Proposal Format

To be considered for selection, vendors must submit a complete response to this RFP as shown in Part V. Proposal Format. Pages must be numbered consecutively, and submitted in ____copies.

Multiple proposals to this RFP are encouraged. We specifically solicit alternative ways of providing service if they can meet the capacity requirements. At least one of the proposals must be complete and comply with all instructions of this RFP, with the other proposals to be evaluated as variations of the original. The additional proposals may be abbreviated following the same format. They need provide only information that differs in any way from that contained in the complete proposal.

Proposals should be prepared simply and economically, providing a straightforward, concise description of vendor capabilities to satisfy the requirements of the RFP. Emphasis should be on clarity and completeness of content.

G. Acceptance of Proposal Content

The successful vendor will be expected to enter into a contract based on proposals made and the terms and conditions of the RFP. Failure to complete negotiations will result in disqualification of the bid.

The contents of the proposal of the selected vendor, specifically including those parts that deal with contractual requirements, purchase prices, and official published specifications will become contractual obligations, and will be made part of the final contract. If any discrepancy arises between the vendor's contract and the proposal, the terms in the proposal shall prevail. Failure of the selected vendor to accept these obligations in a purchase agreement, purchase order, delivery order, or similar acquisition instrument may result in cancellation of the award.

H. Price Increases

Vendors must affirm that the purchase price of the system stated in the proposal will be valid for a minimum of ____days after the proposal is submitted. Vendors shall be obligated to inform Customer of any price increases that occur within ____days of submitting the proposal or until Customer announces selection of a vendor, whichever is less.

I. Performance Bond

Upon execution of a contract for purchase of a system, the vendor shall, if requested, furnish and maintain until the system has been accepted by Customer, a performance bond in the amount of the purchase price of the system. In lieu of a performance bond the parties may agree to withholding of ____% of the purchase price until the system has been operating without major failure for ____days.

J. Insurance

The vendor shall carry liability insurance against any and all claims for damage to person or property that may arise out of performance of this contract in limits of not less than $________ for general liability and not less than $________ for property damage. Certificates of insurance shall be filed with Customer, and shall be subject to its approval as to the adequacy of the protection.

K. Prime Contractor Responsibilities

The successful vendor will be required to assume responsibility for delivery, installation, engineering, and maintenance of all equipment, software, and support services offered in this proposal, whether or not it is the manufacturer or producer of them. Further, Customer will consider the vendor to be the sole point of contact with regard to contractual matters, including payment of any and all charges resulting from lease or purchase of the entire system.

L. Permits, Licenses, Ordinances, and Regulations
Any and all fees that pertain to the installed system and the work of any vendor required by state, county, or city laws will be paid by the vendor. All other applicable permits or fees required by laws, ordinances, and regulations shall also be paid by the vendor. The vendor must give all notices necessary in connection therewith.

All work of the vendor shall comply with local and other governing ordinances, codes, and regulations, including all OSHA regulations and requirements. No claims for additional payment will be approved for changes required to comply with codes, ordinances, and regulations in effect on the date of the installation.

M. Proprietary Information
This RFP contains information that is proprietary to Customer. No use is to be made of this information except for responding to the RFP. Customer will make every effort to keep confidential any information vendors submit in response to this RFP if the information is clearly marked as confidential.

N. Payment to Subcontractors
Before any bills including payment to subcontractors are submitted to Customer for payment, the vendor shall submit documentary evidence that subcontractors have been paid.

II. SYSTEM REQUIREMENTS AND SPECIFICATIONS
This part of the request for proposals presents system requirements for the hardware and software that will comprise the PBX and telecommunications system for Customer. It also presents background information that may be useful to vendors in responding fully to our needs.

A. System Objectives
This project intends to accomplish the following objectives:
Enumerate system objectives

B. General System Characteristics
Describe system characteristics. Samples are listed below.
The system to be acquired by Customer must have the following characteristics:

1. A system architecture that supports simultaneous voice and data communication over twisted pair wire. For the purposes of this RFP, simultaneous voice and data communication is defined as a system that permits both voice and data communication to occur without mutual interference over the same communication path between the station and the PBX. It does not imply that single-pair wire is

required or that the signals cannot be split into separate ports at the PBX.

2. All telephone instruments to be DTMF dialing.
3. System must support optional workstation interfaces that allow access to PBX ports without the use of modems.
4. System must be designed for an objective service level of P.01 on all trunks.
5. Provision of usage statistics on all classes of lines, trunks, common equipment, consoles, and service circuits without special study by vendor.
6. Ease of feature use, both from single line conventional telephones and from proprietary instruments.
7. System management and station message detail recorder features, including a full range of restriction features, and recording of details on all outward calls.
8. Full support by the vendor's maintenance forces with a guaranteed response time in case of major system failure.
9. Ease of administration, including a customer interface that enables access to traffic usage information and supports user administered adds, moves, and changes.

C. System Configuration

The system must be engineered to accommodate the "equipped" requirements at cutover. "Wired" requirements reflect those needed within ten years and should include cabinets, shelves, common equipment and memory, but not the line or trunk cards beyond those equipped. The proposal must also indicate the maximum capacity in lines, trunks, or ports of the type and model of the proposed PBX.

Enter system configuration. Suggested format follows.

	Equipped	*Wired*	*Capacity*
CO Trunks			
DID Trunks			
Paging Trunks			
Recorded Announcement Trunks			
WATS Trunks			
Tie Trunks			
Total Trunks			
Attendant Consoles			
Headsets			
Handsets			
Off Premise Stations			
Single Line Desk Telephones			

	Equipped	Wired	Capacity
Single Line Wall Telephones			
Small Multibutton Desk Telephones*			
Large Multibutton Desk Telephones*			
Display Telephones			
Integrated Voice/Data Terminals			

* Define the number of buttons for feature telephones.

D. Functional Requirements

The following are functional requirements of the PBX. Features are listed as mandatory (M) and desirable (D). Desirable features are weighted on a scale of 1 (low) to 5 (high). Vendors must state the optional cost, if any, of desirable features in the response sheet included in Appendix ____. Feature definitions in this section follow the definitions in Appendix ____. Vendors are requested to describe any features proposed but not meeting this definition.

List features as desirable or mandatory. Weight desirables on a scale of 1 to 5 as shown in samples below. Feature list should conform to the definitions that are included in Appendix D of this book.

Feature	
Add-on Conference	M
Alarm Display on Attendant Console	M
Alphanumeric Display for Attendant Console	D-5
Area Code Restriction	M
Attendant Busy Lamp Field	D-2
Attendant Camp-on	M
Attendant Controlled Conference	M
Attendant Control of Trunk Group Access	D-5
Attendant Direct Paging	D-5
Attendant Direct Station Selection	D-3
Attendant Trunk Group Selection	D-3
Attendant Forced Release	D-1
Attendant Incoming Call Control	D-4
Attendant Override	D-3
Attendant Recall	M
Attendant Release Loop	D-4
Attendant Station Number Display	D-2
Attendant Transfer of Incoming Call	M
Attendant Transfer-Outgoing	D-3
Attendant Trunk Busy Lamp Field	D-5
Attendant Trunk Group Busy Lamps	D-5

Automatic Attendant Recall	D-3
Automatic Callback-Calling	D-4
Automatic Circuit Assurance	D-5
Automatic Recall	M
Automatic Redial	D-3
Battery Backup	M
Busy Verification of Trunks	D-5
Call Forwarding All Calls	M
Call Forwarding Busy Line	M
Call Forwarding Don't Answer	M
Call Hold	M
Call Park	D-5
Call Pickup	M
Call Transfer	M
Call Waiting Service	D-5
Call Warning Tone	D-5
Calling Number Display to Station	D-3
Class of Service Display to Attendant	D-5
Class of Service Intercept	D-3
Code Call	D-3
Code Restriction	M
Conference Calling	D-5
Consultation Hold	M
Customer Maintenance and Administration Panel	M
Deluxe Queuing	D-5
Dial Dictation	D-2
Digital Line Interface	M
Digital Telephones	D-2
Digital Trunk Interface	M
Direct-in Lines	M
Direct Inward Dialing (DID)	M
Direct Inward System Access (DISA)	D-5
Direct Outward Dialing (DOD)	M
Direct Trunk Group Selection	D-3
Distinctive Ringing	D-4
Don't Answer Recall	D-3
Do Not Disturb	D-5
Emergency Access	D-5
Emergency Dialing	M
Executive Override	D-5
Flexible Intercept	D-4
Flexible Night Transfer	D-4
Hands-Free Station (selected stations)	M
Hot Line Service	D-1
Inward Restriction	D-2
Key System Features (selected stations)	M
Last Number Redial	D-5
Least Cost Routing	M
Line Lockout With Warning	M
Listed Directory Number (LDN)	M

Master Number Hunting	D-5
Meet-me Conference	D-5
Message Waiting	M
Modem Pooling	D-5
Night Service Automatic Switching	D-3
Night Service	M
Nonconsecutive Hunting	M
On-hook Dialing	D-5
Outgoing Station Restriction	M
Outgoing Trunk Queuing	M
Power Failure Transfer	M
Priority Trunk Queuing	M
Recorded Announcement Service	D-5
Remote Maintenance	D-5
Remote Traffic Measurement	D-5
Self Test and Fault Isolation	M
Speakerphone	M
Speed Dialing	M
Station Hunting	M
Station Message Detail Recording (SMDR)	M
Station Transfer Security	M
Stutter Dial Tone	M
Temporary Station Disconnection	M
Tie Trunk Access	M
Timed Reminder	D-5
Toll Restriction	M
Total System Paging	M
Traffic Data To Customer	M
Trunk Answer Any Station	M
Trunk Queuing	M
Trunk Reservation	M
Trunk-to-Tie Trunk Connections	D-5
Trunk Verification by Customer	D-5
Vacant Code Intercept	M
Vacant Number Intercept	M
Voice Mail	D-5
Zone Paging	M

E. Technical Requirements

These requirements address the compatibility of the PBX with the Customer's existing systems and address its need for flexibility to accommodate future applications.

Enter technical requirements. Samples shown below.

1. The switching matrix must provide virtually nonblocking access for all station users.

2. The system must be capable of directly interfacing another PBX over a T-1 line without using channel banks, even if it is not initially equipped to do so.
3. The system must be capable of being synchronized from a master clock signal from a distant PBX or central office.
4. The system must have a universal wiring plan with a minimum of four pair twisted pair wire at every user location. This wiring plan shall support both voice and data over a radius of at least 2500 feet. Station wiring shall terminate on _______ type connectors or equivalent at the junction point between station wiring and the PBX. Station wiring shall be terminated in _______ jacks at the station end of the connection. All station wiring and PBX line and trunk appearances shall be designated clearly for customer administration. An initial record of station wiring crossconnections shall be furnished.
5. The PBX shall have a customer maintenance and administration panel or an equivalent method of providing management information reports on trunk, port, and feature usage. The system shall allow Customer personnel to add, change, or delete features and stations as necessary. The method used shall provide hourly traffic measurements. If measurements are not automatically polled, system shall retain the peak value for a 24 hour period until polled.
6. The system must allow asynchronous terminals to communicate with the host computer and with each other without the use of modems over a radius of at least 500 feet. Stations shall be capable of communicating at a minimum of 19.2 kb/s asynchronous, and 56 kb/s synchronous. The system must be capable of connecting to any IBM-compatible personal computer over an RS-232-C interface.
7. The system must include modem pooling.
8. To support direct terminal-to-terminal communications with a minimum amount of time for call setup, the system should support some form of abbreviated dialing plan.
9. The prices quoted shall include all equipment recommended by the manufacturer to protect the system from damage caused by external sources. This equipment shall include, if necessary, gas tube protection on lines and trunks, power line voltage regulation, and surge protection equipment.
10. After installation is complete, the vendor shall perform a complete set of installation and acceptance tests as recommended by the manufacturer. Customer shall be permitted to witness the comple-

tion of these tests, and shall be furnished with a certified copy of the results. For the 30 consecutive days following completion of the test, each system shall operate with at least 99.9 percent availability.

11. PBX shall be equipped with battery backup with a minimum of ____ hours reserve.
12. The vendor shall provide a certified inventory of all station and PBX equipment and software together with the location and identification of the equipment so Customer can maintain an inventory of the equipment.
13. Station message detail recorders shall be provided. Call details shall be stored on floppy disk, tape, or other semi-permanent medium so that power outages do not result in loss of call records. Vendor shall specify a call accounting system that provides the following data:
    a) Distribution of costs to account code based on originating station.
    b) Summary of calls by area code, number of calls, minutes of usage, and cost.
    c) V&H table built in for rating calls.
    d) Ability to track calls placed to specified numbers.
    e) Ability to extract traffic statistics from the call accounting system. Note: If vendor does not handle such a call accounting system, it is not necessary to quote prices. It is necessary only to specify compatible equipment that meets the requirements.
14. The system must be able to display digits on selected telephone instruments or use a voice callback unit to retrieve waiting messages.

F. Maintenance and Support Requirements

It is Customer's intent to manage the PBX itself to the fullest extent possible. However, we expect to use vendor support on certain outage conditions, and therefore the level of maintenance and design support offered are key criteria in our evaluation of bid responses.

The following are mandatory requirements:

1. Maintenance Plan--Each vendor must submit a comprehensive maintenance plan that addresses the following:
    a) Overall maintenance philosophy of the manufacturer and vendor. Describe level of support provided locally and by manufacturer.
    b) Preventive maintenance required for each system element, and extent of impact on system operation.

c) Remote maintenance capabilities including availability and cost of services from remote maintenance and administration center.
d) Parts stocking plan including a guaranteed interval for obtaining emergency replacements and guaranteed continuing availability of replacement parts for a minimum of five years after manufacture of the system is discontinued.
e) Experience and qualifications of vendor's maintenance force. List each person trained on the system, work location, years experience, and average number of systems maintained per technician.
f) Complete warranty information on the system. A copy of the warranty statement for each system element must be included with the response.

2. Ongoing Support--The price quoted for the system must include the cost of warranty support for all new equipment during the first year of operation of the system. Each vendor must also quote the price for on-going support after the expiration of the warranty. Prices should be quoted separately for each of the following items if the service is available. If not available, so note in the response. Quote prices for both a maintenance contract and hourly rates.
    a) Repair of all failed switching equipment. Note: Customer will assume responsibility for dealing with the telephone company for pursuing repairs of commercial telephone service.
    b) Preventive maintenance program if required.
    c) Remote diagnostics of system. Customer personnel will replace failed circuit cards under direction of vendor's maintenance technicians.
    d) Total system administration including monitoring usage and recommending additions of trunks or equipment to maintain objective service levels.
3. Support Plan--Each vendor is required to submit a support plan that addresses at least the following:
    a) Project schedule from time of contract acceptance to turnover. Vendor is expected to provide a critical path chart specifying each major step of the project.
    b) Restoration following catastrophic failures. A catastrophic failure is defined as one or more of the following:
        (1) A total processor failure
        (2) Any failure that affects more than 25 percent of the lines in the system
        (3) Any failure that affects more than 25 percent of the trunks in the system

(4) Any failure that affects more than 25 percent of the service circuits in the system.

c) Documentation provided to Customer --The system must include the following minimum levels of documentation:

(1) Station user's manuals
(2) System test procedures manuals
(3) System maintenance manuals
(4) Attendant console operating manual
(5) System description
(6) System cabling and crossconnect diagrams
(7) System wiring list showing all equipped features and options

4) User training plan--The system must include manuals and training materials for station users and attendants. Vendors must explain how training will be accomplished, and what cost, if any, will be incurred by Customer. Vendors must also specify what type of ongoing training support is available after the cutover. Plan must also outline training program to permit Customer personnel to perform adds, moves, changes, and to replace PBX station and trunk cards after the first year.

5) Dialing plan--The vendor shall include a plan for dialing the following types of calls:

a) Intra company
b) Local telephone numbers
c) Long distance calls to any equal access carrier

G. Other Cost Factors

Customer will consider all initial and recurring costs in evaluating the total economics of each proposal. These are the life cycle costs referred to in Part IV-A In addition to the recurring maintenance costs discussed in Part II-F, the following factors will also be evaluated:

1. Electric power consumed.
2. Cost of spares and of stocking replacement parts.
3. Initial and recurring costs of floor space.
4. Ongoing training and administration.
5. Cost of administering adds, moves, and changes.

III. EVALUATION PROCEDURES

A. Objectives

The objective of the evaluation will be to examine each proposal, to determine if it meets the mandatory requirements, and to judge how well it meets the elements outlined in Part II of this RFP.

B. Changes to Proposals

During the evaluation period no changes in price nor any major changes in service or equipment may be made. Vendors are required to notify Customer of any changes to system hardware or service offerings that could materially affect this proposal during the evaluation period. Any vendor wishing to amend its proposal may do so during this period; however, Customer is under no obligation to consider alterations made in the proposals in making its selection

C. Evaluation Process

The first step in the evaluation process will be to determine the degree to which proposals conform to the mandatory requirements listed in Part II. Those proposals failing to meet mandatory requirements may be rejected without further evaluation; however, Customer reserves the right to waive any mandatory requirements. Any waiver of mandatory requirements will apply equally to all products. Also, incomplete proposals may be rejected as unresponsive. The remaining proposals will be weighted according to how well they meet the mandatory and desirable components. Mandatory criteria will be weighted on how well or with what techniques the vendor meets the criteria. For example, paging is a mandatory feature that can be fulfilled in a variety of ways. Proposals will be judged on how well the paging system meets Customer's objectives of effective paging without intrusiveness.

The selection decision will be based on the following criteria: *Note: Change or add criteria as necessary; total should add to 100 percent*

1. Total life cycle cost including voice and data equipment and services. ____ %
2. Reliability of both voice and data systems ____ %
3. Vendor's support plan ____%
4. Mandatory system features. ____ %
5. Weighting of desirable system features. ____ %

IV. INFORMATION REQUESTS

This part includes specific questions about the proposed system. Responses to these questions will be used in the evaluation to aid in analyzing responses. Vendors may respond to these questions by submitting manufacturer's publications that have been clearly marked to cross reference specific questions. If responses are not identical for all locations, please so indicate.

A. Life Cycle Cost Factors

Please provide the following information about each of the proposed systems:

1. Provide a footprint and specify the floor space required at each location including power equipment, distributing frames, any work area around the system. Is additional floor space required to grow the system to its maximum configuration?

2. What is the average electric power in KWH consumed by the system, both with and without channel banks:
   a) At cutover?
   b) At the ultimate wired capacity?
3. What is the mean time between failures:
   a) Of the central processing unit?
   b) Of an electronic telephone?
   c) Of the system power supply?
   d) Of tape or disk unit used for rebooting the system?
   e) Of tape or disk unit used for storing traffic information and station message detail?
   f) How is mean time between failure from the above units computed?
   g) Has MTBF been validated with actual field results?
4. Are any of the units in 3 above (except electronic telephones) duplicated in the system proposed? If so, describe the redundancy method used (eg., shared load, hot standby, etc.)
5. Is redundancy available for any of the above units? If so, please specify the additional cost.
6. List any spares recommended for this system but not included in the purchase price. Include a list of quantities recommended, their costs, and the guaranteed amount of time required to obtain emergency replacements from another source.
7. What is the operating temperature range of the system?
   a) Is air conditioning required or recommended for its operation in this state?
   b) If air conditioning is not required, what ventilation requirements must be met?
   c) What other environmental requirements must be met in the equipment room?

B. Traffic Usage Information

Describe the types of traffic usage information that are provided by the proposed system.

1. How is traffic usage information collected? (ie., via a dedicated terminal, dial-up line, etc).
2. What equipment, if any, is required to collect and process traffic usage information?
   a) Is this equipment included in the cost of the system proposed?
   b) If not, please indicate equipment recommended and its list price.
3. What service indicators (eg. dial tone speed, trunk group usage, peg count, overflow) are measured by the system? Provide sample

copies of all such reports and a brochure, practice, or narrative describing how to collect and interpret information.

4. Does the vendor or manufacturer provide any support systems for evaluating service provided by the system? If so, please describe and specify the cost.

C. Data Integration

Describe the method of voice/data integration used by this system.

1. How many pairs of station cable are required for both a telephone and work station?
2. How many PBX ports are required for both a telephone and a work station?
3. Does the system support conventional 2500 type telephones?
4. What is thc maximum data rate that can be supported for:
   a) Asynchronous terminals?
   b) Synchronous terminals?

D. System Technical Characteristics

1. What is the maximum radius over which the system can support voice/data integration?
2. What is the maximum station loop range for a conventional 2500 type telephone?
3. What is the maximum station loop range for an electronic telephone?
4. What is the maximum station loop range for an integrated voice/data terminal?
5. Describe the method used for extending the loop beyond the maximum unaided range for all types of telephones the system supports.

E. Messaging Features

Please indicate which, if any, of thc following messaging features are available for the system proposed, and indicate the cost of adding them now or at a later time.

1. Display Telephone
2. Electronic Mail
3. Voice Messaging
4. Other (describe any other messaging features not listed above)

F. Self Diagnostic Capability

Describe the method used by the manufacturer to diagnose troubles in the system.

1. What percentage of the troubles can be diagnosed by the system down to the card level without the use of an external diagnostic center?
2. To what extent is the system capable of diagnosing line and trunk trouble?

3. What kind of man/machine interface is provided with the system for trouble diagnosis?
4. Is the system capable of performing any routine trunk tests? If so, please indicate what tests are performed and the type of responders required at the distant end of the circuit.
5. How are alarms indicated? Does the attendant or key operator have any diagnostic information displayed on the console or on a printout before calling for assistance?

G. System Architecture
1. Describe how system ports are used by lines, trunks, and service circuits.
   a) Are card slots universal--ie. can line cards be intermixed with trunk cards or must they be segregated?
   b) Do some types of terminal devices or trunks require more than one port? If so, specify which.
   c) What is the growth increment for adding station and trunk ports? What is the installed cost of each increment? Please specify both the cost of adding a shelf and adding a card.
   d) Are any elements of the architecture sensitive to load balance? If so, indicate what action is required by system administrator to keep the system in balance.
2. Describe the method used and the time required to restart the system after a power or processor failure.
   a) Are any translations lost during such a failure?
   b) Is a trained technician required to restart the system following power failure?
3. What modulation method is used in the switching network (ie. PCM, PAM, delta modulation, etc)?
4. As the system is configured, indicate what component is the limiting factor in capacity, and how much load can be added to the system as configured before that limit is reached.
5. When the capacity limit described in 4 above is reached, what action is required to increase capacity, and approximately how much does it cost?
6. What is the maximum CCS per port in the system as configured?
7. What is the total capacity in CCS of the switching network?
8. How many simultaneous connections will the system support as configured?

H. Protocol Conversion
1. Please list all data communications protocols optionally supported by the system.
2. Are X.25 gateways to public data networks available? If so, what PDNs are supported?

3. Is terminal emulation software available? If so, what terminals are supported?

I. Work Station Support

1. Are integrated voice/data work stations available? If so, please indicate the cost and features of each type.
2. Do proprietary telephones support the connection of IBM-compatible personal computers? If not, describe the method of connecting a PC to the network.

J. Call Accounting Capabilities

Please describe the methods provided with the system that management can use to control telephone usage and operating costs. Please provide copies of all station message detail recorder output reports available with the system proposed.

1. How are call detail reports processed?
2. Is special equipment that is not included in the price of this system required to process call management reports? If so, what is required and what does it cost?
3. Are compatible systems available to maintain the following:
   a) Telephone Directory
   b) Equipment inventory
   c) Feature list by station

K. System Upgrades and Changes

Please describe the manufacturer's method for keeping the system updated.

1. Please describe your policy for software issue changes?
   a) Can a user skip a software issue?
   b) What is the average charge for a software issue change?
   c) How frequently do software changes occur on the average?
   d) How long does the manufacturer support previous software issues before support is withdrawn?
   e) In the system proposed, what is the generic and issue number?
   f) When was this issue released?
   g) When is the next release scheduled?
2. What is the manufacturer's process for handling hardware changes? How are users notified?
3. Is there an annual maintenance or user's fee for software that is not included in the maintenance price? If so, indicate what the charge is.

L. Status of System Development

The intention of Customer is to obtain technology that is neither so new that it is untested, nor so old that it is likely to become obsolescent soon. If the system being proposed is a new model or modification of an existing product, this will be taken under consideration in evaluating the stability of the technology. These questions are intended to aid in making that determination. Please add any further information you believe is relevant.

1. What was the date of delivery of the first system of this type?
2. Approximately how many systems of this type have been delivered to date?
3. If this system is a modification of previous technology (as opposed to a complete redesign), please answer the following questions:
   a) What was the nomenclature of the previous system?
   b) How many of the previous systems are in service?
   c) Please describe briefly the major changes between the system being proposed and its predecessor.
4. How long will the manufacturer guarantee availability of repair parts? Submit a written statement from the manufacturer to the Customer stating this warranty.

V. PROPOSAL FORMAT

We have structured this RFP to give vendors wide latitude to demonstrate their capabilities to meet the Customer's voice and data communications needs. The system requirements outlined in Part II should be considered minimum standards. Vendors are encouraged to submit any additional information they consider appropriate or helpful for evaluating their proposals.

Please review this entire RFP carefully before responding to ensure that you understand fully all procedural, system, and contractual requirements. The vendor's conference is the appropriate forum for requesting clarification of any elements that remain unclear.

A. Documents Required

Vendors are required to submit three copies each of three documents in response to this RFP:

1. Proposal
2. Pricing Information
3. Reference Materials

B. Proposal

This document is the primary response to this RFP and must be submitted in the format described in this section. All pages must be numbered consecutively. A set of tabs to identify each part of the proposal must be inserted to facilitate quick reference. The response form, Appendix ____must be easily removable so it can be used separately.

1. General
   a) Cover letter to identify proposer and to state duration of firm price guarantee (ie., 30, 60, 90 days) and other general information.

b) Executive Summary of important features of the proposal including conclusions and general recommendations. It will also state that the proposal does or does not fully comply with the requirements of this RFP, and shall be signed by an authorized representative of the vendor.
c) A brief statement of the vendor's capabilities and experience with systems of this type.
d) Any additional information the vendor may wish to submit.

2. General Characteristics (Reference Part II-B)
Respond to each element described in Part II-B. If it would increase clarity or reduce redundancy, the response may be in the form of a reference to other parts of your proposal that address these points.
3. System Configuration
Show the configuration of the PBX. At a minimum, show the following:
   a) Quantities of each component (eg. line cards, trunk cards, DTMF receivers, etc.)
   b) Interface to T-1 lines
   c) Auxiliary apparatus such as call accounting system, etc.
   d) Software generic and issue
4. Functional requirements (Reference Part 2.4)
   a) For every system feature listed in Part 2.4, please indicate on the response form Appendix ____whether the feature is: (i) included with the system at no extra cost; (o) optional at extra cost (specify cost); (n) not available; (e) equivalent feature is available (also indicate whether (i) or (o).); (d) under development (indicate when available)
   b) For additional features you wish to propose, list them separately and code as in V-B4a above.
   c) Include separate descriptions for each feature.
   d) All central system hardware and software to implement mandatory features must be included in the quoted price. Indicate any additional central system hardware and software cost for optional features.
5. Technical Requirements (Reference Part II-E)
   a) Respond to each element described in Part II- E. If it would increase clarity or reduce redundancy, the response may be in the form of a reference to other parts of your proposal that address these points.
   b) Include diagrams and descriptive material that show the system architecture. Identify all hardware components and interfaces.
   c) Provide descriptions of each system component.

6. Maintenance and Support Requirements
   a) Submit a comprehensive maintenance plan following the outline in Part II-F1. Indicate whether this type of support is available after expiration of the warranty and the charges for this support.
   b) Submit a comprehensive plan for ongoing support of the system following the outline in Part II-F2. Indicate whether this type of support is available after expiration of the warranty and the charges for this support.
   c) Submit a support plan following the outline in Part II-F3, including cutover schedule and plan, restoration plan, and a description of documentation that will be provided with the system.
   d) Submit a statement on the guaranteed response time from the time a call is received in your office until a qualified technician reaches our premises for the following types of failures:
      (1) Catastrophic failure as defined in Part II-F3d.
      (2) Other service affecting failure
      If different levels of service support are offered, include the cost of each.
   e) Submit a training plan for users and attendants following the outline in Part II-F3d. Indicate what training is included in the price of the system, what training support is available after the initial cutover, and any charge for such training. Include at least the following for each course:
      - abstract of course
      - length of course
      - location of course
      - intended audience of course
      - number of students that can be trained at once
      - suggested course schedule
      - availability of key instructor training material
   f) Submit a plan to train one or two Customer personnel to make moves, adds, and changes, and to replace PBX station and trunk cords after the first year. Include at least the following for each course:
      - abstract of each course
      - prerequisites
      - length of course
      - location of course
      - cost of course
   g) Submit a description of the dialing plan proposed following the outline in Part II- F3f.

h) Submit a list of documentation that is available for using, installing, administering, and maintaining the system. If the cost of documentation is not included in the proposal price, indicate the price of each publication. Note: Vendors may be requested to loan sample copies of documentation for the duration of the review period.

7. Vendor Qualifications
   a) Complete the vendor qualification form in Appendix ____and return it with your proposal.
   b) Provide a history of the organization, including founding dates and when telecommunications activities began.
   c) Provide annual financial statements audited by a certified public accountant for the last three fiscal years. The vendor's annual report will suffice if the auditor's certification is included.
   d) If available, provide a copy of Dunn and Bradstreet's most current report on the vendor.
   e) Provide references who can be called, including company name, contact person, position in the company, address, and telephone number of at least three users with system configurations similar to that proposed. Include for each reference a description of their system, indicating how long it has been in service.
8. Other Costs (Reference Part II-G)
   Provide other cost factors as requested in Part IV-A.

C. Specific Information Requests (Reference Part IV)
Submit information responding to each specific question asked in this section. Because this RFP does not include every possible feature and benefit of a PBX system, vendors are encouraged to identify any other features or services that are pertinent in evaluating their systems. Costs for any additional features or capabilities must be included in the pricing information.

D. Pricing Information
Multiple proposals are encouraged, and if submitted, should be identified as the basic or alternative system. The configurations for each location must show components as follows:
1. PBX Switching System. Show all hardware and software components required for each configuration proposed. Indicate the PBX name and model, show the "equipped for" and "wired for" quantities for the various type of stations and trunks. Format the components and pricing as shown on the following page.

Description	Part #	Qty	Unit Price	Extended Price
Main Switch Cabinet	99999	1	$10,000	$10,000
Line Card Module	88888	2	$ 5,000	$10,000
Subtotal PBX Switch				
Total System (Of Location)				
Grand Total All Locations				

2. Station Equipment. Show the station equipment required to meet the requirements shown in Part II-C System Configuration. Format the information as in IV-D1 above.
3. Other Equipment. Show battery backup, SMDR, customer administration, channel banks, if needed and other special equipment required in Part II-D in the same format as V-D1 above
4. *Station Wiring*. Include cable, wiring, distributing frames, terminal blocks, station connectors, and all other components to all locations marked in Appendix ____and not presently wired.
5. *Labor.* Provide all labor required to install the systems to meet the specifications. Include the costs of training, programming, and the preparation of the initial cabling records. If these are automatically included as part of the system price, they do not have to be separately detailed.
6. *Totals*. Show grand total purchase prices for each location.

APPENDIX D

FEATURE DEFINITIONS FOR PBX AND KEY TELEPHONE SYSTEMS

Account Codes: Station users can dial or be forced to dial an account code that is printed on the station message detail recorder to identify outgoing calls.

Add-on Conference: Station user can add a third party to a conversation by flashing the switchhook to get second dial tone and dialing the additional number.

Alarm Calls Waiting Indicator: A call-waiting indicator on the attendant's console is illuminated when incoming calls are waiting to be answered.

Alarm Display on Attendant Console: System alarms are displayed on lamps or alphanumeric readout on the attendant console.

Alphanumeric Display for Attendant Console: Displays call-related information and message information in an alphanumeric readout on attendant's console.

Area Code Restriction: Prevents dialing specified area codes from restricted stations.

Attendant Busy Lamp Field: A group of lights displays the busy/idle status of all stations in the system so attendant is not required to dial number and receive busy signal to know station is busy.

Attendant Camp-on: Attendant can temporarily attach an incoming call to a busy station. Called party hears a camp-on tone. When the called party hangs up, the call automatically rings through.

Attendant Control of Trunk Group Access: Places a trunk group under exclusive control of attendant so users cannot access it directly.

Attendant Controlled Conference: Attendant can establish and monitor conference calls through a conference bridge to users both on and off the system.

Attendant Direct Paging: Attendant can access paging system from console by pushing paging access button.

Attendant Direct Station Selection: Attendant can access stations by pushing a button in a station selection field.

Attendant Direct Trunk Group Selection: Attendant can access an idle trunk directly by pressing a button associated with the trunk group.

Attendant Direct Trunk Termination: Incoming trunks are terminated directly on the attendant console as opposed to termination on a switched loop.

Attendant Forced Release: Attendant can disconnect all parties on a circuit.

Attendant Incoming Call Control: Incoming calls are automatically directed to a designated extension after a specified number of rings.

Attendant Override: Attendant can override a busy line to hold a conversation with parties.

Attendant Recall: Enables users on a conference call to recall attendant for assistance.

Attendant Release Loop: Attendant is able to place on hold the switched loop of calls that are camped on to a busy station or ringing.

Attendant Splitting: Attendant can confer privately with the called party on an attendant-handled connection without the calling party overhearing. Attendant can alternate between or conference the called and calling station parties.

Attendant Station Number Display: Displays the extension number of a call placed to the attendant within the system.

Attendant Switched Loop: Incoming trunks are switched to attendant instead of connected directly on the console. When call is connected to called station, loop is released.

Attendant Transfer-Incoming: Enables station user to signal attendant to transfer an incoming trunk call to another station.

Attendant Transfer-Outgoing: Enables station user to signal attendant to transfer an outgoing trunk call to another station.

Attendant Trunk Busy Lamp Field: Provides a lamp indication when all trunks in a particular trunk group are busy.

Attendant Trunk Group Busy Lamps: Provides a visual lamp display when all trunks in a group such as WATS and central office are busy.

Automatic Call Distribution: Provides automatic connection of incoming calls to a specific group of answering stations (split). System generates reports on the status of agents, splits, and trunks.

Automatic Callback-Calling: When a call is placed to a busy station, the caller can activate a feature that automatically rings both telephones when the called number hangs up, providing the calling station is also free. If the callback is not answered within a given time, ringing will stop and callback will be automatically cancelled.

Automatic Circuit Assurance: System records incidence of long or short holding time calls and alerts attendant of possible circuit malfunction.

Automatic Recall: Automatically alerts attendant to camped-on or unanswered call that has exceeded reminder threshold.

Battery Backup: The PBX or key system is operated from storage batteries instead of commercial power. A rectifier keeps the batteries charged. When power fails, system continues to operate without interruption.

Busy Verification of Station Lines: The attendant can enter a busy station connection within the system to determine whether it is in use or off-hook. Before the connection is permitted, a call warning tone is sent to the connected parties, after which they and the attendant will be connected in a three-way bridge.

Busy Verification of Trunks: Allows attendant to make test calls to selected trunks to determine busy/idle status.

Call Forwarding All Calls: Station user can dial a code or press a button to cause all calls to be automatically forwarded to a designated answering position.

Call Forwarding Busy Line: When called number is busy, the system automatically forwards the call to a designated position or receptionist.

Call Forwarding Don't Answer: When called number fails to answer, the system automatically forwards the call to a designated position or receptionist.

Call Forwarding Override: Permits a selected class of service to ring preselected stations that have set up call forwarding-all calls.

Call Forwarding Variable: A station user can individually program any number as a forwarding destination so that calls to the user's number will immediately forward to the programmed number.

Call Hold: Calls can be placed on hold by flashing the switchhook and pressing a button or dialing a code.

Call Park: An incoming call can be placed in a parking orbit by dialing a code. To retrieve the call at another station the called party dials the code and is connected to the caller.

Call Pickup: By dialing a code or pushing a button, a user can pickup a call that is ringing at another telephone within a designated pickup group.

Call Transfer: Incoming, and in some systems outgoing, calls can be transferred to another line by flashing the switchhook, dialing a second number, and hanging up.

Call Waiting: When a call arrives at a busy station, the called party is signaled with a tone and can then place the original call on hold with a switchhook flash and converse with the second caller.

Call Warning Tone: A tone that sounds when a long distance call is about to advance to a higher-cost route. Caller can terminate or allow it to advance.

Called Station Status Display: Provides a visual indication on a display telephone of the status of the called station such as busy or forwarded.

Calling Number Display to Station: Station user sees the number or identity of the calling party displayed on an alphanumeric readout on the telephone instrument.

Camp-on Inquiry: Enables a busy station to receive an incoming call and place the existing call on hold to answer the incoming call. The station can then alternate between both calls.

Centralized Attendant Service (CAS): Allows multiple PBX system to concentrate attendant functions at a central location.

Circular Hunting: Permits a call to be processed through a hunt group of busy extensions in a programmed order. If all stations in the hunt group are busy, the caller receives busy tone. Station numbers may be assigned to a circular hunt group in any order.

Class of Service Day/Night: Station Class of Service can be automatically altered when the system is placed in night transfer mode.

Class of Service Display to Attendant: The user's class of service, indicating restrictions and features, is displayed on the console for calls placed or routed to the attendant.

Class of Service Intercept: If station user places a call prohibited by its class of service, call is routed to attendant.

Class of Service: Permits a station to be assigned a class of service conforming to the feature and restriction level desired.

Code Call: Users within the system can be signaled by a distinctive combination of tones over a code call announcement system.

Code Restriction: The system administrator can exclude certain dialing codes so they cannot be dialed from within the system.

Conference Calling: Permits multiple parties to be interconnected by the station user for simultaneous conversation.

Consultation Hold: Station user can place a line on hold and originate a call to another station. After consultation the user can add the second station to the call.

Data Security: Permits lines used for data transmission to be protected from interruptions such as executive and attendant override, attendant camp-on, and busy verification.

Deluxe Queuing: When all trunks in a group are busy, feature places users in a queue and completes call when facility is available. Feature permits user to hang up and be automatically called by the system, or remain off hook until trunk becomes available.

Dial Dictation Access: Dictation system provided by customer and attached to PBX or KTS is accessed by dialing a special code. DTMF telephone is used to control recorder functions.

Dialed Number Identification System (DNIS) A feature used in conjunction with 800 lines to enable the PBX to route incoming calls to the appropriate answering position based on the digits dialed.

Digital Line Interface: System allows direct connection of a digital telephone to system line circuits.

Digital Telephone: A telephone instrument in which voice is converted to digital form inside the telephone instrument.

Digital Trunk Interface: System supports interface to T-1 digital line without the use of channel banks.

Direct Inward Dialing (DID): Callers from outside the system reach users directly by dialing a seven-digit telephone number. Calls are completed over a dedicated DID trunk group.

Direct Inward System Access (DISA): Callers from outside the system can gain access to station features such as tie lines and WATS lines by dialing a special access code.

Direct Station Selection (DSS): Enables attendant to access a station directly by pressing a button.

Direct Trunk Group Selection: Attendant can access a trunk in a particular group by pressing a button instead of dialing an access code.

Direct-in Lines: "Personal" lines bypass the attendant and terminate directly on user's telephone set. Lines are not available for shared access by other users.

Distinctive Ringing: Different ringing tones are used to alert caller to source of an incoming call. Tones indicate whether the source of calls is inside or outside the system, from tie line or 800 number, etc.

Do Not Disturb: Station user can divert numbers to a secondary answering position by pressing a button that indicates the called party does not want to be disturbed.

Don't Answer Recall: After a prescribed time, a call is automatically placed between a calling and called number that did not answer.

Emergency Access: All phones in a designated emergency group can be simultaneously rung by dialing a code.

Emergency Dialing: A predefined emergency code can connect station users to outside emergency services such as police and fire.

Executive Override: Selected telephone users can interrupt a call in progress and hold a conversation with the interrupted party

Flexible Intercept: Allows customer to change routing or announcement when user dials a disconnected number or vacant code.

Flexible Night Transfer: Attendant can alter night service connections as conditions change without involving vendor.

Forced Authorization Codes: The system can be programmed so that users must dial authorization codes to reach long distance facilities either in a direct inward system access (DISA) mode or a standard internal use mode.

Hands-Free Station : See Speakerphone.

Hot Line Service: Allows station to automatically place a call to a predesignated station by lifting the handset.

Intercepted Call Control: Any call that cannot be completed is automatically directed to the attendant, who is alerted that the arriving call is an intercepted call.

Inward Restriction: Station is restricted from receiving calls from outside the system.

Key System Features: Selected features such as call hold, call pickup, and speed dialing are activated by pressing a button.

Last Number Redial: The telephone instrument stores the last number dialed so the user can redial it by pressing a single button.

Least Cost Routing: Long distance calls are automatically routed over the lowest cost available facility without the caller dialing special access codes.

Line Lockout With Warning: A station permanently off-hook is locked from system access and a warning tone applied after a prescribed interval.

Listed Directory Number (LDN): Calls from outside the system to a listed directory number are automatically connected to the attendant, who can transfer the call to another station.

Maintenance and Administration Terminal: Provides customer with access to traffic usage information and facility for revising station translations that must be changed when adding, moving, or changing stations. Also allows customer to make diagnostic tests for fault isolation.

Master Number Hunting: Any number within a hunt group can be assigned as the master or pilot number without regard to its numeric sequence.

Meet-me Conference: Conference calls are established between users by dialing a conferencing number at a prescribed time. Any user can terminate or hold a connection without disrupting the rest of the conference.

Message Waiting: A lamp on the called party's telephone is illuminated or stutter dial tone is provided when the message center or voice mail is holding a message for the station user.

Modem Pooling: A pool of modems is connected to the system for shared access by any station within the system.

Music-on-hold: Provides music to a party that is placed on hold or parked to indicate that the call is still connected.

Night Service Automatic Switching: Trunks are automatically connected to stations lines if attendant fails to configure system for night service.

Night Service: Incoming trunks are transferred to specified station lines by action of attendant.

Nonconsecutive Hunting: Station numbers can be assigned to a hunting group without being in consecutive numerical order.

On-hook Dialing: Station can dial and monitor progress of telephone calls without lifting the receiver until the called party answers.

Outgoing Station Restriction: Selected stations can be denied permission to place any outgoing calls.

Outgoing Trunk Queuing: See Trunk Queuing.

Power Failure Transfer: When power fails in a system with no battery backup, central office lines are transferred to single line telephones where calls can be placed and received for the duration of the outage.

Priority Trunk Queuing: Station users are queued for access to busy trunks by priority code. High priority lines move to the head of the queue before lines with lower priority within limits programmed into the system.

Recorded Announcement Service: A prerecorded announcement can be reached by all users on the system by dialing an access code.

Remote Access: See Direct Inward System Access (DISA).

Remote Maintenance: System can be accessed from remote location to perform diagnostics and take limited corrective action, or perform station and trunk feature changes.

Remote Traffic Measurement: Traffic usage information can be collected from a remote location by dialing over an access line and downloading the information from a storage device in the PBX.

Route Advance: Completes calls over a series of alternative long distance routes according to variables programmed into the system. Differs from least cost routing in that route advance lacks the intelligence to choose the lowest cost route.

Self Test and Fault Isolation: Performs internal diagnostics and provides diagnostics on probable cause and corrective action recommended.

Speakerphone: User can initiate and receive calls by depressing a button that activates speaker/microphone system in the telephone.

Speed Dialing-Station: Station users can establish a private list of numbers that are accessed by pushing a button or dialing an abbreviated code stored in the telephone.

Speed Dialing-System: System contains lists of frequently called numbers that are dialed by an abbreviated code.

Station Hunting - Terminal: If call is placed to a busy station in a hunt group, caller receives busy tone and does not hunt to the next idle number. Does not apply to the pilot number.

Station Hunting: Calls to a busy station are routed to an idle station within the hunting group.

Station Message Detail Recording (SMDR): A device that registers dialed number, time of day, duration of call, and cost of call. Either prints calls to paper or disk for allocating usage and controlling unauthorized use of telephone service.

Station Transfer Security: When an incoming trunk call is transferred to a station and is not answered within a specified interval, it is automatically routed to the attendant.

Stutter Dial Tone: Special distinctive dial tone is reached when user flashes the switchhook to access a feature such as call transfer or add-on conferencing.

Tandem Trunk Switching: Permits an incoming tie line call to complete a trunk-to-trunk connection without attendant assistance.

Temporary Station Disconnection: Attendant can temporarily disconnect a station from the console.

Tie Trunk Access: Provides dialing level for access to inter-machine tie trunks.

Timed Reminder: Automatically signals attendant about calls camped on but not answered, not answered after transfer by station user, or placed on hold at the console.

Toll Restriction: Limits the access that certain classes of users have to long distance service. Feature can deny access to toll calls altogether, or allows access to low cost services, certain area codes, etc.

Total System Paging: Paging system can access all speakers within the paging system.

Traffic Measurement: Provides traffic usage information for customer access without intervention by centralized facility of the vendor.

Trunk Answer Any Station: Incoming trunk calls can be answered from any authorized telephone by dialing an access code when attendant position is not provided or is unattended.

Trunk Queuing: When all trunks in a group are busy, user can dial an access code to enter a queue and automatically be completed in order of preference. If user does not answer the returned call within a given number of seconds, the station is dropped from the queue.

Trunk Reservation: Allows attendant to reserve a trunk and extend it to a specific station.

Trunk Verification by Customer: Allows customer to test trunks to verify operation.

Trunk-to-Trunk Connections: Enables attendant or any station user to connect together either two central office calls or a central office and tie trunk call. Station user can hang up if desired, and trunks remain connected.

Uniform Call Distribution (UCD): Incoming calls are routed to a group of service positions, usually based on first-come-first-served to distribute the work load more or less evenly.

Uniform Dialing Plan: permits users in a multi-PBX system to reach one another over tie trunks by dialing a uniform station code.

Vacant Code Intercept: Routes calls to vacant codes such as a vacant number level to attendant or recorder.

Vacant Number Intercept: Routes calls to a vacant station number to attendant or recorder.

Voice Mail: Provides storage facility for recording messages to an unanswered station. User can forward calls to mail box or they can be transferred by attendant or other user. User can retrieve calls from DTMF telephone by dialing an access code.

Zone Paging: Paging system is divided into different zones within the PBX or KTS serving area. Zones are accessed by dialing a code or pressing a button.

APPENDIX E

AT&T WATS
SERVICE AREAS - INTERSTATE

SUBSCRIBER'S HOME STATE†	*SERVICE AREAS**					
	1	*2*	*3*	*4*	*5*	*6*
Alabama	FL,MS GA,SC KY,TN LA	AR,NC So IL IN,So OH MO	DC,No OH No IL, OK IA, TX, WV MD,VA	CT,MN,PA DE,NE,RI KS,NJ,WI MI,NY	AZ,MA,NH,NM,ND, CA,ME,OR,PR,SD, CO,MT,UT,VI,VT ID,NV,WA,WY,HI	AK
Arizona	So CA,UT CO NV NM	No CA ID We TX WY	AR,NE,Ea TX, So TX KS,OK,WA MT,OR ND,SD	AL,IN,MO,MS AR,KY,TN IL,LA,WI IA,MN	CT,DC,MI,NH,NJ,NY DE,FL,NC,OH,PA,PR GA,MA,RI,SC,VA MD,ME,VI,VT,WV,HI	AK
Arkansas	LA,TN MS, Ea TX MO OK	AL No TL,So TL KS KY	FL,NC,NE,We TX GA,OH,SC,WV IN,So TX IA,WI	CO,MI,MN DC,NC,NM DE,PA,SD MS,VA,WY	AZ,MA,ME,MT,NH CA,NV,NJ,NY,OR CT,PR,RI,VT ID,UT,VI,WA,HI	AK
California-No	AR,UT ID,WA NV OR	CO,WY MT NE NM	IA,ND KS,OK MN,SD MO,TX	AL,KY,LA AR,MI IL,MS IN,WI	CT,DC,DE,FL,GA,MA MD,NJ,NY,OH,VA,NC ME,PA,PR,RI,WV NH,SC,VI,VT,TN,HI	AK
California-So	AR,UT NV NM OR	CO,WY ID We TX WA	IA,MT,NE,Ea TX KS,NS,So TX MO,OK MN,SD	AL,KY,LA AR,MI IL,MO,MS IN,TN,WI	CT,DC,DE,FL,GA,MA MD,NH,NJ,NY,OH,PA ME,PR,RI,SC,VA NC,VI,VT,WV,HI	AK

*Each greater service area includes the preceding service area states. The subscriber's hcme state is not included.

SUBSCRIBER'S HOME STATE†	SERVICE AREAS*					
	1	2	3	4	5	6
Colorado	AZ,OK,UT KS,WY NE NM	ID,We TX MT NV SD	AR,MO,ND CA,Ea TX IA MN	IL,MS,OR IN,TN,So TX LA,WA MI,WI	AL,FL,GA,KY,MA,MD CT,ME,NC,NH,NJ,NY DC,OH,PA,PR,RI,SC DE,VA,VI,VT,WV,HI	AK
Connectiut	MA,Ne NY NH, Se NY NJ,Ea PA RI,VT	DC,We NY DE No OH,So OH,VA MD,WV ME,We PA	AL,KY,TN GA,MI,WI IL,NC IN,SC	AR,LA,ND,NE FL,MN,OK IA,MO,SD LS,MS	AZ,CA,CO,ID,MT,NM NV,TX,UT,VI OR,WA PR,WY,HI	AK
Delaware	DC,Se NY MD,Ea PA NJ VA	CT Ne NY,VT MA,We NY,WV NH,So OH,NC We PA,RI	AL,IN,KY FL,ME,MI GA,No OH IL,SC,TN	AR,MN,MO IA,MS,NE KS,OK,WI LA,ND,SD	AZ,MT,NM,NV,OR,PR CA,TX,UT,VI,WA CO,WY ID,HI	AK
District of Columbia	DE,VA MD,WV NJ Ea PA,We PA	CT Ne NY, Se NY NC No OH,So OH RI	AL,IN,KY,NH FL,MA,SC,TN GA,ME,VT IL,MI	AR,MN,MO IA,MS,SD KS,ND,WI LA,NE,OK	AZ,MT,PR,TX CA,NM,UT,VI CO,NV,WA ID,OR,WY,HI	AK
Florida	AL,NC GA,SC LA,TN MS	AR,KY,So OH DC,MD DE,VA IN,WV	IL,OK MO,PA NJ,TX No OH	CT,MI,PR,VI IA,NE,RI KS,NH,VT MA,NY,WI	AZ,CA,OR,SD,UT CO,ND,WA,MN ID,NM,WY MT,NV,ME,HI	AK

*Each greater service area includes the preceding service area states. The subscriber's home state is not included.

SUBSCRIBER'S HOME STATE†	SERVICE AREAS*					
	1	2	3	4	5	6
Georgia	AL,NC FL,SC KY,TN MS	AR,So IL IN,So OH LA,WV VA	DC,No IL,No OH DE,MD,PA MI,NJ MO,NY	CT,NE,TX,VT IA,NH,WI KS,OK MA,RI	AZ,ME,NM,ND,WY CA,MN,OR,PR CO,MT,SD,UT ID,NV,VI,WA,HI	AK
Hawaii‡	No CA,So CA OR	AZ,UT ID,WA MT NV	AK,SD CO,WY NE NM	AR,LA,OK IA,MN,TX IL,MO,WI KS,ND	AL,FL,MD CT,GA,ME DC,IN,KY,MI DE,MA,MS NC,NH,NY,PA,PR NJ,OH,RI SC,TN,VI VA,VT,WV	
Idaho	MT,WA NV,WY OR UT	No CA,So CA CO	AZ,MO,ND IA,NE,SD KS,NM,WI MN,OK	AR,LA IN,MI IL,MS KY,TX	AL,CT,DC,DE,FL,GA,MA MD,NH,NJ,NY,OH,PA ME,PR,SC,TN,VA NC,RI,VI,VT,WV,HI	AK
Illinois-No	IA,So MI IN MO WI	AR,No MI KY,TN MN,WV No OH,No OH	AL,MS,PA GA,NC,SC KS,NE,VA LA,OK	CO,ND,NJ DC,NY DE,SD MD,TX	AZ,ID,NH,NM,NV CA,MA,OR,PR,RI CT,ME,UT,VI,WY FL,MT,VT,WA,HI	AK
Illinois-So	IN,So OH IA,TN KY MO	AL,So MI AR,No OH MS WI	GA,No MI,PA KS,NC,SC LA,NE,VA MN,OK,WV	CO,ND,NJ DC,NY DE,SD MD,TX	AZ,CA,ME,MT,NH CT,MA,NM,NV,OR FL,PR,UT,VI,WA ID,RI,VT,WY,HI	AK

*Each greater service area includes the preceding service area states. The subscriber's home state is not included.

SUBSCRIBER'S HOME STATE†	SERVICE AREAS*					
	1	2	3	4	5	6
Indiana	No IL,So IL KY No OH,So OH So MI	IA,We PA MO,WV TN WI	AL,GA,No MI AR,MD,NC,NY DC,MN,Ea PA DE,MS,SC,VA	CT,MA,ND,SD,VT FL,NE,NH KS,NJ,NY LA,OK,RI	AZ,ME,MT,NM,NV CA,OR,PR,TX CO,UT,VI ID,WA,WY,HI	AK
Iowa	No IL,So IL,SD MN,WI MO NE	IN KS No MI,So MI OK	AL,LA,MS AR,ND,WV CO,OH,WY KY,TN	DC,NC,NM GA,PA,VA MD,SC MT,TX	AZ,FL,NH,NJ,NY CA,ID,NV,RI,UT CT,MA,OR,VI,VT DE,ME,PR,WA,HI	AK
Kansas	CO,OK IA,SD MO NE	AR,So IL NM,Ea TX We TX	AL,No IL,TN,WY IN,MN,So TX KY,MS,UT LA,ND,WI	AZ,MT,SC GA,NC,VA ID,NV,WV MI,OH	CA,FL,MA,MD,VT CT,ME,NH,NJ,WA DC,NY,OR,PA DE,PR,RI,VI,HI	AK
Kentucky	So IL,So OH IN,VA MO,WV TN	AL,No IL GA,No OH NC SC	AR,LA,NJ DC,MD,PA DE,MI,WI IA,MS	CT,MN,OK FL,NE,RI KS,NH,TX MA,NY,VT	AZ,ME,ND,NM CA,MT,NV,OR CO,PR,SD,UT ID,VI,WA,WY,HI	AK
Louisiana	AL,Ea TX AR,So TX MS OK	GA,So IL KY,We TX MC TN	FL,No IL,OH IN,NE,SC IA,NC,WV KS,NM	CO,MI,SD DC,MN,VA DE,NJ,WI MD,PA	AZ,MA,MT,ND,VT CA,ME,NH,NV,WA CT,NY,OR,PR,WY ID,RI,UT,VI,HI	AK

*Each greater service area includes the preceding service area states. The subscriber's home state is not included.

SUBSCRIBER'S HOME STATE†	*SERVICE AREAS**					
	1	*2*	*3*	*4*	*5*	*6*
Maine	CT,Ne NY MA,Se NY NH,RI NJ,VT	DC,We NY DE,No OH MD,VA Ea PA,We PA,WV	GA,MI,So OH IL,MN,TN IN,NC,WI KY,SC	AL,KS,ND AR,LA,NE FL,MO,OK IA,MS,SD	AZ,MT,PR,TX,UT CA,NM,VI,WA CO,NV,WY ID,OR,HI	AK
Maryland	DC,Se NY DE,Ea PA,We PA NJ,VA WV	CT,Ne NY MA,We NY NC,RI No OH,So OH	AL,IN,NH FL,KY,SC GA,ME,TN IL,MI,VT	AR,MN,NE IA,MO,OK KS,MS,SD LA,ND,WI	AZ,MT,PR,WA CA,NM,TX,WY CO,NV,UT ID,OR,VI,HI	AK
Massachusetts	CT,Ne NY ME,Se NY NH,Ea PA NJ,RI,VT	DC,No OH,So OH,We PA DE,VA MD,WV We NY	AL,IA,SC GA,KY,TN IL,MI,WI IN,NC	AR,MN,NE FL,MO,OK KS,MS,SD LA,ND	AZ,MT,PR,WA CA,NM,TX,WY CO,NV,UT ID,OR,VI,HI	AK
Michigan-No	No IL,No OH IA MN WI	So IL,We NY IN,ND,So OH KY,We PA MO	CT,MD,Ea PA DC,NC,RI,SD DE,NH,TN,VT, Ne NY MA,NJ,VA,WV,Se NY	AL,GA,MS AR,KS,NE CO,LA,OK FL,ME,SC	AZ,NM,NV,OR,PR CA,TX,UT ID,VI,WA MT,WY,HI	AK
Michigan-So	No IL,We NY IN No OH,So OH WI	DC,So IL IA,Ne NY KY,WV Ea PA,We PA	CT,MN,NJ,Se NY DE,MO,RI,VT MA,NC,SD,VA MD,NH,TN	AL,GA,KS,SC AR,LA,ME CO,MS,ND FL,NE,OK	AZ,NM,TX,UT CA,NV,VI ID,OR,WA MT,PR,WY,HI	AK

*Each greater service area includes the preceding service area states. The subscriber's home state is not included.

SUBSCRIBER'S HOME STATE†	*SERVICE AREAS**					
	1	*2*	*3*	*4*	*5*	*6*
Minnesota	IA,No IL No MI ND,SD NE,WI	So IL So MI IN,MO KS	AR,OH,WV CO,OK,WY KY,PA MT,TN	AL,MD,TX DC,MS,UT ID,NM,VA LA,NY	AZ,FL,GA,OR,RI CA,MA,ME,SC,WA CT,NC,NH,VI,PR DE,NJ,NV,VT,HI	AK
Mississippi	AL,TN,Ea TX AR GA LA	FL,So IL KY,SC MO OK	IA,No IL,So TX IN,OH,We TX KS,VA,WV NC,WI	CO,MI,NM DC,MN,PA DE,NE,SD MD,NJ	AZ,MA,ME,MT,ND CA,NH,NV,NY,WA CT,OR,PR,RI,WY ID,UT,VI,VT,HI	AK
Missouri	AR,KY No IL,So IL,NE IA,OK KS,TN	IN LA MS WI	AL,OH,SD GA,TX MI,WV MN	CO,MD,PA DC,NC,SC DE,ND,VA FL,NM,WY	AZ,MA,NH,PR,VT CA,ME,NJ,RI,WA CT,MT,NV,UT,NY ID,OR,VI,HI	AK
Montana	ID,UT ND,WA OR,WY SD	No CA CO,NV MN NE	AZ,So CA IA,NM KS,OK MO,WI	AR,MI IL,OH IN,TX KY	AL,FL,MD,ME,MS,VT CT,GA,NC,NH,NJ,WV DC,LA,NY,PA,SC,TN DE,MA,PR,RI,VA,VI,HI	AK
Nebraska	CO,SD IA,WY KS MO	No IL MN,WI ND OK	AR,So IL ID,MT,UT IN,NM MI,TX	AL,LA,MS AZ,NV,OH GA,OR,WV KY,TN	CA,FL,NC,PA,RI CT,MA,NH,SC,VA DC,MD,NJ,VI,WA DE,ME,NY,VT,PR,HI	AK

*Each greater service area includes the preceding service area states. The subscriber's home stateis not included.

SUBSCRIBER'S HOME STATE†	SERVICE AREAS*					
	1	2	3	4	5	6
Nevada	AZ,UT No CA,So CA ID OR	CO WA WY	IA,NE,TX KS,NM MT,OK ND,SD	AR,MI,TN IL,MN,WI IN,MO LA,MS	AL,FL,MD,ME,NC,WV CT,GA,NH,NJ,NY DC,KY,OH,PR,SC,VA DE,MA,PA,RI,VI,VT,HI	AK
New Hampshire	CT,Ne NY MA,Se NY ME,RI NJ,VT	DC,We NY DE,Ea PA,We PA MD,VA No OH,So OH,WV	GA,KY,TN IA,MI,WI IL,NC IN,SC	AL,LA,ND AR,MN,NE FL,MO,OK KS,MS,SD	AZ,MT,PR,WA CA,NM,TX,WY CO,NM,UT ID,OR,VI,HI	AK
New Jersey	CT,MD DC,Se NY DE,Ea PA MA,RI	NC,No OH,So OH,VT NH,We PA Ne NY,VA We NY,WV	AL,KY,TN GA,ME,WI IL,MI IN,SC	AR,LA,ND FL,MN,NE IA,MO,OK KS,MS,SD	AZ,MT,PR,TX CA,NM,UT,WY CO,NV,VI ID,OR,WA,HI	AK
New Mexico	AZ,We TX CO OK UT	KS,Ea TX NE,So TX NV WY	AR,LA CA,MO ID,MT IA,SD	AL,MN,TN IL,MS,WA IN,ND,WI KY,OR	CT,GA,MI,NY,RI,SC DC,MA,NC,OH,VA DE,MD,NH,PA,VI,WV FL,ME,NJ,PR,VT,HI	AK
New York-Ne	CT,NH,VT DE,NJ MA,Ea PA,We PA MD,RI	DC,So MI ME,WV No OH,So OH VA	AL,IN,No MI GA,KY,SC IA,MO,TN IL,NC,WI	AR,MN,OK FL,MS,SD KS,ND LA,NE	AZ,MT,PR,WA CA,NM,TX,WY CO,NV,UT ID,OR,VI,HI	AK

*Each greater service area includes the preceding service area states. The subscriber's home state is not included.

SUBSCRIBER'S HOME STATE†	SERVICE AREAS*					
	1	2	3	4	5	6
New York-Se	CT,MD,RI DC,NH,VT DE,NJ MA,Ea PA	ME,VA NC,WV No OH,So OH We PA	AL,IN,SC GA,KY,TN IA,MI,WI IL,MO	AR,MN,OK FL,MS,SD KS,ND LA,NE	AZ,MT,PR,TX CA,NM,UT,WY CO,NV,VI ID,OR,WA,HI	AK
New York-We	DC,No OH DE,Ea PA,We PA MD,WV NJ	CT,So MI ME,So OH MA,VA,NH RI,VT	AL,IN,No MI GA,KY,SC IA,MO,TN IL,NC,WI	AR,MN,OK FL,MS,SD KS,ND LA,NE	AZ,MT,PR,WA CA,NM,TX,WY CO,NV,UT ID,OR,VI,HI	AK
North Carolina	DC,So OH DE,SC,WV GA,TN MD,VA	AL,No OH KY NJ PA	CT,MA,RI FL,MI,VT IL,MS IN,NY	AR,ME,NH IA,MN,OK KS,MO,TX LA,NE,WI	AZ,MT,OR,VI CA,ND,PR,WA CO,NM,SD,WY ID,NV,UT,HI	AK
North Dakota	IA,SD NY,WY MT,MN NE	CO,No IL ID,No MI KS WI	AR,So IL,WA IN,So MI MO,NM,OR NV,OK,UT	AZ,OH CA,TN KY,TX MS,WV	AL,FL,ME,NJ,RI,VT CT,GA,MD,NY,SC DC,LA,NC,PA,VA DE,MA,NH,PR,VI,HI	AK
Ohio-No	IN,So MI KY,We NY We PA VA,WV	DC,Ea PA DE,TN No IL,SO IL,MD	AL,IA,NJ,Ne NY CT,MA,NC,SC No MI,VT GA,MO,WI,Se NY	AR,ME,NE,SD FL,MN,NH KS,MS,OK LA,ND,RI	AZ,MT,OR,VI CA,PR,WA CO,NM,TX,WY ID,NV,UT,HI	AK

*Each greater service area includes the preceding service area states. The subscriber's home state is not included.

SUBSCRIBER'S HOME STATE†	*SERVICE AREAS**					
	1	*2*	*3*	*4*	*5*	*6*
Ohio-So	IN,So MI KY,We NY We PA VA,WV	DC,Ea PA No IL,So IL,TN MD NC	AL,IA,No MI,WI CT,MA,Ne NY DE,MO,Se NY GA,NJ,SC,VT	AR,ME,NE,SD FL,MN,NH KS,MS,OK LA,ND,RI	AZ,MT,PR,WA CA,NV,TX,WY CO,NM,UT ID,OR,VI,HI	AK
Oklahoma	AR,NM CO,Ea TX KS,We TX MO	LA MS NE So TX	AL,KY,UT IL,MM,WI IN,SD,WY IA,TN	AZ,MT,SC FL,NC,VA GA,ND,WV MI,OH	CA,ID,NH,OR,PA CT,MA,NJ,PR,VT DC,MD,NV,RI,WA DE,ME,NY,VI,HI	AK
Oregon	No CA ID,WA NV UT	AZ So CA MT WY	CO,ND,SD IA,NE,TX KS,NM MN,OK	AR,MI IN,MO IL,MS LA,WI	AL,FL,MD,NJ,PR,RI CT,GA,ME,NY,SC,TN DC,KY,NC,OH,VA,VI DE,MA,NH,PA,VT,WV,HI	AK
Pennsylvania-Ea	CT,NJ DC,Ne NY DE,Se NY MD,We NY	MA,RI NC,VA NH,VT No OH,So OH,WV	AL,KY,SC GA,ME,TN IL,MI,WI IN,MO	AR,LA,NE FL,MN,OK IA,MS,SD KS,ND	AZ,MT,PR,WA CA,NM,TX,WY CO,NV,UT ID,OR,VI,HI	AK

*Each greater service area includes the preceding service area states. The subscriber's home state is not included.

SUBSCRIBER'S HOME STATE†	SERVICE AREAS*					
	1	2	3	4	5	6
Pennsylvania-We	DE,We NY DC,No OH,So OH,WV MD,VA Ne NY	CT,So M MA,NC NJ Se NY	AL,KY,NH,TN GA,ME,RI,VT IL,MO,SC,WI IN,No MI	AR,LA,NE FL,MN,OK IA,MS,SD KS,ND	AZ,MT,OR,VI CA,PR,WA CO,NM,TX,WY ID,NV,UT,HI	AK
Rhode Island	CT,Ne NY MA,Se NY NH,VT NJ	DC,We NY DE,No OH MD,Ea PA,We PA, WV ME,VA	AL,KY,SC GA,MI,TN IL,NC,WI IN,So OH	AR,LA,ND,SD FL,MN,NE,VI IA,MS,OK KS,MO,PR	AZ,MT,TX CA,NM,UT CO,NV,WA ID,OR,WY,HI	AK
South Carolina	AL,TN GA,VA KY,WV NC	DC,MD DE,No OH,So OH FL,We PA IN	AR,MO,Ea PA CT,MS IL,NJ LA,NY	IA,MI,TX KS,NH,VT MA,OK,WI ME,RI	AZ,MN,NM,SD,WY CA,MT,NV,UT CO,ND,OR,VI ID,NE,PR,WA,HI	AK
South Dakota	IA,NE MN,WY MT ND	CO,No IL KS MO WI	AR,So IL ID,NM,UT IN,OK MI,TX	AL,MS,TN AZ,NV,WA KY,OH,WV LA,OR	CA,FL,ME,NC,NH,NJ CT,GA,NY,PA,VI DC,MA,PR,RI,VT DE,MD,SC,VA,HI	AK
Tennessee	AL,MS AR,MO GA,NC KY,VA	No IL,So IL,WV IN No OH,So OH SC	DC,MD,WI FL,MI IA,OK LA,PA	CT,NE,TX DE,NJ KS,NY MN,RI	AZ,MA,NH,NM,NV CA,ME,OR,PR,SD CO,MT,UT,VI,VT ID,ND,WA,WY,HI	AK

*Each greater service area includes the preceding service area states. The subscriber's home state is not included.

SUBSCRIBER'S HOME STATE†	*SERVICE AREAS**					
	1	*2*	*3*	*4*	*5*	*6*
Texas-Ea	AR,MO KS,OK LA MS	AL,NM CO,TN NE So IL	AZ,No IL FL,IN,UT GA,KY,WY IA,SD	ID,NC,SC MI,ND,WV MN,NV,WI MT,OH	CA,ID,NH,NJ,NY CT,MA,OR,PA,PR DC,MD,RI,VI,VA DE,ME,VT,WA,HI	AK
Texas-So	AR,NM KS,OK LA MS	AL,TN AZ, CO MO	FL,IN,UT GA,KY,WI IA,NE,WY IL,SD	ID,NC,SC MI,ND,WV MN,NV MT,OH	CA,MA,NJ,NY,OR CT,MD,PA,PR,RI DC,ME,VA,VI,VT DE,NH,WA,HI	AK
Texas-We	AR,OK CO KS NM	AZ,NE LA,UT MS MO	AL,IL,TN FL,IN,WY GA,KY IA,SD	ID,NC,SC MI,ND,WV MN,NV,WI MT,OH	CA,MA,MD,ME;NH CT,NJ,OR,PA,PR DC,RI,VA,VI,VT DE,WA,NY,HI	AK
Utah	AZ,NM CO,WY ID NV	No CA,So CA MT	IA,OK,WA KS,OR ND,SD NE,TX	AR,MI,TN IL,MN,WI IN,MO LA,MS	AL,CT,DC,DE,FL,GA KY,MA,MD,ME,NC,WV NH,NJ,NY,OH,PA PR,RI,SC,VA,VI,VT,HI	AK

*Each greater service area includes the preceding service area states. The subscriber's home state is not included.

SUBSCRIBER'S HOME STATE†	SERVICE AREAS*					
	1	2	3	4	5	6
Vermont	CT,NJ,Ne NY MA,Se NY ME,RI NH	DC,No OH,So OH,WV DE,PA MD,VA We NY	GA,KY,TN IA,MI,WI IL,NC IN,SC	AL,LA,ND AR,MN,NE FL,MO,OK KS,MS,SD	AZ,MT,PR,TX CA,NM,UT,VI CO,NV,WA ID,OR,WY,HI	AK
Virginia	DC,NC,WV DE,So OH KY,We PA MD,TN	NJ,Ea PA Se NY,SC We NY No OH	AL,IL,MS,VT CT,IN,NH FL,MA,Ne NY GA,MI,RI	AR,ME,OK IA,MN,SD KS,MO,WI LA,NE	AZ,MT,ND CA,NM,NV,OR CO,PR,TX,UT ID,VI,WA,WY,HI	AK
Washington	No CA ID,OR MT NV	So CA ND UT WY	AZ,MN,OK CO,MO,SD IA,NE,WI KS,NM	AR IL MI TX	AL,CT,DC,DE,FL,GA,WV IN,KY,LA,MA,ME,MS NC,NH,NJ,NY,OH,PA,MD RI,PR,SC,TN,VA,VI,VT,HI	AK
West Virginia	DC,No OH,So OH KY,We PA MD,VA NC	DE Ea PA IN,SC NJ,TN We NY	AL,MA,Ne NY CT,MI,Se NY GA,MO,RI IL,NH,VT,WI	AR,LA,NE FL,ME,OK IA,MN,SD KS,MS	AZ,MT,OR,TX CA,ND,PR,UT CO,NM,VI,WY ID,NV,WA,HI	AK
Wisconsin	IA No IL No MI,So MI MN	So IL IN,No OH,So OH MO,SD NE	AR,MD,PA DC,ND,TN KS,NY,VA KY,OK,WV	AL,DE,GA,NJ,RI CO,LA,MA,SC, VT CT,MS,MT,WY NC,NH	AZ,ME,PR,WA CA,NM,TX FL,NV,UT ID,OR,VI,HI	AK
Wyoming	CO,SD ID,UT MT NE	AZ,NV KS,OR ND NM	CA,OK IA,WA NM,WI MO	AR,LA IN,MI IL,MS KY,TX	AL,CT,DC,DE,FL,GA MA,MD,ME,NC NH,NJ,NY,OH,PA,PR RI,SC,TN,VA,VI,VT,WV,HI	AK

*Each greater service area includes the preceding service area states. The subscriber's home state is not included.

APPENDIX F

INTERNATIONAL DIALING PLAN

(International Numbering Plan Listed by Country Code)

Country Code	*Country*
201	Egypt
202	Egypt, Cairo
203	Egypt, Alexandria
204	Egypt
205	Egypt
206	Egypt
207	Egypt
208	Egypt
209	Egypt
210	Spanish Sahara
212	Morocco
213	Algeria
216	Tunisia
217	Tunisia
218	Libya
219	Libya
220	Gambia
221	Senegal
222	Mauritania
223	Mali
224	Guinea
225	Ivory Coast
226	Burkina Faso
227	Niger
228	Togo
229	Benin
230	Mauritius
231	Liberia
232	Sierra Leone
233	Ghana
234	Nigeria
235	Chad
236	Central African Republic

Country Code	*Country*
237	Cameroon
239	Sao Tome & Principe
240	Equatorial Guinea
241	Gabon
242	Congo
243	Zaire
244	Angola
245	Guinea-Bissau
247	Niger
248	Seychelles
249	Sudan
250	Rwanda
251	Ethiopia
253	Afars and Issas (Djibouti)
254	Kenya
255	Tanzania
257	Burundi
258	Mozambique
259	Zanzibar
260	Zambia
261	Madagascar
262	Namibia
263	Zimbabwe
264	Reunion
265	Malawi
266	Lesotho
267	Botswana
268	Swaziland
269	Comorro Islands
270	South Africa
271	South Africa, Pretoria
272	South Africa, Cape Town

Country Code	*Country*
273	South Africa
274	South Africa
275	South Africa
276	South Africa
277	South Africa
278	South Africa
279	South Africa
297	Aruba
300	Greece
301	Greece, Athens
302	Greece, Rhodes
303	Greece
304	Greece
305	Greece
306	Greece
307	Greece
308	Greece
309	Greece
310	Netherlands
311	Netherlands, Rotterdam
312	Netherlands, Amsterdam
313	Netherlands
314	Netherlands
315	Netherlands
316	Netherlands
317	Netherlands, The Hague
318	Netherlands
319	Netherlands
320	Belgium
321	Belgium
322	Belgium, Brussels
323	Belgium, Antwerp
324	Belgium
325	Belgium
326	Belgium
327	Belgium
328	Belgium
329	Belgium
330	Andorra
331	France, Paris
332	France
333	France
334	France
335	France, Bordeaux
336	France

Country Code	*Country*
337	France
338	France
339	France - Monaco
340	Spain
341	Spain, Madrid
342	Spain - Canary Islands
343	Spain, Barcelona
344	Spain
345	Spain, Seville
346	Spain
347	Spain
348	Spain
349	Spain
350	Gibraltar
351	Portugal
352	Luxembourg
353	Ireland
354	Iceland
355	Albania
356	Malta
357	Cyprus
358	Finland
359	Bulgaria
360	Hungary
361	Hungary, Budapest
362	Hungary
363	Hungary
364	Hungary
365	Hungary
366	Hungary
367	Hungary
368	Hungary
369	Hungary
370	East Germany
371	East Germany
372	East Germany, Berlin
373	East Germany
374	East Germany, Leipzig
375	East Germany, Dresden
376	East Germany
377	East Germany
378	East Germany
379	East Germany
380	Yugoslavia
381	Yugoslavia, Belgrade
382	Yugoslavia

Country Code	*Country*
383	Yugoslavia
384	Yugoslavia
385	Yugoslavia
386	Yugoslavia
387	Yugoslavia
388	Yugoslavia, Titograd
389	Yugoslavia
390	Italy
391	Italy
392	Italy, Milan
393	Italy
394	Italy, Venice
395	Italy, Florence - San Marino
396	Italy, Rome - Vatican City
397	Italy
398	Italy, Naples
399	Italy
400	Romania, Bucharest
401	Romania
402	Romania
403	Romania
404	Romania
405	Romania
406	Romania
407	Romania
408	Romania
409	Romania
410	Switzerland
411	Switzerland, Zurich
412	Switzerland, Geneva
413	Switzerland, Berne
414	Switzerland, Lucerne
415	Switzerland
416	Switzerland
417	Liechtenstein
418	Switzerland
419	Switzerland
420	Czechoslovakia
421	Czechoslovakia
422	Czechoslovakia
423	Czechoslovakia
424	Czechoslovakia
425	Czechoslovakia
426	Czechoslovakia
427	Czechoslovakia

Country Code	*Country*
428	Czechoslovakia
429	Czechoslovakia
430	Austria
431	Austria
432	Austria, Vienna
433	Austria
434	Austria
435	Austria, Innsbruck
436	Austria
437	Austria
438	Austria
439	Austria
440	United Kingdom
441	United Kingdom, London
442	United Kingdom, N. Ireland - Wales
443	United Kingdom, Edinburgh
444	United Kingdom, Glasgow
445	United Kingdom, Liverpool
446	United Kingdom
447	United Kingdom
448	United Kingdom
449	United Kingdom
450	Denmark
451	Denmark, Copenhagen
452	Denmark, Copenhagen
453	Denmark
454	Denmark
455	Denmark, Esbjerg
456	Denmark
457	Denmark
458	Denmark, Aalborg
459	Denmark
460	Sweden
461	Sweden
462	Sweden
463	Sweden, Goteborg
464	Sweden
465	Sweden
466	Sweden
467	Sweden
468	Sweden, Stockholm
469	Sweden

Country Code	*Country*
470	Norway
471	Norway
472	Norway, Oslo
473	Norway
474	Norway
475	Norway, Bergen
476	Norway
477	Norway
478	Norway
479	Norway
480	Poland
481	Poland
482	Poland. Warsaw
483	Poland
484	Poland
485	Poland
486	Poland
487	Poland
488	Poland
489	Poland
490	West Germany
491	West Germany
492	West Germany, Bonn
493	West Germany, Berlin
494	West Germany, Hamburg
495	West Germany
496	West Germany, Frankfurt
497	West Germany
498	West Germany, Munich
499	West Germany
501	Belize
502	Guatemala
503	El Salvador
504	Honduras
505	Nicaragua
506	Costa Rica
507	Panama
509	Haiti
510	Peru
511	Peru, Lima
512	Peru
513	Peru
514	Peru
515	Peru, Arequipa
516	Peru
517	Peru
518	Peru
519	Peru
520	Mexico, Mexico City
521	Mexico, Chihuahua
522	Mexico, Puebla
523	Mexico, Guadalahara
524	Mexico, Leon
525	Mexico, Mexico City
526	Mexico, Tijuana
527	Mexico, Acapulco
528	Mexico, Monterrey
529	Mexico, Cancun
530	Cuba
531	Cuba
532	Cuba
533	Cuba
534	Cuba
535	Cuba
536	Cuba
537	Cuba
538	Cuba
539	Cuba
540	Argentina
541	Argentina, Buenos Aires
542	Argentina
543	Argentina
544	Argentina
545	Argentina, Cordoba
546	Argentina, San Juan
547	Argentina
548	Argentina
549	Argentina
550	Brazil
551	Brazil, Sao Paulo
552	Brazil, Rio de Janiero
553	Brazil
554	Brazil
555	Brazil
556	Brazil, Brasillia
557	Brazil
558	Brazil
559	Brazil
560	Chile
561	Chile
562	Chile, Santiago
563	Chile, Valparso

Country Code	*Country*
564	Chile, Concepcion
565	Chile
566	Chile
567	Chile
568	Chile
569	Chile
570	Colombia
571	Colombia
572	Colombia
573	Colombia
574	Colombia
575	Colombia
576	Colombia
577	Colombia
578	Colombia
579	Colombia
580	Venezuela
582	Venezuela, Caracas
583	Venezuela
584	Venezuela
586	Venezuela, Maracaibo
587	Venezuela
588	Venezuela
589	Venezuela
590	Guadaloupe
591	Bolivia
592	Guyana
593	Ecuador
594	French Guinea
595	Paraguay
596	French Antille
597	Surinam
598	Uraguay
599	Netherland Antille
600	Malaysia
601	Malaysia
602	Malaysia
603	Malaysia
604	Malaysia, Alor Star
605	Malaysia
606	Malaysia, Port Dickson
607	Malaysia
608	Malaysia
609	Malaysia
610	Australia
611	Australia
612	Australia, Sidney

Country Code	*Country*
613	Australia, Melbourne
614	Australia
615	Australia, Brisbane
616	Australia
618	Australia
619	Australia, Perth
620	Indonesia
621	Indonesia
622	Indonesia
623	Indonesia
624	Indonesia
625	Indonesia
626	Indonesia
627	Indonesia
628	Indonesia
629	Indonesia
630	Philippines
631	Philippines
632	Philippines, Manila
633	Philippines
634	Philippines
635	Philippines
636	Philippines
637	Philippines
638	Philippines
639	Philippines
640	New Zealand
641	New Zealand
642	New Zealand
643	New Zealand, Christchurch
644	New Zealand, Wellington
645	New Zealand
646	New Zealand
647	New Zealand
648	New Zealand
649	New Zealand, Aukland
650	Singapore
651	Singapore
652	Singapore
653	Singapore
654	Singapore
655	Singapore
656	Singapore
657	Singapore
658	Singapore

Country Code	*Country*
659	Singapore
660	Thailand
661	Thailand
662	Thailand, Bangkok
663	Thailand
664	Thailand
665	Thailand
666	Thailand
667	Thailand
668	Thailand
669	Thailand
671	Guam
672	Portuguese Timor
673	Brunei
674	Nauru
674	Papua New Guinea
676	Tonga
677	Solomon Islands
678	New Hebrides
679	Fiji
681	Wallis Futuna
682	Cook Islands
683	Nieuw Island
684	American Samoa
685	Western Samoa
686	Kiribati
687	New Caledonia
688	Tuvalu
689	French Polynesia
700—799	USSR
810	Japan
811	Japan
812	Japan
813	Japan, Tokyo
814	Japan, Yokohama
815	Japan
816	Japan, Osaka
817	Japan, Kyoto
818	Japan, Hiroshima
819	Japan
820	Korea
821	Korea
822	Korea, Seoul
823	Korea
824	Korea
825	Korea, Pusan

Country Code	*Country*
826	Korea
827	Korea
828	Korea
829	Korea
840	Viet Nam
841	Viet Nam
842	Viet Nam
843	Viet Nam
844	Viet Nam
845	Viet Nam
846	Viet Nam
847	Viet Nam
848	Viet Nam
849	Viet Nam
852	Hong Kong
853	Macao
855	Kampuchea
856	Laos
860	China
861	China, Beijing
862	Taiwan
863	China
864	China
865	China
866	China
867	Taiwan
868	China
869	China
870	Maritime Mobile
880	Bangladesh
886	Taiwan
900	Turkey
901	Istanbul Turkey
902	Turkey
903	Turkey
904	Turkey, Ankara
905	Turkey, Izmir
906	Turkey
907	Turkey
908	Turkey
909	Turkey
910	India
911	India
912	India
913	India, Calcutta
914	India
915	India

Country Code	*Country*
916	India
917	India
918	India
919	India
920	Pakistan
921	Pakistan
922	Pakistan, Karachi
923	Pakistan
924	Pakistan
925	Pakistan
926	Pakistan
927	Pakistan
928	Pakistan
929	Pakistan
930	Afganistan
931	Afganistan
932	Afganistan
933	Afganistan
934	Afganistan
935	Afganistan
936	Afganistan
937	Afganistan
938	Afganistan
939	Afganistan
941	Sri Lanka, Colombo
942	Sri Lanka
943	Sri Lanka
944	Sri Lanka, Matara
945	Sri Lanka
946	Sri Lanka
947	Sri Lanka
948	Sri Lanka
949	Sri Lanka
950	Burma
951	Burma
952	Burma
953	Burma
954	Burma
955	Burma
956	Burma
957	Burma
958	Burma
960	Maldives
961	Lebanon
962	Jordan
963	Syria
964	Iraq
965	Kuwait
966	Saudi Arabia
967	Yeman Arab Republic
968	Oman
969	Yeman Peoples Democratic Republic
971	United Arab Emirates
972	Israel
973	Bahrain
974	Qatar
976	Mongolia - Bhutan
977	Nepal
978	Dubai
979	Abu Dhabi
980	Iran
981	Iran
982	Iran
983	Iran
984	Iran
985	Iran
986	Iran
987	Iran
988	Iran
989	Iran

(International Numbering Plan Listed by Country)

Country	Country Code
Abu Dhabi	979
Afars and Issas (Djibouti)	253
Afganistan	930
Afganistan	931
Afganistan	932
Afganistan	933
Afganistan	934
Afganistan	935
Afganistan	936
Afganistan	937
Afganistan	938
Afganistan	939
Albania	355
Algeria	213
American Samoa	684
Andorra	330
Angola	244
Argentina	540
Argentina	542
Argentina	543
Argentina	544
Argentina	547
Argentina	548
Argentina	549
Argentina, Buenos Aires	541
Argentina, Cordoba	545
Argentina, San Juan	546
Aruba	297
Australia	610
Australia	611
Australia	614
Australia	616
Australia	618
Australia, Brisbane	615
Australia, Melbourne	613
Australia, Perth	619
Australia, Sidney	612
Austria	430
Austria	431
Austria	433
Austria	434
Austria	436
Austria	437
Austria	438
Austria	439
Austria, Innsbruck	435
Austria, Vienna	432
Bahrain	973
Bangladesh	880
Belgium	320
Belgium	321
Belgium	324
Belgium	325
Belgium	326
Belgium	327
Belgium	328
Belgium	329
Belgium, Antwerp	323
Belgium, Brussels	322
Belize	501
Benin	229
Bolivia	591
Botswana	267
Brazil	550
Brazil	553
Brazil	554
Brazil	555
Brazil	557
Brazil	558
Brazil	559
Brazil, Brasillia	556
Brazil, Rio de Janiero	552
Brazil, Sao Paulo	551
Brunei	673
Bulgaria	359
Burkina Faso	226
Burma	950
Burma	951
Burma	952
Burma	953
Burma	954
Burma	955
Burma	956
Burma	957
Burma	958
Burundi	257
Cameroon	237
Central African Republic	236
Chad	235

Country	Country Code
Chile	560
Chile	561
Chile	565
Chile	566
Chile	567
Chile	568
Chile	569
Chile, Concepcion	564
Chile, Santiago	562
Chile, Valparso	563
China	860
China	863
China	864
China	865
China	866
China	868
China	869
China, Beijing	861
Colombia	570
Colombia	571
Colombia	572
Colombia	573
Colombia	574
Colombia	575
Colombia	576
Colombia	577
Colombia	578
Colombia	579
Comorro Islands	269
Congo	242
Cook Islands	682
Costa Rica	506
Cuba	530
Cuba	531
Cuba	532
Cuba	533
Cuba	534
Cuba	535
Cuba	536
Cuba	537
Cuba	538
Cuba	539
Cyprus	357
Czechoslovakia	420
Czechoslovakia	421
Czechoslovakia	422
Czechoslovakia	423

Country	Country Code
Czechoslovakia	424
Czechoslovakia	425
Czechoslovakia	426
Czechoslovakia	427
Czechoslovakia	428
Czechoslovakia	429
Denmark	450
Denmark	453
Denmark	454
Denmark	456
Denmark	457
Denmark	459
Denmark, Aalborg	458
Denmark, Copenhagen	451
Denmark, Copenhagen	452
Denmark, Esbjerg	455
Dubai	978
East Germany	370
East Germany	371
East Germany	373
East Germany	376
East Germany	377
East Germany	378
East Germany	379
East Germany, Berlin	372
East Germany, Dresden	375
East Germany, Leipzig	374
Ecuador	593
Egypt	201
Egypt	204
Egypt	205
Egypt	206
Egypt	207
Egypt	208
Egypt	209
Egypt, Alexandria	203
Egypt, Cairo	202
El Salvador	503
Equatorial Guinea	240
Ethiopia	251
Fiji	679
Finland	358
France	332
France	333
France	334
France	336
France	337

Country	Country Code
France	338
France - Monaco	339
France, Bordeaux	335
France, Paris	331
French Antille	596
French Guinea	594
French Polynesia	689
Gabon	241
Gambia	220
Ghana	233
Gibraltar	350
Greece	300
Greece	303
Greece	304
Greece	305
Greece	306
Greece	307
Greece	308
Greece	309
Greece, Athens	301
Greece, Rhodes	302
Guadaloupe	590
Guam	671
Guatemala	502
Guinea	224
Guinea-Bissau	245
Guyana	592
Haiti	509
Honduras	504
Hong Kong	852
Hungary	360
Hungary	362
Hungary	363
Hungary	364
Hungary	365
Hungary	366
Hungary	367
Hungary	368
Hungary	369
Hungary, Budapest	361
Iceland	354
India	910
India	911
India	912
India	914
India	915
India	916

Country	Country Code
India	917
India	918
India	919
India, Calcutta	913
Indonesia	620
Indonesia	621
Indonesia	622
Indonesia	623
Indonesia	624
Indonesia	625
Indonesia	626
Indonesia	627
Indonesia	628
Indonesia	629
Iran	980
Iran	981
Iran	982
Iran	983
Iran	984
Iran	985
Iran	986
Iran	987
Iran	988
Iran	989
Iraq	964
Ireland	353
Israel	972
Istanbul Turkey	901
Italy	390
Italy	391
Italy	393
Italy	397
Italy	399
Italy, Florence - San Marino	395
Italy, Milan	392
Italy, Naples	398
Italy, Rome - Vatican City	396
Italy, Venice	394
Ivory Coast	225
Japan	810
Japan	811
Japan	812
Japan	815
Japan	819
Japan, Hiroshima	818

Country	Country Code
Japan, Kyoto	817
Japan, Osaka	816
Japan, Tokyo	813
Japan, Yokohama	814
Jordan	962
Kampuchea	855
Kenya	254
Kiribati	686
Korea	820
Korea	821
Korea	823
Korea	824
Korea	826
Korea	827
Korea	828
Korea	829
Korea, Pusan	825
Korea, Seoul	822
Kuwait	965
Laos	856
Lebanon	961
Lesotho	266
Liberia	231
Libya	218
Libya	219
Liechtenstein	417
Luxembourg	352
Macao	853
Madagascar	261
Malawi	265
Malaysia	600
Malaysia	601
Malaysia	602
Malaysia	603
Malaysia	605
Malaysia	607
Malaysia	608
Malaysia	609
Malaysia, Alor Star	604
Malaysia, Port Dickson	606
Maldives	960
Mali	223
Malta	356
Maritime Mobile	870
Mauritania	222
Mauritius	230
Mexico, Acapulco	527

Country	Country Code
Mexico, Cancun	529
Mexico, Chihuahua	521
Mexico, Guadalahara	523
Mexico, Leon	524
Mexico, Mexico City	520
Mexico, Mexico City	525
Mexico, Monterrey	528
Mexico, Puebla	522
Mexico, Tijuana	526
Mongolia - Bhutan	976
Morocco	212
Mozambique	258
Namibia	262
Nauru	674
Nepal	977
Netherland Antille	599
Netherlands	310
Netherlands	313
Netherlands	314
Netherlands	315
Netherlands	316
Netherlands	318
Netherlands	319
Netherlands, Amsterdam	312
Netherlands, Rotterdam	311
Netherlands, The Hague	317
New Caledonia	687
New Hebrides	678
New Zealand	640
New Zealand	641
New Zealand	642
New Zealand	645
New Zealand	646
New Zealand	647
New Zealand	648
New Zealand, Aukland	649
New Zealand, Christchurch	643
New Zealand, Wellington	644
Nicaragua	505
Nieuw Island	683
Niger	227
Niger	247
Nigeria	234
Norway	470

Country	*Country Code*
Norway	471
Norway	473
Norway	474
Norway	476
Norway	477
Norway	478
Norway	479
Norway, Bergen	475
Norway, Oslo	472
Oman	968
Pakistan	920
Pakistan	921
Pakistan	923
Pakistan	924
Pakistan	925
Pakistan	926
Pakistan	927
Pakistan	928
Pakistan	929
Pakistan, Karachi	922
Panama	507
Papua New Guinea	674
Paraguay	595
Peru	510
Peru	512
Peru	513
Peru	514
Peru	516
Peru	517
Peru	518
Peru	519
Peru, Arequipa	515
Peru, Lima	511
Philippines	630
Philippines	631
Philippines	633
Philippines	634
Philippines	635
Philippines	636
Philippines	637
Philippines	638
Philippines	639
Philippines, Manila	632
Poland	480
Poland	481
Poland	483
Poland	484
Poland	485
Poland	486
Poland	487
Poland	488
Poland	489
Poland. Warsaw	482
Portugal	351
Portuguese Timor	672
Qatar	974
Reunion	264
Romania	401
Romania	402
Romania	403
Romania	404
Romania	405
Romania	406
Romania	407
Romania	408
Romania	409
Romania, Bucharest	400
Rwanda	250
Sao Tome & Principe	239
Saudi Arabia	966
Senegal	221
Seychelles	240
Sierra Leone	232
Singapore	650
Singapore	651
Singapore	652
Singapore	653
Singapore	654
Singapore	655
Singapore	656
Singapore	657
Singapore	658
Singapore	659
Solomon Islands	677
South Africa	270
South Africa	273
South Africa	274
South Africa	275
South Africa	276
South Africa	277
South Africa	278
South Africa	279
South Africa, Cape Town	272

Country	*Country Code*
South Africa, Pretoria	271
Spain	340
Spain	344
Spain	346
Spain	347
Spain	348
Spain	349
Spain - Canary Islands	342
Spain, Barcelona	343
Spain, Madrid	341
Spain, Seville	345
Spanish Sahara	210
Sri Lanka	942
Sri Lanka	943
Sri Lanka	945
Sri Lanka	946
Sri Lanka	947
Sri Lanka	948
Sri Lanka	949
Sri Lanka, Colombo	941
Sri Lanka, Matara	944
Sudan	249
Surinam	597
Swaziland	268
Sweden	460
Sweden	461
Sweden	462
Sweden	464
Sweden	465
Sweden	466
Sweden	467
Sweden	469
Sweden, Goteborg	463
Sweden, Stockholm	468
Switzerland	410
Switzerland	415
Switzerland	416
Switzerland	418
Switzerland	419
Switzerland, Berne	413
Switzerland, Geneva	412
Switzerland, Lucerne	414
Switzerland, Zurich	411
Syria	963
Taiwan	862
Taiwan	867
Taiwan	886

Country	*Country Code*
Tanzania	255
Thailand	660
Thailand	661
Thailand	663
Thailand	664
Thailand	665
Thailand	666
Thailand	667
Thailand	668
Thailand	669
Thailand, Bangkok	662
Togo	228
Tonga	676
Tunisia	216
Tunisia	217
Turkey	900
Turkey	902
Turkey	903
Turkey	906
Turkey	907
Turkey	908
Turkey	909
Turkey, Ankara	904
Turkey, Izmir	905
Tuvalu	688
United Arab Emirates	971
United Kingdom	440
United Kingdom	446
United Kingdom	447
United Kingdom	448
United Kingdom	449
United Kingdom, Edinburgh	443
United Kingdom, Glasgow	444
United Kingdom, Liverpool	445
United Kingdom, London	441
United Kingdom, N. Ireland - Wales	442
Uraguay	598
USSR	700–799
Venezuela	580
Venezuela	583
Venezuela	584
Venezuela	587

Country	*Country Code*
Venezuela	588
Venezuela	589
Venezuela, Caracas	582
Venezuela, Maracaibo	586
Viet Nam	840
Viet Nam	841
Viet Nam	842
Viet Nam	843
Viet Nam	844
Viet Nam	845
Viet Nam	846
Viet Nam	847
Viet Nam	848
Viet Nam	849
Wallis Futuna	681
West Germany	490
West Germany	491
West Germany	495
West Germany	497
West Germany	499
West Germany, Berlin	493
West Germany, Bonn	492
West Germany, Frankfurt	496
West Germany, Hamburg	494
West Germany, Munich	498
Western Samoa	685
Yeman Arab Republic	967
Yeman Peoples Democratic Republic	969
Yugoslavia	380
Yugoslavia	382
Yugoslavia	383
Yugoslavia	384
Yugoslavia	385
Yugoslavia	386
Yugoslavia	387
Yugoslavia	389
Yugoslavia, Belgrade	381
Yugoslavia, Titograd	388
Zaire	243
Zambia	260
Zanzibar	259
Zimbabwe	263

APPENDIX G

NORTH AMERICAN AREA CODES

North American Area Codes by State or Province

State or Province	*Area Code*
Alabama	205
Alaska	907
Arizona	602
Arkansas	501
Bermuda, Puerto Rico, Virgin Islands. & other Caribbean Islands.	809
California	209
California	213
California	408
California	415
California	619
California	707
California	714
California	805
California	818
California	916
Canada:	
Alberta	403
British Columbia	604
Manitoba	204
New Brunswick	506
Newfoundland	709
Nova Scotia & Prince Edward Island.	902
Ontario	416
Ontario	519
Ontario	613
Ontario	705
Ontario	807
Quebec	418
Quebec	514
Quebec	819
Saskatchewan	306
Colorado	303
Colorado	719
Connecticut	203
Delaware	302
DIAL IT Services	900
District of Columbia	202
Florida	305
Florida	407
Florida	813
Florida	904
Georgia	404
Georgia	912
Govt. Spec. Svc.	710
Hawaii	808
IC Services	700
Idaho	208
Illinois	217
Illinois	309
Illinois	312
Illinois	618
Illinois	815
Indiana	219
Indiana	317
Indiana	812
800 Service	800
Iowa	319
Iowa	515
Iowa	712
Kansas	316
Kansas	913

State or Province	Area Code
Kentucky	502
Kentucky	606
Louisiana	318
Louisiana	504
Maine	207
Maryland	301
Massachusetts	413
Massachusetts	508
Massachusetts	617
Mexico:	
Mexico City	905
Northwest Mexico	706
Michigan	313
Michigan	517
Michigan	616
Michigan	906
Minnesota	218
Minnesota	507
Minnesota	612
Mississippi	601
Missouri	314
Missouri	417
Missouri	816
Montana	406
Nebraska	308
Nebraska	402
Nevada	702
New Hampshire	603
New Jersey	201
New Jersey	609
New Mexico	505
New York	212
New York	315
New York	516
New York	518
New York	607
New York	716
New York	718
New York	914

State or Province	Area Code
North Carolina	704
North Carolina	919
North Dakota	701
Ohio	216
Ohio	419
Ohio	513
Ohio	614
Oklahoma	405
Oklahoma	918
Oregon	503
Pennsylvania	215
Pennsylvania	412
Pennsylvania	717
Pennsylvania	814
Rhode Island	401
South Carolina	803
South Dakota	605
Tennessee	615
Tennessee	901
Texas	214
Texas	409
Texas	512
Texas	713
Texas	806
Texas	817
Texas	915
TWX	610
Utah	801
Vermont	802
Virginia	703
Virginia	804
Washington	206
Washington	509
West Virginia	304
Wisconsin	414
Wisconsin	608
Wisconsin	715
Wyoming	307

North American Area Codes by Area Code

Area Code	*State or Province*
201	New Jersey
202	District of Columbia
203	Connecticut
204	Manitoba
205	Alabama
206	Washington
207	Maine
208	Idaho
209	California
212	New York
213	California
214	Texas
215	Pennsylvania
216	Ohio
217	Illinois
218	Minnesota
219	Indiana
301	Maryland
302	Delaware
303	Colorado
304	West Virginia
305	Florida
306	Saskatchewan
307	Wyoming
308	Nebraska
309	Illinois
312	Illinois
313	Michigan
314	Missouri
315	New York
316	Kansas
317	Indiana
318	Louisiana
319	Iowa
401	Rhode Island
402	Nebraska
403	Alberta
404	Georgia
405	Oklahoma
406	Montana
407	Florida
408	California
409	Texas
412	Pennsylvania
413	Massachusetts
414	Wisconsin
415	California
416	Ontario
417	Missouri
418	Quebec
419	Ohio
501	Arkansas
502	Kentucky
503	Oregon
504	Louisiana
505	New Mexico
506	New Brunswick
507	Minnesota
508	Massachusetts
509	Washington
512	Texas
513	Ohio
514	Quebec
515	Iowa
516	New York
517	Michigan
518	New York
519	Ontario
601	Mississippi
602	Arizona
603	New Hampshire
604	British Columbia
605	South Dakota
606	Kentucky
607	New York
608	Wisconsin
609	New Jersey
610	TWX
612	Minnesota
613	Ontario
614	Ohio
615	Tennessee
616	Michigan
617	Massachusetts
618	Illinois
619	California
700	IC Services
701	North Dakota
702	Nevada
703	Virginia
704	North Carolina
705	Ontario

Area Code	*State or Province*
706	Northwest Mexico
707	California
709	Newfoundland
710	Govt. Spec. Svc.
712	Iowa
713	Texas
714	California
715	Wisconsin
716	New York
717	Pennsylvania
718	New York
719	Colorado
800	800 Service
801	Utah
802	Vermont
803	South Carolina
804	Virginia
805	California
806	Texas
807	Ontario
808	Hawaii
809	Bermuda, Puerto Rico, Virgin Islands and Other Caribboan Islands.

Area Code	*State or Province*
812	Indiana
813	Florida
814	Pennsylvania
815	Illinois
816	Missouri
817	Texas
818	California
819	Quebec
900	"DIAL IT" Services
901	Tennessee
902	Nova Scotia & Prince Edward Is.
904	Florida
905	Mexico City
906	Michigan
907	Alaska
912	Georgia
913	Kansas
914	New York
915	Texas
916	California
918	Oklahoma
919	North Carolina

APPENDIX H

TRADE PUBLICATIONS

BCR Handbook of Telecommunications Management,. 950 York Rd., Hinsdale, IL 60521, BCR Enterprises, Inc., (312) 986-1432.

BOC Week, Suite 444, 1101 King St, Alexandria, VA 22314, (703) 683-4100.

Cellular Business. P.O. Box 12901, Overland Park, KS 66212-0930, (913) 888-4664.

CO. 12 West 21 St., New York, NY 10010, 212-206-6660.

Common Carrier Week. 1836 Jefferson Pl. N.W., Washington, DC 20036, (202) 872-9200.

Communications Consultant. 352 Park Ave. South, New York, NY 10010-1709, Jobson Publishing Co., (212) 685-4848.

Communications Daily. 1836 Jefferson Pl. N.W., Washington, DC 20036, (202) 872-9200.

Communications News. 124 S.W. 1st St., Geneva, Illinois 60134, Harcourt Brace Jovanovich: (312) 232-1401.

Communications Week. 600 Community Dr., Manhasset, NY 11030, CMP Publications, Inc., (516) 365-4600.

Data Communications. 1221 Avenue of the Americas, New York, NY 10020, McGraw-Hill, Inc., (212) 512-2000.

Edge on and About AT&T. 103 Washington St., Morristown, NJ 07960, (201) 292-1877.

FCC Week. Suite 444, 1101 King St, Alexandria, VA 22314, (703) 683-4100.

International Communications Week. Suite 444, 1101 King St, Alexandria, VA 22314, (703) 683-4100.

ISDN Strategies. 950 York Rd., Hinsdale, IL 60521, (312) 986-1432.

LAN: The Local Area Network Magazine. 12 West 21 St., New York, NY 10010, 212-206-6660.

Lightwave. 235 Bear Hill Rd., Waltham, MA 02154, (617) 890-2700.

Mobile Communications Business. P.O. Box 2000-141, Mission Viejo, CA 92690, (714) 859-5502.

Network World. Box 9171, 375 Cocituate Rd., Framingham, MA 01701-9171, (617) 879-0700.

Pay Phone News. P.O. Box 1218, McLean, VA 22101, (703) 734-7050.

Report on AT&T. Suite 444, 1101 King St, Alexandria, VA 22314, (703) 683-4100.

Satellite Week. 1836 Jefferson Pl. N.W., Washington, DC 20036, (202) 872-9200.

State Telephone Regulation Report. Suite 444, 1101 King St, Alexandria, VA 22314, (703) 683-4100.

Telecom Manager. Suite 444, 1101 King St, Alexandria, VA 22314, (703) 683-4100.

Telecommunications. Horizon House-Microwave, Inc., 685 Canton St., Norwood MA 02062.

Telecommunications Products + Technology. P.O. Box 1425, 119 Russell St., Littleton, MA 01460-4425, Pennwell Publishing Co.: (617) 486-9501.

Telecommunications Reports. 1036 National Press Bldg., Washington, DC 20045, (202) 347-2654.

Teleconnect. 12 Wcst 21 St., New York, NY 10010, 212-691-8215.

Telephone Angles. Suite 700N, 4550 Montgomery Ave., Bethesda, MD 20814, (301) 656-6666.

Telephone Engineer and Management. 124 S. First St., Geneva, IL 60134, Harcourt Brace Jovanovich Publications, Inc.

Telephone News. 7811 Montrose Rd., Potomac, MD 20854, (301) 424-3700.

Telephony. 55 E Jackson Blvd., Chicago, IL 60604, Telephony Publishing Corporation, (312) 922-2435.

Transnational Data & Communications Report. P.O. Box 2039, Springfield, VA 22152, (202) 488-3434.

APPENDIX I

V AND H COORDINATES OF CATEGORY A CITIES

Alabama

Anniston	7406	2304
Birmingham	7518	2446
Decatur	7324	2585
Huntsville	7267	2535
Mobile	8167	2367
Montgomery	7692	2247
Troy	7771	2136

Arizona

Flagstaff	8746	6760
Phoenix	9135	6748
Prescott	8917	6872
Tucson	9345	6485
Yuma	9385	7171

Arkansas

Fayetteville	7600	3872
Forrest City	7555	3232
Hot Springs	7827	3554
Jonesboro	7389	3297
Pine Bluff	7803	3358
Searcy	7581	3407

California

Anaheim	9250	7810
Bakersfield	8947	8060
Chico	8057	8668
Eureka	7856	9075
Fresno	8669	8239
Gardena	9250	7879
Hayward	8513	8660
Long Beach	9217	7856
Los Angeles	9213	7878
Oakland	8486	8695
Point Reyes	8420	8794
Redwood City	8556	8682
Sacramento	8304	8580
Salinas	8722	8560
San Bernardino	9172	7710
San Diego	9468	7629
San Francisco	8492	8719
San Jose	8583	8619
San Luis Obispo	9005	8349
Santa Monica	9227	7920
Santa Rosa	8354	8787
Stockton	8435	8530
Sunnyvale	8576	8643
Ukiah	8206	8885
Van Nuys	9197	7919

Colorado

Colorado Springs	7679	5813
Denver	7501	5899
Fort Collins	7331	5965
Fort Morgan	7335	5739
Glenwood Springs	7651	6263
Grand Junction	7804	6438
Greeley	7345	5895
Montrose	7898	6292
Pueblo	7787	5742

Connecticut

Bridgeport	4841	1360
Hartford	4687	1373
New Haven	4792	1342
New London	4700	1242
Stamford	4897	1388

Delaware

Wilmington	5326	1485

District Of Columbia

Washington	5622	1583

Florida

Chipley	7927	1958
Clearwater	8203	1206
Cocoa	7925	0903
Crestview	8025	2128
Daytona Beach	7791	1052
Fort Lauderdale	8282	0557
Fort Myers	8359	0904
Fort Pierce	8054	0737
Fort Walton Beach	8097	2097
Gainsville	7838	1310
Jackconville	7649	1276
Key West	8745	0668
Lake City	7768	1419
Miami	8351	0527
Ocala	7909	1227
Orlando	7954	1031
Panama City	8057	1914
Pensacola	8147	2200
St. Petersburg	8224	1159
Sarasota	8295	1049
Tallahassee	7877	1716
Tampa	8173	1147
West Palm Beach	8166	0607
Winter Garden	7970	1069
Winter Haven	8084	1034

Georgia

Albany	7649	1817
Atlanta	7260	2083
Augusta	7089	1674
Brunswick	7476	1340
Columbus	7556	2045
Conyers	7243	2016
Dublin	7354	1718
Fitzgerald	7539	1684
Macon	7364	1865
Rome	7234	2662
Savannah	7266	1379
Thomasville	7773	1709
Waycross	7550	1485

Idaho

Boise	7096	7869
Pocatello	7146	7250
Twin Falls	7275	7557

Illinois

Centralia	6744	3311
Champaign-Urbana	6371	3336
Chicago	5986	3426
Collinsville	6781	3455
De Kalb	6061	3591
Hinsdale	6023	3461
Joliet	6088	3454
Marion	6882	3202
Newark	6123	3527
Peoria	6362	3592
Rockford	6022	3675
Rock Island	6276	3816
Springfield	6539	3518
Woodstock	5964	3587

Indiana

Bloomington	6417	2984
Evansville	6729	3019
Fort Wayne	5942	2982
Indianapolis	6272	2992
Muncie	6130	2925
New Albany	6525	2786
South Bend	5918	3206
Terre Haute	6428	3145

Iowa

Boone	6394	4355
Burlington	6449	3829
Cedar Rapids	6261	4021
Davenport	6273	3817
Des Moines	6471	4275
Dubuque	6088	3925
Iowa City	6313	3972

City		
Sioux City	6468	4768
Waterloo	6208	4167
Kansas		
Dodge City	7640	4958
Hutchison	7452	4644
Kansas City	7028	4212
Manhattan	7143	4520
Salina	7275	4656
Topeka	7110	4369
Wichita	7489	4520
Kentucky		
Danville	6558	2561
Frankfort	6462	2634
Madisonville	6845	2942
Paducah	6982	3088
Winchester	6441	2509
Louisiana		
Alexandria	8409	3168
Baton Rouge	8476	2874
Lafayette	8587	2996
Lake Charles	8679	3202
Monroe	8148	3218
New Orleans	8483	2638
Shreveport	8272	3495
Maine		
Augusta	3961	1870
Lewiston	4042	1391
Portland	4121	1384
Maryland		
Baltimore	5510	1575
Massachusetts		
Boston	4422	1249
Brockton	4465	1205
Cambridge	4425	1258
Fall River	4543	1170
Framingham	4472	1284
Lawrence	4373	1811
Springfield	4620	1408
Worchester	4513	1330
Michigan		
Detroit	5536	2828
Flint	5461	2993
Grand Rapids	5628	3261
Houghton	5052	4088
Iron Mountain	5255	3993
Jackson	5663	3009
Kalamazoo	5749	3177
Lansing	5584	3081
Petoskey	5120	3425
Plymouth	5562	2891
Fontiac	5498	2895
Saginaw	5404	3074
Sault	4863	3471
Traverse City	5284	3447
Minnesota		
Duluth	5352	4530
Minneapolis	5777	4513
St. Cloud	5721	4705
St. Paul	5776	4498
Virginia	5234	4657
Wadena	5606	4915
Willmar	5864	4790
Mississippi		
Biloxi	8296	2481
Columbus	7657	2704
Greenville	7888	3126
Gulfport	8317	2511
Hattiesburg	8152	2636
Jackson	8035	2880
Laurel	8066	2645
McComb	8262	2823
Meridian	7899	2639
Tupelo	7535	2825
Missouri		
Cape Giradeau	7013	3251
Joplin	7421	4015
Kansas City	7027	4203
St. Joseph	6913	4301
St. Louis	6807	3482
Sikeston	7099	3221
Springfield	7310	3836

Montana

Billings	6391	6790
Glendive	5961	6313
Helena	6336	7348
Missoula	6336	7650

Nebraska

Grand Island	6901	4936
Omaha	6687	4595
Sidney	7112	5671

Nevada

Carson City	8139	8306
Las Vegas	8665	7411
Reno	8064	8323

New Hampshire

Concord	4326	1426
Dover	4261	1344
Manchester	4354	1388
Nashua	4394	1356

New Jersey

Atlantic City	5284	1284
Camden	5249	1453
Hackensack	4976	1432
Morristown	5035	1478
Newark	5015	1430
New Brunswick	5085	1434
Trenton	5164	1440

New Mexico

Albuquerque	8549	5887
Las Cruces	9132	5742
Roswell	8787	5413
Santa Fe	8389	5804

New York

Albany	4639	1629
Binghamton	4943	1837
Buffalo	5076	2326
Huntington	4918	1349
Nassau	4961	1355
New York City	4977	1406
Potsdam	4404	2054
Poughkeepsie	4821	1526
Rochester	4913	2195
Syracuse	4798	1990
Troy	4616	1633
Westchester	4912	1330

North Carolina

Asheville	6749	2001
Charlotte	6657	1698
Fayetteville	6501	1385
Gastonia	6683	1754
Greensboro	6400	1638
Greenville	6350	1226
Laurinburg	6610	1437
New Bern	6307	1119
Raleigh	6344	1436
Rocky Mount	6232	1329
Wilmington	6559	1143
Winston-Salem	6440	1710

North Dakota

Bismarck	5840	5736
Casselton	5633	5241
Dickinson	5922	6024
Fargo	5615	5182
Grand Forks	5420	5300

Ohio

Akron	5637	2472
Canton	5676	2419
Cincinnati	6263	2679
Cleveland	5574	2543
Columbus	5872	2555
Dayton	6113	2705
Findlay	5828	2766
Mansfield	5783	2575
Toledo	5704	2820
Youngstown	5557	2353

Oklahoma

Enid	7783	4505
Lawton	8178	4451
Muskogee	7746	4042
Oklahoma City	7947	4373
Tulsa	7707	4173

Oregon

Medford	7503	8892
Pendleton	6707	8326
Portland	6799	8914

Pennsylvania

Allentown	5166	1585
Altoona	5460	1972
Harrisburg	5363	1733
Philadelphia	5257	1501
Pittsburgh	5621	2185
Pottsville	5221	1695
Reading	5258	1612
Scranton	5042	1715
State College	5360	1933
Williamsport	5200	1873

Rhode Island

Providence	4550	1219

South Carolina

Charleston	7021	1281
Columbia	6901	1589
Florence	6744	1417
Greenville	6873	1894
Orangeburg	6980	1502
Spartanburg	6811	1833

South Dakota

Aberdeen	5992	5308
Huron	6201	5183
Sioux Falls	6279	4900

Tennessee

Chattanooga	7098	2366
Clarksville	6988	2837
Jackson	7282	2976
Johnson City	6595	2050
Kingspoint	6570	2107
Knoxville	6801	2251
Memphis	7471	3125
Morristown	6699	2183
Nashville	7010	2710

Texas

Abilene	8698	4513
Amarillo	8266	5076
Austin	9005	3996
Beaumont	8777	3344
Corpus Christi	9475	3739
Dallas	8436	4034
El Paso	9231	5655
Fort Worth	8479	4122
Freeport	9096	3466
Harlingen	9820	3663
Houston	8938	3563
Laredo	9681	4099
Longview	8348	3660
Lubbock	8598	4962
Midland	8934	4888
San Angelo	8944	4563
San Antonio	9225	4062
Sweetwater	8737	4632
Waco	8706	3993

Utah

Logan	7367	7102
Ogden	7480	7100
Provo	7680	7006
Salt Lake City	7576	7065

Vermont

Burlington	4270	1808
White River Junction	4327	1585

Virginia

Blacksburg	6247	1867
Leesburg	5634	1685
Lynchburg	6093	1703
Newport News	5908	1260
Norfolk	5918	1223
Petersburg	5961	1429
Richmond	5906	1472
Roanoke	6196	1801

Washington

Bellingham	6087	8933
Kennewick	6595	8391
North Bend	6354	8815
Seattle	6336	8896
Spokane	6247	8180
Yakima	6533	8607

West Virginia

Blacksburg	6247	1867
Charleston	5631	1747
Clarksburg	5865	2095
Fairmont	5808	2091
Huntington	6212	2299
Morgantown	5764	2083
Parkersburg	5976	2268
Wheeling	5755	2241

Wisconsin

Appleton	5589	3776
Dodgeville	5963	3390
Eau Claire	5698	4261
Green Bay	5512	3747
La Crosse	5874	4133
Madison	5887	3796
Milwaukee	5788	3589
Racine	5837	3535
Stevens Point	5622	3964

Wyoming

Casper	6918	6297
Cheyenne	7203	5958

APPENDIX J

CODE OF ETHICS OF THE SOCIETY OF TELECOMMUNICATIONS CONSULTANTS*

CODE OF ETHICS

A. Members shall maintain the highest standards of honesty and fair dealing toward their clients past and present, other members, and the general public.
B. Members shall treat all information relating to the affairs of clients, obtained in the course of a consulting assignment, as confidential.
C. Members shall not knowingly place themselves in a position in which their interests are, or may be, in conflict with those of any client.
D. Members are required to terminate any business or organization relationship which would require them to act in a manner inconsistent with the principles laid down in this code.
E. Members shall inform their clients of any business connections, affiliations, or interests of which the client would have a reasonable expectation to be made aware.
F. In performing services for a client, no member shall accept any fee, commission, or other valuable consideration in connection with those services from anyone other than the client.
G. Prior to the commencement of services, members shall make clients fully aware of the fee structure and all associated costs.
H. Prior to the commencement of services, members shall take all reasonable steps to ensure that the client has a clear understanding of the scope and objectives of the work to be performed.

---

*Reprinted by permission

I. No member or vendor advisory council member shall intentionally injure the professional reputation of another member, but members shall be required to inform the society's Professional Conduct Committee of any violation or apparent violation of this code or of the society's bylaws by any member.

J. No member or vendor advisory council member shall attempt to influence the professional judgement of any telecommunications consultant through the payment or offering of any fee, commission, or valuable consideration.

APPENDIX K

LIST OF SELECTED MANUFACTURERS AND VENDORS OF TELECOMMUNICATIONS PRODUCTS AND SERVICES

This appendix lists manufacturers and vendors of principal telecommunications equipment and services. Because addresses and telephone numbers are changing frequently, it is advisable to check these against sources that are kept updated. Readers are advised to consult one of the following sources for current and more complete manufacturer information:

Data Communications Buyer's Guide Issue, Data Communications Magazine. McGraw-Hill Building, 1221 Avenue of the Americas, New York, NY 10020, (212) 512-2000.

Datapro Reports on Telecommunications and *Datapro Reports on Data Communications.* 1805 Underwood Blvd. Delran, NJ 08075, Data Pro Research Corporation: (609) 764-0100.

Telecommunications Reference Data and Buyers Guide. 610 Washington St. Dedham, MA 02026, Telecommunications: (617) 326-8220.

Teleconnect, Buyers Guide issue (July). 12 West 21 St. New York, NY 10010, 212-691-8215.

Telephone Engineer and Management Directory, Telephone Engineer and Management Magazine. 124 South First St. Geneva, IL 60134, (312) 232-1400.

Telephony Directory and Buyers Guide. Telephony Publishing Co.: 55 E. Jackson Blvd. Chicago, IL 60604.

Manufacturers Listed by Product

Automatic Call Distributors

- Fuijitsu Business Communications
- IBM/Rolm
- Lanier/Harris
- NEC America
- Northern Telecom
- Rockwell International
- Siemens Information Systems

Cable, Hardware, and Wiring Supplies

- 3M Tel Comm Products
- ADC Telecommunications
- Alpha Wire
- Amp Products
- Amphenol
- Anaconda-Ericsson Inc.
- AT&T
- Belden Corp.
- Cook Electric/Northern Telecom
- Communication Mfg. Co.
- General Cable Company
- Lanier/Harris
- Newton Instrument Co.
- Nevada Western
- Siecor
- Simplex Wire And Cable Company
- Standard Wire and Cable Co.
- Thomas & Betts

Call Accounting Systems

- Account-A-Call Corp.
- American Telecommunications
- Com Dev
- Infortext Systems
- Lanier/Harris
- Moscom
- NEC Information Systems
- Summa Four
- Sykes
- Telco Research
- Telemon-Tecom
- Western Telematic
- Xerox
- Xiox
- Xtend Communications

Data Communications Equipment

- Amdahl Communications Systems Division
- AT&T
- BBL Industries
- Codex Corp.
- Case Communications
- Equinox Systems
- Gandalph Technologies
- General Datacomm, Inc.
- Micom Systems
- NEC America
- Northern Telecom
- Paradyne Corp.
- Racal-Milgo
- Redcom Laboratories
- Rockwell International
- Timeplex
- Universal Data Systems Inc.

Data PBXs

- Equinox Systems
- Gandalf Technologies
- Micom Systems
- Solid State Systems
- Toshiba

Distributing Frames

- ADC Telecommunications
- AT&T
- Cook Electric/Northern Telecom
- Porta Systems

Facsimile

AT&T
Canon USA
Fujitsu
Gamma Link
Harris/3M
Hitachi America
Murata Business Systems
NEC America
Omnifax
Panafax
Panasonic
Pitney Bowes
Ricoh
Sanyo
Sharp
Toshiba
Xerox

Key Telephone Equipment/Hybrid PBX

AT&T
Candela Electronics
Comdial Corporation
Eagle Telephonics, Inc.
Fujitsu
GAltronics
IBM/Rolm
Isotec
Lanier/Harris
Melco Labs.
NEC America
Panasonic
Redcom Laboratories
Solid State Systems
Tadiran Electronic Industries
Telrad
TIE/communications

Local Area Networks

3COM
Allen Bradley
AT&T
Banyan Systems
Bridge Communications
Concord Data Systems
Corvus Systems, Inc.
Excelan
Gandalph Technologies
IBM
Lanier/Harris
Northern Telecom
Novell
Proteon
Redcom Laboratories
Siemens Information Systems
Sytek, Inc.
Tadiran Electronic Industries
Ungermann-Bass
Wang Laboratories, Inc.
Xerox
Xtend Communications

Power Equipment

AT&T
Lorain Products
Northern Telecom
Reliable Electric/Utility Products

Private Branch Exchanges (PBXs)

AT&T
CXC
Ericsson
Fujitsu Business Communications
GTE Communication Systems
Harris Digital
Hitachi America
IBM/Rolm
InteCom
ITT Business Communications
Mitel, Inc.
NEC America
Redcom Laboratories
Northern Telecom

Siemens Information Systems
Solid State Systems
SRX
Toshiba America
Tie/communications

Private Payphones

ACT Communications
American Coin Telephone
American Paytel
Extrom Communications
Intellicall
Network Paystations
Northern Telecom
Protel
R-Tec Systems
Stromberg-Carlson

Protection Equipment

AT&T
Cook Electric/Northern Telecom
Lorain Products
Porta Systems

Test Equipment

AT&T
Atlantic Research Corporation
Biddle Instruments
Cushman Electronics
Dynatech Corporation
Dynatel 3M
Frederick Electronics Corp.
Halcyon Communications, Inc.
Harris/Dracon
Hekimian Laboratories, Inc.
Hewlett Packard Co.
John Fluke
Lear Siegler, Inc.
Marconi Instruments
Northern Telecom
Plantronics Wilcom
Scientific-Atlanta Inc.
Sierra Electronic Division, Lear Siegler, Inc.
Systron-Donner Corp.
Tektronix, Inc.
W & G Instruments Inc.
Triplett Corporation
Wilcom Products
Wiltron Company

T-1 Carrier Equipment

AT&T
Aydin Monitor System Digital Communications Group
Coastcom
Granger Associates
ITT Corp.
Kentrox Industries
Lynch Communications Systems Inc.
Micom Systems
NEC America
Northern Telecom
Telco Systems
Tellabs Inc.
Wescom, Telephone Products Div., Rockwell Telecommunications Inc.

Transmission Services

ADP Autonet
American Satellite Corporation
AT&T
CompuServe, Inc.
GTE Telenet Communications Corp.
MCI Telecommunications Corporation
RCA Global Communications
Satellite Business Systems
Tymnet, Inc.
Western Union Telegraph Co.

Voice Mail

- Alston/Conrac
- AT&T
- BBL Industries
- Genesis Electronics
- IBM/Rolm
- Octel Communications
- Opcom
- Optelcom
- VMX
- Voice Messaging Systems

Names and Addresses of Manufacturers

3COM
1365 Shorebird Way
Mountain View, CA 97043
(415) 961-9602

3M Company, Telcomm Products Division
3M Center, Bldg. 225-4S-06
St. Paul, MN 55144
(612) 733-9646

3M Tel Comm Products
1909 W Braker Ln. #504
Austin, TX 78769
(512) 834-3862

Account-A-Call Corporation
4450 Lakeside Dr.
Burbank, CA 91505
(818) 846-3340

ACT Communications
109 N. Main St.
Eaton, TX 75439
(214) 961-2300

ADC Telecommunications
4900 W. 78th St.
Minneapolis, MN 55435
(612) 835-6800

ADP Autonet
175 Jackson Plaza
Ann Arbor, MI 4810
(313) 769-6800

Allen Bradley
747 Alpha Dr.
Cleveland, OH 44143
(216) 449-6700

Alpha Wire
Box 711
Elizabeth, NJ 07207
(201) 925-8000

Alston Div. Conrac Corporation
1724 S. Mountain Ave.
Duarte, CA 91010
(818) 357-2121

Amdahl Communications Systems Division
2500 Walnut Ave.
Marina del Rey, CA 90291
(213) 822-3202

American Coin Telephone
4203 Woodstock Dr. #252
San Antonio, TX 78228
(512)736-4401

American PayTel
9642 W. Farragut Ave.
Rosemont, IL 60018
(312) 678-6681

American Satellite Corporation
20301 Century Blvd.
Germantown, Md. 20767
(301) 428-6040

American Telecommunications
15851 Dallas Parkway #1120
Dallas, TX 75248
(214) 934-9500

AMP Telecom Division
P. O. Box 1776
Southeastern, PA 19399
(215) 647-1000

Amphenol
2122 York Rd.
Oak Brook, IL 60521
(312) 986-2300

Anaconda-Ericsson Inc.
Wire & Cable Div.-Telcom Cable
P.O. Box 4405
Overland Park, KS. 66204
(913) 677-7500

AT&T
Bedminister, NJ 07921
(201) 234-4000

AT&T
1 Speedwell Ave.
Morristown, N.J. 07960
(800) 247-1212

AT&T Network Systems
P.O. Box 1278R
Morristown, NJ 07960

AT&T Technologies Consumer Products
5 Woodhollow Rd.
Parsippany, NJ 07054
(201) 581-3000

Atlantic Research Corporation
5390 Cherokee Ave.
Alexandria, VA 22312
(703) 642-4000

Aydin Monitor System Digital Communications Group
502 Office Center Dr.
Ft. Washington, PA 19034
(215) 646-8100

Banyan Systems
115 Flanders Rd.
Westboro, MA 01581
(617) 898-1000

BBL Industries
2935 Northeast Pkwy.
Atlanta, GA 30360
(404) 449-7740

Belden Corp.
2000 S. Batavia Ave.
Geneva IL 60134
(312) 232-8900

Biddle Instruments
510 Township Line Rd.
Blue Bell, PA 19422
(215) 646-9200

Bridge Communications
1345 Shoebird Way
Mountain View, CA 94043
(415) 969-4400

Candela Electronics
550 Del Rey Ave.
P.O. Box 461
Sunnyvale, CA 94088
(408) 738-3800

Canon USA
One Canon Plaza
Lake Success, NY 11042
(212) 688-1200

Case Communications
7200 Riverwood Dr.
Columbia, MD 21046
(301) 290-7710

Coastcom
2312 Stanwell Dr.
Concord, CA 94520
(415) 825-7500

Code-A-Phone Corporation
P.O. Box 5656
Portland, OR 97228
(503) 655-894

Codex Corp.
20 Cabot Blvd.
Mansfield, MA 02408
(617) 364-2000

Com Dev
2006 Whitefield Industrial Way
Sarasota, FL 34243
(813) 753-6411

Comdial Corporation
P.O. Box 7266
Charlottesville, VA 22906
(804) 978-2458

Communication Mfg. Co.
P.O. Box 2708
Long Beach, CA 90801
(213) 426-8345

CompuServe, Inc.
5000 Arlington Centre Blvd.
Columbus, OH 43220
(614) 457-8600

Concord Data Systems
303 Bear Hill Rd.
Waltham, MA 02154
(617) 890-1394

Cook Electric Division of Northern
Telecom
6201 Oakton St.
Morton Grove, IL 60059
(312) 967-6600

Corvus Systems, Inc.
2100 Corvus Dr.
San Jose, CA 95124
(408) 559-7000

Cushman Electronics
2450 N. First St.
San Jose, CA 95131
(408) 263-8100

Dynatech Data Systems Division of
Dynatech Corporation
7644 Dynatech Ct.
Springfield, VA 22153
(703) 569-9000

Dynatel 3M
380 N. Pastoria Ave.
Sunnyvale, CA 95086
(408) 733-4300

Eagle Telephonics, Inc.
375 Oser Ave.
Hauppauge, NY 11788
(516) 273-6700

Equinox Systems
14260 S.W. 119th Ave.
Miami, FL 33186
(800) 328-2729

Ericsson Inc.
Communications Division
7465 Lampson Ave.
Garden Grove, CA 92641
(714) 895-3962

Excelan
2180 Fortune Dr.
San Jose, CA 95131
(408) 434-2300

Extrom Communications
137 Express St.
Plainview, NY 11803
(516) 935-5311

Frederick Electronics Corp.
Hayward Rd.
P.O. Box 502
Frederick, MD 21701
(301) 662-5901

Fuijitsu
Corporate Drive Commerce Park
Danbury, CT 06810
(203) 796-5400

Fuijitsu America, Inc. Data
Communications Div.
3055 Orchard Dr.
San Jose, CA 95134
408-946-8777

Fuijitsu Business Communications
3190 MiraLoma Ave.
Anaheim, CA 92806
(714) 630-7721

GAltronics
P.O. Box 31
Reading, PA 19603
(215) 777-1374

Gamma Link
2452 Embarcadero Way
Palo Alto, CA 94303
(415) 856-7421

Gandalf Technologies, Inc.
350 E. Dundee Rd.
Wheeling, IL 60090
(312) 541-6060

General Cable Co.
Fiber Optics Div.
160 Fieldcrest Ave.
Raritan Center, Edison, NJ 08810
(201) 225-4780

General Cable Company
500 W. Putnam Ave.
Greenwich CT 06830
(203) 661-0100

General Datacomm, Inc.
One Kennedy Ave.
Danbury, CT 06762-1299
(203) 797-0711

Genesis Electronics
103 Woodmere Rd.
Folsom, CA 95630
(916) 985-4050

Granger Associates
3101 Scott Blvd.
Santa Clara, CA 95051
(408) 727-3101

GTE Communication Systems
2500 W. Utopia Rd.
Phoenix, AZ 85027 (602)
582-7000

GTE Corporation Business
Communications Systems, Inc.
12502 Sunrise Valley Dr.
Reston, VA 22090
(703) 435-7643

GTE Telenet Communications
Corporation
8229 Boone Blvd.
Vienna, VA 22180
(703) 442-1000

Halcyon Communications, Inc.
2121 Zanker Rd.
San Jose, CA 95131
(408) 293-9970

Harris Corporation, Digital
Telephone Systems Div.
1 Digital Dr.
P.O. Box 1188
Novato, CA 94947
(415) 472-2500

Harris/3M Center
2300 Parklake Dr. NE
Atlanta, GA 30345-2979
(404) 873-9500

Hekimian Laboratories, Inc.
9298 Gaither Rd.
Gaithersburg, MD 20877
(301) 840-1217

Hewlett-Packard Co.
1501 Page Mill Rd.
Palo Alto, CA 94304
(415) 493-1501

Hitachi America, LTD.,
Telecommunications Research
and Sales Div.
2990 Gateway Dr., Suite 1000
Norcross, GA 30071
(404) 446-8820

IBM Corporation
Old Orchard Rd.
Armonk, NY 10504
(914) 765-1900

IBM Corporation
P.O. Box 12195
Research Triangle Park, NC 27709
(919) 543-5221

IBM/Rolm Corporation
4900 Old Ironsides Dr.
Santa Clara, CA 95050
(408) 986-1000

Infortext Systems
1067 E. State Pkwy.
Schaumburg, IL 60173
(312) 490-1155

Intecom, Inc.
601 Intecom Dr.
Allen TX 75002
(214) 727-9141

Intellicall
2155 Chanault Dr. #410
Carrollton, TX 75006
(214) 416-0022

Isotec
6 Thorndal Circle
Darien CT 06820
(203) 655-6500

ITT Telecom, Network Systems Division
3100 Highwoods Blvd.
Raleigh, NC 27604
(919) 872-3359

John Fluke Mfg. Co., Inc.
P.O. Box C9090
Everett, WA 98206
(206) 365-5400

Kentrox Industries, Inc.
P.O. Box 10704
Portland, OR 97201
(503) 643-1681

Lanier/Harris Business Products Inc.
1700 Chantilly Dr. N.E.
Atlanta, GA 30324
(404) 329-8000

Lear Siegler, Inc., Electronic Instrumentation Div.
714 N. Brookhurst St.
Anaheim, CA 92803
(714) 774-1010

Lynch Communications Systems Inc.
204 Edison Way
Reno, NV 89520
(702) 786-4020

Marconi Instruments, Division of Marconi Electronics, Inc. 100
Stonehurst Ct.
Northvale NJ 07647
(201) 767-7250

MCI International
International Dr.
Rye Brook, NY 10573
(914) 937-3444

MCI Telecommunications Corporation
1133 19th St. N.W.
Washington, DC 20036
(202) 872-1600

Melco Labs
14408 N.E. 20th St.
Bellevue, WA 98007
(206) 643-3400

Micom Systems, Inc.
20151 Nordhoff St.
Chatsworth,, CA 91311
(213) 998-8844

Mitel, Inc.
5400 Broken Sound Blvd.
N.W. Boca Raton, FL 33431
(305) 994-8500

Moscom
300 Main St.
E. Rochester, NY
(716) 385-6440

Murata Business Systems
4801 Spring Valley Rd.
Building 108-B
Dallas, TX 75244
(214) 392-1622

NEC Telephones, Inc.
532 Broad Hollow Rd.
Melville, NY 11747
(800) 626-4952

Network Paystations
124 W 66th St.
Cincinnati, OH 45216
(513) 242-7483

Nevada Western
930 W. Maude Ave.
Sunnyvale, CA 94086
(408) 737-1600

Newton Instrument Co., Inc.
111 East A St.
Butner, NC 27509
(919) 575-6426

Northern Telcom Inc.
Northeast Electronics Div.
Airport Rd., P.O. Box 649
Concord, NH 03301
(603) 224-6511

Northern Telcom Inc., Digital Switching Systems
4001 E. Chapel Hill-Nelson Hwy.
Research Triangle Pk., NC 27709
(919) 549-5000

Northern Telcom Inc., Network Systems
1201 E. Arapaho Rd.
Richardson, TX 75081
(214) 234-7500

Northern Telecom, Inc.
2150 Lakeside Blvd.
Richardson, TX 75081
(214) 437-8000

Novell, Inc.
1170 N. Industrial Park Dr.
Orem, UT 84057
(801) 226-8202

Octel Communications
890 Tasman Dr.
Milpitas, CA 95035
(408) 942-6500

Optelcom
15930 Luanne Dr.
Gaithersburg, MD 20877
(301) 840-2121

Panafax Corp.
10 Melville Park Rd.
Melville, NY 11747
(516) 420-0055

Panasonic Co. Telephone Products Div.
1 Panasonic Way
Secaucus, NJ 07094
(201) 348-7000

Paradyne Corporation
8550 Ulmerton Rd.
Largo, FL 33540
(813) 536-4771
213) 618-9910

Pitney Bowes Facsimile Systems
1515 Summer St. 5th Floor
Stamford, CT 06926
(203) 356-7178

Plantronics Wilcom
P.O. Box 508
Laconia, N.H. 03246
(603) 524-2622

Porta Systems Corp.
575 Underhill Blvd.
Syosset, NY 11791
(516) 364-9300

Protel
P.O. Box 1052
Boulder, CO 80306
(303) 444-0971

Racal-Milgo, Inc.
8600 N.W. 41st St.
Miami, FL 33166
(305) 592-8600

RCA Global Communications
60 Broad St.
New York, NY 10004
(212) 806-7000

Redcom Laboratories
1 Redcom Ctr.
Victor, NY 14564
(716) 924-7550

Reliable Electric/Utility Products
11333 Addison St.
Franklin Park, IL 60131
(312)455-8010

Reliance Comm/Tec, R-Tec Systems
2100 Reliance Pkwy.
P.O. Box 919
Bedford, TX 76021
(817) 267-3141

Ricoh Corp.
5 Didrick Pl.
West Caldwell, NJ
(201) 882-2000

Rockwell International Corp.,
Collins Transmission Systems
Div.
1200 N. Alma Rd.
Richardson, TX 75081
(214) 996-5000

Rockwell International Corp.,
Switching Systems Div.
1431 Opus Pl.
Downers Grove, IL 60515
(312) 852-5700

Sanyo
51 Joseph St.
Moonachie, NJ 07074
(201) 440-9300

Satellite Business Systems
8283 Greensboro Dr.
McLean, VA 22102
(703) 442-5000

Scientific-Atlanta Inc.
Box 105600
Atlanta, GA 30348
(404) 411-4000

Sharp
Sharp Plaza
Mahwah, NJ 07430
(201) 529-9500

Siecor Corporation
489 Siecor Park
Hickory, NC 28603
(704) 328-2171

Siecor Fiberlan
Box 12726
Research Triangle Park, NC 27709
(919) 544-3791

Siemens Corporation, Telephone
Division
N.W. 58th St.
Boca Raton, FL 33431
(201) 494-1000

Sierra Electronic Division, Lear
Siegler, Inc.
3885 Bohannon Dr.
Menlo Park, CA 94025
(415) 321-5374

Simplex Wire And Cable Company
P.O. Box 479
Portsmouth, NH 03801
(603) 436-6100

Solid State Systems
1300 Shiloh Rd. NW
Kennecow, GA 30144
(404) 423-2200

SRX
15926 Midway Rd.
Dallas, TX 75244
(214) 934-9111

Standard Wire and Cable Co.
2345 Alaska Ave.
El Segundo, CA 90245
(213) 536-0006

Stromberg Carlson Corp.
400 Rinehart Rd.
Lake Mary, FL 32746
(305) 849-3000

Summa Four
2456 Brown Ave.
Manchester, NH 03103
(603) 625-4050

Sykes Datatronics, Inc.
159 Main St. E.
Rochester, NY 14604
(716) 325-9000

Systron-Donner Corp.
2727 Systron Dr.
Concord, CA 94518
(415) 671-6589

Sytek, Inc.
1225 Charleston Rd.
Mountain View, CA 94039
(415) 966-7333

Tadiran Electronic Industries
5733 Myerlake Circle
Clearwater, FL 33520
(813) 536-3222

Tektronix, Inc.
P.O. Box 500
Beaverton OR 97077
(503) 644-0161

Telamon-Tecom
8134 Zionsville Rd.
Indianapolis, IN 46268
(317) 875-0045

Telco Research
1207 17th Ave. S
Nashville, TN 37212
(615) 329-0031

Telco Systems Fiber Optics Corp.
33 Boston-Providence Highway
Norwood, MA 02062
(617) 769-7510

Telco Systems, Inc.
1040 Marsh Rd.
Suite 100
Menlo Park, CA 94025
(415) 324-4300

Tellabs Inc.
4951 Indiana Ave.
Lisle, IL 60532
(312) 969-8800

Telrad
510 Broad Hollow Rd.
Melville, NY 11747
(516) 420-1350

Thomas & Betts
920 Rt. 202
Raritan, NJ 08869
(201) 685-1600

TIE/communications, Inc.
8 Progress Dr.
Shelton, CT 06484
(203) 929-7373

Timeplex Inc.
400 Chestnut Ridge Rd.
Woodcliff Lake, NJ 07675
(201) 930-4600

Toshiba Facsimile Product Group
9740 Irvine Blvd.
Irvine, CA 92718
(714) 380-3000

Toshiba Telecom
2441 Michelle Dr.
Tustin, CA 92680
(714) 730-5000

Triplett Corporation
One Triplett Dr.
Buffton, OH 45817
(419) 358-5015

Tymnet, Inc.
2710 Orchard Parkway
San Jose, CA 95134
(408) 946-4900

Ungermann Bass Inc.
2560 Mission College Blvd.
Santa Clara, CA 95050
(408) 496-0111

Universal Data Systems Inc.
5000 Bradford Dr.
Huntsville, AL 35805
(205) 837-8100

VMX
17217 Waterview Pkwy.
Dallas, TX 75244
(214) 386-0300

Voice Messaging Systems
21 Franklin St.
Quincy, MA 02169
(617) 770-4301

W & G Instruments, Inc.
119 Naylon Ave.
Livingston, NJ 07039
(201) 994-0854

Wang Laboratories, Inc.
One Industrial Ave
Lowell, MA 01851
(617) 459-5000

Wescom, Telephone Products Div., Rockwell Telecommunications, Inc.
8245 S. Lemont Rd.
Downers Grove, IL 60515
(312) 985-9000

Western Telematic
5 Sterling St.
Irvine, CA 92718
(714) 586-9950

Western Union Telegraph Co.
1 Lake St.
Upper Saddle River, NJ 07458
(201) 825-5000

Wiltron Company
805 E. Middlefield Rd.
P.O. Box 7290
Mountain View, CA 94042
(415) 969-6500

Xerox Corporation, Information Products Div.
1301 Ridgeview Dr.
Lewisville, TX 75067
(214) 412-7200

Xiox
577 Airport Blvd #700
Burlingame, CA 94010
(415) 375-8188

Xtend Communications
171 Madison Ave.
New York, NY 10016
(212) 725-2010

APPENDIX L

TELECOMMUNICATIONS SYSTEM OPERATIONS REVIEW

Check marks in boxes below indicate a discrepancy or need for further investigation.

Costs

- □ Are all telecom cost components being tracked?
- □ Does someone regularly watch cost trends?
- □ Is there a budget for each telecom component?
- □ Are all components within the budget? If not, why?
- □ Is there evidence that someone investigates budgetary discrepancies?
- □ Are telecom costs allocated to cost centers?
- □ Are telecom costs treated as a total company overhead?
- □ What has been the trend in telecom costs over the past three years?
- □ Is there a logical explanation for cost variances?
- □ Are toll messages regularly analyzed to determine feasibility of alternate long distance services?

Equipment Room

- □ Are physical aspects of equipment room satisfactory?
  - □ Floor space?
  - □ Lighting?
  - □ Heating?
  - □ Air conditioning and air circulation?
  - □ Fire prevention?
  - □ Accident prevention?
- □ Is housekeeping satisfactory?
- □ Is the room inaccessible to unauthorized persons?
- □ Are cables properly secured?

- ☐ Are cables properly tagged and designated?
- ☐ Is equipment adequately grounded?

Equipment

- ☐ Is equipment clearly labeled and designated?
- ☐ Are equipment cabinets locked?
- ☐ Is the system equipped with a call accounting system? If so, what use is made of the output? If not, could one be profitably employed?
- ☐ Is call accounting equipment in a secured location?
- ☐ Is an equipment maintenance log maintained?

Records

- ☐ Are the following records available, accurate, current, and legible?
  - ☐ Wiring diagrams?
  - ☐ Interconnect records?
  - ☐ Cable assignment records?
  - ☐ Station records?
  - ☐ Station restriction records?
  - ☐ Answering location records?
  - ☐ Equipment property records?
  - ☐ Power failure transfer records?

Wiring Records

- ☐ Are cables clearly labeled?
- ☐ Are cables securely terminated?
- ☐ Are cable pairs labeled on distributing blocks?
- ☐ Are wiring closets secure?
- ☐ Are cords clear of traffic patterns?
- ☐ Is wiring properly secured in supporting structures?

Trouble Handling Procedures

- ☐ Are all trouble reports fully documented?
  - ☐ Is there a standard trouble report format?
  - ☐ Are trouble reports complete?
  - ☐ Are reports retained?
- ☐ Are trouble reports being analyzed?
- ☐ Is there evidence of unusual trouble patterns?
- ☐ Is there evidence of repeated troubles of the same type?
- ☐ Does someone follow up on trouble patterns?

- ☐ Are troubles tracked by major service and equipment category?
- ☐ Is clearing time within objectives?
- ☐ Does the company test trouble before referring to vendor?
- ☐ How does the company decide which vendor to call?
- ☐ Has anyone been trained in trouble testing?
- ☐ Is there evidence of costs incurred for calling the wrong vendor?

System Administration

- ☐ Does the system have any provisions for measuring service and load?
- ☐ If so, is data being collected regularly?
- ☐ If not, are special studies being made?
- ☐ What usc is made of traffic data?
- ☐ Are attendant console statistics being evaluated?
- ☐ Do trunk groups appear to be properly sized?
- ☐ Have objectives been set for trunk overflows?
  - ☐ Are objectives reasonable?
  - ☐ Are overflows within the objective range?
- ☐ Are outgoing trunk overflows out of limits?
- ☐ Are ARS patterns logically assigned?
- ☐ Is feature usage regularly checked?
- ☐ Are restriction levels logically assigned?
- ☐ Is data terminal response time regularly measured?
- ☐ Are data performance records retained and analyzed?
- ☐ Are administrative changes logged?

Operational Performance

- ☐ Are quality checks made?
  - ☐ Transmission quality
  - ☐ Trunk group blockage
  - ☐ Message handling
- ☐ Are user satisfaction surveys made?
- ☐ Is there evidence of lost time attributable to telecom services?
- ☐ Are dialing plans devised to take full advantage of system capabilities?
- ☐ Does the company have a policy on personal use of telecommunications services? Is it enforced?
- ☐ What percent of station calls are answered by the fourth ring?
- ☐ Are station users making full use of features such as call forwarding and call pickup?
- ☐ Have station users been trained?
- ☐ Are messages relayed efficiently?

- ☐ Do telecommunications employees have documented job duties?
- ☐ Do console attendants present a professional image for the company?
- ☐ Are attendants fully trained?
- ☐ Has relief training been administered?
- ☐ What is the average number of rings before telephones are answered?
- ☐ Are abandoned call statistics available? If so, are they trended? Do they coincide with peak console load periods?
- ☐ Does the attendant have duties that interfere with answering incoming calls?

Security

- ☐ Is remote access (DISA) used?
- ☐ Are statistics on its use available?
- ☐ Is there evidence of abuse?
- ☐ Who has access to the password for the system administration console?
  - ☐ Is the password posted?
  - ☐ Is it ever changed?
  - ☐ Who covers in absence of the custodian of the password?
- ☐ Is equipment and cable secure from tapping?
- ☐ Are terminal passwords guarded?
- ☐ Are passwords changed when personnel are replaced?

GLOSSARY

Note: Readers are also referred to Appendix D, which lists descriptions for PBX and key telephone features.

Access The capability of terminals to be interconnected with one another for the purpose of exchanging traffic.

Adaptive Differential Pulse Code Modulation (ADPCM) A method approved by CCITT for coding voice channels at 32 kb/s to increase the capacity of T-1 to either 44 or 48 channels.

Adaptive equalizer Circuitry in a modem that allows the modem to compensate automatically for circuit conditions that impair high speed data transmission.

Algorithm A set of processes in a computer program used to solve a problem with a given set of steps.

Alternate routing The ability of a switching machine to establish a path to another machine over more than one circuit group.

American Standard Code for Information Interexchange (ASCII) A seven bit (plus one parity bit) coding system used for encoding characters for transmission over a data network.

Amplitude distortion Any variance in the level of frequencies within the passband of a communications channel.

Analog A transmission mode in which information is transmitted by converting it to a continuously variable electrical signal.

Answer supervision A signal that is sent from a switching system through the trunking network to the originating end of a call to signal that a call has been answered.

Asynchronous transmission A means of transmitting data over a network wherein each character contains a start and stop bit to keep the transmitting and receiving terminals in synchronism with each other.

Automatic Call Distributor (ACD) A switching system which automatically distributes incoming calls to a group of answering positions without going through an operator. If all answering positions are busy, the calls are held until one becomes available.

Availability The ratio of circuit up time to total elapsed time.

Backplane The wiring in a card carrier that interconnects circuit packs and equipment.

Balance The degree of electrical match between the two sides of a cable pair or between a two-wire circuit and the matching network in a four-wire terminating set.

Balun A device that converts the unbalanced wiring of a coaxial terminal system to a balanced twisted pair system.

Bandwidth The range of frequencies a communications channel is capable of carrying without excessive attenuation.

Baseband A form of modulation in which data signals are pulsed directly on the transmission medium without frequency division.

Baud The number of data signal elements per second a data channel is capable of carrying.

Bearer channel A 64 kb/s information-carrying channel that furnishes integrated services digital network (ISDN) services to end users.

Binary A numbering system consisting of two digits, zero and one.

Binary Synchronous Communications (BSC or Bisync) An IBM byte-controlled half-duplex protocol using a defined set of control characters and sequences for data transmission.

Bit Error Rate (BER) The ratio of bits transmitted in error to the total bits transmitted on the line.

Bit The smallest unit of information that can be processed or transported over a circuit; contraction of the words ''BInary digiT.''

Bit rate The speed at which bits are transmitted on a circuit; usually expressed in bits per second.

Bit stream A continuous string of bits transmitted serially in time.

Block error rate In a given unit of time, measures the number of blocks that must be retransmitted because of error

Blocked Calls Delayed (BCD) A variable used in queuing theory to describe the behavior of the input process when the user is held in queue when encountering blockage.

Blocked Calls Held (BCH) A variable used in queuing theory to describe the behavior of the input process when the user immediately redials when encountering blockage.

Blocked Calls Released (BCR) A variable used in queuing theory to describe the behavior of the input process when the user waits for a time before redialing when encountering blockage.

Blocking A switching system condition in which no circuits are available to complete a call, and a busy signal is returned to the caller.

Bridge Circuitry used to interconnect networks with a common set of higher level protocols.

Bridged tap Any section of a cable pair that is not on the direct electrical path between the central office and the user's premises.

Broadband A form of modulation in which multiple channels are formed by dividing the transmission medium into discrete frequency segments.

Busy hour The composite of various peak load periods selected for the purpose of designing network capacity.

Byte A set of eight bits of information equivalent to a character. Also sometimes called "octet."

C message weighting A factor used in noise measurements to describe the lesser annoying effect on the human ear of high and low frequency noise compared to mid-range noise.

Call-back queueing A trunk queueing system in which the switching system signals the users that all trunks are busy, and calls them back when a trunk is available.

Call sequencer An electronic device similar to an automatic call distributor that can answer calls, distribute them to agent positions, hold callers in queue, and provide statistical information.

Call warning tone A tone placed on a circuit to indicate that the call is about to route to a high-cost facility.

Capacity The number of call attempts and busy hour load that a switching machine is capable of supporting.

Carrier (1) A type of multiplexing equipment used to derive several channels from one communications link by combining signals on the basis of time or frequency division. (2) A card cage used in an apparatus cabinet to contain multiple circuit packs. (3) A company that carries telecommunications messages and private channels for a fee.

Carrier Sense Multiple Access with Collision Detection (CSMA/CD) A system used in contention networks where the network interface unit listens for the presence of a carrier before attempting to send, and detects the presence of a collision by monitoring for a distorted pulse.

Causal forecast A forecasting method that assumes the variable being forecast is directly related to and caused by another variable.

Central Processing Unit (CPU) The control logic element used to execute instructions and manipulate data in a computer.

Centralized Attendant Service (CAS) A PBX feature that allows the using organization to route all calls from a multi-PBX system to a central answering location where attendants have access to features as if they were colocated with the PBX.

Centrex A class of central office service that provides the equivalent of PBX service from a telephone company switching machine. Incoming calls can be dialed directly to extensions without operator intervention.

Channel A path in a communications system between two or more points, furnished by a wire, radio, lightwave, satellite, or a combination of media.

Channel bank Apparatus that encodes multiple voice frequency signals into frequency or time division multiplexed signals for transmitting over a transmission medium.

Channel Service Unit (CSU) Apparatus that interfaces DTE to a line connecting to a dataport channel unit to enable digital communications without a modem. Used with DSU when DTE lacks complete digital line interface capability.

Circuit A transmission path between two points in a telecommunications system.

Circuit pack A plug-in electronic device that contains the circuitry to perform a specific function. A circuit pack is not capable of stand alone operation, but functions only as an element of the parent device.

Circuit switching A method of network access in which terminals are connected by switching together the circuits to which they are attached. In a circuit switched network the terminals have full real time access to each other up to the bandwidth of the circuit.

Class of service The service classification within a telecommunications system that controls the features, calling privileges, and restrictions the user is assigned.

Clear channel A 64 kb/s digital channel that uses external signaling and therefore permits all 64 kb/s to be used for data transmission.

Coder/Decoder (CODEC) A device used to convert analog signals to digital and vice versa.

Collision A condition that occurs when two or more terminals on a contention network attempt to acquire access to the network simultaneously.

Common control switching A switching system that uses shared equipment to establish, monitor, and disconnect paths through the network. The equipment is called into the connection to perform a function and then released to serve other users.

Compounding The growth in capital that occurs when the principal is increased by the amount of interest that accumulates each period.

Computer Branch Exchange (CBX) A computer-controlled PBX.

Concentrator A data communications device that subdivides a channel into a larger number of data channels. Asynchronous channels are fed into a high speed synchronous channel via a concentrator to derive several lower speed channels.

Conditioning Special treatment given to a transmission facility to make it acceptable for high speed data communication.

Cost of money The composite cost of capital that is used as the discount rate in a financial study.

Consultative Committee on International Telephone and Telegraph (CCITT) An international committee that sets telephone, telegraph, and data communications standards.

Contention A form of multiple access to a network in which the network capacity is allocated on a "first come first served" basis.

Critical path The longest series of tasks in a PERT network. The critical path determines the overall completion time required.

Crossconnect A wired connection between two or more elements of a telecommunications circuit.

Crosstalk The unwanted coupling of a signal from one transmission path into another.

Current planning Planning that is concerned with an organization's near term (one to three years) requirements.

Customer Premise Equipment (CPE) Telephone apparatus mounted on the user's premises and connected to the telephone network.

Cutover Any change from an existing to a new telecommunications system.

Cyclicality The assumption in forecasting that the variable being forecast changes with some observable cycle.

Cyclical Redundancy Check (CRC) A data error-detecting system wherein an information block is subjected to a mathematical process designed to ensure that errors cannot occur undetected.

Data Circuit-Terminating Equipment (DCE) Equipment designed to establish a connection to a network, condition the input and output of DTE for transmission over the network, and terminate the connection when completed.

Data line monitor A data line impairment-measuring device that bridges the data line and observes the condition of data, addressing, and protocols.

Data Service Unit (DSU) Apparatus that interfaces DTE to a line connecting to a dataport channel unit to enable digital communications without a modem. Used with CSU when DTE lacks complete digital line interface capability or alone when DTE includes digital line interface capability.

Data Terminal Equipment (DTE) Any form of computer, peripheral, or terminal that can be used for originating or receiving data over a communication channel.

DBm A measure of signal power as compared to one milliwatt (1/1000 watt) of power. It is used to express power levels. For example, a signal power of -10 dBm is 10 dB lower than one milliwatt.

DBrn A measure of noise power relative to a reference noise of -90 dBm.

DBrnc0 A measure of C message noise referred to a zero test level point.

Decibel (dB) A measure of relative power level between two points in a circuit.

Dedicated circuit A communications channel assigned for the exclusive use of an organization.

Delphi A forecasting method in which several knowledgeable individuals make forecasts. The analyst derives the forecast from a weighted average.

Delta modulation A system of converting analog to digital signals by transmitting a single bit indicating the direction of change in amplitude from the previous sample.

Demarcation point The point at which customer-owned wiring and equipment interfaces with the telephone company.

Demodulation The process of retrieving an original signal from a modulated carrier wave.

Diagnostic Test programs used for error and fault detection in apparatus or a circuit.

Dial-One WATS A long distance service that carries a fixed monthly fee that entitles the user to purchase service at a discount.

Digital A mode of transmission in which information is coded in binary form for transmission on the network.

Digital Access Crossconnect System (DACS) A specialized digital switch that enables crossconnection of channels at the digital line rate.

Digital Multiplexed Interface (DMI) A T-1 interface between a PBX and a computer.

Digital Service Crossconnect (DSX) A physically wired crossconnect frame to enable connecting digital transmission equipment at a standard bit rate.

Discount rate See *Cost of Money.*

Discounted payback period The number of years in which a stream of cash flows, discounted at an organization's cost of money, repays an initial investment.

Discounting The process of computing the present worth of a future cash flow by reducing it by a factor equivalent to the organization's cost of money and the time until the cash flow occurs.

Distributing frame A framework holding terminal blocks that are used to interconnect cable and equipment and to provide test access.

Downlink The radio path from a satellite to an earth station.

Dual Tone Multi Frequency (DTMF) A signaling method used between the station and the central office consisting of a push button dial that emits dual tone encoded signals. The generic name for Touchtone, which is a trademark of AT&T.

E and M signaling A common designation for the transmit and receive leads of signaling equipment at the point of interface with connected equipment.

Earth station The assembly of radio equipment, antenna, and satellite communication control circuitry that is used to provide access from terrestrial circuits to satellite capacity.

Echo canceler An electronic device that processes an echo signal and cancels it out to prevent annoyance to the talker.

Echo return loss The weighted return loss of a circuit across a band of frequencies from 500 to 2500 Hz.

Echo suppressor A device that opens the receive path of a circuit when talking power is present in the transmit path.

Echo The reflection of a portion of a signal back to its source.

Electronic mail A service that allows text-form messages to be stored in a central file, and retrieved over a data terminal by dialing access and identification codes.

Envelope delay The difference in propagation speed of different frequencies within the pass band of a telecommunications channel.

Equal access A central office feature that allows all interexchange carriers to have identical access to the trunk side of the switching network in an end office.

Erlang A unit of network load. One Erlang equals 36 CCS and represents 100 percent occupancy of a circuit or piece of equipment. Also used to define the input process under the BCC (Erlang B) and BCD (Erlang C) blockage conditions.

Error Any discrepancy between a received data signal and the signal as it was transmitted.

Error free seconds The number of seconds per unit of time that a circuit vendor guarantees the circuit will be free of errors.

Ethernet A proprietary contention bus network developed by Xerox, Digital Equipment Corporation, and Intel. Ethernet formed the basis for the IEEE 802.3 standard.

Explanatory forecast A forecasting method that attempts to discover an underlying pattern without necessarily discovering the fundamental cause.

Extended Super Frame (ESF) T-1 carrier framing format that provides 64 kb/s clear channel capability, error checking, 16 state signaling, and other data transmission features.

Facility Any set of transmission paths that can be used to transport voice or data. Facilities can range from a cable to a carrier system or a microwave radio system.

Facsimile A system for scanning a document, encoding it, transmitting it over a telecommunication circuit, and reproducing it in its original form at the receiving end.

Feasibility analysis The process of evaluating the technical and economic effectiveness of one or more alternatives.

Flow control The process of protecting network service by denying access to additional traffic that would add further to congestion.

Foreign Exchange (FEX) A special service that connects station equipment located in one telephone exchange with switching equipment located in another.

Four-wire circuit A circuit that uses separate paths for each direction of transmission.

Front end processor An auxiliary computer attached to a network to perform control operations and relieve the host computer for data processing.

Full-duplex A data communication circuit over which data can be sent in both directions simultaneously.

Gantt chart A bar chart display of tasks in a network and calendar dates during which they will be executed.

Gas tube protector A protector containing an ionizing gas that conducts external voltages to ground when they exceed a designed threshold level.

Gateway Circuitry used to interconnect networks by converting the protocols of each network to that used by the other.

Gauge The physical size of an electrical conductor, specified by American Wire Gauge (AWG) standards.

Generic program The operating system in a SPC central office that contains logic for call processing functions and controls the overall machine operation.

Glare A condition that exists when both ends of a circuit are simultaneously seized.

Grade of service (1) The percentage of time or probability that a call will be blocked in a network. (2) A quality indicator used in transmission measurements to specify the quality of a circuit based on both noise and loss.

Ground start A method of circuit seizure between a central office and a PBX that transmits an immediate signal by grounding the tip of the line.

Half-duplex A data communications circuit over which data can be sent in only one direction at a time.

Handshaking Signaling between two DCE devices on a link to set up communications between them.

Hang-on queueing A trunk queueing system in which the switching system signals users that all trunks are busy and allows them to remain off-hook while they are held in queue until the call can be completed.

Heat coil A protection device that opens a circuit and grounds a cable pair when operated by stray currents.

High usage groups Trunk groups established between two switching machines to serve as the first choice path between the machines, and thus handle the bulk of the traffic.

Holding time The average length of time per call that calls in a group of circuits are off hook.

Hundred Call Seconds (CCS) A measure of network load. Thirty-six CCS represents 100 percent occupancy of a circuit or piece of equipment.

Hybrid (1) A multi-winding coil or electronic circuit used in a four-wire terminating set or switching system line circuits to separate the four wire and two wire paths. (2) A key telephone system that has many of the features of a PBX. Such features as pooled trunk access characterize a hybrid.

Impedance The opposition to flow of alternating current in an electrical circuit.

Impulse noise Short bursts of short duration, high amplitude interference.

Independent Telephone Company (IC) A non-Bell telephone company.

Inside wiring The wiring on the customer's premises between the telephone set and the telephone company's demarcation point.

Integrated voice/data The combination of voice and data signals from a work station over a communication path to the PBX.

Interexchange Carrier (IEC) A common carrier that provides long distance service between LATAs.

Interface The connection between two systems. Usually hardware and software connecting a computer terminal with peripherals such as DCE, printers, etc.

Intermediate Distributing Frame (IDF) A crossconnection point between the main distributing frame and station wiring.

Intermodulation distortion Distortion or noise generated in electronic circuits when the power carried is great enough to cause non-linear operation.

Internal Rate of Return (IRR) In a financial study is the discount rate at which the net present values of two alternatives are equal.

Intertoll trunks Trunks interconnecting Class 4 and higher switching machines in the AT&T network.

Jitter The phase shift of digital pulses over a transmission medium.

Jumper Wire used to interconnect equipment and cable on a distributing frame.

Key Telephone System (KTS) A method of allowing several central office lines to be accessed from multiple telephone sets.

Leased line An unswitched telecommunications channel leased to an organization for its exclusive use.

Least Cost Routing (LCR) A PBX service feature that chooses the most economical route to a destination based on cost of the terminated services and time of delay.

Level The signal power at a given point in a circuit.

Life cycle analysis A financial study in which all cash flows are identified over the life of the project and discounted to their present value at the organization's cost of money.

Line conditioning A service offered by telephone companies to reduce envelope delay, noise, and amplitude distortion to enable transmission of higher speed data.

Link A circuit or path joining two communications channels in a network.

Loading The process of inserting fixed inductors in series with both wires of a cable pair to reduce voice frequency loss.

Local Access Transport Area (LATA) The geographical boundaries within which Bell Operating Companies are permitted to offer long distance traffic.

Local Area Network (LAN) A form of local network using one of the non-switched multiple access technologies.

Long range planning Planning that is concerned with an organization's requirements over a period greater than two or three years.

Loop start A method of circuit seizure between a central office and station equipment that operates by bridging the tip and ring of the line through a resistance.

Loopback test A test applied to a full-duplex circuit by connecting the receive leads to the transmit leads, applying a signal, and reading the returned test signal at the near end of the circuit.

Loss The drop in signal level between points on a circuit.

Main distributing frame (MDF) The cable rack used to terminate all distribution and trunk cables in a central office or PBX.

Mean Time Between Failures (MTBF) The average time a device or system operates without failing.

Mean Time To Repair (MTTR) The average time required for a qualified technician to repair a failed device or system.

Message switching A form of network access in which a message is forwarded from a terminal to a central switch where it is stored and forwarded to the addressee after some delay.

Message Telephone Service (MTS) A generic name for the switched long distance telephone service offered by all interexchange carriers.

Messenger A metallic strand attached to a pole line to support aerial cable.

Microwave A high frequency, high capacity radio system, usually used to carry multiple voice channels.

Milliwatt One one-thousandth of a watt. Used as a reference power for signal levels in telecommunications circuits.

Modeling The process of designing a network from a series of mathematical formulas that describe the behavior of network elements.

Modem pool A centralized pool of modems accessed through a PBX to provide off-net data transmission from modemless terminals.

Modem turnaround time The time required for a half-duplex modem to reverse the direction of transmission.

Modem A contraction of the terms ''MOdulator/DEModulator.'' A modem is used to convert analog signals to digital form and vice versa.

Modulation The process by which some characteristic of a carrier signal, such as frequency, amplitude, or phase is varied by a low frequency information signal.

Moving averages A method of developing a forecast by adding the most recent period, dropping the oldest period, and dividing by the number of forecast periods.

Multi-drop A circuit dedicated to communication between multiple terminals that are connected to the same circuit.

Multiline hunt The ability of a switching machine to connect calls to another number in a group when other numbers in the group are busy.

Multiple access The capability of multiple terminals connected to the same network to access one another by means of a common addressing scheme and protocol.

Multiple regression A forecasting method that predicts a variable based on actual results or predictions from multiple sources.

Net Present Value (NPV) The algebraic sum of all discounted cash flows less the initial investment.

Multiplexer A device used for combining several lower speed channels into a higher speed channel.

Network A set of communications points connected by channels.

Network administration The process of monitoring loads and service results in a network and making adjustments needed to maintain service and costs at the design objective level.

Network Channel Terminating Equipment (NCTE) Apparatus mounted on the user's premises that is used to amplify, match impedance, or match network signaling to the interconnected equipment.

Network design The process of determining quantities and architecture of circuit and equipment to achieve a cost/service balance.

Node A major point in a network where lines from many sources meet and may be switched.

Noise Any unwanted signal in a transmission path.

Octet A group of eight bits; also known as a "byte."

Off-hook A signaling state in a line or trunk when it is working or busy.

Off Premise Extension (OPX) Any extension telephone that uses public facilities to connect to the main telephone service.

On-hook A signaling state in a line or trunk when it is non-working or idle.

Open Network Architecture (ONA) A telephone architecture that provides the interfaces to enable service providers to connect to the public switched telephone network.

Open Systems Interconnect (OSI) A seven-layer data communications protocol model that specifies standard interfaces which all vendors can adapt to their own designs.

Operational planning Telecommunications planning that is concerned with day-to-day provision of service.

Overhead Any noninformation bits such as headers, error checking bits, start and stop bits, etc., used for controlling a network.

Packet Assembler/Disassembler (PAD) A device used on a packet switched network to assemble information into packets and to convert received packets into a continuous data stream.

Packet switching A method of allocating network time by forming data into packets and relaying it to the destination under control of processors at each major node. The network determines packet routing during transport of the packet.

Packet A unit of data information consisting of header, information, error detection and trailer records.

Parity A bit or series of bits appended to a character or block of characters to ensure that either an odd or even number of bits are transmitted. Parity is used for error detection.

Patch The temporary interconnection of transmission and signaling paths. Used for temporary rerouting and restoral of failed facilities or equipment.

Pay back period In financial analysis, the original investment divided by periodic savings, which yields the number of periods required to pay back the investment.

Peak-to-Average Ratio (P/AR) An analog test that provides an index of data circuit quality by sending a pulse into one end of a circuit and measuring its envelope at the distant end of the circuit.

Peg count The number of times a specified event occurs; derived from an early method of counting the number of busy lines in a manual switchboard.

Personal Identification Number (PIN) A billing identification number dialed by the user to enable the switching machine to identify the calling party.

Point-to-point circuit A telecommunications circuit that is exclusively assigned to the use of two devices.

Poisson A curve that describes the distribution of arrival times at the input to a service queue.

Poke-through A wiring method in which a hole is drilled through the floor and wiring is routed through the hole to a termination in the space above.

Poll cycle time The amount of time required for a multi-drop data communications controller to make one complete polling cycle through all devices on the network.

Polling A network sharing method in which remote terminals send traffic upon receipt of a polling message from the host. The host accesses the terminal, determines if it has traffic to send, and causes traffic to be uploaded to the host.

Power fail transfer A unit in KTS that transfers one or more telephone instruments to central office lines during a power failure.

Present value In financial analysis, the value of a future cash flow discounted to the present time at the organization's cost of money.

Primary channel A 1.544 mb/s information-carrying channel that furnishes integrated services digital network (ISDN) services to end users. Consists of 23 bearer channels and one signaling channel.

Private Automatic Branch Exchange (PABX) A term often used synonymously for PBX. A PABX is always automatic, whereas switching is manual in some PBXs.

Private Branch Exchange (PBX) A switching system dedicated to telephone and data use in a private communication network.

Program Evaluation and Review Technique (PERT) A method of displaying relationships between tasks in a project in graphic form.

Propagation delay The absolute time delay of a signal from the sending to the receiving terminal.

Propagation speed The speed at which a signal travels over a transmission medium.

Protector A device that prevents hazardous voltages or currents from injuring a user or damaging equipment connected to a cable pair.

Protocol The conventions used in a network for establishing communications compatibility between terminals, and for maintaining the line discipline while they are connected to the network.

Protocol analyzer A data communications test set that enables an operator to observe bit patterns in a data transmission, trap specific patterns, and simulate network elements.

Protocol converter A device that converts one communications protocol to another.

Public data network A data transmission network operated by a private telecommunications company for public subscription and use.

Public Switched Telephone Network (PSTN) A generic term for the interconnected networks of operating telephone companies.

Pulse Amplitude Modulation (PAM) A digital modulation method that operates by varying the amplitude of a stream of pulses in accordance with the instantaneous amplitude of the modulating signal.

Pulse Code Modulation (PCM) A digital modulation method that encodes a PAM signal into an eight-bit digital word representing the amplitude of each pulse.

Queuing The holding of calls in queue when a trunk group is busy and completing them in turn when an idle circuit is available.

Redundancy The provision of more than one circuit element to assume call processing when the primary element fails.

Reference noise (rn) The threshold of audibility to which noise measurements are referred, -90 dBm.

Registration The process the FCC follows in certifying that customer premise equipment will not cause harm to the network or personnel.

Regression analysis The use of one collection of data to forecast another variable.

Remote access The ability to dial into a switching machine over a local telephone number in order to complete calls over a private network from a distant location.

Remote Call Forwarding (RCF) A service offered by most telephone companies that allows a user to obtain a telephone number in a local calling area and have calls automatically forwarded at the user's expense to another telephone number.

Remote Maintenance and Testing System (RMATS) A service offered by PBX manufacturers and vendors that enables the vendor to access a PBX over the PTSN and perform testing and administrative functions.

Reorder A fast busy tone used to indicate equipment or circuit blockage.

Repeater A bi-directional signal regenerator (digital) or amplifier (analog). Repeaters are available to work on analog or digital signals from audio to radio frequency.

Request for Information (RFI) A request issued by prospective purchasers to determine if a product or service is available and what its major features are; normally does not request a specific price quotation.

Request for Proposals (RFP) A request for specific proposals on a product or service in which there are significant differences among products, which makes it necessary to choose the product on the basis of several different factors.

Request for Quotation (RFQ) A request for price quotations, usually on a product or service in which there is little difference in quality or features among competitors.

Reseller An interexchange carrier who purchases long distance service from another carrier and resells it to the end user.

Response time The interval between the terminal operator's sending the last character of a message and the time the first character of the response from the host arrives at the terminal.

Restriction Limitations to a station on the use of PBX features or trunks on the basis of service classification.

Return loss The degree of isolation, expressed in dB, between the transmit and the receive ports of a four wire terminating set.

Ring The designation of the side of a telephone line that carries talking battery to the user's premises.

Riser cable Cabling connecting a main distributing frame to intermediate distributing frames. Although a riser cable is vertical in buildings, it may be horizontal in campus environments.

RJ-11 A standard four-conductor jack and plug arrangement typically used for connecting a standard telephone to inside wiring.

Routing The path selection made for a telecommunications signal through the network to its destination.

Seasonality The characteristic of data that causes it to vary by season of the year, day of the week, or hour of the day.

Sensitivity analysis The process of rerunning a financial study to determine the degree to which changing the assumptions changes the result of the analysis.

Serial interface Circuitry used in DTE to convert parallel data to serial data for transmission on a network.

Shared Tenant Telecommunications Services (STTS) A telecommunications service provided by a building or industrial park owner or service provider to tenants within the facility. The service usually consists of local and long distance services, adds, moves, and changes, maintenance service, and other telecommunications-related services.

Short A circuit impairment that exists when two conductors of the same pair are connected at an unintended point.

Sidetone The sound of a talker's voice audible in the handset of the telephone instrument.

Signal-to-noise ratio The ratio between signal power and noise power in a circuit.

Simulation The process of designing a network by simulating the events and facilities that represent network load and capacity.

Singing The tendency of a circuit to oscillate when the return loss is too low.

Software Defined Network (SDN) A private network that is implemented using the carrier's switched network. The SDN is defined in software as opposed to consisting of discrete circuits.

Split A designated group of answering stations in an automatic call distributor (ACD).

Split channel modem A modem that divides a communication channel into separate send and receive directions.

Station Message Detail Recording (SMDR) The use of equipment in a PBX to record called station, time of day, and duration on trunk calls.

Station range The number of feet or ohms over which a telephone instrument can signal and transmit voice and data.

Station review The process of evaluating each user's needs for features, instrument type, and restrictions in preparation for a major cutover.

Statistical multiplexing A form of data multiplexing in which the time on a communications channel is assigned to terminals only when they have data to transport.

Store-and-forward A method of switching messages in which a message or packet is sent from the originating terminal to a central unit where it is held for retransmission to the receiving terminal.

Strategic planning Planning that is concerned with an organization's long range market position vis-a-vis its competitors.

Subscriber carrier A multi-channel device that enables several subscribers to share a single facility in the local loop.

Subscriber loop The circuit that connects a user's premises to the telephone central office.

Supervision The process of monitoring the busy/idle status of a circuit to detect changes of state.

Synchronous A method of transmitting data over a network wherein the sending and receiving terminals are kept in synchronism with each other by a clock signal embedded in the data.

Systems Network Architecture (SNA) An IBM data communications architecture that includes structure, formats, protocols, and operating sequences.

T-1 multiplexer An intelligent device that divides a 1.544 mb/s facility into multiple voice and data channels.

Terminal (1) A fixture attached to distribution cable to provide access for making connections to cable pairs. (2) Any device meant for direct operation over a telecommunications circuit by an end user.

Telecommunications The electronic movement of information.

Text messaging The use of a computer-based network of terminals to store and transmit messages among users.

Thermal noise Noise created in an electronic circuit by the movement and collisions of electrons.

Throughput The effective rate of transmission of information between two points excluding noninformation (overhead) bits.

Tie trunk A privately owned or leased trunk used to interconnect PBXs in a private switching network.

Time division multiplexing A method of combining several communication channels by dividing a channel into time increments and assigning each channel to a time slot. Multiple channels are interleaved when each channel is assigned the entire bandwidth of the backbone channel for a short period of time.

Time series forecast A forecast consisting of data that varies over time.

Tip The designation of the side of a telephone line that serves as the return path to the central office.

Token passing A method of allocating network access wherein a terminal can send traffic only after it has acquired the network's token.

Token A software mark or packet that circulates among network nodes.

Topology The architecture of a network, or the way circuits are connected to link the network nodes.

Traffic Usage Recorder (TUR) Hardware or software that monitors traffic-sensitive circuits or apparatus and records usage, usually in terms of CCS and peg count.

Transceiver A device that has the capability of both transmitting and receiving information.

Transducer Any device that changes energy from one state to another; examples are microphones, speakers, and telephone handsets.

Translations Software in a switching system that establishes the characteristics and features of lines and trunks.

Transmission Level Point (TLP) A dcsignated measurement point in a circuit where the transmission level has been specified by the designer.

Transmission The process of transporting voice or data over a network or facility from one point to another.

Trunk A communications channel between two switching systems equipped with terminating and signaling equipment.

Uplink The radio path from an earth station to a satellite.

Value added network A data communication network that adds processing services such as error correction and storage to the basic function of transporting data.

Videotex An interactive information retrieval service that usually employs the telephone network as the transmission medium.

Virtual banding In a WATS systems, the ability of trunks to carry traffic to all WATS bands, with billing based on the terminating points of the call instead of the band over which traffic is carried.

Virtual circuit A circuit that is established between two terminals by assigning a logical path over which data can flow. A virtual circuit can either be permanent, in which terminals are assigned a permanent path, or switched, in which the circuit is reestablished each time a terminal has data to send.

Voice mail A service that allows voice messages to be stored digitally in secondary storage and retrieved remotely by dialing access and identification codes.

Voice store-and-forward (see voice mail)

Wide Area Telephone Service (WATS) A bulk-rated long distance telephone service that carries calls at a cost based on usage and the state in which the calls terminate.

BIBLIOGRAPHY

AT&T Bell Laboratories. *Engineering and Operations in the Bell System,* Second Ed. Murray Hill, NJ: Bell Telephone Laboratories, 1983.

Bartee, Thomas C., Ed. *Digital Communications.* Indianapolis, IN: Howard W. Sams Co., 1986.

Belitsos, Byron and Jay Misra. *Business Telematics.* Homewood, IL: Dow Jones-Irwin, 1986.

Bellamy, John C. *Digital Telephony.* New York: John Wiley & Sons, 1982.

Brown, Robert J. and Rudolph R. Yanuck. *Introduction to Life Cycle Costing,* Atlanta: The Fairmount Press, Inc., 1985.

Camrass, Roger and Smith, Ken. *Wiring Up the Workplace.* Hinsdale, IL: BCR Enterprises, 1987.

Dordick, Herbert S. and Frederick Williams. *Innovative Management Using Telecommunications.* New York: John Wiley & Sons, 1986.

Ellis, Robert L. *Designing Data Networks.* Englewood Cliffs, NJ: Prentice-Hall, Inc., 1986.

Freeman, Roger L. *Reference Manual for Telecommunications Engineering.* New York: John Wiley & Sons, 1985.

Freeman, Roger L. *Telecommunications System Engineering.* New York: John Wiley & Sons, 1980.

Gilbreath, Robert D. *Winning at Project Management.* New York: John Wiley & Sons, 1986.

Goeller, L. and J. Goldstone. *The BCR Manual of PBXs. Hinsdale, IL: BCR Enterprises, Inc., 1986.*

Gouin, Michelle D. and Thomas B. Cross. *Intelligent Buildings.* Homewood, IL: Dow Jones-Irwin, 1986.

Green, James H. *The Dow Jones-Irwin Handbook of Telecommunications.* Homewood, IL: Dow Jones-Irwin, 1986.

Green, James H. *Local Area Networks.* Glenview, IL: Scott, Foresman Co., 1985.

Held, Gilbert. *Data Communications Networking Devices.* New York: John Wiley & Sons, 1986.

Horn, Frank W. *Cable Inside and Out.* Geneva, IL: ABC Teletraining, Inc.
Jewett, J; J. Shirago, and B. Yomtov. *Designing Optimal Voice Networks for Businesses, Government, and Telephone Companies.* Chicago: Telephony Publishing Corp., 1980.
Kepner, Charles H. and Benjamin B. Tregoe. *The Rational Manager.* New York, McGraw-Hill, 1965.
Madron, Thomas W. *Local Area Networks: The Second Generation.* New York: John Wiley & Sons, Inc., 1988.
Marrus, Stephanie K. *Building the Strategic Plan.* New York: John Wiley & Sons, 1984.
Meadow, Charles T., and Albert S. Teedesco. *Telecommunications For Management.* New York: McGraw-Hill, 1985.
Naisbitt, John. *Megatrends.* New York: Warner Book, Inc., 1982.
National Fire Prevention Association. *1987 National Electic Code,* Section 800-3. Quincy, MA: National Fire Prevention Association.
Stuck, B.W. and E. Arthurs. *A Computer and Communications Network Performance Analysis Primer.* Englewood Cliffs, NJ: Prentice-Hall, Inc., 1985.
Tofler, Alvin. *The Third Wave. New York:* William Morrow and Co., Inc., 1980.
Wheelwright, Steven C. and Spyros Makridakis. *Forecasting Methods for Management,* 4th Ed. New York: John Wiley & Sons, 1985.
Wiseman, Charles. *Strategy and Computers: Information Systems as Competitive Weapons.* Homewood, IL: Dow Jones-Irwin, 1987.

INDEX

A

B

D

E

F

I

J

K

L

N

O

P

S

X